Lecture Notes in Computer Scier

T0237963

Commenced Publication in 1973
Founding and Former Series Editors:
Gerhard Goos, Juris Hartmanis, and Jan van Leeuwen

Editorial Board

Shlomo Geva   Jaap Kamps
Andrew Trotman (Eds.)

Advances
in Focused Retrieval

7th International Workshop of the Initiative
for the Evaluation of XML Retrieval, INEX 2008
Dagstuhl Castle, Germany, December 15-18, 2008
Revised and Selected Papers

 Springer

Volume Editors

Shlomo Geva
Queensland University of Technology
Faculty of Science and Technology
GPO Box 2434, Brisbane Qld 4001, Australia
E-mail: s.geva@qut.edu.au

Jaap Kamps
University of Amsterdam
Archives and Information Studies/Humanities
Turfdraagsterpad 9, 1012 XT Amsterdam, The Netherlands
E-mail: kamps@uva.nl

Andrew Trotman
University of Otago
Department of Computer Science
P.O. Box 56, Dunedin 9054, New Zealand
E-mail: andrew@cs.otago.ac.nz

Library of Congress Control Number: 2009933190

CR Subject Classification (1998): H.3, H.3.3, H.3.4, H.2.8, H.2.3, H.2.4, E.1

LNCS Sublibrary: SL 3 – Information Systems and Application, incl. Internet/Web and HCI

ISSN   0302-9743
ISBN-10   3-642-03760-7 Springer Berlin Heidelberg New York
ISBN-13   978-3-642-03760-3 Springer Berlin Heidelberg New York

springer.com

© Springer-Verlag Berlin Heidelberg 2009
Printed in Germany

Typesetting: Camera-ready by author, data conversion by Scientific Publishing Services, Chennai, India
Printed on acid-free paper   SPIN: 12727117   06/3180   5 4 3 2 1 0

Foreword

I write with pleasure this foreword to the proceedings of the 7th workshop of the Initiative for the Evaluation of XML Retrieval (INEX). The increased adoption of XML as the standard for representing a document structure has led to the development of retrieval systems that are aimed at effectively accessing XML documents. Providing effective access to large collections of XML documents is therefore a key issue for the success of these systems. INEX aims to provide the necessary methodological means and worldwide infrastructures for evaluating how good XML retrieval systems are.

Since its launch in 2002, INEX has grown both in terms of number of participants and its coverage of the investigated retrieval tasks and scenarios. In 2002, INEX started with 49 registered participating organizations, whereas this number was more than 100 for 2008. In 2002, there was one main track, concerned with the ad hoc retrieval task, whereas in 2008, seven tracks in addition to the main ad hoc track were investigated, looking at various aspects of XML retrieval, from book search to entity ranking, including interaction aspects.

INEX follows the predominant approach in information retrieval of evaluating retrieval approaches using a test collection constructed specifically for that purpose and associated effectiveness measures. A test collection usually consists of a set of documents, user requests (topics), and relevance assessments, which specify the set of "right answers" for the requests. Throughout the years, INEX faced a range of challenges regarding the evaluation of XML retrieval systems, as the consideration of the structure led to many complex issues, which were not always identified at the beginning (e.g., the issue of overlap and a counting-based effectiveness measure, the difficulty in consistently assessing elements using a four-graded two-dimensional scale). In addition, limited funding was available, for example, to pay assessors. This led to a research problem in itself, namely, how to elicit quality assessments in order for the test collections to be reusable. As a result, different theories and methods for the evaluation of XML retrieval were developed and tested at INEX, e.g., the definition of relevance, a plethora of effectiveness measures, leading to a now-stable evaluation setup and a rich history of learned lessons. This is now allowing researchers worldwide to make further progress in XML retrieval, including the investigation of other research questions in XML retrieval, for instance, efficiency in 2008.

What I have greatly enjoyed with INEX is the people working together not only to develop approaches for XML retrieval, but also methodologies for evaluating XML retrieval. Many of the people that actually joined INEX to test their XML retrieval approaches got *hooked* on the problem of properly evaluating XML retrieval approaches. Three of them are the current INEX organizers, Shlomo Geva, Jaap Kamps, and Andrew Trotman. I am very glad that they got hooked, as they are dedicated and, even more importantly, enthusiastic people with extensive expertise in both building XML retrieval systems and evaluating them. INEX is in very safe hands with them. Looking at the current proceedings, I am delighted to see so many excellent works from people from all over the world. Many of them met at the annual workshop, and I heard it was a great success. Well done to all.

April 2009   Mounia Lalmas

Preface

Welcome to the 7th workshop of the Initiative for the Evaluation of XML Retrieval (INEX)! Now, in its seventh year, INEX is one of the established evaluation forums in information retrieval (IR), with 150 organizations worldwide registering and over 50 groups participating actively in the different tracks. INEX aims to provide an infrastructure, in the form of a large structured test collection and appropriate scoring methods, for the evaluation of focused retrieval.

Information on the Web is a mixture of text, multimedia, and metadata, with a clear internal structure, usually formatted according to the eXtensible Markup Language (XML) standard, or another related W3C standard. While many of today's information access systems still treat documents as single large (text) blocks, XML offers the opportunity to exploit the internal structure of documents in order to allow for more precise access, thus providing more specific answers to user requests. Providing effective access to XML-based content is therefore a key issue for the success of these systems.

INEX 2008 was an exciting year for INEX, and brought a lot of changes. Seven research tracks were included, which studied different aspects of focused information access: Ad Hoc, Book, Efficiency, Entity Ranking, Interactive (iTrack), Link the Wiki, and XML Mining. The aim of the INEX 2008 workshop was to bring together researchers who participated in the INEX 2008 campaign. During the past year, participating organizations contributed to the building of a large-scale XML test collection by creating topics, performing retrieval runs, and providing relevance assessments. The workshop concluded the results of this large-scale effort, summarized and addressed issues encountered, and devised a work plan for the future evaluation of XML retrieval systems. These proceedings report the final results of INEX 2008. We accepted a total of 49 out of 53 papers, yielding a 92% acceptance rate.

This was also the seventh INEX Workshop to be held at the *Schloss Dagstuhl – Leibniz Center for Informatics*, providing a unique setting where informal interaction and discussion occurs naturally and frequently. This has been essential to the growth of INEX over the years, and we feel honored and privileged that *Dagstuhl* housed the INEX 2008 Workshop. Finally, INEX was run for, but especially by, the participants. It was a result of tracks and tasks suggested by participants, topics created by participants, systems built by participants, and relevance judgments provided by participants. So the main thank you goes to each of these individuals!

April 2009

Shlomo Geva
Jaap Kamps
Andrew Trotman

Organization

Steering Committee

Charlie Clarke	University of Waterloo, Canada
Norbert Fuhr	University of Duisburg-Essen, Germany
Shlomo Geva	Queensland University of Technology, Australia
Jaap Kamps	University of Amsterdam, The Netherlands
Mounia Lalmas	University of Glasgow, UK
Stephen Robertson	Microsoft Research Cambridge, UK
Andrew Trotman	University of Otago, New Zealand
Ellen Voorhees	NIST, USA

Chairs

Shlomo Geva	Queensland University of Technology, Australia
Jaap Kamps	University of Amsterdam, The Netherlands
Andrew Trotman	University of Otago, New Zealand

Track Organizers

Ad Hoc

Shlomo Geva	General, Queensland University of Technology, Australia
Jaap Kamps	General, University of Amsterdam, The Netherlands
Andrew Trotman	General, University of Otago, New Zealand
Ludovic Denoyer	Document Collection, University Paris 6, France
Ralf Schenkel	Document Exploration, Max-Planck-Institut für Informatik, Germany
Martin Theobald	Document Exploration, Stanford University, USA

Book

Antoine Doucet	University of Caen, France
Gabriella Kazai	Microsoft Research Limited, Cambridge, UK
Monica Landoni	University of Strathclyde, UK

Efficiency

Ralf Schenkel   Max-Planck-Institut für Informatik, Germany
Martin Theobald   Stanford University, USA

Entity Ranking

Gianluca Demartini   L3S, Leibniz Universität Hannover, Germany
Tereza Iofciu   L3S, Leibniz Universität Hannover, Germany
Arjen de Vries   CWI, The Netherlands
Jianhan Zhu   University College London, UK

Interactive (iTrack)

Nisa Fachry   University of Amsterdam, The Netherlands
Ragnar Nordlie   Oslo University College, Norway
Nils Pharo   Oslo University College, Norway

Link the Wiki

Shlomo Geva   Queensland University of Technology,
                               Australia
Wei-Che (Darren) Huang   Queensland University of Technology,
                               Australia
Andrew Trotman   University of Otago, New Zealand

XML Mining

Ludovic Denoyer   University Paris 6, France
Patrick Gallinari   University Paris 6, France

Table of Contents

Book Track

Efficiency Track

Entity Ranking Track

Interactive Track

Link the Wiki Track

XML Mining Track

Overview of the INEX 2008 Ad Hoc Track

Jaap Kamps[1], Shlomo Geva[2], Andrew Trotman[3],
Alan Woodley[2], and Marijn Koolen[1]

[1] University of Amsterdam, Amsterdam, The Netherlands
{kamps,m.h.a.koolen}@uva.nl
[2] Queensland University of Technology, Brisbane, Australia
{s.geva,a.woodley}@qut.edu.au
[3] University of Otago, Dunedin, New Zealand
andrew@cs.otago.ac.nz

Abstract. This paper gives an overview of the INEX 2008 Ad Hoc
Track. The main goals of the Ad Hoc Track were two-fold. The first goal
was to investigate the value of the internal document structure (as provided by the XML mark-up) for retrieving relevant information. This is
a continuation of INEX 2007 and, for this reason, the retrieval results
are liberalized to arbitrary passages and measures were chosen to fairly
compare systems retrieving elements, ranges of elements, and arbitrary
passages. The second goal was to compare focused retrieval to article
retrieval more directly than in earlier years. For this reason, standard
document retrieval rankings have been derived from all runs, and evaluated with standard measures. In addition, a set of queries targeting
Wikipedia have been derived from a proxy log, and the runs are also
evaluated against the clicked Wikipedia pages. The INEX 2008 Ad Hoc
Track featured three tasks: For the *Focused Task* a ranked-list of non-
overlapping results (elements or passages) was needed. For the *Relevant
in Context Task* non-overlapping results (elements or passages) were returned grouped by the article from which they came. For the *Best in
Context Task* a single starting point (element start tag or passage start)
for each article was needed. We discuss the results for the three tasks,
and examine the relative effectiveness of element and passage retrieval.
This is examined in the context of content only (CO, or Keyword) search
as well as content and structure (CAS, or structured) search. Finally, we
look at the ability of focused retrieval techniques to rank articles, using
standard document retrieval techniques, both against the judged topics
as well as against queries and clicks from a proxy log.

1 Introduction

This paper gives an overview of the INEX 2008 Ad Hoc Track. There are two
main research question underlying the Ad Hoc Track. The first main research
question is that of the value of the internal document structure (mark-up) for
retrieving relevant information. That is, does the document structure help in
identify where the relevant information is within a document? This question,
first studied at INEX 2007, has attracted a lot of attention in recent years.

S. Geva, J. Kamps, and A. Trotman (Eds.): INEX 2008, LNCS 5631, pp. 1–28, 2009.
© Springer-Verlag Berlin Heidelberg 2009

Trotman and Geva [11] argued that, since INEX relevance assessments are not bound to XML element boundaries, retrieval systems should also not be bound to XML element boundaries. Their implicit assumption is that a system returning passages is at least as effective as a system returning XML elements. This assumption is based on the observation that elements are of a lower granularity than passages and so all elements can be described as passages. The reverse, however is not true and only some passages can be described as elements. Huang et al. [4] implement a fixed window passage retrieval system and show that a comparable element retrieval ranking can be derived. In a similar study, Itakura and Clarke [5] show that although ranking elements based on passage-evidence is comparable, a direct estimation of the relevance of elements is superior. Finally, Kamps and Koolen [6] study the relation between the passages highlighted by the assessors and the XML structure of the collection directly, showing reasonable correspondence between the document structure and the relevant information.

Up to now, element and passage retrieval approaches could only be compared when mapping passages to elements. This may significantly affect the comparison, since the mapping is non-trivial and, of course, turns the passage retrieval approaches effectively into element retrieval approaches. To study the value of the document structure through direct comparison of element and passage retrieval approaches, the retrieval results were liberalized to arbitrary passages. Every XML element is, of course, also a passage of text. At INEX 2008, a simple passage retrieval format was introduced using file-offset-length (FOL) triplets, that allow for standard passage retrieval systems to work on content-only versions of the collection. That is, the offset and length are calculated over the text of the article, ignoring all mark-up. The evaluation measures are based directly on the highlighted passages, or arbitrary best-entry points, as identified by the assessors. As a result it is now possible to fairly compare systems retrieving elements, ranges of elements, or arbitrary passages. These changes address earlier requests to liberalize the retrieval format to ranges of elements [2] and later requests to liberalize to arbitrary passages of text [11].

The second main question is to compare focused retrieval directly to traditional article retrieval. Throughout the history of INEX, participating groups have found that article retrieval—a system retrieving the whole article by default—resulted in fairly competitive performance [e.g., 7, 10]. Note that every focused retrieval system also generates an underlying article ranking, simply by the order is which results from different articles are ranked. This is most clear in the Relevant in Context and Best in Context tasks, where the article ranking is an explicit part of the task description. To study the importance of the underlying article ranking quality, we derived article level judgments by treating every article with some highlighted text as relevant, derived article rankings from every submission on a first-come, first-served basis, and evaluated with standard measures. This will also shed light on the value of element or passage level evidence for document retrieval [1]. In addition to this, we also include queries derived from a proxy log in the topic set, and can derive judgments from the later clicks in the same proxy log, treating all clicked articles as relevant for the query at hand. All submissions are also evaluated against these clicked Wikipedia pages, giving some insight in the differences between an IR test collection and real-world searching of Wikipedia.

The INEX 2008 Ad Hoc Track featured three tasks:

1. For the *Focused Task* a ranked-list of non-overlapping results (elements or passages) must be returned. It is evaluated at early precision relative to the highlighted (or believed relevant) text retrieved.
2. For the *Relevant in Context Task* non-overlapping results (elements or passages) must be returned, these are grouped by document. It is evaluated by mean average generalized precision where the generalized score per article is based on the retrieved highlighted text.
3. For the *Best in Context Task* a single starting point (element's starting tag or passage offset) per article must be returned. It is also evaluated by mean average generalized precision but with the generalized score (per article) based on the distance to the assessor's best-entry point.

We discuss the results for the three tasks, giving results for the top 10 participating groups and discussing the best scoring approaches in detail. We also examine the relative effectiveness of element and passage runs, and with content only (CO) queries and content and structure (CAS) queries.

The rest of the paper is organized as follows. First, Section 2 describes the INEX 2008 ad hoc retrieval tasks and measures. Section 3 details the collection, topics, and assessments of the INEX 2008 Ad Hoc Track. In Section 4, we report the results for the Focused Task (Section 4.2); the Relevant in Context Task (Section 4.3); and the Best in Context Task (Section 4.4). Section 5 details particular types of runs (such as CO versus CAS, and element versus passage), and on particular subsets of the topics (such as topics with a non-trivial CAS query). Section 6 looks at the article retrieval aspects of the submissions, both in terms of the judged topics treating any article with highlighted text as relevant, and in terms of clicked Wikipedia pages for queries derived from a proxy log. Finally, in Section 7, we discuss our findings and draw some conclusions.

2   Ad Hoc Retrieval Track

In this section, we briefly summarize the ad hoc retrieval tasks and the submission format (especially how elements and passages are identified). We also summarize the measures used for evaluation.

2.1   Tasks

Focused Task. The scenario underlying the Focused Task is the return, to the user, of a ranked list of elements or passages for their topic of request. The Focused Task requires systems to find the most focused results that satisfy an information need, without returning "overlapping" elements (shorter is preferred in the case of equally relevant elements). Since ancestor elements and longer passages are always relevant (to a greater or lesser extent) it is a challenge to choose the correct granularity.

The task has a number of assumptions:

Display: the results are presented to the user as a ranked-list of results.
Users: view the results top-down, one-by-one.

Relevant in Context Task. The scenario underlying the Relevant in Context Task is the return of a ranked list of articles and within those articles the relevant information (captured by a set of non-overlapping elements or passages). A relevant article will likely contain relevant information that could be spread across different elements. The task requires systems to find a set of results that corresponds well to all relevant information in each relevant article. The task has a number of assumptions:

Display: results will be grouped per article, in their original document order, access will be provided through further navigational means, such as a document heat-map or table of contents.
Users: consider the article to be the most natural retrieval unit, and prefer an overview of relevance within this context.

Best in Context Task. The scenario underlying the Best in Context Task is the return of a ranked list of articles and the identification of a best-entry-point from which a user should start reading each article in order to satisfy the information need. Even an article completely devoted to the topic of request will only have one best starting point from which to read (even if that is the beginning of the article). The task has a number of assumptions:

Display: a single result per article.
Users: consider articles to be natural unit of retrieval, but prefer to be guided to the best point from which to start reading the most relevant content.

2.2   Submission Format

Since XML retrieval approaches may return arbitrary results from within documents, a way to identify these nodes is needed. At INEX 2008, we allowed the submission of three types of results: XML elements; ranges of XML elements; and file-offset-length (FOL) text passages.

Element Results. XML element results are identified by means of a file name and an element (node) path specification. File names in the Wikipedia collection are unique so that the next example identifies 9996.xml as the target document from the Wikipedia collection (with the .xml extension removed).

```
<file>9996</file>
```

Element paths are given in XPath, but only fully specified paths are allowed. The next example identifies the first "article" element, then within that, the first "body" element, then the first "section" element, and finally within that the first "p" element.

```
<path>/article[1]/body[1]/section[1]/p[1]</path>
```

Importantly, XPath counts elements from 1 and counts element types. For example if a section had a title and two paragraphs then their paths would be: `title[1]`, `p[1]` and `p[2]`.

A result element, then, is identified unambiguously using the combination of file name and element path, as shown in the next example.

```
<result>
  <file>9996</file>
  <path>/article[1]/body[1]/section[1]/p[1]</path>
  <rsv>0.9999</rsv>
</result>
```

Ranges of Elements. To support ranges of elements, elemental passages are given in the same format.[1] As a passage need not start and end in the same element, each is given separately. The following example is equivalent to the element result example above since it starts and ends on an element boundary.

```
<result>
  <file>9996</file>
  <passage start="/article[1]/body[1]/section[1]/p[1]"
      end="/article[1]/body[1]/section[1]/p[1]"/>
  <rsv>0.9999</rsv>
</result>
```

Note that this format is very convenient for specifying ranges of elements, e.g., the following example retrieves the first three sections.

```
<result>
  <file>9996</file>
  <passage start="/article[1]/body[1]/section[1]"
      end="/article[1]/body[1]/section[3]"/>
  <rsv>0.9999</rsv>
</result>
```

FOL passages. Passage results can be given in file-offset-length (FOL) format, where offset and length are calculated in characters with respect to the textual content (ignoring all tags) of the XML file. A special text-only version of the collection is provided to facilitate the use of passage retrieval systems. File offsets start counting a 0 (zero).

---

[1] At INEX 2007, and in earlier qrels, an extended format allowing for optional character-offsets was used that allowed these passages to start or end in the middle of element or text-nodes. This format is superseded with the clean file-offset-length (FOL) passage format.

The following example is effectively equivalent to the example element result above.

```
<result>
  <file>9996</file>
  <fol offset="461" length="202"/>
  <rsv>0.9999</rsv>
</result>
```

The paragraph starts at the 462th character (so 461 characters beyond the first character), and has a length of 202 characters.

2.3   Evaluation Measures

We briefly summarize the main measures used for the Ad Hoc Track. Since INEX 2007, we allow the retrieval of arbitrary passages of text matching the judges ability to regard any passage of text as relevant. Unfortunately this simple change has necessitated the deprecation of element-based metrics used in prior INEX campaigns because the "natural" retrieval unit is no longer an element, so elements cannot be used as the basis of measure. We note that properly evaluating the effectiveness in XML-IR remains an ongoing research question at INEX.

The INEX 2008 measures are solely based on the retrieval of highlighted text. We simplify all INEX tasks to highlighted text retrieval and assume that systems return all, and only, highlighted text. We then compare the characters of text retrieved by a search engine to the number and location of characters of text identified as relevant by the assessor. For best in context we use the distance between the best entry point in the run to that identified by an assessor.

Focused Task. Recall is measured as the fraction of all highlighted text that has been retrieved. Precision is measured as the fraction of retrieved text that was highlighted. The notion of rank is relatively fluid for passages so we use an interpolated precision measure which calculates interpolated precision scores at selected recall levels. Since we are most interested in what happens in the first retrieved results, the INEX 2008 official measure is interpolated precision at 1% recall (iP[0.01]). We also present interpolated precision at other early recall points, and (mean average) interpolated precision over 101 standard recall points $(0.00, 0.01, 0.02, ..., 1.00)$ as an overall measure.

Relevant in Context Task. The evaluation of the Relevant in Context Task is based on the measures of generalized precision and recall [9], where the per document score reflects how well the retrieved text matches the relevant text in the document. Specifically, the per document score is the harmonic mean of precision and recall in terms of the fractions of retrieved and highlighted text in the document. We use an F_β score with $\beta = 1/4$ making precision four times as important as recall (at INEX 2007, F_1 was used). We are most interested in overall performances so the main measure is mean average generalized precision (MAgP). We also present the generalized precision scores at early ranks (5, 10, 25, 50).

Best in Context Task. The evaluation of the Best in Context Task is based on the measures of generalized precision and recall where the per document score reflects how well the retrieved entry point matches the best entry point in the document. Specifically, the per document score is a linear discounting function of the distance d (measured in characters)

$$\frac{n - d(x, b)}{n}$$

for $d < n$ and 0 otherwise. We use $n = 500$ which is roughly the number of characters corresponding to the visible part of the document on a screen (at INEX 2007, $n = 1,000$ was used). We are most interested in overall performance, and the main measure is mean average generalized precision (MAgP). We also show the generalized precision scores at early ranks (5, 10, 25, 50).

3 Ad Hoc Test Collection

In this section, we discuss the corpus, topics, and relevance assessments used in the Ad Hoc Track.

3.1 Corpus

The document collection was the Wikipedia XML Corpus based on the English Wikipedia in early 2006 [3]. The Wikipedia collection contains 659,338 Wikipedia articles. On average an article contains 161 XML nodes, where the average depth of a node in the XML tree of the document is 6.72.

The original Wiki syntax has been converted into XML, using both general tags of the layout structure (like *article, section, paragraph, title, list* and *item*), typographical tags (like *bold, emphatic*), and frequently occurring link-tags. For details see Denoyer and Gallinari [3].

3.2 Topics

The ad hoc topics were created by participants following precise instructions. Candidate topics contained a short CO (keyword) query, an optional structured CAS query, a one line description of the search request, and narrative with a details of the topic of request and the task context in which the information need arose. Figure 1 presents an example of an ad hoc topic. Based on the submitted candidate topics, 135 topics were selected for use in the INEX 2008 Ad Hoc Track as topic numbers 544–678.

In addition, 150 queries were derived from a proxy-log for use in the INEX 2008 Ad Hoc Track as topic numbers 679–828. For these topics, as well as the candidate topics without a ⟨castitle⟩ field, a default CAS-query was added based on the CO-query: `//*[about(., "`*CO-query*`")]`.

```
<topic id="544" ct_no="6">
  <title>meaning of life</title>
  <castitle>
    //article[about(., philosophy)]//section[about(., meaning of life)]
  </castitle>
  <description>What is the meaning of life?</description>
  <narrative>
    I got bored of my life and started wondering what the meaning of
    life is. An element is relevant if it discusses the meaning of life
    from different perspectives, as long as it is serious. For example,
    Socrates discussing meaning of life is relevant, but something like
    "42" from H2G2 or "the meaning of life is cheese" from a comedy is
    irrelevant. An element must be self contained. An  element that is a
    list of links is considered irrelevant because it is not
    self-contained in the sense that I don't know in which context the
    links are given.
  </narrative>
</topic>
```

Fig. 1. INEX 2008 Ad Hoc Track topic 544

3.3 Judgments

Topics were assessed by participants following precise instructions. The assessors used the new GPXrai assessment system that assists assessors in highlight relevant text. Topic assessors were asked to mark all, and only, relevant text in a pool of documents. After assessing an article with relevance, a separate best entry point decision was made by the assessor. The Focused and Relevant in Context Tasks were evaluated against the text highlighted by the assessors, whereas the Best in Context Task was evaluated against the best-entry-points.

The relevance judgments were frozen on October 22, 2008. At this time 70 topics had been fully assessed. Moreover, 11 topics were judged by two separate assessors, each without the knowledge of the other. All results in this paper refer to the 70 topics with the judgments of the first assigned assessor, which is typically the topic author.

- The 70 assessed topics were: 544–547, 550–553, 555–557, 559, 561, 562–563, 565, 570, 574, 576–582, 585–587, 592, 595–598, 600–603, 607, 609–611, 613, 616–617, 624, 626, 628, 629, 634–637, 641–644, 646–647, 649–650, 656–657, 659, 666–669, 673, 675, and 677.

In addition, there are clicked Wikipedia pages available in the proxy log for 125 topics:

- The 125 topics with clicked articles are numbered: 679–682, 684–685, 687–693, 695–704, 706–708, 711–727, 729–732, 734–751, 753–776, 778, 780–782, 784, 786–787, 789–790, 792–793, 795–796, 799–804, 806–807, 809–810, 812–813, 816–819, 821–824, and 826–828.

Table 1. Statistics over judged and relevant articles per topic

	total		# per topic				
	topics	number	min	max	median	mean	st.dev
judged articles	70	42,272	588	618	603	603.9	5.6
articles with relevance	70	4,887	2	376	49	69.8	68.9
highlighted passages	70	6,908	3	897	56	98.7	124.6
highlighted characters	70	11,471,649	1,419	1,113,578	99,569	163,880.7	202,757.2
Unique articles with clicks	125	225	1	10	1	1.8	1.5
Total clicked articles	125	532	1	24	3	4.3	3.8

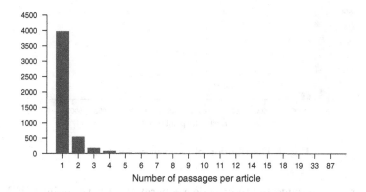

Fig. 2. Distribution of passages over articles

Table 1 presents statistics of the number of judged and relevant articles, and passages. In total 42,272 articles were judged. Relevant passages were found in 4,887 articles. The mean number of relevant articles per topic is 70, but the distribution is skewed with a median of 49. There were 6,908 highlighted passages. The mean was 99 passages and the median was 56 passages per topic.[2]

Table 1 also includes some statistics of the number of clicked articles in the proxy log. There are in total 225 clicked articles (unique per topic) over in total 125 topics, with a mean of 1.8 and a median of 1 clicked article per topic. We filtered the log for queries issued by multiple persons, and can also count the total number of clicks. Here, we see a total of 532 clicks (on the same 225 articles before), with a mean of 4.3 and a median of 3 clicks per topic. It is clear that the topics and clicked articles from the log are very different in character from the ad hoc topics.

Figure 2 presents the number of articles with the given number of passages. The vast majority of relevant articles (3,967 out of 4,887) had only a single highlighted passage, and the number of passages quickly tapers off.

---

[2] Recall from above that for the Focused Task the main effectiveness measures is precision at 1% recall. Given that the average topic has 99 relevant passages in 70 articles, the 1% recall roughly corresponds to a relevant passage retrieved—for many systems this will be accomplished by the first or first few results.

Table 2. Statistics over best entry point judgement

	# topics number		min	max	median	mean	st.dev
best entry point offset	70	4,887	1	87,982	14	1,738.1	4,814.3
first relevant character offset	70	4,887	1	87,982	20	1,816.1	4,854.2
fraction highlighted text	70	4,850	0.0005	1.000	0.583	0.550	0.425

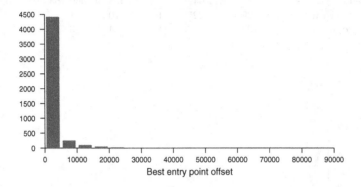

Fig. 3. Distribution of best entry point offsets

Assessors where requested to provide a separate best entry point (BEP) judgment, for every article where they highlighted relevant text. Table 2 presents statistics on the best entry point offset, on the first highlighted or relevant character, and on the fraction of highlighted text in relevant articles. We first look at the BEPs. The mean BEP is well within the article with offset 1,738 but the distribution is very skewed with a median BEP offset of only 14. Figure 3 shows the distribution of the character offsets of the 4,887 best entry points. It is clear that the overwhelming majority of BEPs is at the beginning of the article.

The statistics of the first highlighted or relevant character (FRC) in Table 2 give very similar numbers as the BEP offsets: the mean offset of the first relevant character is 1,816 but the median offset is only 20. This suggests a relation between the BEP offset and the FRC offset. Figure 4 shows a scatter plot the BEP and FRC offsets. Two observations present themselves. First, there is a clear diagonal where the BEP is positioned exactly at the first highlighted character in the article. Second, there is also a vertical line at BEP offset zero, indicating a tendency to put the BEP at the start of the article even when the relevant text appears later on.

Finally, the statistics on the fraction of highlighted text in Table 2 show that amount of relevant text varies from almost nothing to almost everything. The mean fraction is 0.55, and the median is 0.58, indicating that typically over half the article is relevant. Given that the majority of relevant articles contain such a large fraction of relevant text plausibly explains that BEPs being frequently positioned on or near the start of the article.

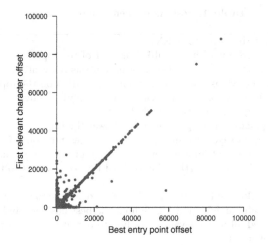

Fig. 4. Scatter plot of best entry point offsets versus the first relevant character

Table 3. Candidate Topic Questionnaire

B1	How familiar are you with the subject matter of the topic?
B2	Would you search for this topic in real-life?
B3	Does your query differ from what you would type in a web search engine?
B4	Are you looking for very specific information?
B5	Are you interested in reading a lot of relevant information on the topic?
B6	Could the topic be satisfied by combining the information in different (parts of) documents?
B7	Is the topic based on a seen relevant (part of a) document?
B8	Can information of equal relevance to the topic be found in several documents?
B9	Approximately how many articles in the whole collection do you expect to contain relevant information?
B10	Approximately how many relevant document parts do you expect in the whole collection?
B11	Could a relevant result be (check all that apply): a single sentence; a single paragraph; a single (sub)section; a whole article
B12	Can the topic be completely satisfied by a single relevant result?
B13	Is there additional value in reading several relevant results?
B14	Is there additional value in knowing all relevant results?
B15	Would you prefer seeing: only the best results; all relevant results; don't know
B16	Would you prefer seeing: isolated document parts; the article's context; don't know
B17	Do you assume perfect knowledge of the DTD?
B18	Do you assume that the structure of at least one relevant result is known?
B19	Do you assume that references to the document structure are vague and imprecise?
B20	Comments or suggestions on any of the above (optional)

Table 4. Post Assessment Questionnaire

C1	Did you submit this topic to INEX?
C2	How familiar were you with the subject matter of the topic?
C3	How hard was it to decide whether information was relevant?
C4	Is Wikipedia an obvious source to look for information on the topic?
C5	Can a highlighted passage be (check all that apply): a single sentence; a single paragraph; a single (sub)section; a whole article
C6	Is a single highlighted passage enough to answer the topic?
C7	Are highlighted passages still informative when presented out of context?
C8	How often does relevant information occur in an article about something else?
C9	How well does the total length of highlighted text correspond to the usefulness of an article?
C10	Which of the following two strategies is closer to your actual highlighting: (I) I located useful articles and highlighted the best passages and nothing more, (II) I highlighted all text relevant according to narrative, even if this meant highlighting an entire article.
C11	Can a best entry point be (check all that apply): the start of a highlighted passage; the sectioning structure containing the highlighted text; the start of the article
C12	Does the best entry point correspond to the best passage?
C13	Does the best entry point correspond to the first passage?
C14	Comments or suggestions on any of the above (optional)

3.4 Questionnaires

At INEX 2008, all candidate topic authors and assessors were asked to complete a questionnaire designed to capture the context of the topic author and the topic of request.

The candidate topic questionnaire (shown in Table 3) featured 20 questions capturing contextual data on the search request.

The post-assessment questionnaire (shown in Table 4) featured 14 questions capturing further contextual data on the search request, and the way the topic has been judged (a few questions on GPXrai were added to the end).

The responses to the questionnaires show a considerable variation over topics and topic authors in terms of topic familiarity; the type of information requested; the expected results; the interpretation of structural information in the search request; the meaning of a highlighted passage; and the meaning of best entry points. There is a need for further analysis of the contextual data of the topics in relation to the results of the INEX 2008 Ad Hoc Track.

4   Ad Hoc Retrieval Results

In this section, we discuss, for the three ad hoc tasks, the participants and their results.

4.1 Participation

A total of 163 runs were submitted by 23 participating groups. Table 5 lists the participants and the number of runs they submitted, also broken down over

Table 5. Participants in the Ad Hoc Track

Id Participant	Focused	Relevant in Context	Best in Context	CO query	CAS query	Element results	Passage results	FOL results	# valid runs	# submitted runs
4 University of Otago	0	6	0	6	0	3	3	0	6	6
5 Queensland University of Technology	6	6	6	15	3	9	9	0	18	18
6 University of Amsterdam	6	6	3	9	6	13	0	2	15	15
9 University of Helsinki	3	0	0	3	0	3	0	0	3	3
10 Max-Planck-Institut Informatik	3	1	1	5	0	5	0	0	5	5
12 University of Granada	3	3	3	9	0	9	0	0	9	9
14 University of California, Berkeley	2	0	1	3	0	3	0	0	3	3
16 University of Frankfurt	1	3	3	0	7	7	0	0	7	9
22 ENSM-SE	2	0	0	2	0	0	0	2	2	9
25 Renmin University of China	3	0	1	2	2	4	0	0	4	4
29 INDIAN STATISTICAL INSTITUTE	3	0	0	3	0	3	0	0	3	3
37 Katholieke Universiteit Leuven	6	0	0	3	3	6	0	0	6	6
40 IRIT	0	0	2	1	1	2	0	0	2	6
42 University of Toronto	2	0	0	0	2	2	0	0	2	3
48 LIG	3	2	0	5	0	5	0	0	5	5
55 Doshisha University	0	0	1	0	1	1	0	0	1	3
56 JustSystems Corporation	3	3	3	6	3	9	0	0	9	9
60 Saint Etienne University	3	0	0	3	0	3	0	0	3	9
61 Universit Libre de Bruxelles	0	0	0	0	0	0	0	0	0	2
68 University Pierre et Marie Curie - LIP6	2	0	0	2	0	2	0	0	2	2
72 University of Minnesota Duluth	2	2	2	6	0	6	0	0	6	6
78 University of Waterloo	3	3	4	10	0	8	2	0	10	13
92 University of Lyon3	5	5	5	15	0	15	0	0	15	15
Total runs	61	40	35	108	28	118	14	4	136	163

the tasks (Focused, Relevant in Context, or Best in Context); the used query (Content-Only or Content-And-Structure); and the used result type (Element, Passage or FOL). Unfortunately, no less than 27 runs turned out to be invalid and will only be evaluated with respect to their "article retrieval" value in Section 6.

Participants were allowed to submit up to three element result-type runs per task and three passage result-type runs per task (for all three tasks). This totaled to 18 runs per participant.[3] The submissions are spread well over the ad hoc retrieval tasks with 61 submissions for Focused, 40 submissions for Relevant in Context, and 35 submissions for Best in Context.

---

[3] As it turns out, two groups submitted more runs than allowed: *University of Lyon3* submitted 6 extra element runs, and *University of Amsterdam* submitted 4 extra element runs. At this moment, we have not decided on any repercussions other than mentioning them in this footnote.

Table 6. Top 10 Participants in the Ad Hoc Track Focused Task

Participant	iP[.00]	iP[.01]	iP[.05]	iP[.10]	MAiP
p78-FOERStep	0.7660	0.6897	0.5714	0.4908	0.2076
p10-TOPXCOarti	0.6808	0.6799	0.5812	0.5372	0.2981
p48-LIGMLFOCRI	0.7127	0.6678	0.5223	0.4229	0.1446
p92-manualQEin*	0.6664	0.6664	0.6139	0.5583	0.3077
p9-UHelRun394	0.7109	0.6648	0.5558	0.5044	0.2268
p60-JMUexpe142	0.6918	0.6640	0.5800	0.5071	0.2347
p14-T2FBCOPARA	0.7319	0.6427	0.4908	0.4036	0.1399
p29-LMnofb020	0.6855	0.6365	0.5566	0.5152	0.2868
p25-weightedfi	0.6553	0.6346	0.5495	0.5263	0.2661
p5-GPX1COFOCe	0.6818	0.6344	0.5693	0.5180	0.2592

4.2 Focused Task

We now discuss the results of the Focused Task in which a ranked-list of non-overlapping results (elements or passages) was required. The official measure for the task was (mean) interpolated precision at 1% recall (iP[0.01]). Table 6 shows the best run of the top 10 participating groups. The first column gives the participant, see Table 5 for the full name of group. The second to fifth column give the interpolated precision at 0%, 1%, 5%, and 10% recall. The sixth column gives mean average interpolated precision over 101 standard recall levels (0%, 1%, ..., 100%).

Here we briefly summarize what is currently known about the experiments conducted by the top five groups (based on official measure for the task, iP[0.01]).

University of Waterloo. Element retrieval run using the CO query. Description: the run uses the Okapi BM25 model in Wumpus to score all content-bearing elements such as sections and paragraphs using Okapi BM25. In addition, scores were boosted by doubling the tf values of the first 10 words of an element.

Max-Planck-Institut für Informatik. Element retrieval run using the CO query. Description: The TopX system retrieving only article elements, using a linear combination of a BM25 content score with a BM25 proximity score that also takes document structure into accout.

LIG Grenoble. An element retrieval run using the CO query. Description: Based on a language Model using a Dirichlet smoothing, and equally weighting element score and its context score, where the context score are based on the collection-links in Wikipedia.

University of Lyon3. A *manual* element retrieval run using the CO query. Description: Using indri search engine in Lemur with manually expanded queries from CO, description and narrative fields. The run is retrieving only articles.

University of Helsinki. An element retrieval run using the CO query. Description: A special phrase index was created based on the detection of phrases in the collection, where the phrases are replication three times—effectively

Table 7. Top 10 Participants in the Ad Hoc Track Relevant in Context Task

Participant	gP[5]	gP[10]	gP[25]	gP[50]	MAgP
p78-RICBest	0.4100	0.3454	0.2767	0.2202	0.2278
p92-manualQEin*	0.4175	0.3589	0.2692	0.2095	0.2106
p5-GPX1CORICe	0.3759	0.3441	0.2677	0.2151	0.2106
p10-TOPXCOallA	0.3681	0.3108	0.2386	0.1928	0.1947
p4-WHOLEDOC	0.3742	0.3276	0.2492	0.1962	0.1929
p6-inex08artB	0.3510	0.3008	0.2216	0.1741	0.1758
p72-UMDRic2	0.3853	0.3361	0.2357	0.1894	0.1724
p12-p8u3exp511	0.2966	0.2726	0.2169	0.1621	0.1582
p56-VSMRIP05	0.3281	0.2647	0.2113	0.1616	0.1500
p48-LIGMLRIC4O	0.3634	0.3115	0.2327	0.1721	0.1497

boosting query word occurrences in phrases. In addition, a standard keyword index was used. The run using BM25 is a combination of the retrieval
status value on the word-index (94% of weight) and the phrase-index (6% of
weight).

Saint Etienne University. An element retrieval run using the CO query. Description: A probabilistic model used to evaluate a weight for each tag: "the
probability that tags distinguishes terms which are the most relevant", i.e.
based on the fact that the tag contains relevant or non relevant passages.
The resulting tag weights are incorporated into an element-level run with
BM25 weighting.

Based on the information from these and other participants

- All ten runs use the CO query. The fourth run, *p92-manualQEin*, uses a
  manually expanded query using words from the description and narrative
  fields. The eighth run, *p29-LMnofb020*, is an automatic run using the title
  and description fields. All other runs use only the CO query in the title field.
- All runs retrieve elements as results.
- The systems at rank second (*p10-TOPXCOarti*), fourth (*p92-manualQEin*),
  and eighth (*p29-LMnofb020*), are retrieving only full articles.

4.3   Relevant in Context Task

We now discuss the results of the Relevant in Context Task in which non-
overlapping results (elements or passages) need to be returned grouped by the
article they came from. The task was evaluated using generalized precision where
the generalized score per article was based on the retrieved highlighted text. The
official measure for the task was mean average generalized precision (MAgP).

Table 7 shows the top 10 participating groups (only the best run per group is
shown) in the Relevant in Context Task. The first column lists the participant,
see Table 5 for the full name of group. The second to fifth column list generalized
precision at 5, 10, 25, 50 retrieved articles. The sixth column lists mean average
generalized precision.

Here we briefly summarize the information available about the experiments conducted by the top five groups (based on MAgP).

University of Waterloo. Element retrieval run using the CO query. Description: the run uses the Okapi BM25 model in Wumpus to score all content-bearing elements such as sections and paragraphs using Okapi BM25, and grouped the results by articles and ranked the articles by their best scoring element.

University of Lyon3. A manual element retrieval run using the CO query. Description: the same as the Focused run above. In fact it is literally the same article ranking as the Focused run. Recall that the run is retrieving only whole articles.

Queensland University of Technology. Element retrieval run using the CO query. Description: GPX run using a //*[about(.,keywords)] query, serving non-overlapping elements grouped per article, with the articles ordered by their best scoring element.

Max-Planck-Institut für Informatik. Element retrieval run using the CO query. Description: An element retrieval run using the new BM25 scoring function (i.e., considering each element as "document" and then computing a standard BM25 model), selecting non-overlapping elements based on score, and grouping them per article with the articles ranked by their highest scoring element.

University of Otago. Element retrieval run using the CO query. Description: BM25 is used to select and rank the top 1,500 documents and whole documents are selected as the passage. The run is retrieving only whole articles.

Based on the information from these and other participants

- The runs ranked sixth (*p6-inex08artB*) and ninth (*p56-VSMRIP05*) are using the CAS query. The run ranked second, *p92-manualQEin*, is using a manually expanded query based on keywords in the description and narrative. All other runs use only the CO query in the topic's title field.
- All runs retrieve elements as results.
- Solid article ranking seems a prerequisite for good overall performance, with second best run, *p92-manualQEin*, the fifth best run, *p4-WHOLEDOC*, and the ninth best run, *p56-VSMRIP05*, retrieving only full articles.

4.4   Best in Context Task

We now discuss the results of the Best in Context Task in which documents were ranked on topical relevance and a single best entry point into the document was identified. The Best in Context Task was evaluated using generalized precision but here the generalized score per article was based on the distance to the assessor's best-entry point. The official measure for the task was mean average generalized precision (MAgP).

Table 8 shows the top 10 participating groups (only the best run per group is shown) in the Best in Context Task. The first column lists the participant,

Table 8. Top 10 Participants in the Ad Hoc Track Best in Context Task

Participant	gP[5]	gP[10]	gP[25]	gP[50]	MAgP
p78-BICER	0.3896	0.3306	0.2555	0.2019	0.2238
p92-manualQEin*	0.4144	0.3688	0.2834	0.2244	0.2197
p25-weightedfi	0.3510	0.3058	0.2531	0.2042	0.2037
p5-GPX1COBICp	0.3711	0.3395	0.2605	0.2046	0.1989
p6-submitinex	0.3475	0.2898	0.2236	0.1706	0.1709
p10-TOPXCOallB	0.2417	0.2374	0.1913	0.1550	0.1708
p12-p8u3exp501	0.2546	0.2331	0.1952	0.1503	0.1468
p72-UMDBIC1	0.3192	0.2752	0.1891	0.1474	0.1455
p56-VSMRIP08	0.2269	0.2038	0.1748	0.1403	0.1317
p55-KikoriBest	0.2041	0.1958	0.1552	0.1210	0.0960

see Table 5 for the full name of group. The second to fifth column list generalized precision at 5, 10, 25, 50 retrieved articles. The sixth column lists mean average generalized precision.

Here we briefly summarize the information available about the experiments conducted by the top five groups (based on MAgP).

University of Waterloo. Element retrieval run using the CO query. Description: the run uses the Okapi BM25 model in Wumpus to score all content-bearing elements such as sections and paragraphs using Okapi BM25, and kept only the best scoring element per article.

University of Lyon3. A manual element retrieval run using the CO query. Description: the same as the Focused and Relevant in Context runs above. In fact all three runs have literally the same article ranking. This run is retrieving the start of the whole article as best entry point, in other words an article retrieval run.

Renmin University of China. Element retrieval run using the CO query. Description: using language model to compute RSV at leaf level combined with aggregation at retrieval time, assuming independence.

Queensland University of Technology. Run retrieving ranges of elements using the CO query. The run is always returning a whole article, setting the BEP at the very start of the article. Description: GPX run using a //*[about(.,keywords)] query, ranking articles by their best scoring element, but transformed to return the complete article as a passages. This is effectively an article level GPX run.

University of Amsterdam. Run retrieving FOL passages using the CO query. Description: language model with local indegree prior, setting the BEP always at the start of the article. Since the offset is always zero, this is similar to an article retrieval run.

Based on the information from these and other participants

– As for the Relevant in Context Task, we see again that solid article ranking is very important. In fact, we see runs putting the BEP at the start

Table 9. Statistical significance (t-test, one-tailed, 95%)

	1	2	3	4	5	6	7	8	9	10
p78		-	-	-	-	-	-	-	-	-
p10			-	-	-	-	-	-	-	-
p48				-	-	-	-	-	-	-
p92					-	-	-	-	-	-
p9						-	-	-	-	-
p60							-	-	-	-
p14								-	-	-
p29									-	-
p25										-
p5										

	1	2	3	4	5	6	7	8	9	10
p78		-	⋆	⋆	⋆	⋆	⋆	⋆	⋆	⋆
p92				-	-	-	⋆	⋆	⋆	⋆
p5				⋆	-	⋆	⋆	⋆	⋆	⋆
p10					-	-	-	⋆	⋆	⋆
p4						-	-	⋆	⋆	⋆
p6							-	-	-	⋆
p72								-	-	⋆
p12									-	-
p56										-
p48										

	1	2	3	4	5	6	7	8	9	10
p78		-	-	⋆	⋆	⋆	⋆	⋆	⋆	⋆
p92			-	-	⋆	⋆	⋆	⋆	⋆	⋆
p25				-	⋆	⋆	⋆	⋆	⋆	⋆
p5					⋆	-	⋆	⋆	⋆	⋆
p6						-	-	-	⋆	⋆
p10							-	-	⋆	⋆
p12								-	-	⋆
p72									-	⋆
p56										⋆
p55										

of all the retrieved articles at rank two (*p92-manualQEin*), rank four (*p5-GPX1COBICp*), and rank five (*p6-submitinex*).

- The fourth ranked run, *p5-GPX1COBICp*, uses ranges of elements, albeit a degenerate case where always the full article is selected. The fifth run, *p6-submitinex*, uses fol passages, albeit again a degenerate case where the BEP is always the zero offset.
- With the exception of the runs at rank nine (*p56-VSMRIP08*) and ten (*p55-KikoriBest*), which used the CAS query, all the other best runs per group use the CO query.

4.5 Significance Tests

We tested whether higher ranked systems were significantly better than lower ranked system, using a t-test (one-tailed) at 95%. Table 9 shows, for each task, whether it is significantly better (indicated by "⋆") than lower ranked runs. For example, For the Focused Task, we see that the early precision (at 1% recall) is a rather unstable measure and none of the runs are significantly different. Hence we should be careful when drawing conclusions based on the Focused Task results. For the Relevant in Context Task, we see that the top run is significantly better than ranks 3 through 10, the second best run better than ranks 6 through 10, the third ranked system better than ranks 4 and 6 through 10, and the fourth and fifth ranked systems better than ranks 8 through 10. For the Best in Context Task, we see that the top run is significantly better than ranks 4 through 10, the second and third runs significantly better than than ranks 5 to 10. The fourth ranked system is better than the systems ranked 5 and 7 to 10, and the fifth ranked system better than ranks 9 and 10.

5 Analysis of Run and Topic Types

In this section, we will discuss relative effectiveness of element and passage retrieval approaches, and on the relative effectiveness of systems using the keyword and structured queries.

Table 10. Ad Hoc Track: Runs with ranges of elements or FOL passages

(a) Focused Task

Participant	iP[.00]	iP[.01]	iP[.05]	iP[.10]	MAiP
p5-GPX2COFOCp	0.6311	0.6305	0.5365	0.4719	0.2507
p22-EMSEFocuse*	0.6757	0.5724	0.4487	0.3847	0.1555

(b) Relevant in Context Task

Participant	gP[5]	gP[10]	gP[25]	gP[50]	MAgP
p4-WHOLEDOCPA	0.3742	0.3276	0.2492	0.1962	0.1929
p5-GPX1CORICp	0.3566	0.3220	0.2430	0.1875	0.1900

(c) Best in Context Task

Participant	gP[5]	gP[10]	gP[25]	gP[50]	MAgP
p5-GPX1COBICp	0.3711	0.3395	0.2605	0.2046	0.1989
p6-submitinex	0.3475	0.2898	0.2236	0.1706	0.1709
p78-BICPRplus	0.2651	0.2252	0.1666	0.1268	0.1254

5.1   Elements versus Passages

We received 18 submissions using ranges of elements of FOL-passage results, from in total 5 participating groups. We will look at the relative effectiveness of element and passage runs.

As we saw above, in Section 4, for all three tasks the best scoring runs used elements as the unit of retrieval. Table 10 shows the best runs using ranges of elements or FOL passages for the three ad hoc tasks. All these runs use the CO query. As it turns out, the best focused run using passages ranks outside the top scoring runs in Table 6; the best relevant in context run using passages is ranked fifth among the top scoring runs in Table 7; and the best best in context run using passages is ranked fourth among the top scoring runs in Table 8. This outcome is consistent with earlier results using passage-based element retrieval, where passage retrieval approaches showed comparable but not superior behavior to element retrieval approaches [4, 5].

However, looking at the runs in more detail, their character is often unlike what one would expect from a "passage" retrieval run. For Focused, *p5-GPX2COFOCp* is an article run using ranges of elements; and *p22-EMSEFocuse* is a manual query run using FOL passages. For Relevant in Context, both *p4-WHOLEDOCPA* and *p5-GPX1CORICp* are article runs using ranges of elements. For Best in Context, *p5-GPX1COBICp* is an article runs using ranges of elements; *p6-submitinex* is an article run using FOL passages; and *p78-BICPRplus* is an element retrieving run using ranges of elements. So, all but two of the runs retrieve only articles. Hence, this is not sufficient evidence to warrant any conclusion on the effectiveness of passage level results. We hope and expect that the test collection and the passage runs will be used for further research into the relative effectiveness of element and passage retrieval approaches.

Table 11. CAS query target elements over all 135 topics

Target Element	Frequency
*	51
section	39
article	30
p	11
figure	3
body	1

Table 12. Ad Hoc Track CAS Topics: CO runs (left-hand side) versus CAS runs (right-hand side)

(a) Focused Task

Participant	iP[.00]	iP[.01]	iP[.05]	iP[.10]	MAiP	Participant	iP[.00]	iP[.01]	iP[.05]	iP[.10]	MAiP
p60-JMUexpe136	0.7321	0.7245	0.6416	0.5936	0.2934	p6-inex08artB	0.6514	0.6379	0.5901	0.5248	0.2261
p48-LIGMLFOCRI	0.7496	0.7209	0.5307	0.4440	0.1570	p56-VSMRIP02	0.7515	0.6333	0.4781	0.3667	0.1400
p78-FOER	0.7263	0.7089	0.6084	0.5485	0.2225	p5-GPX3COSFOC	0.6232	0.6220	0.5509	0.4626	0.2137
p5-GPX1COFOCe	0.7168	0.6972	0.6416	0.5616	0.2616	p25-RUCLLP08	0.5969	0.5969	0.5761	0.5545	0.2491
p29-LMnofb020	0.7193	0.6766	0.5926	0.5611	0.2951	p37-kulcaselem	0.6824	0.5626	0.3532	0.2720	0.1257
p10-TOPXCOallF	0.7482	0.6680	0.5555	0.4871	0.1925	p42-B2U0visith	0.6057	0.5364	0.4830	0.4449	0.1739
p25-weightedfi	0.6665	0.6634	0.5907	0.5646	0.2671	p16-001RunofUn	0.3111	0.2269	0.1675	0.1206	0.0365
p6-inex08artB	0.6689	0.6571	0.5570	0.4961	0.2104						
p9-UHelRun394	0.7024	0.6567	0.5602	0.5221	0.2255						
p72-UMDFocused	0.7259	0.6491	0.4947	0.3812	0.1115						

(b) Relevant in Context Task

Participant	gP[5]	gP[10]	gP[25]	gP[50]	MAgP	Participant	gP[5]	gP[10]	gP[25]	gP[50]	MAgP
p78-RICBest	0.4808	0.3818	0.2994	0.2274	0.2485	p6-inex08artB	0.3757	0.3113	0.2334	0.1847	0.1937
p5-GPX1CORICe	0.3946	0.3518	0.2670	0.2169	0.2166	p5-GPX3COSRIC	0.3482	0.3232	0.2381	0.1923	0.1764
p4-WHOLEDOC	0.4020	0.3534	0.2508	0.2009	0.2125	p56-VSMRIP05	0.3401	0.2796	0.2143	0.1616	0.1501
p10-TOPXCOallA	0.3892	0.3220	0.2366	0.1910	0.1967	p16-009RunofUn	0.0153	0.0156	0.0123	0.0095	0.0023
p92-manualQEin*	0.3818	0.3395	0.2515	0.1970	0.1933						
p6-inex08artB	0.3762	0.3140	0.2293	0.1790	0.1900						
p72-UMDRic2	0.3952	0.3434	0.2289	0.1868	0.1745						
p12-p8u3exp511	0.3229	0.2880	0.2245	0.1631	0.1680						
p48-LIGMLRIC4O	0.3818	0.3408	0.2461	0.1832	0.1583						
p56-VSMRIP04	0.2315	0.2031	0.1675	0.1368	0.1275						

(c) Best in Context Task

Participant	gP[5]	gP[10]	gP[25]	gP[50]	MAgP	Participant	gP[5]	gP[10]	gP[25]	gP[50]	MAgP
p78-BICER	0.3935	0.3386	0.2544	0.1956	0.2172	p5-GPX3COSBIC	0.3109	0.2883	0.2235	0.1780	0.1661
p25-weightedfi	0.3342	0.3065	0.2390	0.1958	0.2004	p40-xfirmcos07	0.2381	0.1794	0.1348	0.1078	0.0908
p5-GPX1COBICp	0.3663	0.3358	0.2504	0.1926	0.1983	p55-KikoriBest	0.1817	0.1721	0.1422	0.1123	0.0803
p92-manualQEin*	0.3728	0.3383	0.2599	0.2082	0.1952	p16-006RunofUn	0.0307	0.0347	0.0307	0.0261	0.0128
p10-TOPXCOallB	0.2424	0.2419	0.1788	0.1457	0.1727						
p6-submitinex	0.3505	0.3062	0.2278	0.1713	0.1716						
p12-p8u3exp501	0.2586	0.2397	0.1934	0.1425	0.1448						
p72-UMDBIC1	0.3222	0.2751	0.1757	0.1377	0.1369						
p56-VSMRIP09	0.1562	0.1537	0.1377	0.1127	0.1038						
p40-xfirmbicco	0.1594	0.1546	0.1367	0.1137	0.0661						

5.2   CO versus CAS

We now look at the relative effectiveness of the keyword (CO) and structured (CAS) queries. As we saw above, in Section 4, one of the best runs per group for the Relevant in Context Task, and two of the top 10 runs for the Best in Context Task used the CAS query.

Table 13. Top 10 Participants in the Ad Hoc Track: Article retrieval

Participant	P5	P10	1/rank	map	bpref
p78-BICER	0.6286	0.5343	0.8711	0.3789	0.3699
p92-manualQEin*	0.6429	0.5886	0.8322	0.3629	0.3924
p10-TOPXCOarti	0.5943	0.5443	0.8635	0.3516	0.3628
p5-GPX1COBICe	0.5743	0.5257	0.7868	0.3413	0.3588
p37-kulcoeleme	0.5286	0.4557	0.7468	0.3268	0.3341
p25-weightedfi	0.4971	0.4657	0.7192	0.3255	0.3355
p29-VSMfbElts0	0.5543	0.4857	0.7955	0.3195	0.3388
p60-JMUexpe136	0.5457	0.4857	0.7843	0.3192	0.3383
p9-UHelRun293	0.5829	0.5029	0.7766	0.3144	0.3323
p4-SWKL200	0.5714	0.5000	0.7950	0.3107	0.3297

All topics have a CAS query since artificial CAS queries of the form

`//*[about(., keyword title)]`

were added to topics without CAS title. Table 11 show the distribution of target elements. In total 86 topics had a non-trivial CAS query.[4] These CAS topics are numbered 544–550, 553–556, 564, 567, 568, 572, 574, 576–578, 580, 583, 584, 586–591, 597–605, 607, 608, 610, 615–625, 627, 629–633, 635–640, 646, 651–655, 658, 659, 661–670, 673, and 675–678. As it turned out, 39 of these CAS topics were assessed. The results presented here are restricted to the 39 CAS topics.

Table 12 lists the top 10 participants measured using just the 39 CAS topics and for the Focused Task (a), the Relevant in Context Task (b), and the Best in Context Task (c). For the Focused Task the CAS runs score lower than the CO query runs. For the Relevant in Context Task, the best CAS run would have ranked fifth among the CO runs. For the Best in Context Task, the best CAS run would rank seventh among the CO runs. Overall, we see the that teams submitting runs with both types of queries have higher scoring CO runs, with participant 6 as a notable exception for Relevant in Context.

6   Analysis of Article Retrieval

In this section, we will look in detail at the effectiveness of Ad Hoc Track submissions as article retrieval systems. We look first at the article rankings in terms of the Ad Hoc Track judgments—treating every article that contains highlighted text as relevant. Then, we look at the article rankings in terms of the clicked pages for the topics from the proxy log—treating every clicked article as relevant.

6.1   Article Retrieval: Relevance Judgments

We will first look at the topics judged during INEX 2008, the same topics as in earlier sections, but now using the judgments to derive standard document-level

---

[4] Note that some of the wild-card topics (using the "*" target) in Table 11 had non-trivial about-predicates and hence have not been regarded as trivial CAS queries.

Table 14. Top 10 Participants in the Ad Hoc Track: Article retrieval per task over judged topics (left) and clicked pages (right)

(a) Focused Task

Participant	P5	P10	1/rank	map	bpref	Participant	P5	P10	1/rank	map	bpref
p92-manualQEin*	0.6429	0.5886	0.8322	0.3629	0.3924	p5-Terrier	0.1594	0.0877	0.5904	0.5184	0.8266
p10-TOPXCOarti	0.5943	0.5443	0.8635	0.3516	0.3628	p6-inex08artB	0.1623	0.0870	0.5821	0.5140	0.8150
p5-GPX1COFOCp	0.5743	0.5257	0.7868	0.3413	0.3588	p92-autoindri0	0.1565	0.0884	0.5601	0.4853	0.8211
p37-kulcoeleme	0.5286	0.4557	0.7468	0.3268	0.3341	p60-JMUexpe142	0.1536	0.0862	0.5624	0.4853	0.8250
p78-FOER	0.5800	0.5043	0.7995	0.3259	0.3277	p48-LIGMLFOCRI	0.1449	0.0833	0.5191	0.4596	0.7153
p29-VSMfbElts0	0.5543	0.4857	0.7955	0.3195	0.3388	p10-TOPXCOarti	0.1522	0.0841	0.5164	0.4538	0.8167
p25-weightedfi	0.4971	0.4657	0.7192	0.3195	0.3324	p78-FOER	0.1304	0.0819	0.4979	0.4404	0.8136
p60-JMUexpe136	0.5457	0.4857	0.7843	0.3192	0.3383	p40-xfirmcos07	0.1217	0.0717	0.4301	0.3748	0.7184
p9-UHelRun293	0.5829	0.5029	0.7766	0.3144	0.3323	p55-KikoriFocu	0.1261	0.0732	0.4334	0.3727	0.7785
p6-inex08artB	0.5514	0.4800	0.7851	0.3010	0.3109	p22-EMSEFocus*	0.1203	0.0783	0.4233	0.3704	0.8105

(b) Relevant in Context Task

Participant	P5	P10	1/rank	map	bpref	Participant	P5	P10	1/rank	map	bpref
p92-manualQEin*	0.6429	0.5886	0.8322	0.3629	0.3924	p5-Terrier	0.1594	0.0877	0.5904	0.5184	0.8266
p5-GPX1CORICp	0.5743	0.5257	0.7868	0.3413	0.3588	p6-inex08artB	0.1623	0.0870	0.5821	0.5140	0.8150
p78-RICBest	0.5886	0.5029	0.8161	0.3404	0.3422	p60-JMUexpe150	0.1536	0.0862	0.5624	0.4853	0.8167
p10-TOPXCOallA	0.5314	0.4843	0.8226	0.3122	0.3279	p92-autoindri0	0.1565	0.0884	0.5601	0.4853	0.8211
p60-JMUexpe150	0.5886	0.4900	0.8266	0.3119	0.3185	p48-LIGMLRIC4O	0.1464	0.0841	0.5238	0.4647	0.7081
p4-SWKL200	0.5714	0.5000	0.7950	0.3107	0.3297	p78-RICBest	0.1348	0.0812	0.4979	0.4422	0.8126
p6-inex08artB	0.5514	0.4800	0.7851	0.3010	0.3109	p10-TOPXCOallA	0.1333	0.0775	0.5139	0.4397	0.7863
p56-VSMRIP05	0.5486	0.4543	0.7752	0.2880	0.3045	p72-UMDRic2	0.1275	0.0717	0.4560	0.4088	0.7526
p72-UMDRic2	0.6000	0.5200	0.8579	0.2739	0.3048	p4-SWKL200	0.1159	0.0732	0.4168	0.3701	0.8007
p22-EMSERICStr*	0.5057	0.4543	0.7079	0.2728	0.3064	p55-KikoriRele	0.1232	0.0710	0.4125	0.3501	0.7712

(c) Best in Context Task

Participant	P5	P10	1/rank	map	bpref	Participant	P5	P10	1/rank	map	bpref
p78-BICER	0.6286	0.5343	0.8711	0.3789	0.3699	p5-Terrier	0.1594	0.0877	0.5904	0.5184	0.8266
p92-manualQEin*	0.6429	0.5886	0.8322	0.3629	0.3924	p6-submitinex	0.1594	0.0862	0.5673	0.4976	0.8164
p5-GPX1COBICe	0.5743	0.5257	0.7868	0.3413	0.3588	p92-autoindri0	0.1565	0.0884	0.5601	0.4853	0.8211
p10-TOPXCOallB	0.5314	0.4843	0.8226	0.3290	0.3344	p60-JMUexpe151	0.1536	0.0855	0.5624	0.4844	0.8214
p25-weightedfi	0.4971	0.4657	0.7192	0.3255	0.3355	p78-BICPRplus	0.1522	0.0841	0.5432	0.4673	0.7799
p60-JMUexpe157	0.5714	0.5000	0.8215	0.3098	0.3176	p10-TOPXCOallB	0.1333	0.0775	0.5139	0.4398	0.8205
p6-submitinex	0.5486	0.4657	0.7793	0.2984	0.3086	p72-UMDBIC1	0.1275	0.0710	0.4482	0.4011	0.7398
p56-VSMRIP08	0.5486	0.4543	0.7752	0.2880	0.3045	p40-xfirmcos07	0.1217	0.0717	0.4301	0.3748	0.7160
p72-UMDBIC2	0.5914	0.5171	0.8511	0.2761	0.3022	p55-KikoriBest	0.1261	0.0732	0.4334	0.3727	0.7785
p12-p8u3exp501	0.4829	0.4371	0.7044	0.2723	0.3061	p56-VSMRIP08	0.1130	0.0659	0.3943	0.3445	0.7258

relevance by regarding an article as relevant if some part of it is highlighted by the assessor. Throughout this section, we derive an article retrieval run from every submission using a first-come, first served mapping. That is, we simply keep every first occurrence of an article (retrieved indirectly through some element contained in it) and ignore further results from the same article.

We use `trec_eval` to evaluate the mapped runs and qrels, and use mean average precision (map) as the main measure. Since all runs are now article retrieval runs, the differences between the tasks disappear. Moreover, runs violating the task requirements—most notably non-overlapping results for all tasks, and having scattered results from the same article in relevant in context—are now also considered, and we work with all 163 runs submitted to the Ad Hoc Track.

Table 13 shows the best run of the top 10 participating groups. The first column gives the participant, see Table 5 for the full name of group. The second and third column give the precision at ranks 5 and 10, respectively. The fourth column gives the mean reciprocal rank. The fifth column gives mean average precision. The sixth column gives binary preference measures (using the top R judged non-relevant documents). Recall from the above that second ranked run (*p92-manualQEin*) is a manual article retrieval run submitted to all three tasks. Also the run ranked three (*p10-TOPXCOarti*) and the run ranked seven

Table 15. Top 10 Participants in the Ad Hoc Track: Clicked articles

Participant	P5	P10	1/rank	map	bpref
p5-Terrier	0.1594	0.0877	0.5904	0.5184	0.8266
p6-inex08artB	0.1623	0.0870	0.5821	0.5140	0.8150
p60-JMUexpe150	0.1536	0.0862	0.5624	0.4853	0.8167
p92-autoindri0	0.1565	0.0884	0.5601	0.4853	0.8211
p78-BICPRplus	0.1522	0.0841	0.5432	0.4673	0.7799
p48-LIGMLRIC4O	0.1464	0.0841	0.5238	0.4647	0.7081
p10-TOPXCOarti	0.1522	0.0841	0.5164	0.4538	0.8167
p72-UMDRic2	0.1275	0.0717	0.4560	0.4088	0.7526
p40-xfirmcos07	0.1217	0.0717	0.4301	0.3748	0.7184
p55-KikoriFocu	0.1261	0.0732	0.4334	0.3727	0.7785

(*p60-JMUexpe136*) retrieve exclusively articles. The relative effectiveness of these article retrieval runs in terms of their article ranking is no surprise. Furthermore, we see submissions from all three ad hoc tasks. Most notably runs from the Best in Context task at ranks 1, 2, 4, and 6; runs from the Focused task at ranks 2, 3, 5, 7, 8, and 9; and runs from the Relevant in Context task at ranks 2 and 10.

If we break-down all runs over the original tasks, shown on the left-hand side of Table 14, we can compare the ranking to Section 4 above. We see some runs that are familiar from the earlier tables: three Focused runs correspond to Table 6, five Relevant in Context runs correspond to Table 7, and seven Best in Context runs correspond to Table 8. More formally, we looked at how the two system rankings correlate using Kendall's Tau.

- Over all 61 Focused task submissions the system rank correlation is 0.517 between iP[0.01] and map, and 0.568 between MAiP and map.
- Over all 40 Relevant in Context submissions the system rank correlation between MAgP and map is 0.792.
- Over all 35 Best in Context submissions the system rank correlation is 0.795

Overall, we see a reasonable correspondence between the rankings for the ad hoc tasks in Section 4 and the rankings for the derived article retrieval measures. The correlation with the Focused task runs is much lower than with the Relevant in Context and Best in Context tasks. This makes sense, since the ranking of articles is an important part of the two "in context" tasks.

6.2 Article Retrieval: Clicked Pages

In addition to the topics created and assessed by INEX participants, we also included 150 queries derived from a proxy log, and can also construct pseudo-relevance judgments by regarding every clicked Wikipedia article as relevant.

Table 15 shows the best run of the top 10 participating groups. The first column gives the participant, see Table 5 for the full name of group. The second and third column give the precision at ranks 5 and 10, respectively. The fourth column gives the mean reciprocal rank. The fifth column gives mean average precision. The sixth column gives binary preference measures (using the top R judged non-relevant documents). Compared to the judged topics, we immediately see much lower scores for the early precision measures (precision at 5 and 10, and reciprocal ranks), while at the same time higher scores for the overall measures (map and bpref). This is a result of the very low numbers of relevant documents, 1.8 on average, that make it impossible to get a grips on recall aspects. The runs ranked first (*p5-Terrier*), fourth (*p92-autoindri0*), and seventh (*p10-TOPXCOarti*) retrieve exclusively full articles. Again, it is no great surprise that these runs do well for the task of article retrieval.

The resulting ranking is quite different from the article ranking based on the judged ad hoc topics in Table 13. They have only one run in common, although they agree on five of the ten participants. Looking, more formally, at the system rank correlations between the two types of article retrieval we see the following.

- Over all 163 submissions, the system rank correlation is 0.357.
- Over the 76 Focused task submissions, the correlation is 0.356.
- Over the 49 Relevant in task submissions, the correlation is 0.366.
- Over the 38 Best in Context task submissions, the correlation is 0.388.

Hence the judged topics above and the topics derived from the proxy log vary considerable. A large part of the explanation is the dramatic difference between the numbers of relevant articles, with 70 on average for the judged topics and 1.8 on average for the proxy log topics.

7   Discussion and Conclusions

In this paper we provided an overview of the INEX 2008 Ad Hoc Track that contained three tasks: For the *Focused Task* a ranked-list of non-overlapping results (elements or passages) was required. For the *Relevant in Context Task* non-overlapping results (elements or passages) grouped by the article that they belong to were required. For the *Best in Context Task* a single starting point (element's starting tag or passage offset) per article was required. We discussed the results for the three tasks, and analysed the relative effectiveness of element and passage runs, and of keyword (CO) queries and structured queries (CAS). We also look at effectiveness in term of article retrieval, both using the judged topics and using queries and clicks derived from a proxy log.

When examining the relative effectiveness of CO and CAS we found that for all tasks the best scoring runs used the CO query. This is in contrast with earlier results showing that structural hints can help promote initial precision [8]. Part of the explanation may be in the low number of CAS submissions (28) in comparison with the number of CO submissions (108). Only 39 of the 70 judged topics had a non-trivial CAS query, and the majority of those CAS queries made only reference to particular tags and not on their structural relations. This may have diminished the value of the CAS query in comparison with earlier years.

Given the efforts put into the fair comparison of element and passage retrieval approaches, the number of passage and FOL submissions was disappointing.

Eighteen submissions used ranges of elements or FOL passage results, whereas 118 submissions used element results. In addition, many of the passage or FOL submissions used exclusively full articles as results. Although we received too few non-element runs to draw clear conclusions, we saw that the passage based approaches were competitive, but not superior to element based approaches. This outcome is consistent with earlier results in [4, 5].

As in earlier years, we saw that article retrieval is reasonably effective at XML-IR: for each of the ad hoc tasks there were three article-only runs among the best runs of the top 10 groups. When looking at the article rankings inherent in all Ad Hoc Track submissions, we saw that again three of the best runs of the top 10 groups in terms of article ranking (across all three tasks) were in fact article-only runs. This suggests that element-level or passage-level evidence is still valuable for article retrieval. When comparing the system rankings in terms of article retrieval with the system rankings in terms of the ad hoc retrieval tasks, over the exact same topic set, we see a reasonable correlation especially for the two "in context" tasks. The systems with the best performance for the ad hoc tasks, also tend to have the best article rankings. Since finding the relevant articles can be considered a prerequisite for XML-IR, this should not come as a surprize. In addition, the Wikipedia's encyclopedic structure with relatively short articles covering a single topic results in relevant articles containing large fractions of relevant text (with a mean of 55% of text being highlighted). While it is straightforward to define tasks and measures that strongly favor precision over recall, a more natural route would be to try to ellicit more focused information needs that have natural answers in short excerpts of text.

When we look at a different topic set derived from a proxy log, and a shallow set of clicked pages rather than a full-blown IR test collection, we see notable differences. Given the low number of relevant articles (1.8 on average) compared to the ad hoc judgments (70 on average), the clicked pages focus exclusively on precision aspects. This leads to a different system ranking, although there is still some agreement on the best groups. The differences between these two sets of topics require further analysis.

Finally, the Ad Hoc Track had two main research questions. The first main research question was the comparative analysis of element and passage retrieval approaches, hoping to shed light on the value of the document structure as provided by the XML mark-up. We found that the best performing system used predominantly element results, although the number of non-element retrieval runs submitted is too low to draw any definite conclusions. The second main research question was to compare focused retrieval directly to traditional article retrieval. We found that the best scoring Ad Hoc Track submissions also tend to have the best article ranking, and that the best article rankings were generated using element-level evidence. For both main research questions, we hope and expect that the resulting test collection will prove its value in future use. After all, the main aim of the INEX initiative is to create bench-mark test-collections for the evaluation of structured retrieval approaches.

Acknowledgments. Jaap Kamps was supported by the Netherlands Organization for Scientific Research (NWO, grants 612.066.513, 639.072.601, and 640.001.501).

References

[1] Callan, J.P.: Passage-level evidence in document retrieval. In: Proceedings of the 17th Annual International ACM SIGIR Conference, pp. 302–310 (1994)

[2] Clarke, C.L.A.: Range results in XML retrieval. In: Proceedings of the INEX 2005 Workshop on Element Retrieval Methodology, pp. 4–5 (2005)

[3] Denoyer, L., Gallinari, P.: The Wikipedia XML Corpus. SIGIR Forum 40, 64–69 (2006)

[4] Huang, W., Trotman, A., O'Keefe, R.A.: Element retrieval using a passage retrieval approach. In: Proceedings of the 11th Australasian Document Computing Symposium (ADCS 2006), pp. 80–83 (2006)

[5] Itakura, K.Y., Clarke, C.L.A.: From passages into elements in XML retrieval. In: Proceedings of the SIGIR 2007 Workshop on Focused Retrieval, pp. 17–22 (2007)

[6] Kamps, J., Koolen, M.: On the relation between relevant passages and XML document structure. In: Proceedings of the SIGIR 2007 Workshop on Focused Retrieval, pp. 28–32 (2007)

[7] Kamps, J., Marx, M., de Rijke, M., Sigurbjörnsson, B.: The importance of morphological normalization for XML retrieval. In: Proceedings of the First INEX Workshop, pp. 41–48 (2003)

[8] Kamps, J., Marx, M., de Rijke, M., Sigurbjörnsson, B.: Articulating information needs in XML query languages. Transactions on Information Systems 24, 407–436 (2006)

[9] Kekäläinen, J., Järvelin, K.: Using graded relevance assessments in IR evaluation. Journal of the American Society for Information Science and Technology 53, 1120–1129 (2002)

[10] Thom, J.A., Pehcevski, J.: How well does best in context reflect ad hoc XML retrieval. In: Pre-Proceedings of INEX 2007, pp. 124–125 (2007)

[11] Trotman, A., Geva, S.: Passage retrieval and other XML-retrieval tasks. In: Proceedings of the SIGIR 2006 Workshop on XML Element Retrieval Methodology, pp. 43–50 (2006)

A   Appendix: Full Run Names

Group	Run	Label	Task	Query	Results	Notes
4	151	p4-SWKL200	RiC	CO	Pas	
4	152	p4-WHOLEDOC	RiC	CO	Ele	Article-only
4	153	p4-WHOLEDOCPA	RiC	CO	Pas	Article-only
5	122	p5-Terrier	BiC	CO	Pas	Article-only
5	123	p5-Terrier	Foc	CO	Pas	Article-only
5	124	p5-Terrier	RiC	CO	Pas	Article-only
5	133	p5-GPX2COFOCp	Foc	CO	Pas	Article-only
5	138	p5-GPX1COBICe	BiC	CO	Ele	

Continued on Next Page...

Group	Run	Label	Task	Query	Results	Notes
5	139	p5-GPX1COFOCe	Foc	CO	Ele	
5	140	p5-GPX1CORICe	RiC	CO	Ele	
5	141	p5-GPX3COSBIC	BiC	CAS	Ele	
5	142	p5-GPX3COSFOC	Foc	CAS	Ele	
5	143	p5-GPX3COSRIC	RiC	CAS	Ele	
5	144	p5-GPX1COBICp	BiC	CO	Pas	Article-only
5	145	p5-GPX1COFOCp	Foc	CO	Pas	Article-only
5	146	p5-GPX1CORICp	RiC	CO	Pas	Article-only
6	255	p6-submitinex	BiC	CO	FOL	Article-only
6	264	p6-inex08artB	RiC	CAS	Ele	
6	265	p6-inex08artB	RiC	CO	Ele	
6	268	p6-inex08artB	RiC	CAS	Ele	
6	269	p6-inex08artB	RiC	CO	Ele	
6	270	p6-inex08artB	Foc	CAS	Ele	
6	271	p6-inex08artB	Foc	CO	Ele	
6	274	p6-inex08artB	Foc	CO	Ele	
6	276	p6-inex08artB	Foc	CO	Ele	
9	174	p9-UHelRun293	Foc	CO	Ele	
9	176	p9-UHelRun394	Foc	CO	Ele	
10	91	p10-TOPXCOallF	Foc	CO	Ele	
10	92	p10-TOPXCOallB	BiC	CO	Ele	
10	93	p10-TOPXCOallA	RiC	CO	Ele	
10	207	p10-TOPXCOarti	Foc	?	Ele	Article-only
12	97	p12-p8u3exp501	BiC	CO	Ele	
12	100	p12-p8u3exp511	RiC	CO	Ele	
14	205	p14-T2FBCOPARA	Foc	CO	Ele	
16	233	p16-009RunofUn	RiC	CAS	Ele	
16	234	p16-006RunofUn	BiC	CAS	Ele	
16	244	p16-001RunofUn	Foc	CAS	Ele	
22	62	p22-EMSEFocuse	Foc	CO	Ele	Manual Invalid
22	66	p22-EMSEFocuse	Foc	CO	FOL	Manual
22	68	p22-EMSERICStr	RiC	CO	Ele	Manual Invalid
25	30	p25-RUCLLP08	Foc	CAS	Ele	
25	278	p25-weightedfi	Foc	CO	Ele	
25	282	p25-weightedfi	BiC	CO	Ele	
29	238	p29-VSMfbElts0	Foc	CO	Ele	
29	253	p29-LMnofb020	Foc	CO	Ele	Article-only
37	227	p37-kulcaselem	Foc	CAS	Ele	
37	230	p37-kulcoeleme	Foc	CO	Ele	
40	54	p40-xfirmbicco	BiC	CO	Ele	
40	296	p40-xfirmcos07	BiC	CAS	Ele	
40	297	p40-xfirmcos07	Foc	CAS	Ele	Invalid
42	299	p42-B2U0visith	Foc	CAS	Ele	
48	59	p48-LIGMLFOCRI	Foc	CO	Ele	

Continued on Next Page...

Group	Run	Label	Task	Query	Results	Notes
48	72	p48-LIGMLRIC4O	RiC	CO	Ele	
55	279	p55-KikoriFocu	Foc	CAS	Ele	Invalid
55	280	p55-KikoriRele	RiC	CAS	Ele	Invalid
55	281	p55-KikoriBest	BiC	CAS	Ele	
56	190	p56-VSMRIP02	Foc	CAS	Ele	
56	197	p56-VSMRIP04	RiC	CO	Ele	Article-only
56	199	p56-VSMRIP05	RiC	CAS	Ele	Article-only
56	202	p56-VSMRIP08	BiC	CAS	Ele	
56	224	p56-VSMRIP09	BiC	CO	Ele	
60	11	p60-JMUexpe136	Foc	CO	Ele	Article-only
60	53	p60-JMUexpe142	Foc	CO	Ele	
60	81	p60-JMUexpe150	RiC	CO	Ele	Invalid
60	82	p60-JMUexpe151	BiC	CO	Ele	Invalid
60	175	p60-JMUexpe157	BiC	CO	Ele	Invalid
72	106	p72-UMDFocused	Foc	CO	Ele	
72	154	p72-UMDBIC1	BiC	CO	Ele	
72	155	p72-UMDBIC2	BiC	CO	Ele	
72	277	p72-UMDRic2	RiC	CO	Ele	
78	156	p78-FOER	Foc	CO	Ele	
78	157	p78-FOERStep	Foc	CO	Ele	
78	160	p78-BICER	BiC	CO	Ele	
78	163	p78-BICPRplus	BiC	CO	Pas	
78	164	p78-RICBest	RiC	CO	Ele	
92	177	p92-autoindri0	BiC	CO	Ele	Article-only
92	178	p92-autoindri0	Foc	CO	Ele	Article-only
92	179	p92-autoindri0	RiC	CO	Ele	Article-only
92	183	p92-manualQEin	BiC	CO	Ele	Manual Article-only
92	184	p92-manualQEin	Foc	CO	Ele	Manual Article-only
92	185	p92-manualQEin	RiC	CO	Ele	Manual Article-only

Experiments with Proximity-Aware Scoring for XML Retrieval at INEX 2008

Andreas Broschart[1,2], Ralf Schenkel[1,2], and Martin Theobald[1]

[1] Max-Planck-Institut für Informatik, Saarbrücken, Germany
[2] Saarland University, Saarbrücken, Germany
{abrosch,schenkel,mtb}@mpi-inf.mpg.de

Abstract. Proximity enhanced scoring models significantly improve retrieval quality in text retrieval. For XML IR, we can sometimes enhance the retrieval efficacy by exploiting knowledge about the document structure combined with established text IR methods. This paper elaborates on our approach used for INEX 2008 which modifies a proximity scoring model from text retrieval for usage in XML IR and extends it by taking the document structure information into account.

1 Introduction

Term proximity has been a common means to improve effectiveness for text retrieval, passage retrieval, and question answering, and several proximity scoring functions have been developed in recent years (for example, [4–7]). For XML retrieval, however, proximity scoring has not been similarly successful. To the best of our knowledge, there is only a single existing proposal for proximity-aware XML scoring [1] that computes, for each text position in an element, a fuzzy score for the query, and then computes the overall score for the element as average score over all its positions.

Our proximity score for content-only queries on XML data [2] extends the existing proximity score by Büttcher et al. [4], taking into account the document structure when computing the distance of term occurrences.

2 Proximity Scoring for XML

To compute a proximity score for an element e with respect to a query $q = \{t_1 \ldots t_n\}$ with multiple terms, we first compute a linear representation of e's content that takes into account e's position in the document, and then apply a variant of the proximity score by Büttcher et al. [4] on that linearization.

Figure 1 shows an example for the linearization process. We start with the sequence of terms in the element's content. Now, as different elements often discuss different topics or different aspects of a topic, we aim at giving a higher weight to terms that occur together in the same element than to terms occurring close together, but in different elements. To reflect this in the linearization, we introduce virtual gaps at the borders of certain elements, whose sizes depend on the element's tag (or, more generally, on the tags of the path from the document's

S. Geva, J. Kamps, and A. Trotman (Eds.): INEX 2008, LNCS 5631, pp. 29–32, 2009.

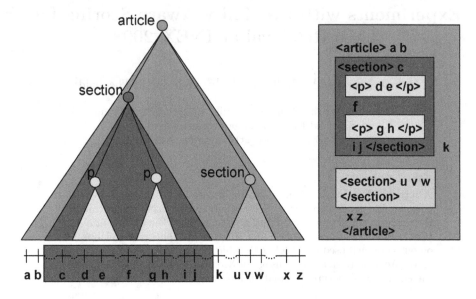

Fig. 1. An XML document and its linearization

root to the element). In the example, gaps of section elements may be larger than those of p (paragraph) elements, because the content of two adjacent p elements within the same section element may be considered related, whereas the content of two adjacent section elements could be less related. Some elements (like those used purely for layout purposes such as bold or for navigational purposes such as link) may get a zero gap size. The best choice for gaps depends on the collection. Gap sizes are currently chosen manually; an automated selection of gap sizes is subject to future work.

Based on the linearization, we apply the proximity scoring model of Büttcher et al. [4] for each element in the collection to find the best matches for a query $q = \{t_1, \ldots, t_n\}$ with multiple terms. This model linearly combines, for each query term, a BM25 content score and a BM25-style proximity score into a proximity-aware score. Note that unlike the original, we compute these scores for elements, not for documents, so the query-independent term weights in the formulas are inverse *element* frequencies $ief(t) = \log_2 \frac{N - ef(t) + 0.5}{ef(t) + 1}$, where N is the number of elements in the collection and $ef(t)$ is the number of elements that contain the term t. Similarly, average and actual lengths are computed for elements. The BM25 score of an element e for a query q is

$$score_{\text{BM25}}(e, q) = \sum_{t \in q} ief(t) \frac{tf(e, t) \cdot (k_1 + 1)}{tf(e, t) + K}$$

To compute the proximity part of the score, Büttcher et al. first compute an accumulated interim score $acc(t_i)$ for each query term t_i that depends on the distance of this term's occurrences in the element to other, adjacent query term occurrences. Formally, for each adjacent occurrence of a term t_j at distance d to an occurrence of t_i, $acc(t_i)$ grows by $ief(t_j)/d^2$. The proximity part of an element's score is then computed by plugging the acc values into a BM25-style scoring function:

$$score_{prox}(e,q) = \sum_{t \in q} min\{1, ief(t)\} \frac{acc(t) \cdot (k_1 + 1)}{acc(t) + K}$$

where, $K=k \cdot [(1-b) + b \cdot \frac{|e|}{avgel}]$ (analogously to the BM25 formula) and b, k_1, and k are configurable parameters that are set to $b = 0.5$ and $k = k_1 = 1.2$, respectively. In our modified version we consider *every* query term occurrence, not only adjacent query terms which allows for index precomputation without knowing the query load. The overall score is then the sum of the BM25 score and the proximity score:

$$score(e,q) = score_{BM25}(e,q) + score_{prox}(e,q)$$

3 AdHoc Track Results

3.1 Results for Focused Task

Our recent development of TopX focused on improving its retrieval quality. For the Focused Task, we submitted the following three runs:

- TopX-CO-Baseline-articleOnly: a CO run that considered the non-stem-med terms in the title of a topic (including the terms in phrases, but not their sequence) except terms in negations and stop words. We restricted the collection to the top-level article elements and computed the 1,500 articles with the highest $score_{BM25}$ value as described in Section 2. Note that this approach corresponds to standard document-level retrieval.
- TopX-CO-Proximity-articleOnly: a CO run that reranked the results of the TopX-CO-Baseline-articleOnly run by adding the $score_{prox}$ part described in Section 2. We used gaps of size 30 for section and p elements. (Due to the limited number of runs we could not evaluate different gap sizes; see [2] for a more thorough study with older INEX topics.)
- TopX-CO-Focused-all: an element-level CO run that considered the terms in the title of a topic without phrases and negations, allowing all tags for results. Note that, unlike our runs in previous years, we did not use a tag-specific ief score, but a single global ief value per term; we demonstrated in [3] that this gives better results for CO queries than tag-specific inverse element frequencies.

Table 1 shows the official results for these runs. It is evident that element-level retrieval generally yields a higher early precision than article-level retrieval, but

Table 1. Results for the Focused Task: interpolated precision at different recall levels (ranks for iP[0.01] are in parentheses) and mean average interpolated precision

run	iP[0.00]	iP[0.01]	iP[0.05]	iP[0.10]	MAiP
TopX-CO-Proximity-articleOnly	0.6808	0.6799 (3)	0.5812	0.5372	0.2981
TopX-CO-Baseline-articleOnly	0.6705	0.6694 (4)	0.5944	0.5474	0.2963
TopX-CO-Focused-all	0.7480	0.6471 (11)	0.5339	0.4687	0.1857

the quality quickly falls behind that of article-level retrieval. This is reflected in the official results where our article-level runs are at positions 3 and 4, whereas the element-level run is at position 11. Proximity scoring with gaps can in general help to improve early precision with article-level retrieval, at the cost of a slightly reduced recall. However, the MAiP average of the proximity-based run slightly improves over the baseline without proximity.

3.2   Other Tasks

We submitted a run to each of the other two tasks in the AdHoc track, where each of them was based on the CO titles of topics and the BM25-style element-level score shown in Section 2. To produce the runs for the RelevantInContext task, we ran TopX in document mode which generated a list of documents ordered by the highest score of any element within the document, together with a list of elements and their scores for each document. This yielded reasonable results with a MAgP value of 0.19470553, corresponding to rank 6 of all runs; this is a good improvement over 2007, which we mainly attribute to the better performance of the new scoring function.

To compute the best entry point for a document, we post-processed the RelevantInContext runs by simply selecting the element with highest score from each document and ordered them by score. This yielded reasonable results as well, with a MAgP value of 0.17081437, corresponding to rank 13 among all runs.

4   Conclusions and Future Work

This paper presented a structure-aware proximity score for XML retrieval that helps to improve the retrieval effectiveness of gap-free approaches for article-level retrieval. Our future work will focus on automatic methods to determine good gap sizes for elements, determining characteristics for queries where proximity boosts performance, and extending proximity scoring to queries with structural constraints.

References

[1] Beigbeder, M.: ENSM-SE at INEX 2007: Scoring with proximity. In: Preproceedings of the 6th INEX Workshop, pp. 53–55 (2007)
[2] Broschart, A., Schenkel, R.: Proximity-aware scoring for XML retrieval. In: SIGIR, pp. 845–846 (2008)
[3] Broschart, A., Schenkel, R., Theobald, M., Weikum, G.: TopX @ INEX 2007. In: Fuhr, N., Kamps, J., Lalmas, M., Trotman, A. (eds.) INEX 2007. LNCS, vol. 4862, pp. 49–56. Springer, Heidelberg (2008)
[4] Büttcher, S., Clarke, C.L.A., Lushman, B.: Term proximity scoring for ad-hoc retrieval on very large text collections. In: SIGIR, pp. 621–622 (2006)
[5] de Kretser, O., Moffat, A.: Effective document presentation with a locality-based similarity heuristic. In: SIGIR, pp. 113–120 (1999)
[6] Rasolofo, Y., Savoy, J.: Term proximity scoring for keyword-based retrieval systems. In: Sebastiani, F. (ed.) ECIR 2003. LNCS, vol. 2633, pp. 207–218. Springer, Heidelberg (2003)
[7] Song, R., Taylor, M.J., Wen, J.-R., Hon, H.-W., Yu, Y.: Viewing term proximity from a different perspective. In: Macdonald, C., Ounis, I., Plachouras, V., Ruthven, I., White, R.W. (eds.) ECIR 2008. LNCS, vol. 4956, pp. 346–357. Springer, Heidelberg (2008)

Finding Good Elements for Focused Retrieval

Carolyn J. Crouch, Donald B. Crouch, Salil Bapat,
Sarika Mehta, and Darshan Paranjape

Department of Computer Science
University of Minnesota Duluth
Duluth, MN 55812
(218) 726-7607
ccrouch@d.umn.edu

Abstract. This paper describes the integration of our methodology for the dynamic retrieval of XML elements [2] with traditional article retrieval to facilitate the Focused and the Relevant-in-Context Tasks of the INEX 2008 Ad Hoc Track. The particular problems that arise for dynamic element retrieval in working with text containing both tagged and untagged elements have been solved [3]. The current challenge involves utilizing its ability to produce a rank-ordered list of elements in the context of focused retrieval. Our system is based on the Vector Space Model [8]; basic functions are performed using the Smart experimental retrieval system [7]. Experimental results are reported for the Focused, Relevant-in-Context, and Best-in-Context Tasks of both the 2007 and 2008 INEX Ad Hoc Tracks. These results indicate that the goal of our 2008 investigations—namely, finding good focused elements in the context of the Wikipedia collection--has been achieved.

1 Introduction

Our work for INEX 2008 centers on producing good elements in a focused retrieval environment. Dynamic element retrieval—i.e., the dynamic retrieval of elements at the desired degree of granularity—has been the focus of our investigations at INEX for some time [2, 3]. We have demonstrated that our method works well for both structured [1] and semi-structured text [3] and that it produces a result identical to that produced by the search of the same query against the corresponding all-element index [5]. In [3], we show that dynamic element retrieval (with terminal node expansion) produces a result considerably higher than that reported by the top-ranked participant for the INEX 2006 Thorough task. The picture changes, however, when overlap is no longer allowed—i.e., when the task changes to focused retrieval. A review of our INEX 2007 Ad Hoc results for all three tasks in [3] shows that in each case, our results rank in the mid-range of participant scores. In 2008, our goal is to improve those results. Since the Ad Hoc tasks for INEX 2008 are identical to those of INEX 2007 and the evaluation procedures remain largely unchanged as well, we are able to compare our 2008 results not only to those of other participants but also to our own earlier (2007) results for the same tasks.

S. Geva, J. Kamps, and A. Trotman (Eds.): INEX 2008, LNCS 5631, pp. 33–38, 2009.
© Springer-Verlag Berlin Heidelberg 2009

2 Experiments with the INEX 2007 and 2008 Collections

In this section, we include results produced by our current methods for the three INEX Ad Hoc tasks, namely, Focused, Relevant-in-Context (RiC), and Best-in-Context (BiC). To produce its best results, our system needs tuning to establish appropriate values for *Lnu-ltu* term weighting with respect to a metric (e.g., iP[0.01] for the Focused task); these results reflect that tuning. (See [3] for details.)

2.1 Focused Task Methodology

There are two important issues which arise with respect to the Focused and Relevant-in-Context (RIC) tasks. The first relates to how the documents of interest (i.e., with respect to a specific query) are identified. The second is the method by which the focused elements are selected from those documents. The results reported in [2] for 2007 compare the values achieved by dynamic element retrieval to (base case) all-element retrieval. The focused elements themselves are selected based on correlation (i.e., the highest-correlating element along a path is chosen). Thus, for these experiments, the documents of interest were determined by the process of dynamic element retrieval (see [2] for details), and the focused elements were selected based on correlation.

To improve these results, in 2008 we revised our approach to focused retrieval by incorporating dynamic element retrieval with article retrieval as follows. For each query, we retrieve n articles or documents. We then use dynamic element retrieval to produce the rank-ordered list of all elements (from these n documents) having a positive correlation with the query. A parameter, m, represents the upper bound on the number of focused elements that are reported after overlap is removed by one of the three strategies described below. This is the method by which focused elements are produced for the 2007 and 2008 Focused and RiC tasks reported here. Negative terms (those preceded by a minus) are removed from both query sets.

Three focusing or overlap removal strategies were investigated in these experiments. The *section strategy* chooses as the focused element the highest correlating element along a path which is not a body. (Most of these turn out in fact to be sections. A body element appears in this list of focused elements if and only if none of its child elements are ranked within the top m.). The *correlation strategy* chooses the highest correlating element along a path as the focused element, without restriction on element type, And the *child strategy* chooses the terminal element along a path as the focused element (i.e., ignores correlation and always gives precedence to the child rather than the parent).

All 2008 runs are evaluated against the most recent (i.e., corrected) relevance assessments provided by INEX.

2.2 Ad Hoc Focused Task

The results produced for the INEX 2007 Focused Task are given in Table 1. In these experiments, overlap is removed using the *child strategy*. Best results are produced when at most 25 documents are retrieved and 500 elements reported ($n = 25$, $m = 500$); at 0.5386 this value would place at rank 1 in the 2007 rankings. (If negative terms are not omitted from the query set, the best value is 0.5293 at $n = 50$ and $m = 750$, which still exceeds the first place value in the rankings). Details of these experiments are reported in [1, 6].

The results produced by the identical methodology (but using the *section* focusing strategy) for the INEX 2008 Focused Task are given in Table 2. Best case results here are produced at $n = 25$ and $m = 100$, where the value of iP[0.01] is 0.6709 (equivalent to rank 4 in the 2008 rankings). Our INEX 2008 submission (which included negative query terms and utilized the child strategy for overlap removal) produced a value of 0.6309 at iP[0.01] for a rank of 19.

Table 1. iP[0.01] Results for 2007 Focused Task (Child Strategy)

NUMBER OF DOCUMENTS	NUMBER OF ELEMENTS										
	50	100	150	200	250	500	1000	1500	2000	3000	4000
25	0.4972	0.4734	0.5089	0.5144	0.5383	0.5386	0.5381	0.5381	0.5381	0.5381	0.5381
50	0.4751	0.4736	0.4747	0.4943	0.5338	0.5367	0.5343	0.5343	0.5343	0.5343	0.5343
100	0.4510	0.4429	0.4315	0.4538	0.4962	0.5242	0.5273	0.5238	0.5238	0.5238	0.5238
150	0.4398	0.4435	0.4351	0.4369	0.4557	0.5016	0.5066	0.5002	0.4964	0.4964	0.4964
200	0.4379	0.4445	0.4477	0.4361	0.4480	0.4829	0.4894	0.4928	0.4932	0.4893	0.4893
250	0.4697	0.4376	0.4513	0.4361	0.4603	0.4920	0.5031	0.5148	0.5136	0.5098	0.5098
500	0.4664	0.4244	0.4508	0.4444	0.4589	0.4746	0.4990	0.4900	0.4995	0.5052	0.5058

Table 2. iP[0.01] Results for 2008 Focused Task (Section Strategy)

NUMBER OF DOCUMENTS	NUMBER OF ELEMENTS										
	50	100	150	200	250	500	1000	1500	2000	3000	4000
25	0.6589	0.6457	0.6451	0.6440	0.6458	0.6455	0.6455	0.6455	0.6455	0.6455	0.6455
50	0.6698	0.6471	0.6580	0.6486	0.6533	0.6521	0.6521	0.6521	0.6521	0.6521	0.6521
100	0.6709	0.6572	0.6443	0.6371	0.6384	0.6450	0.6446	0.6443	0.6443	0.6443	0.6443
150	0.6702	0.6572	0.6487	0.6412	0.6332	0.6372	0.6411	0.6409	0.6410	0.6410	0.6410
200	0.6679	0.6613	0.6518	0.6445	0.6339	0.6313	0.6348	0.6329	0.6327	0.6327	0.6327
250	0.6697	0.6748	0.6555	0.6488	0.6511	0.6357	0.6353	0.6356	0.6355	0.6355	0.6355
500	0.6662	0.6689	0.6544	0.6480	0.6517	0.6269	0.6336	0.6334	0.6317	0.6337	0.6333

Tables 3 and 4 show the results of the corresponding experiments for the 2008 Focused Task when *the correlation* and *child* focusing strategies, respectively, are used. Although the *correlation strategy* shows relatively decent results on the whole, the *child strategy* is clearly not competitive.

Table 3. iP[0.01] Results for 2008 Focused Task (Correlation Strategy)

NUMBER OF DOCUMENTS	NUMBER OF ELEMENTS										
	50	100	150	200	250	500	1000	1500	2000	3000	4000
25	0.6563	0.6599	0.6610	0.6620	0.6620	0.6621	0.6621	0.6621	0.6621	0.6621	0.6621
50	0.6576	0.6598	0.6603	0.6605	0.6607	0.6611	0.6611	0.6611	0.6611	0.6611	0.6611
100	0.6568	0.6592	0.6592	0.6592	0.6595	0.6595	0.6599	0.6599	0.6599	0.6599	0.6599
150	0.6563	0.6582	0.6583	0.6585	0.6585	0.6586	0.6592	0.6593	0.6593	0.6593	0.6593
200	0.6539	0.6554	0.6554	0.6556	0.6558	0.6558	0.6565	0.6565	0.6565	0.6565	0.6565
250	0.6553	0.6564	0.6565	0.6568	0.6568	0.6570	0.6571	0.6576	0.6576	0.6576	0.6576
500	0.6533	0.6544	0.6546	0.6549	0.6550	0.6551	0.6554	0.6554	0.6554	0.6560	0.6561

Table 4. iP[0.01] Results for 2008 Focused Task (Child Strategy)

NUMBER OF DOCUMENTS	NUMBER OF ELEMENTS										
	50	100	150	200	250	500	1000	1500	2000	3000	4000
25	0.6252	0.6213	0.6155	0.6151	0.6266	0.6150	0.6150	0.6150	0.6150	0.6150	0.6150
50	0.6107	0.5961	0.6083	0.5800	0.6015	0.6155	0.6057	0.6057	0.6057	0.6057	0.6057
100	0.6162	0.5822	0.5698	0.5755	0.5778	0.5890	0.6035	0.6003	0.6003	0.6003	0.6003
150	0.6160	0.5975	0.5757	0.5630	0.5643	0.5751	0.5903	0.5907	0.5906	0.5906	0.5906
200	0.6107	0.5958	0.5742	0.5734	0.5661	0.5699	0.5840	0.5754	0.5819	0.5820	0.5820
250	0.6119	0.6065	0.5955	0.5846	0.5782	0.5733	0.5779	0.5904	0.5784	0.5849	0.5786
500	0.6082	0.5993	0.6060	0.5870	0.5941	0.5516	0.5762	0.5690	0.5759	0.5679	0.5717

2.3 Ad Hoc Retrieval-in-Context Task

The RiC results are produced using the elements produced by the Focused task and grouping them by article. Results are reported in document-rank order. Table 5 shows the 2007 best case result with an MAgP value of 0.1415 at $n = 250$, $m = 4000$ (rank 9 for the 2007 rankings). (See [1, 6] for details.) 2008 RiC results are shown in Table 6. The best value based on the *section* focusing strategy is achieved at $n = 500$, $m = 2000$, where MAgP = 0.1771 (which would rank at 15 for 2008). The *correlation* focusing strategy produces very similar but slightly higher values (e.g., 0.1783 at $n = 500$, $m = 2000$) across the table; it is not shown here. Our submitted run (which includes negative terms) produced a MAgP value of 0.1723 and ranked 18.

Table 5. MAgP Results for 2007 RIC Task (Child Strategy)

NUMBER OF DOCUMENTS	NUMBER OF ELEMENTS										
	50	100	150	200	250	500	1000	1500	2000	3000	4000
25	0.0767	0.0821	0.0865	0.0915	0.0951	0.0978	0.0977	0.0977	0.0977	0.0977	0.0977
50	0.0772	0.0889	0.0946	0.0980	0.1034	0.1129	0.1155	0.1154	0.1154	0.1154	0.1154
100	0.0772	0.0908	0.0989	0.1021	0.1098	0.1190	0.1305	0.1310	0.1315	0.1313	0.1313
150	0.0741	0.0913	0.0991	0.1048	0.1084	0.1197	0.1288	0.1345	0.1343	0.1346	0.1343
200	0.0750	0.0918	0.0982	0.1041	0.1088	0.1203	0.1299	0.1350	0.1375	0.1371	0.1374
250	0.0771	0.0921	0.0987	0.1040	0.1113	0.1243	0.1330	0.1385	0.1408	0.1408	0.1415
500	0.0772	0.0911	0.0985	0.1032	0.1112	0.1230	0.1326	0.1373	0.1393	0.1391	0.1405

Table 6. MAgP Results for 2008 RIC Task (Section Strategy)

NUMBER OF DOCUMENTS	NUMBER OF ELEMENTS										
	50	100	150	200	250	500	1000	1500	2000	3000	4000
25	0.1004	0.1047	0.1050	0.1048	0.1048	0.1038	0.1038	0.1038	0.1038	0.1038	0.1038
50	0.1141	0.1231	0.1260	0.1294	0.1296	0.1279	0.1271	0.1271	0.1271	0.1271	0.1271
100	0.1153	0.1343	0.1385	0.1418	0.1445	0.1506	0.1478	0.1471	0.1471	0.1471	0.1471
150	0.1139	0.1362	0.1437	0.1478	0.1499	0.1580	0.1586	0.1571	0.1570	0.1565	0.1565
200	0.1143	0.1367	0.1444	0.1488	0.1524	0.1607	0.1650	0.1633	0.1627	0.1617	0.1617
250	0.1146	0.1353	0.1464	0.1499	0.1533	0.1635	0.1689	0.1674	0.1665	0.1659	0.1652
500	0.1153	0.1350	0.1468	0.1551	0.1561	0.1662	0.1734	0.1771	0.1768	0.1752	0.1728

2.4 Ad Hoc Best-in-Context Task

Finding the Best Entry Point (BEP) is a task which is not related to focused retrieval. In 2007, we examined a number of factors which might be useful in determining the

BEP, including correlation, tag set membership, physical location and combinations of these factors [4]. For 2007, best results were obtained based purely on physical position. In this case, a MAgP value of 0.1729, based on RiC input, was generated; this value would appear at rank 10 in the 2007 rankings. A very similar result (MAgP = 0.1722) was produced by a combination of two factors (tag set membership and location). In 2008, our experiments with BEP were based purely on physical location; they use the name tag as BEP. Best results are achieved when 1500 articles are retrieved, with MAgP = 0.1875 (which would rank at 9 in the 2008 rankings).

3 Conclusions

Our 2008 agenda has centered on producing good focused elements. Our basic method (which used the earlier, less effective *child strategy* focusing technique) nevertheless performs very well for the 2007 Focused task, exceeding all ranked results. When we apply the same methodology to the same task in 2008 (changing only to the more effective *section strategy* for focusing), the best results rank near the top of the 2008 rankings. Ongoing experiments (not reported here) encourage us to believe that we can substantially improve on these results.

With respect to the Relevant-in-Context task experiments, our current results (falling at rank 9 for 2007 and rank 15 for 2008) could clearly be improved. Best-in-Context results are acceptable but would benefit from more experimentation.

References

[1] Bapat, S.: Improving the results for focused and relevant-in-context tasks. M.S. Thesis, Department of Computer Science, University of Minnesota Duluth (2008),
http://www.d.umn.edu/cs/thesis/bapat.pdf
[2] Crouch, C.: Dynamic element retrieval in a structured environment. ACM TOIS 24(4), 437–454 (2006)
[3] Crouch, C., Crouch, D., Kamat, N., Malik, V., Mone, A.: Dynamic element retrieval in the Wikipedia collection. In: Fuhr, N., Kamps, J., Lalmas, M., Trotman, A. (eds.) INEX 2007. LNCS, vol. 4862, pp. 70–79. Springer, Heidelberg (2008)
[4] Mehta, S.: Finding the best entry point. M.S. Thesis, Department of Computer Science, University of Minnesota Duluth (2008), http://www.d.umn.edu/cs/thesis/mehta.pdf
[5] Mone, A.: Dynamic element retrieval for semi-structured documents. M.S. Thesis, Department of Computer Science, University of Minnesota Duluth (2007),
http://www.d.umn.edu/cs/thesis/mone.pdf
[6] Paranjape, D.: Improving focused retrieval. M.S. Thesis, Department of Computer Science, University of Minnesota Duluth (2007),
http://www.d.umn.edu/cs/thesis/paranjape.pdf
[7] Salton, G. (ed.): The Smart Rretrieval System—Experiments in Automatic Document Processing. Prentice-Hall, Englewood Cliffs (1971)
[8] Salton, G., Wong, A., Yang, C.S.: A vector space model for automatic indexing. Comm. ACM 18(11), 613–620 (1975)

New Utility Models for the Garnata Information Retrieval System at INEX'08

Luis M. de Campos, Juan M. Fernández-Luna, Juan F. Huete,
Carlos Martín-Dancausa, and Alfonso E. Romero

Departamento de Ciencias de la Computación e Inteligencia Artificial
E.T.S.I. Informática y de Telecomunicación, Universidad de Granada,
18071 – Granada, Spain
{lci,jmfluna,jhg,cmdanca,aeromero}@decsai.ugr.es

Abstract. In this work we propose new utility models for the structured information retrieval system Garnata, and expose the results of our participation at INEX'08 in the AdHoc track using this system.

1 Introduction

Garnata [5] is a Structured Information Retrieval System for XML documents, based on probabilistic graphical models [8,9], developed by members of the research group "Uncertainty Treatment in Artificial Intelligence" at the University of Granada. Garnata has already been tested at two editions of the INEX Workshop [4,6], and its theoretical basis is explained in more detail in [1,2].

Garnata computes the relevance degree of each component or structural unit in a document by combining two different types of information. On the one hand, the specificity of the component with respect to the query: the more terms in the component appear in the query, the more relevant becomes the component, that is to say, the more clearly the component is only about (at least a part of) the topic of the query. On the other hand, the exhaustivity of the component with respect to the query: the more terms in the query match with terms in the component, the more relevant the component is, i.e., the more clearly the component comprises the topic of the query. The components that best satisfy the user information need expressed by means of the query should be, simultaneously, as specific and exhaustive as possible.

These two dimensions of the relevance of a component with respect to the query are calculated in a different way. To compute the specificity, the probability of relevance of each component is obtained through an inference process in a Bayesian network representing the structured document collection. The exhaustivity is obtained by first defining the utility of each component as a function of the proportion of the terms in the query that appear in this component. Then the Bayesian network is transformed into an influence diagram which computes the expected utility of each component, by combining the probabilities of relevance and the utilities in a principled way.

S. Geva, J. Kamps, and A. Trotman (Eds.): INEX 2008, LNCS 5631, pp. 39–45, 2009.

In this work we propose a modification of the system by defining the utility in a different manner, in such a way that those components that do not contain most of the query terms are penalized more heavily. By defining a parametric model, it is possible to adjust the degree of utility to make the system behave more similarly to a strict AND (if not all or almost all the query terms are in the considered component, this one will be scarcely relevant) or to a less strict AND.

2 Utility Models in the Garnata System

As we focus in this work on the utility component of the Garnata system, we will not enter into details of the Bayesian network model representing the document collection. This model is able to efficiently compute the posterior probabilities of relevance of all the structural units U of all the documents, given a query Q, $p(U|Q)$. These probabilities represent the specificity component of each structural unit U: the more terms indexing U also belong to Q, the more probable is U.

The Bayesian network is then enlarged by including decision variables R_U, representing the possible alternatives available to the decision maker (retrieve unit U), and utility variables V_U, thus transforming it into an influence diagram. The objective is to compute the expected utility of each decision given Q, $EU(R_U|Q)$.

In Garnata the utility value V_U of each structural unit U is made of a component which depends on the involved unit, other component which depends only on the kind of tag associated to that unit, and another component independent on the specific unit (these three components are multiplied in order to form the utility value, see [4]).

The part depending on the involved unit, which is the only one we are going to modify, is defined as the sum of the inverted document frequencies of those terms contained in U that also belong to the query Q, normalized by the sum of the idfs of the terms contained in the query: a unit U will be more useful (more exhaustive), with respect to a query Q, as more terms of Q also belong to U:

$$nidf_Q(U) = \frac{\sum_{T \in An(U) \cap Q} idf(T)}{\sum_{T \in Q} idf(T)} \tag{1}$$

$An(U)$ in the previous equation represents the set of terms contained (either directly or indirectly) in the structural unit U.

3 New Utility Models

As it can be observed from Eq. (1), the utility or exhaustivity of a structural unit U with respect to a query Q grows linearly with the number of query terms appearing in U (reaching a maximum equal to 1 when all the terms of the query appear in the unit). In our experience with the system in different applications [3,4], we have observed that this linear growing, when combined with the probabilities computed from the Bayesian network (which measure specificity), can cause that small structural units, which only match with a fraction of the query terms, become more relevant that other, greater structural units that contain more terms from the query. In many cases this behaviour is not the expected one, because probably a user who employs several terms to express his/her query is expecting to find most of these terms in the structural units obtained as the answer of the system to this query. For that reason we believe that it is interesting to define other utility models which give more importance (in a non-linear way) to the appearance of most of the terms in the query.

In this work we propose a parametric non-linear utility model that, as the parameter grows, the more terms from the query must be contained in a structural unit in order to get a high utility value for this unit. A way of obtaining this behaviour is through the use of the following transformation:

$$nidf_{Q,n}(U) = nidf_Q(U)\frac{e^{(nidf_Q(U))^n} - 1}{e - 1} \tag{2}$$

In this way, when $n = 0$ we have $nidf_{Q,0}(U) = nidf_Q(U)$, that is to say, we reproduce the original model, and the greater the value of the integer parameter n, we obtain a behaviour more similar to a strict AND operator. In Figure 1 we can observe several plots of the function $x\frac{e^{x^n}-1}{e-1}$ for different values of n.

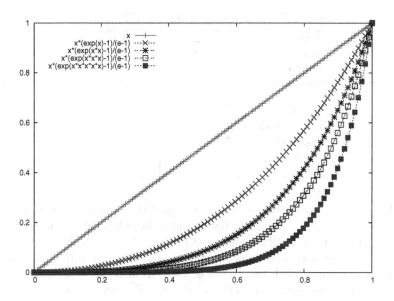

Fig. 1. Function $x\frac{e^{x^n}-1}{e-1}$, for $n = 0, 1, 2, 3, 5$

4   Experimental Results

In this INEX 2008 edition, we have participated submitting nine runs in the AdHoc track (content only). More specifically, three in each of the *Focused*, *Relevant in Context* and *Best in Context* sub-tasks. Table 1 shows the positions in the ranking according to the official evaluation measures (*MAgP* for *Best in Context* and *Relevant in Context*, and $iP[0.01]$ for *Focused*), the sub-task and finally the run identifier.

Table 1. Runs submitted to the INEX'2008 AdHoc tasks and positions in the rankings

Position	Value	Sub-task	RunId
52	0.468856	Focused	p8_u3_exp_5_1110
53	0.467071	Focused	p8_u3_exp_3_1110
54	0.448733	Focused	p15_u3_exp_5_1110
25	0.158177	Relevant in Context	p8_u3_exp_5_1110
26	0.158177	Relevant in Context	p8_u3_exp_5_0100
27	0.152320	Relevant in Context	p8_u3_exp_3_1110
18	0.146799	Best in Context	p8_u3_exp_5_0100
19	0.146536	Best in Context	p8_u3_exp_3_0100
22	0.138141	Best in Context	p15_u3_exp_3_0100

Table 2. Importance of the different types of units used in the official runs

Tag	Weight file 8	Weight file 15
name	20	200
title	20	50
caption	10	30
collectionlink	10	30
emph2	10	30
emph3	10	30
conversionwarning	0	0
languagelink	0	0
template	0	0
default value	1	1

With respect to the parameters, we have used the weight files 8 and 15 (p8 and p15 as prefixes of the run identifiers), and utility file 3 (u3, contained in the identifiers), with the first values presented in Table 2 and in Table 3 the second ones (see [4] for details about these parameters and their use within the model). We have experimented with two values of the parameter n in Eq. (2), 3 and 5 (exp_3 and exp_5, also contained in the identifiers). These values were selected by means of experimentation with previous INEX collections. Finally, the suffix of the run identifier corresponds to the values of each of the four configurations of the component of the utility function independent on the involved unit (see [4]).

Table 3. Relative utility values of the different types of units used in the official runs

Tag	Utility file 3	Tag	Utility file 3
conversionwarning	0	section	1.25
name	0.85	p	1.5
title	0.85	body	2.0
collectionlink	0.75	emph2	1.0
languagelink	0.0	emph3	1.0
article	2.5	default value	1.0

Table 4. Comparison between runs with and without applying the transformation in Eq. (2)

With $nidf_Q(U)$	With $nidf_{Q,n}(U)$	%Change	Sub-tasks	Run Id.
0.366249	0.468856	28.01	Focused	p8_u3_exp_5_1110
0.366249	0.467071	27.53	Focused	p8_u3_exp_3_1110
0.341804	0.448733	31.28	Focused	p15_u3_exp_5_1110
0.083034	0.158177	90.50	Relevant in Context	p8_u3_exp_5_1110
0.067706	0.158177	133.62	Relevant in Context	p8_u3_exp_5_0100
0.083034	0.152320	83.44	Relevant in Context	p8_u3_exp_3_1110
0.075842	0.146799	93.56	Best in Context	p8_u3_exp_5_0100
0.075842	0.146536	93.21	Best in Context	p8_u3_exp_3_0100
0.078910	0.138141	75.06	Best in Context	p15_u3_exp_3_0100

Although there has been a significant reduction of runs submitted in this 2008 edition – measured as focused retrieval – (*Focused*: from 79 last year to 61 this edition; *Relevant in Context*: from 66 to 40; *Best in context*: from 71 to 35), we could say that in terms of the percentiles of the positions in the rankings, we are improving our results in *Relevant in Context* (from percentiles 66-74 last year, to 62-67 this year) and *Best in Context* (from 63-70 to 51-62), and keeps more or less the same positions in *Focused* (from 84-89 to 85-88).

It is noticeable that within the *Focused* task, Garnata's performance is relatively low, and keeps more or less the same positions than last year, and how the methods described in [4] for adjusting the output for the requirements of the other two tasks make a good job from the raw results generated by Garnata. Clearly *Best in Context* is the sub-task where the performance is higher, and where the best improvement is achieved.

In order to better determine the improvement obtained by the new utility model presented in this paper, we have run an experiment without using the transformation presented in Eq. (2), but applying instead the original Eq. (1), $nidf_Q(U)$. Table 4 shows the values of the official evaluation measures with the old utility model used in previous editions (first column), this year with the new model (second column) and the percentage of change (third column). As noticed, the percentages of change are generally quite large, and this fact confirms our initial hypothesis that the new transformation could improve the results.

We have carried out another series of experiments, motivated by the following fact: we realised that among the systems obtaining the best results in the official competition at INEX'08 [7], there are many systems that do not return any possible structural unit as a result but only some of them, typically only content-bearing elements like section, paragraphs or the complete article. In contrast, our official runs retrieved almost any elements, and this may be a source of poor behaviour specially when removing overlapping elements. So, we have repeated our official experiments but filtering the results in order to retrieve only article, or only article, body, section and paragraph elements. This can be easily done by using an utility file giving weight zero to all the structural units except the selected ones (with weight equal to one). The results of these experiments are displayed in Table 5.

Table 5. Runs retrieving only content-bearing elements and positions in the rankings

article+section+...		only article			
Position	Value	Position	Value	Sub-task	RunId
48	0.517808	52	0.482262	Focused	p8_u3_exp_5_1110
46	0.524948	52	0.478478	Focused	p8_u3_exp_3_1110
52	0.474641	54	0.455649	Focused	p15_u3_exp_5_1110
20	0.171119	27	0.157455	Relevant in Context	p8_u3_exp_5_1110
24	0.164420	27	0.157455	Relevant in Context	p8_u3_exp_5_0100
22	0.168308	27	0.155347	Relevant in Context	p8_u3_exp_3_1110
20	0.146501	14	0.168893	Best in Context	p8_u3_exp_5_0100
22	0.140705	14	0.167468	Best in Context	p8_u3_exp_3_0100
24	0.131170	18	0.148391	Best in Context	p15_u3_exp_3_0100

We can observe that this strategy of retrieving only the more general elements is useful for the *Focused* and *Relevant in Context* tasks, where we would obtain better positions in the ranking (going from percentiles 85-88 to 75-85 in *Focused* and from 62-67 to 50-60 in *Relevant in Context*, when using the four elements selected). However, the results are slightly worse for the *Best in Context* task (going from percentiles 51-63 to 57-68) in the case of using the four elements but better when using only the article element. These results point out that the choice of the structural elements to be retrieved has a non-negligible impact on the performance of an XML retrieval system.

5 Concluding Remarks

In this paper we have presented the participation of the University of Granada group in the 2008 INEX edition in the AdHoc tasks. This is based on the work developed in previous years, but introducing a new utility model which gives more importance (in a non-linear way) to the appearance of most of the terms in the query. We have shown in the previous section that this new approach considerably improves the results with respect to not using it.

With respect to the comparison of our results with the rest of participants, we could say that we are in the middle of the rankings, improving with respect to the last edition of INEX.

Regarding future research in the context of INEX, we have to work in the improvement of the raw results of Garnata, as they are the base for the different sub-tasks, and in the filtering strategy used to remove overlapping elements. Also, we have designed an approach to answer CAS queries, which will be evaluated in the next edition of the evaluation campaign.

Acknowledgments. This work has been jointly supported by the Spanish Consejería de Innovación, Ciencia y Empresa de la Junta de Andalucía, Ministerio de Ciencia de Innovación and the research programme Consolider Ingenio 2010, under projects TIC-276, TIN2008-06566-C04-01 and CSD2007-00018, respectively.

References

1. de Campos, L.M., Fernández-Luna, J.M., Huete, J.F.: Using context information in structured document retrieval: An approach using Influence diagrams. Information Processing & Management 40(5), 829–847 (2004)
2. De Campos, L.M., Fernández-Luna, J.M., Huete, J.F.: Improving the context-based influence diagram for structured retrieval. In: Losada, D.E., Fernández-Luna, J.M. (eds.) ECIR 2005. LNCS, vol. 3408, pp. 215–229. Springer, Heidelberg (2005)
3. de Campos, L.M., Fernández-Luna, J.M., Huete, J.F., Martín, C., Romero, A.E.: An information retrieval system for parliamentary documents. In: Pourret, O., Naim, P., Marcot, B. (eds.) Bayesian Networks: A Practical Guide to Applications, pp. 203–223. Wiley, Chichester (2008)
4. de Campos, L.M., Fernández-Luna, J.M., Huete, J.F., Martín-Dancausa, C.J., Romero, A.E.: The Garnata information retrieval system at INEX 2007. In: Fuhr, N., Kamps, J., Lalmas, M., Trotman, A. (eds.) INEX 2007. LNCS, vol. 4862, pp. 57–69. Springer, Heidelberg (2008)
5. de Campos, L.M., Fernández-Luna, J.M., Huete, J.F., Romero, A.E.: Garnata: An information retrieval system for structured documents based on probabilistic graphical models. In: Proceedings of the Eleventh International Conference of Information Processing and Management of Uncertainty in Knowledge-Based Systems (IPMU), pp. 1024–1031 (2006)
6. de Campos, L.M., Fernández-Luna, J.M., Huete, J.F., Romero, A.E.: Influence diagrams and structured retrieval: Garnata implementing the SID and CID models at INEX 2006. In: Fuhr, N., Lalmas, M., Trotman, A. (eds.) INEX 2006. LNCS, vol. 4518, pp. 165–177. Springer, Heidelberg (2007)
7. Geva, S., Kamps, J., Trotman, A. (eds.): INEX 2008 Workshop Pre-proceedings (2008)
8. Jensen, F.V.: Bayesian Networks and Decision Graphs. Springer, Heidelberg (2001)
9. Pearl, J.: Probabilistic Reasoning in Intelligent Systems: Networks of Plausible Inference. Morgan and Kaufmann, San Mateo (1988)

UJM at INEX 2008: Pre-impacting of Tags Weights

Mathias Géry, Christine Largeron, and Franck Thollard

Université de Lyon, F-42023, Saint-Étienne, France
CNRS UMR 5516, Laboratoire Hubert Curien
Université de Saint-Étienne Jean Monnet, F-42023, France
{mathias.gery,christine.largeron,franck.thollard}@univ-st-etienne.fr

Abstract. This paper[1] addresses the impact of structure on terms weighting function in the context of focused Information Retrieval (IR). Our model considers a certain kind of structural information: tags that represent logical structure (title, section, paragraph, etc.) and tags related to formatting (bold, italic, center, etc.). We take into account the tags influence by estimating the probability that a tag distinguishes relevant terms. This weight is integrated in the terms weighting function. Experiments on a large collection during INEX 2008 IR competition showed improvements for focused retrieval.

1 Introduction

The focused information retrieval (IR) aims at exploiting the documents structure (e.g. HTML or XML markup) in order to retrieve the relevant elements (parts of documents) for a user information need. The structure can be used to emphasize some particular words or some parts of the document: the importance of a term depends on its formatting (e.g. bold font, italic, etc.), and also on its position in the document (e.g., title terms versus text body).

Different approaches have been proposed to integrate the structure at the step of querying or at the step of indexing. Following [2], we propose to integrate the structure in the weighting function: the weights of terms are based not only on the terms frequencies in the documents and in the collection, but also on the terms position in the documents. This position can be defined by XML tags. This approach raises two questions: how to choose the structural weights? How to integrate them in the classical models?

Some works propose to choose empirically the tags and their weights [5] or to learn them automatically using genetic algorithms [8]. These approaches use generally less than five tags. We propose to learn automatically the tags weights, without limit on the number of tags.

Concerning the integration of the structure weights, Robertson et al. suggests to preserve the non linearity of the BM25 weighting function by pre-impacting

---

[1] This work has been partly funded by the Web Intelligence project (région Rhône-Alpes, cf. http://www.web-intelligence-rhone-alpes.org)

S. Geva, J. Kamps, and A. Trotman (Eds.): INEX 2008, LNCS 5631, pp. 46–53, 2009.

structure on the terms frequencies instead of impacting it directly on the global terms weights [6]. We propose to apply this approach in the context of focused XML IR.

The main contribution of this paper is a formal framework integrating structure, introduced in the next section. We present in section 3 our experiments and in section 4 our results in the INEX 2008 competition.

2 A Structured Document Model

We consider in this paper the problem of extending the classical probabilistic model [7] that aims at estimating the relevance of a document for a given query through two probabilities: the probability of finding a relevant information and the probability of finding a non relevant information.

Our model takes into account the structure at two levels. Firstly, the logical structure (e.g. tags section, paragraph, table, etc.) is used in order to select the XML elements that are handled at the indexing step. These elements are the only ones that can be indexed, ranked and returned to the user. Secondly, the formatting structure (e.g. bold font, italic, etc.) and the logical structure are integrated into the terms weighting function. For a given tag we can estimate if it emphasizes terms in relevant documents or term in non relevant part of documents. A learning step computes a weight for each tag, based on the probability, to distinguish relevant terms and non relevant ones. At querying step, the relevance of an element is estimated based on the weights of the terms it contains, combined with the weights of the tags labeling those terms.

2.1 Term Based Score of XML Elements

The relevance of an element e_j for a query Q is function of the weights of the query terms t_i that appear in the element. We use the weighting function BM25 [7]:

$$w_{ji} = \frac{tf_{ji} * (k_1 + 1)}{k_1 * ((1 - b) + (b * ndl)) + tf_{ji}} * \log \frac{N - df_i + 0.5}{df_i + 0.5} \tag{1}$$

With tf_{ji}: frequency of t_i in e_j; N: number of elements in the collection; df_i: number of elements containing the term t_i; ndl: ratio between the length of e_j and the average element length; k_1 and b: classical BM25 parameters.

2.2 Tag Based Score of XML Elements

The relevance of an element e_j relatively to the tags is based on the weights, noted w'_{ik}, of each term t_i labelled by a tag b_k. We used a learning set LS in which the relevant elements for a given query are known. Given the set R (resp. NR) that contains the relevant (resp. non relevant) elements, a contingency table can be built:

	R	NR	$LS = R \cup NR$
$t_{ik} \in e_j$	r_{ik}	$nr_{ik} = n_{ik} - r_{ik}$	n_{ik}
$t_{ik} \notin e_j$	$R - r_{ik}$	$N - n_{ik} - R + r_{ik}$	$N - n_{ik}$
Total	R	$NR = N - R$	N

With R: number of relevant terms; NR: number of non relevant terms. r_{ik}: number of times term t_i labelled by b_k is relevant; $\sum_i r_{ik}$: number of relevant terms labelled by b_k; n_{ik}: number of times term t_i is labelled by b_k; $nr_{ik} = n_{ik} - r_{ik}$: number of times term t_i labelled by b_k is not relevant.

Then, w'_{ik} can be used to distinguish relevant terms from non relevant ones according to the tags that mark them. This is closely related to probabilistic IR model, but in our approach tags are considered instead of terms and terms instead of documents.

$$w'_{ik} = \frac{P(t_{ik}|R)(1 - P(t_{ik}|NR))}{P(t_{ik}|NR)(1 - P(t_{ik}|R))} = \frac{r_{ik} \times (NR - nr_{ik})}{nr_{ik} \times (R - r_{ik})} \tag{2}$$

Moreover, we hypothesize that the property for a tag to distinguish relevant terms does not depend on terms, *i.e.* the weight of a tag b_k should be the same for all terms. We finally estimate for each tag b_k a weight w'_k:

$$w'_k = \frac{\sum_{t_i \in T} w'_{ik}}{|T|} \tag{3}$$

2.3   Global Score of XML Elements

In order to compute a global score, we propose a linear combination f_{claw}[2] between the weight w_{ji} of a term t_i and the average of the weights w'_k of the tags b_k that mark the term[3]:

$$f_{claw}(e_j) = \sum_{t_{ik} \in e_j / t_i \in Q} w_{ji} \times \frac{\sum_{k/t_{ik}=1} w'_k}{|\{k/t_{ik} = 1\}|} \tag{4}$$

In previous experiments [3], f_{claw} slightly improved recall but the results were not convincing. Even if the estimation of the tag weights must be carefully addressed, it appears that the way such weights are integrated into the final score is essential. Following [6], we take advantage of the non linearity of BM25 by pre-impacting the tags weights at the term frequency level. More precisely, tf is replaced by ttf[4]in BM25:

$$ttf_{ji} = tf_{ji} \times \frac{\sum_{k/t_{ik}=1} w'_k}{|\{k/t_{ik} = 1\}|} \tag{5}$$

---

[2] CLAW: Combining Linearly Average tag-Weights.
[3] w_{ji}: the BM25 weight of term t_i in element e_j, cf. eq. 1.
[4] TTF: Tagged Term Frequency. $t_{ik} = 1$ means that t_i is labelled by b_k.

3   Experiments

We have experimented these models during the INEX 2008 IR competition in a classic IR way (granularity: full articles) as well as in a focused IR way (granularity: XML elements). The English Wikipedia XML corpus [1] contains 659,388 strongly structured articles, which are composed of 52 millions of XML elements (i.e. 79 elements on average; with an average depth of 6.72). The whole articles (textual content + XML structure) represent 4.5 Gb while the textual content only 1.6 Gb. The original Wiki syntax has been converted into XML, using both general tags of the logical structure (article, section, paragraph, title, list and item), formating tags (like bold, emphatic) and frequently occurring link-tags.

3.1   Experimental Protocol

The corpus enriched by the INEX 2006 assessments on 114 queries has been used as a training set in order to estimate the tags weights w'_k. We have evaluated our approach using the 70 queries of INEX 2008.

Our evaluation is based on the main INEX measures ($iP[x]$ the precision value at recall x, AiP the *interpolated average precision*, and $MAiP$ the *interpolated mean average precision* [4]). Note that the main ranking of INEX competition is based on $iP[0.01]$ instead of the overall measure $MAiP$, in order to take into account the importance of precision at low recall levels.

Each run submitted to INEX is a ranked list containing at most 1 500 XML elements for each query. Some runs retrieve all the relevant elements among the first 1 500 XML returned elements, and some others retrieve only part of them. Note that a limit based on a number of documents (instead of *e.g.* a number of bytes) allows to return more information and therefore favours runs composed by full articles. We have calculated $R[1500]$ (the recall at 1 500 elements) and $S[1500]$ (the size of these 1 500 elements in Mbytes).

3.2   Tags Weighting

We have manually selected 16 tags (*article, cadre, indentation1, item, li, normallist, numberlist, p, row, section, table, td, template, th, title, tr*) in order to define the XML elements to consider. These logical structure tags will be considered during the indexing step and therefore those will define the elements the system will be able to return.

Regarding the other tags (namely the formatting tags), we first selected the 61 tags that appear more than 300 times in the 659,388 documents. We then manually removed 6 tags: article, body (they mark the whole information), br, hr, s and value (considered not relevant).

The weights of the 55 remaining tags were computed according to equation w'_k in equation 3. Table 1 presents the top 6 tags and their weights, together with the weakest 6 ones and their weights. Their frequencies in the whole collection is also given.

Table 1. Weight w'_k of the 6 strongest and 6 weakest tags

\multicolumn{3}{Top strongest weights}			Top weakest weights		
tag	weight	freq.	tag	weight	freq.
h4	12,32	307	emph4	0,06	940
ul	2,70	3'050	font	0,07	27'117
sub	2,38	54'922	big	0,08	3'213
indentation1	2,04	135'420	em	0,11	608
section	2,01	1'610'183	b	0,13	11'297
blockquote	1,98	4'830	tt	0,14	6'841

4   Results: Focused Task

Our aim was firstly to obtain a strong baseline, secondly to experiment focused retrieval (i.e. elements granularity) against classic retrieval (i.e. full articles granularity), and thirdly to experiment the impact of tags weights in the BM25 weighting function. Table 2 presents the 3 runs that we have submitted to INEX 2008 Ad-Hoc in focused task. The structure is not taken into account in R1, where the documents are returned to the user (articles granularity) as well as in R2 where the elements are returned (elements granularity), while in R3 the tags weights are integrated in BM25 in a focused retrieval (elements granularity - TTF)

Table 2. Our 3 runs submitted to INEX 2008 Ad-Hoc, focused task

Run (name)	Granularity	Tags weights
R1 (JMU_expe_136)	articles	-
R2 (JMU_expe_141)	elements	-
R3 (JMU_expe_142)	elements	TTF

4.1   Parameters

The parameters of the chosen weighting functions (namely BM25) were tuned in order to improve classic retrieval (articles granularity) and focused retrieval (elements granularity). Among the parameters studied to improve the baseline, we can mention the use of a stoplist, the optimization of BM25 parameters ($k_1 = 1.1$ and $b = 0.75$), etc. Regarding the queries, we set up a better "andish" mode and consider *or* and *and, etc* .... Some specific parameters (*e.g.* the minimum size of the returned elements) were also tuned for focused retrieval.

Our baseline and all other runs have been obtained automatically, and using only the query terms (*i.e* the *title* field of INEX topics). We thus do not use fields *description, narrative* nor *castitle*.

4.2   INEX Ranking: $iP[0.01]$

Our system gives very interesting results compared to the best INEX systems. Our runs are compared on the figure 1 against *FOERStep*, the best run submitted to INEX 2008 according to $iP[0.01]$ ranking, on 61 runs yet evaluated in the focused task. This run outperforms our runs at very low recall levels. Our run $R1$ gives the best results at recall levels higher than 0.05. This is also shown by the $MAiP$ presented in table 3.

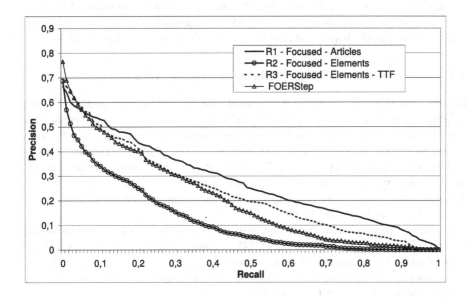

Fig. 1. Recall / Precision of 3 runs on 61 runs yet evaluated in the focused task

Table 3 presents the results of our 3 runs submitted to the track Ad-Hoc (focused task).

Table 3. Our 3 runs compared to 61 "focused" runs

Run (rank)	$iP[0.01]$	$MAiP$	$R[1500]$	$S[1500]$
FOERStep (winner)	**0.6897**	0.2076	0.4468	**78**
R1: articles (14)	0.6412	**0.2801**	**0.7871**	390
R2: elements (37)	0.5697	0.1208	0.2761	**51**
R3: elements+TTF (9)	**0.6640**	0.2347	0.6097	234

4.3   Articles Versus Elements

Our second aim was to compare classic retrieval of full articles versus focused retrieval of XML elements. We therefore indexed either the whole articles or the elements, and the parameters of the system were tuned also for focused retrieval.

It is interesting to notice that the BM25 model applied on full articles ($R1$) outperforms our focused retrieval results ($R2$) considering $MAiP$, despite the fact that BM25 parameter ndl is designed to take into account different documents lengths and thus documents granularities. Classic IR weighting functions, indexing and querying process, are undoubtedly not well adapted to focused retrieval. However, this is consistent with other results obtained during the INEX 2007 campaign where some top ranked systems only consider (and therefore return) full articles.

On the other hand, the focused run $R2$ returns a smallest quantity of information. Indeed, the total size of the 1 500 XML elements returned (for each query) is reduced to 51 Mb instead of 390 Mb for classic retrieval of full articles.

4.4   Pre-impacting of Tags Weights on Terms Weights

Finally, our third aim was to experiment the impact of tag weights in term weighting function in a focused retrieval scheme. In order to understand the pro and cons of our structured model, the weighting functions and the same parameters used for the baseline runs were also used with our structured model.

The figure 1 shows that our TTF strategy ($R3$) improves dramatically the focused retrieval at low recall levels (from 0.5697 to 0.6640 following $iP[0.01]$ ranking). However, it does not improve focused retrieval enough to reach better results than classic retrieval.

These results confirm also that, according to Robertson and al. [6], it is important to keep the non linearity of the BM25 weighting function by "pre-impacting" term position in the structure of document (in other terms, tags weights) on the terms frequencies (strategy TTF) instead of "post-impacting" it directly on the terms weights (strategy CLAW, cf. [3]).

5   Conclusion

We proposed in [3] a new way of integrating the XML structure in the classic probabilistic model. We consider both the logical structure and the formatting structure. The logical structure is used at indexing step to define elements that correspond to part of documents. These elements will be indexed and potentially returned to the user. The formatting structure is integrated in the document model itself. During a learning step using the INEX 2006 collection, a weight is computed for each formatting tag, based on the probability that this tag distinguishes relevant terms. During the querying step, the relevance of an element is evaluated using the weights of the terms it contains, but each term weight is modified by the weights of the tags that mark the term.

The baselines are rather strong as the score of the BM25 run on article (run $R1$) is ranked seven of the competition according to the $iAP[0.01]$ ranking.

Our strategy TTF gives better results than focused retrieval ($R2$) and classic retrieval ($R1$) at low recall levels ($iP[0.01]$). That shows the interest of focused IR ($R3$ vs $R1$), and the interest of using structure ($R3$ vs $R2$). Pre-impacting the structure on terms frequencies (TTF, *R3*) gives also better results than "post-impacting" it on final terms weights (CLAW, [3]). Actually, TTF changes significantly the performances of the methods when considering the $iP[0.01]$ or the $MAiP$ measure.

TTF (*R3*) gives also good recall results ($MAiP = 0.2347$; $R[1500] = 0.6097$). Focused IR eliminates more non relevant elements than relevant elements (*R3* vs *R1*): $R[1500]$ decreases by 16% while S decreases by 40%.

We have presented a document model integrating explicitly the structural information in the weighting function, and a learning process of tags weights. We reach the same conclusions than [6] about the interest of pre-impacting structure, with a very different collection, a more heterogeneous one that contains a much larger set of tags (> 1 thousand).

In previous experiments, a basic average function, that considers all the tags equally (CLAW), gives better results than other combining functions (multiplication, only the closest tag, etc.). But, we think that a finest combining function (e.g. taking into account the distance between terms and tags) should improve the results.

References

1. Denoyer, L., Gallinari, P.: The wikipedia XML corpus. In: SIGIR forum, vol. 40, pp. 64–69 (2006)
2. Fuller, M., Mackie, E., Sacks-Davis, R., Wilkinson, R.: Coherent answers for a large structured document collection. In: SIGIR, pp. 204–213 (1993)
3. Géry, M., Largeron, C., Thollard, F.: Integrating structure in the probabilistic model for information retrieval. In: Web Intelligence, pp. 763–769 (2008)
4. Kamps, J., Pehcevski, J., Kazai, G., Lalmas, M., Robertson, S.: INEX 2007 evaluation measures. In: Fuhr, N., Kamps, J., Lalmas, M., Trotman, A. (eds.) INEX 2007. LNCS, vol. 4862, pp. 24–33. Springer, Heidelberg (2008)
5. Rapela, J.: Automatically combining ranking heuristics for html documents. In: WIDM, pp. 61–67 (2001)
6. Robertson, S., Zaragoza, H., Taylor, M.: Simple BM25 extension to multiple weighted fields. In: CIKM, New York, USA, pp. 42–49 (2004)
7. Robertson, S.E., Sparck Jones, K.: Relevance weighting of search terms. JASIST 27(3), 129–146 (1976)
8. Trotman, A.: Choosing document structure weights. IPM 41(2), 243–264 (2005)

Use of Multiword Terms and Query Expansion for Interactive Information Retrieval

Fidelia Ibekwe-SanJuan[1] and Eric SanJuan[2]

[1] ELICO, Université de Lyon 3
4, Cours Albert Thomas, 69008 Lyon, France
ibekwe@univ-lyon3.fr
[2] LIA & IUT STID, Université d'Avignon
339, chemin des Meinajaries, Agroparc BP 1228,
84911 Avignon Cedex 9, France
eric.sanjuan@univ-avignon.fr

Abstract. This paper reports our participation in the INEX 2008 Ad-Hoc Retrieval track. We investigated the effect of multiword terms on retrieval effectiveness in an interactive query expansion (IQE) framework. The IQE approach is compared to a state-of-the-art IR engine (in this case Indri) implementing a bag-of-word query and document representation, coupled with pseudo-relevance feedback (automatic query expansion(AQE)). The performance of multiword query and document representation was enhanced when the term structure was relaxed to accept the insertion of additional words while preserving the original structure and word order. The search strategies built with multiword terms coupled with QE obtained very competitive scores in the three Ad-Hoc tasks: Focused retrieval, Relevant-in-Context and Best-in-Context.

1 Introduction

The INEX Ad-Hoc track evaluates the capacity of IR systems to retrieve relevant passages from structured documents (XML elements) rather than whole documents. As this is our first participation in INEX, we tested two basic ideas: (i) evaluate the performance of a state-of-art IR engine designed for full document retrieval; (ii) evaluate the effectiveness of multiword terms for representing queries and documents coupled with query expansion (QE) and compare it to a bag-of-word approach coupled with the same QE mechanism. Here, a multiword term is taken to mean a syntactic construct usually associated with a noun phrase. Multiword terms are undeniably richer in information content and are less ambiguous than lone words. Moreover, recent experiments in IR in the biomedical domain, especially the TREC Genomic Track [1] showed that multiword terms and NLP processing hold promise for IR when applied to a corpus from a technical domain with a more homogeneous content. The hypotheses we wished to test were the following:

1. Can multiword terms gathered interactively from the from top n ranked documents returned by an initial query improve retrieval effectiveness?

S. Geva, J. Kamps, and A. Trotman (Eds.): INEX 2008, LNCS 5631, pp. 54–64, 2009.

2. More importantly, can a language model that preserves the structure of noun phrases coupled with a QE mechanism perform better than a bag-of-word model coupled with the same QE mechanism?

To implement our different search strategies, we used the Indri search engine in the Lemur package[1]. The rest of the paper is organized as follows: section 2 describes the Ad-Hoc retrieval tasks; section 3 presents our approach for multiword term selection and the different search strategies implemented; section 4 analyzes results and finally section 5 draws some conclusions from our research experiments.

2   Ad-Hoc Retrieval Tasks

The official corpus for Ad-Hoc retrieval is the 2006 version of the English Wikipedia comprising 659,388 articles without images [2]. Participants were asked to submit query topics corresponding to real life information need. A total of 135 such topics were collected, numbered from 544-678. A topic consists of four fields: content only field (<CO> or <Title>) with a multiword term expression of the topic; a content only + structure version of the topic (<CAS>) which is the title with indication of XML structure where the relevant elements may be found; a <description> field which is a slightly longer version of the title field; and a <narrative> field comprising a summary with more details about the expected answers. Typically, the narrative would indicate things to eliminate from relevant documents and draw boundaries that can be geographic, spatial, genre or historical in nature. Some title fields contained boolean signs that required systems to explicitly exclude (-) or include (+) certain terms in the relevant answer elements.

<topic id="546" ct_no="8">
<title> 19th century imperialism </title>
<castitle>article[about(., history)]
section[about(., 19th century imperialism)]</castitle>
<description>Describe the imperialism around the 19th century.</description>
<narrative>I am writing a thesis on 19th century imperialism. I am interested in which countries and why they practiced imperialism and how it affected the rest of the world. An element describing earlier or later than 19th century is acceptible if it supports the context of 19th century imperialism. But an element that describes post ww2 imperialism is far off. An element that describes about a history book/theory on the topic is also acceptable, but an element describing a person who is not directly related to the topic is not. E.g. An article about Hitler is acceptable, but not a novelist who fought in ww1.</narrative>
</topic>

Fig. 1. Example of a topic in the Ad-Hoc retrieval track

---

[1] http://www.lemurproject.org/lemur/IndriQueryLanguage.php

The Ad-Hoc track has 3 tasks

1. Focused retrieval: this requires systems to return a ranked list of relevant non-overlapping elements or passages.
2. The Relevant-in-Context (RiC) task builds on the results of the focused task. Systems are asked to select, within relevant articles, several non-overlapping elements or passages that are specifically relevant to the topic.
3. The Best-in-Context (BiC) task is aimed at identifying the best entry point (BEP) to start reading a relevant article.

3   Multiword Term Selection and Query Expansion

We first describe the document representation model in section 3.1, then the query representation (3.2) and finally our multiword term selection process (3.3). Section 3.4 describes the different search strategies we implemented using both automatic Indri search as a baseline and different parameters of the Indri QE feature.

3.1   Document Representation

The Wikipedia corpus was indexed using the Indri engine. No pre-processing was performed on the corpus. In particular, no lemmatization was performed and no stop word lists were used. The idea was to test the performance of an existing IR engine on raw texts without using any lexical resources. A nice feature of the Indri index is that word occurrences and positions in the original texts are recorded. A multiword term t is represented as an ordered list of nouns, adjectives and/or prepositions, $t = w_n...w_0$, where w_0 is necessarily a noun. Thus, a multiword term is not simply a sequence of nominals (nouns and adjectives) but a syntactic construct corresponding to noun phrases in English where the last element is compulsorily a noun and the order of the words must be preserved. These noun phrases should ideally be interpretable out of context, thus correspond to concepts or objects of the real world. Multiword terms are encoded in Indri language using the "#4" operator. Therefore t is encoded as $\#4(w_n...w_0)$. This operator will match any sequence of words in documents with at most 4 optional words inserted into it.

3.2   Query Representation

Given a query Q, the user selects some (possibly all) multiword terms in Q. If several terms are selected, we use the indri belief operator "#combine" to combine these terms. Hence, the initial query Q is translated by the user in an indri query Q' of the form

$$\#combine(\#4(w_{1,n_1}...w_{1,0})...\#4(w_{i,n_i}...w_{i,0}))$$

where:

- i and n_i are integers with $i > 0$.
- $w_{i,k}$ can be a noun, and adjective or a preposition.

We did not make use of the "+, -" boolean operators included in the initial topic description. We also tested the belief operators "#or" that is implemented as the complement of fuzzy conjunction, but its behavior appeared to be more confusing for the document ranking task. For more details on the Indri query language, see[2].

3.3 Interactive Multiword Term Selection and Query Expansion

Following an initial query Q to the Indri search engine using only the title field, we consider the top 20 ranked documents based on Q query. The user selects up to 20 multiword terms appearing in these documents. This leads to acquiring synonyms, abbreviations, hypernyms, hyponyms and associated terms with which to expand the original query term. The selected multiword terms are added to the initial Indri query Q using the syntax described in §3.2. This gives rise to a manually expanded query Q' which will be automatically expanded in a Q'' query using Indri QE feature with the following parameters: the number N of added terms is limited to 50 and are all extracted from the $D = 4$ top ranked documents using the query Q'. Moreover, in the resulting automatic expanded query Q'', Q' is weighted to $w = 10\%$. Figure 2 gives an example of multiword query terms used to expand topic 544. These multiword terms were acquired from the top 20 ranked document following the initial query from the title field. This interactive query expansion process required on the average 1 hour for each topic.

These three parameters ($D = 4, N = 50, w = 10$) were optimized on the TREC Enterprise 2007 data on the CSIRO website corpus[3]. Hence, the QE parameters were optimized on a different corpus than the one on which it is being tested now, i.e., Wikipedia.

3.4 Search Strategies

We first determined a baseline search which consisted in submitting the text in the title field of queries to Indri, without stop word removal, without attempting to extract any kind of terms, single or multiword. We then devised more elaborate search strategies, including the interactive multiword term selection process described in §3.3. The different search strategies mainly involved using the expanded set of multiword terms with other features of the Indri search engine such as QE and term weighting. These two features were combined with various possibilities of multiword term representation: bag-of-word, fixed structure, term relaxation (allowing insertion of n additional words). The precise

---

[2] http://www.lemurproject.org/lemur/IndriQueryLanguage.php
[3] Australian Commonwealth Scientific and Industrial Research Organisation, http://www.csiro.au/

```
#combine( #band(#1(nature of life) philosophy)
#1(significance of life)
#1(meaning of life)
#combine(#1(meaning of life) #or(socrates plato aristotle))
#band(#1(meaning of life) philosophy)
#band(#1(meaning of life) existence)
#band(#1(meaning of life) metaphysics)
#band(#1(existence) existentialism)
#band(#2(purpose life) religion)
#band(#2(purpose life) philosophy)
#band(#3(purpose life) religion)
#band(#3(purpose life) philosophy)
#band(#1(reflection of life) philosophy)
#1(philosophy of life)
#1(philosophy of existence)
#combine(#1(philosopher of life) #or(socrates plato aristotle))
#band(#1(source of life) philosophy)
#band(#2(life wheel) philosophy)
#band(#1(center of life) philosophy)
#band(#1(direction of life) philosophy) )
```

Fig. 2. Example of an expanded query with multiword terms for topic 544 on the "Meaning of life"

parameters for each implemented search strategy is detailed hereafter. In the official INEX conference, we submitted five different runs for the three Ad-Hoc retrieval tasks. Thus our runs were not differentiated by task. We carried out additional experiments after the INEX's official evaluation in order to further test the effect of term relaxation on the performance of our search strategies. The different search strategies are summarized in table 1.

Table 1. Ad-hoc runs

RunID	Approach
ID92_manual	multiword term with Indri with #1, #2 and #or operators
manualExt	*multiword term with Indri with #4 and #combine operators*
ID92_auto	automatic one word query with Indri #combine operator
autoQE	*ID92_auto with automatic Indri Query expansion (QE)*
ID92_manualQE	ID92_manual with QE
manualExtQE	*manualExt with QE*
ID92_manual_weighting	multiword term with Indri term weighting (TW)
ID92_manual_weightingQE	multiword term with Indri TW and QE

Only strategies whose ID begin by "ID92..." were submitted to the official INEX Ad-Hoc Retrieval evaluation. The search strategies in italics were performed after the official evaluation.

Baseline bag-of-word search. We carried out two automatic search strategies labeled "ID92_auto" and "autoQE" respectively, using only the text from the title field of the topic, without stopword removal. These constitute our baseline. "ID92_auto" was submitted to INEX, meanwhile it appeared after evaluation that its scores could be slightly improved using the QE function with default parameters. We thus carried out the additional strategy labelled *"autoQE"*.

Multiword terms with Query Expansion. In "ID92_manual", the multiword terms gathered during the process described in section 3.3 were combined with operators $\#n$ with $n \leq 2$ ($n = 1$ requires an exact match of the term, $n = 2$ allows for one insertion in the term) and linked by the "#or" operator. In "ID92_manualQE", we combined the above parameters with the QE mechanism. Note that only the selection of multiwords from the initial Indri ranked documents is manual. The QE function in Indri is automatic once the parameters are fixed. After the official evaluation, we ran additional experiments using the same principle but further relaxed the number of words that can be inserted into the multiword terms ($n = 4$). This gave rise to search strategies labeled *"manualExt"* and *"manualExtQE"* respectively. In both cases, we used the belief operator "#combine".

Query term weighting. Here, we experimented with "scrapping" the multiword term structure. In "ID92_manual_weighting", the multiword terms in "ID92_manual" were converted into a bag of weighted words in the following way:

1. each word w occurring in at least one query term is used.
2. its weight is set to $c + 0.1 \times m$ where c is the number of query terms with w as head word (for example *"teacher"* in *"head teacher"*) and m the number of terms where it appears as a modifier word (for example *"head"* in *"head teacher"*).
3. we then used the Indri operator "weight" to combine these words and their weights.

An additional strategy added the QE function to this vector space model representation of the queries thus giving rise to the "ID92_ manual_weightingQE" run.

4   Results

Two types of evaluation were provided in the Ad-Hoc retrieval tasks: (i) XML element or passage retrieval, (ii) full article retrieval.

4.1   Evaluation Protocol

For the focused task, the official measure is interpolated precision at 1% recall (iP[0.01]). However, results are also calculated for interpolated precision at other early recall points (0.00, 0.01, 0.05 and 0.10). Mean average interpolated precision [MAiP] over 101 standard recall points (0.00, 0.01, 0.02, ..., 1.00) is given as an overall measure.

4.2 Focused Retrieval Evaluation

Table 2 shows the scores obtained by all our runs in all three tasks. For each task, a first column shows the score obtained in the official measure while the second column gives the run's rank out of all submitted runs for that task. We will analyze the results of the focused search here. The analysis of the RiC and BiC results is done in sections 4.3 and 4.4 respectively. For the runs done after the evaluation, we can only provide the scores but not their ranks.

Table 2. Scores at INEX 2008 ad-hoc tasks

Task	Focus		RiC			BiC		
Measure	iP[0.01]	Rank	gP[1]	MAgP	Rank	gP[1]	MAgP	Rank
manualExtQE	0.693	-	0.61	0.215	-	0.61	0.225	-
ID92_manualQE	0.666	6th	0.55	0.211	3rd	0.56	0.220	2nd
ID92_manual	0.642	13th	0.55	0.158	24th	0.55	0.166	14th
ID92_manual_weightingQE	0.622	24th	0.52	0.185	12th	0.48	0.195	6th
ID92_manual_weighting	0.589	30th	0.47	0.148	32nd	0.42	0.153	17th
autoQE	0.574	-	0.46	0.197	-	0.43	0.201	-
ID92_auto	0.566	38th	0.44	0.171	19th	0.40	0.175	10th

For the focused task, 61 runs from 19 different institutions were submitted. Three systems retrieving full articles, including ours were amongst the 10 topmost systems. Four of our search strategies were ranked in the first half of all submitted runs. Our "ID92_manualQE" strategy that combined manual multiword term selection with automatic QE was persistently better than the other four at all levels of recall. It was ranked 4th by institutions and 6th when considering all submitted runs. However one must be cautious when drawing any conclusion from these results as iP[0.01] corresponds roughly to the precision after 1 relevant document has been retrieved. The term weighting strategies which transformed the multiword query terms into a vector space model obtained lower scores although the variant with QE (ID92_manual_weightingQE) performed significantly better than the variant without QE (ID_manual_weighting). The lowest scores were observed for the baseline Indri on single words with or without automatic QE (autoQE, ID92_auto). The additional experiments carried out after official evaluation showed that multiword term relaxation (manualExt) improved our official scores, and that when QE is added (manualExtQE), the score significantly increases from an iP[0.01]=0.674 to iP[0.01]=0.693, slightly surpassing the score obtained by the best system in the focused task with an iP[0.01]=0.690.

Relaxing the multiword term structure. Figure 3 takes a closer look at the precision/recall for our search strategies implementing multiword terms with QE. More precisely, this figure compares:

1. a state of art automatic IR system (Indri) using automatic QE features (autoQE),

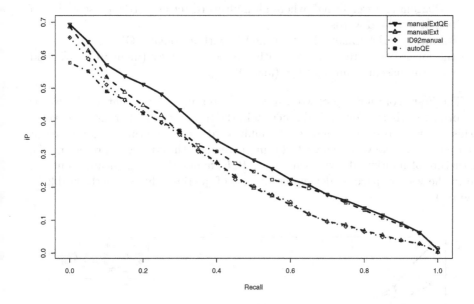

Fig. 3. Impact of multiword terms, query expansion and term relaxation on precision/recall for the focused task

2. IR with manually selected multiword terms where term structure and word order are preserved (ID92_manualQE),
3. the same strategy as in (2) but using a relaxed structure of terms by allowing insertion of additional words into the terms (manualExt).
4. the same strategy as in (3) but with automatic QE (manualExtQE).

For low recall levels (iP[0.05] and lower), all strategies with manually selected multiword terms have similar scores and clearly outperform their baseline counterpart. We can see from figure 3 that the two strategies using a more relaxed term structure (manualExtQE, manualExt) performed better than all the others. At iP[0.15], "manualExtQE" implementing the combination of the two features - QE with a relaxed term structure, clearly outperformed all other three runs and consequently all official INEX 2008 evaluated runs. In fact t-Tests with significance level $\alpha=0.05$ show that average score of manualExtQE between iP[0.0] and iP[0.25] is significantly higher than the average score of any of our other search strategies. It follows from these results that a relaxed multiword term structure combined with QE works better than a crisp one.

Multiword *vs.* bag-of-words representation of queries. We now study the behaviour of the strategies that implement a vector space model representation of multiword terms combined with term weighting. For that we plot in figure 4 the precision/recall for:

- "ID92_manual_weighting" where all multiword terms were represented by a bag of weighted words;
- its variant "ID92_manual_weightingQE" with automatic QE;
- the former two are compared with our best strategy (manualExtQE) and with the baseline run with QE (autoQE).

The best score for bag-of-word model was obtained by weighting the words according to their grammatical function in the term, i.e., head or modifier word. This is a way to project some of the multiword term structure onto the vector space model. However, even with this improvement, the strategies preserving the structure of multiword terms (manualExtQE, manualExt) significantly outperform the vector space model representation of queries. This is clearly visible in figure 4.

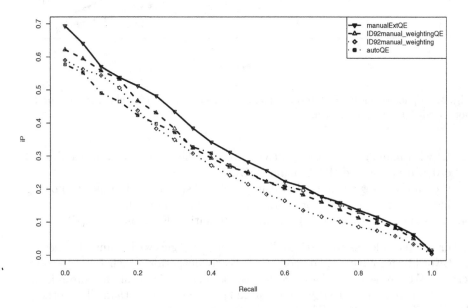

Fig. 4. Bag of word query representation *vs* multiword term structure

It appears that the bag-of-word representation of multiword terms without QE (ID92_manual_weighting) is competitive with the scores obtained by the baseline run (autoQE) on top ranked documents. When we consider higher recall levels (0.25), it performs worse than the baseline.

4.3   Relevant-in-Context Task

A total of 40 runs were submitted for this task by all participating groups. The official INEX evaluation once again showed that systems retrieving full articles instead of XML elements or passages were very competitive [3]. Table 2 shows the scores obtained by all our runs at different recall levels, their MagP and overall ranks.

Our "ID92_manualQE" run was ranked at the 3rd position out of all submitted runs and outperformed all our other runs. This is followed by the "ID92_manualQE". Surprisingly, the additional baseline approach with QE (autoQE) with a MAgP of 0.197 outperformed both the multiword term approach without QE (ID92_manual, MAgP=0.158) and the same approach with weighting and QE (ID92_manual_weightingQE, MAgP=0.185) whereas these two runs had higher precision values at early recall levels (gP[1-5]). It follows that for the Relevant-in-Context measure that combines several levels of recall, multiword terms used alone for queries are not sufficient. It is necessary to enrich them using top ranked documents to increase recall. In fact this phenomenon was also observed in the results of the focused task. Multiword terms queries without QE obtained lower scores than the baseline at higher levels of recall.

4.4   Best-in-Context Task

Our search strategies basically conserve the same order of performance as in the RiC task with all runs moving forward to higher ranks (see table 2). Particularly noticeable is the good performance of the "ID92_manualQE" run, ranked 2nd out of 35 submitted runs. The relaxed version "manualExtQE" does even better with a MAgP=0.225, thereby slightly outperforming the best system in the official evaluation (MAgP=0.224) at this task. Surprisingly again, the score of "ID92_auto" is among the 10 best systems (MAgP=0.175). When the QE mechanism is added (autoQE), it obtains a MAgP score of 0.201 thereby outperforming the system ranked 5th in the official evaluation (MAgP=0.120).

4.5   Document Retrieval Evaluation

INEX official evaluation also provided judgements full article retrieval. Retrieved elements or passages were ranked by descending order of relevance and judged on a first-come, first-served basis. Hence an element or passage represents the first occurrence of the document from which it was taken. For runs retrieving full articles, it was the classical case of document ranking. Evaluation was carried out over all submitted runs irrespective of task. A total of 163 submitted runs were ranked. Precision scores were calculated also at early recall levels of 5, 10 while mean average precision (MAP) was used as the official measure.

Table 3 shows the evaluation scores for our best strategies. Among the 163 runs that were submitted by participating groups, our "manual ID_92manualQE" strategy with a map of 0.3629 was ranked at the 3rd position. Also, this same strategy with relaxed term structure "manualExtQE" gives a score (map=0.3796) slightly better than the best ranked system (map=0.3789) and significantly outperforms our baseline "autoQE" (map=0.3700) from P5-P30 recall levels.

The reason for this very good performance of the "autoQE" run could be because qrels have been simply derived from those for focused task by considering

Table 3. Scores for full document retrieval. Total runs submitted: 163.

Participant	Rank	P5	P10	1/rank	map	bpref
manualExtQE	-	0.6580	0.5942	0.8742	0.3796	0.4076
autoQE	-	0.6171	0.5471	0.8055	0.3700	0.3724
p92-manualQEin	3rd	0.6371	0.5843	0.8322	0.3629	0.3917

that any document with a single relevant passage is relevant regardless of the size of the relevant passage within the document. On the contrary, the Focused and RiC measures takes the portion of the relevant passages into consideration.

5   Concluding Remarks

In this study, we tested the assumption that query and document representation with multiword terms, combined with query expansion (QE) can yield very competitive results. We tested this hypothesis against two baseline strategies implementing the bag-of-word representation using the Indri search engine with QE feature. The results obtained on the Wikipedia corpus in the three Ad-Hoc Retrieval tasks are very promising. All the search strategies implementing a multiword representation of queries and documents with QE were consistently ranked among the top five systems in the official INEX evaluation and outperformed the baseline strategies adopting a bag-of-word representation, even combined with QE. On the whole, our experiments have shown that using manually expanded multiword terms which are further expanded automatically with a query expansion mechanism is a promising research direction for IR when dealing with topically homogenous collection of texts such as Wikipedia articles. In the future, we intend to address how the interactive multiword term selection process may be automated.

References

1. Ruch, F., Tbahriti, I., Gobeill, J., Aronson, A.: Argumentative feedback: A linguistically-motivated term expansion for information retrieval. In: Proceedings of the Joint Conference COLING-ACL 2006, Sydney, July 17-21 (2006)
2. Denoyer, L., Gallinari, P.: The wikipedia xml corpus. In: SIGIR Forum, p. 6 (2006)
3. Kamps, J., Geva, S., Trotman, A., Woodley, A., Koolen, M.: Overview of the inex 2008 ad hoc track. In: PreProceedings of the 15th Text Retrieval Conference (INEX 2008), Dagstuhl, Germany, December 15-18, pp. 1–27 (2008)

Enhancing Keyword Search with a Keyphrase Index

Miro Lehtonen[1] and Antoine Doucet[1,2]

[1] Department of Computer Science
P.O. Box 68 (Gustaf Hällströmin katu 2b)
FI–00014 University of Helsinki
Finland
{Miro.Lehtonen,Antoine.Doucet}~@cs.helsinki.fi
[2] GREYC CNRS UMR 6072,
University of Caen Lower Normandy
F-14032 Caen Cedex
France
Antoine.Doucet~@info.unicaen.fr

Abstract. Combining evidence of relevance coming from two sources —
a keyword index and a keyphrase index — has been a fundamental part
of our INEX-related experiments on XML Retrieval over the past years.
In 2008, we focused on improving the quality of the keyphrase index and
finding better ways to use it together with the keyword index even when
processing non-phrase queries. We also updated our implementation of
the word index which now uses a state-of-the-art scoring function for
estimating the relevance of XML elements. Compared to the results from
previous years, the improvements turned out to be successful in the INEX
2008 ad hoc track evaluation of the focused retrieval task.

1 Introduction

The interest in developing methods for keyphrase search has decreased recently
in the INEX community partly because most of the queries are not keyphrase
queries [1]. However, we believe that indexing interesting phrases found in the
XML documents can be useful even when processing non-phrase queries. As
the XML version of the Wikipedia is full of marked-up phrases, we have been
motivated to work on the quality of the phrase index, as well, in order to capture
those word sequences that document authors really intended to be phrases.

In the previous years, our ad hoc track results have not been at the same
level with the best ad hoc track results. We believed that the reason lay in
the keyword index and the tfidf scoring function because the top results were
achieved with the probabilistic retrieval model. Lesson learned: we introduced
BM25 as the new scoring function for the keyword index. The latest results of
the INEX 2008 evaluation show great improvement from previous years. How
much the improvement is due to the state-of-the-art scoring function and how
much to the improved phrase index is still unclear, though.

S. Geva, J. Kamps, and A. Trotman (Eds.): INEX 2008, LNCS 5631, pp. 65–70, 2009.

This article is organised as follows. Section 2 describes our IR system as it was implemented in 2008. In Section 3, we show how the keyphrases are extracted from the document collection into a keyphrase index. Section 4 details the scoring methods for both the word index and the keyphrase index. The results of the evaluation are presented in Section 5, and finally, we draw conclusions and directions for future work in Section 6.

2   System Description

Our system was built on the EXTIRP architecture [2]. With one pass of the whole collection XML documents, we select a set of disjoint XML fragments which are indexed as an atomic unit of text. We can apply the paradigm of document retrieval to the fragments because they do not overlap. The indexed fragments of XML are entire XML elements in a predefined size range of 150–7,000 XML characters. The total number of indexed fragments is 1,022,062 which is just over 1.5 fragments per article. The number is relatively low because most of the Wikipedia articles are too small to be further divided into smaller fragments.

Two separate inverted indices are built for the fragments. A *word index* is created after punctuation and stopwords are removed, case folded, and the remaining words are stemmed with the Porter algorithm [3]. The *phrase index* where a phrase is defined as a Maximal Frequent Sequence (MFS) [4] is described in Section 3.2.

3   The Anatomy of a Keyphrase Index

Building a keyphrase index starts from finding or detecting the word sequences that should be considered keyphrases. As we are indexing hypertextual XML documents, it is natural to use the characteristics of hypertext documents and the markup language in the analysis as we detect passages that are potentially indexed keyphrases. The analysis is followed by a text mining method for extracting the Maximal Frequent Sequences from the word sequence.

3.1   Phrase Detection and Replication

Most of the XML markup in the Wikipedia articles describes either the presentation of the content or the hyperlink structure of the corpus, both of which show as mixed content with inline level XML elements. In these cases, the start and end tags of the inline level elements denote the start and the end of a word sequence that we call an *inline phrase*. These phrases include the anchor texts of hyperlinks as well as phrases with added emphasis, e.g., italicized passages. An exact definition for the XML structures that qualify was presented at the INEX 2007 workshop [5]. Intuitively, the inline phrases are highly similar to the multi-word sequences that text mining algorithms extract from plain text documents. Therefore, the tags of the inline elements are strong markers of potential phrase boundaries. Because phrase extraction algorithms operate on word sequences without XML, we incorporate the explicit phrase marking tags into the word sequence by replicating the qualified inline phrases.

Considering the effect of replication, we only look at the character data (CDATA) as the tags and other XML markup are parsed and removed before phrase extraction. The most obvious effect is the increase in phrase frequency of the replicated inline phrases with a similar side effect on the individual words they compose of. Moreover, the distance between the words preceding and following the phrase increases, which makes the phrase boundaries more explicit to those phrase extraction algorithms that allow gaps in the multiword sequences.

Duplicating the inline phrases lead to a 10–15% improvement in the MAiP on the INEX 2007 topics [6], but more recent experiments where the phrases were replicated three times have shown even further improvement when tested on the same topics. Note that these results depend on the phrase extraction algorithm and that other algorithms than ours may lead to different figures. Anyway, we chose to see if the triplication of the inline phrases works on the INEX 2008 topics, as well, and built the phrase index correspondingly.

3.2   MFS Extraction

The *phrase index* is based on Maximal Frequent Sequences (MFS) [4]. A sequence is said to be frequent if it occurs more often than a given sentence frequency threshold. It is said to be maximal if no other word can be inserted into the sequence without reducing the frequency below the threshold. This permits to obtain a compact set of document descriptors, that we use to build a phrase index of the collection.

The frequency threshold is decided experimentally, because of the computational complexity of the algorithm. Although lower values for the threshold produce more MFSs, the computation itself would take too long to be practical.

To be able to extract more descriptors, we clustered the XML fragments of the Wikipedia collection into 250 disjoint clusters. This permits to fasten the extraction process and to locally lower frequency threshold values. The result is a phrasal description of the document collection that is enhanced both in terms of quality and quantity of the descriptors. The drawback of this approach is a less compact document description. To perform this divide-and-conquer extraction of MFS, we used the *MFS_MineSweep* algorithm which is discussed in full detail in [7].

3.3   Arguments for Two Phrase Extraction Methods

Extracting the marked-up phrases and computing the frequent word sequences are both adequate methods for finding interesting phrases in hypertext documents. However, as our method that utilizes XML markup ignores the tag names, it generates a substantial amount of noise even though most marked phrases are captured. Examples of this noise include infrequent phrases that occur in few documents and various passages where the typeface differs from the surrounding content. Moreover, if words are inserted in the middle of a phrase, it shows as multiple marked up words instead of a single phrase with gaps in it. Therefore, the markup is not as reliable indicator of a phrase as the statistical occurrences of word sequences.

Maximal Frequent Sequences is a rather stable definition for an indexed phrase. First, the extracted word sequences are statistically frequent. Second, natural variation in the sequences is allowed in the form of gaps within the phrase. Replication of the marked up phrases changes the word sequence where the maximal sequences are computed. Incorporating the markup-based component in the text mining algorithm further stabilizes the method, which on the whole improves the quality of the phrase index [6].

4   Scoring XML Fragments

When processing the queries, we compute two separate RSV values that are later combined: a Word_RSV value based on a word index, and an MFS_RSV value based on the phrase index.

The Word_RSV is calculated using Okapi BM25 as implemented in the Lemur Toolkit [8], while the MFS_RSV is computed through loose phrase matching, in an identical way as in earlier versions of our system [9]. An exact match of the query phrase is not required, but gaps between query words and a different order of the query terms do contribute less to the score than an exact match.

The combination of both RSV values is done as follows. First, both values are normalized into the [0,1] range, using *Max Norm*, as presented by Lee [10]. Following this step, both RSVs are aggregated into a single RSV through linear interpolation, so that the aggregated RSV $= \alpha *$ Word_RSV $+ \beta *$ MFS_RSV.

In previous INEX participations, α was the number of distinct query terms and β was the number of distinct query terms in the query phrases. Post INEX 2007 experiments showed better performance with absolute values throughout the topic set, and we have decided to rely on such a new setting for our 2008 experiments as well, with α ranging between 92 and 94 and $\beta = 100 - \alpha$.

The relatively low value of β is due to the fact that the phrase index only contains words that are frequent enough in phrasal context, that is, frequent enough in conjunction with at least one other word. Important words that do not co-occur sequentially do not appear in the phrasal index. For this reason, the phrasal RSV not self-sufficient and should be perceived as a complement of the word RSV.

5   Results

We submitted three runs for the ad hoc track task of focused retrieval. The configurations of the submitted runs were based on experiments on the ad hoc track topics of INEX 2007, according to which the best proportion of weight given to terms and phrases would be around 92:8–94:6. The weight is given to the word index component is part of the Run ID. The initial results including 70 topics with assessments are shown in Table 1.

Table 1. Evaluation of our three official runs submitted for the focused retrieval task

Run ID	iP[0.00]	iP[0.01]	iP[0.05]	iP[0.10]	MAiP
UHel-Run1-92	0.6920	0.6534	0.5568	0.4996	0.2256
UHel-Run2-93	0.7030	0.6645	0.5583	0.5028	0.2271
UHel-Run3-94	0.7109	0.6648	0.5558	0.5044	0.2268

None of the submitted runs is significantly better than the other two runs although the interpolated precision does show moderately different figures at the lowest levels of recall. However, the results are similar to those of our earlier experiments on INEX 2007 topics where the precision peaks when α is set between 92 and 94. Compared to the peak values, if α is set to 0, precision drops by over 30%, whereas setting α to 100 (BM25 baseline) results in a modest decline of 1–5% depending on the recall point. However, we have not yet conducted this experiment on the 2008 topics and can thus not confirm the previous observations.

6   Conclusion and Future Work

The biggest change in our system from 2007 took place in the scoring function that contributes over 90% of the total relevance score of each XML fragment. We discarded tfidf and replaced it with BM25 which assumes the probabilistic model for information retrieval. Thanks to that update, our results are now comparable with the best results overall. The results also confirm that our phrase index slightly improves precision from a baseline where BM25 is the only scoring function as the optimal weight given to the phrase score is around 7%. Investigating whether the weights should be different for different types of queries is part of our future work.

References

1. Doucet, A., Lehtonen, M.: Let's phrase it: INEX topics need keyphrases. In: Proceedings of the SIGIR 2008 Workshop on Focused Retrieval, pp. 9–14 (2008)
2. Lehtonen, M., Doucet, A.: Extirp: Baseline retrieval from wikipedia. In: Fuhr, N., Lalmas, M., Trotman, A. (eds.) INEX 2006. LNCS, vol. 4518, pp. 115–120. Springer, Heidelberg (2007)
3. Porter, M.F.: An algorithm for suffix stripping. Program 14, 130–137 (1980)
4. Ahonen-Myka, H.: Finding all frequent maximal sequences in text. In: Mladenic, D., Grobelnik, M. (eds.) Proceedings of the 16th International Conference on Machine Learning ICML 1999 Workshop on Machine Learning in Text Data Analysis, Ljubljana, Slovenia, pp. 11–17. J. Stefan Institute (1999)
5. Lehtonen, M., Doucet, A.: Phrase detection in the Wikipedia. In: Fuhr, N., Kamps, J., Lalmas, M., Trotman, A. (eds.) INEX 2007. LNCS, vol. 4862, pp. 115–121. Springer, Heidelberg (2008)

6. Lehtonen, M., Doucet, A.: XML-aided phrase indexing for hypertext documents. In: SIGIR 2008: Proceedings of the 31st annual international ACM SIGIR conference on Research and development in information retrieval, pp. 843–844. ACM, New York (2008)
7. Doucet, A., Ahonen-Myka, H.: Fast extraction of discontiguous sequences in text: a new approach based on maximal frequent sequences. In: Proceedings of IS-LTC 2006, "Information Society, Language Technology Conference", pp. 186–191 (2006)
8. Lemur: Lemur toolkit for language modeling and ir (2003)
9. Doucet, A., Aunimo, L., Lehtonen, M., Petit, R.: Accurate Retrieval of XML Document Fragments using EXTIRP. In: INEX 2003 Workshop Proceedings, Schloss Dagstuhl, Germany, pp. 73–80 (2003)
10. Lee, J.H.: Combining multiple evidence from different properties of weighting schemes. In: Proceedings of the 18th annual international ACM SIGIR conference on Research and development in information retrieval, pp. 180–188. ACM Press, New York (1995)

CADIAL Search Engine at INEX

Jure Mijić[1], Marie-Francine Moens[2], and Bojana Dalbelo Bašić[1]

[1] Faculty of Electrical Engineering and Computing, University of Zagreb,
Unska 3, 10000 Zagreb, Croatia
{jure.mijic,bojana.dalbelo}@fer.hr
[2] Department of Computer Science, Katholieke Universiteit Leuven,
Celestijnenlaan 200A, 3001 Heverlee, Belgium
sien.moens@cs.kuleuven.be

Abstract. Semi-structured document retrieval is becoming more popular with the increasing quantity of data available in XML format. In this paper, we describe a search engine model that exploits the structure of the document and uses language modelling and smoothing at the document and collection levels for calculating the relevance of each element from all the documents in the collection to a user query. Element priors, CAS query constraint filtering, and the +/- operators are also used in the ranking procedure. We also present the results of our participation in the INEX 2008 Ad Hoc Track.

Keywords: Focused retrieval, Index database, Language model, Search engine.

1 Introduction

Information retrieval has become a part of our everyday lives. With the growing amount of available information, it has become challenging to satisfy a specific information need. We expect the retrieval procedure to find the smallest and most relevant information unit available, especially if our information need is very specific. Information retrieval procedures usually return whole documents as a result of a user query, but with the increasing number of semi-structured XML data sources, the information unit size can be varied from whole documents to sections, paragraphs, or even individual sentences. The choice of an appropriate information unit size is left to the retrieval procedure, which determines which portions of a document are considered relevant. If the search procedure is returning parts of a document, it is necessary to eliminate overlapping content so that the user does not have to inspect duplicate content. This reduces the time it takes for the user to browse through the results.

The outline structure of the documents we are searching could also be known, so the user could specify additional structural constraints in the query, i.e., to return only relevant paragraphs or images. Such queries are called content-and-structure (CAS) queries, as opposed to content-only (CO) queries, which do not have those structural constraints and contain only the keywords from the

S. Geva, J. Kamps, and A. Trotman (Eds.): INEX 2008, LNCS 5631, pp. 71–78, 2009.

query. Using CAS queries, the user can generate a more specific query that could improve the retrieved results.

The structure of the document could also be exploited in the ranking of the document elements. The location of the returned element in the document could indicate its potential relevance. For example, elements at the beginning of the document could be considered to be more relevant in most cases. Also, elements that are nested deep in the document structure could be considered less relevant.

In Section 2, we give an overview of our search engine and how it is modelled. In Section 3, we describe the ranking method used in our search engine and the data model that the method requires. In Section 4, we present and discuss the ad hoc results for our runs, and in Section 5, we give our concluding remarks.

2 System Overview

Our search engine [5] was developed for the CADIAL project [1]. The search engine provides access to a collection of Croatian legislative documents and has built-in support for morphological normalization for Croatian language [9]. All of the documents have a similar structure, which consists of a title, introduction, body, and signature. Furthermore, the body is divided into articles, i.e., articles of a law, not articles as documents, and each article into paragraphs. This document structure can prove to be useful, as it can be exploited by the retrieval procedures. This was the main motivation for our participation in the INEX Ad Hoc Track, as the documents from the Wikipedia collection used in that track also have a similar tree like structure and are written in XML format. The documents from the Wikipedia collection contain tags such as article, body, section, paragraph, table, and figure.

For the purpose of text processing, we use the Text Mining Tools (TMT) library [8]. The most basic text processing operation is tokenization, which is implemented for use with the UTF-8 character set that we use for internal text representation. Input documents are in XML format, and any part of the document can be indexed. The search engine can also use an inflectional lexicon for morphological normalization, but we did not have a lexicon built for the English language, so we instead used stemming, specifically the Porter stemmer [7].

At the core of our search engine is an index database containing all words found in the document collection, along with their respective positions in the documents. Words are stored in their normalized form if morphological normalization is used, or stems of words are stored if stemming is used. The index database also contains additional statistical data needed for the new ranking method we implemented for the INEX Ad Hoc Track: see remark further, as well as the structure of the elements for each document in the collection. The list of document elements to index is defined during the process of indexing, so we can choose which elements to index, i.e., article, body, section, paragraph, and figure, or to index the documents without their structure, i.e., only the article root tag. A document collection index database is built using an index builder tool, and than saved to a file in binary format. Serialization and deserialization procedures used are also implemented in the TMT library.

3 Ranking Method and Underlying Data Model

We implemented a new ranking method in our search engine that can exploit the document structure and be relatively straightforward and efficient to use on a large document collection. We also added support for CAS queries and the +/- keyword operators.

3.1 Language Model

For our ranking method, we used language modelling [6]. The basic idea behind this method is to estimate a language model for each element, and then rank the element by the likelihood of generating a query with the given language model. Therefore, we can calculate the relevance of every element e to the specified query Q:

$$P(e|Q) = P(e) \cdot P(Q|e), \tag{1}$$

where $P(e)$ defines the probability of element e being relevant in the absence of a query; and $P(Q|e)$ is the probability of the query Q, given an element e. We estimated the element priors in the following way:

$$P(e) = \frac{1}{1 + e_{location}} \cdot \frac{1}{1 + e_{depth}}, \tag{2}$$

where $e_{location}$ is the local order of an element, ignoring its path; and e_{depth} is the number of elements in the path, including e itself. For example, for an element /article[1]/body[1]/p[5], the location value is 5, and its depth is 3. A similar formula for calculating element priors was used in previous work by Huang et al. [3]. We experimented with this formula, and found that changing the coefficients in the formula does not improve the results any further. The formula, in this simple form, yields noticeable improvements in the retrieval performance.

For a query $Q = (q_1, q_2, ..., q_m)$, assuming the query terms to be independent, $P(Q|e)$ can be calculated according to a mixture language model:

$$P(Q|e) = \prod_{i=1}^{m}(1 - \lambda_d - \lambda_c)P_{elem}(q_i|e) + \lambda_d P_{doc}(q_i|D) + \lambda_c P_{col}(q_i|C), \tag{3}$$

where λ_d is the smoothing factor for the document level; λ_c is the smoothing factor for the collection level; and $P_{elem}(q_i|e)$, $P_{doc}(q_i|D)$, $P_{col}(q_i|C)$ are probabilities of the query term q_i given the element, document, and collection, respectively. The smoothing is done on two levels: the document and the collection levels, with the restriction that $\lambda_d, \lambda_c \in [0, 1]$ and $\lambda_d + \lambda_c < 1$. Wang et al. [10] found that smoothing on both the document and collection levels produced significantly better results than just smoothing on the whole collection. They used the Two-Stage smoothing method, and compared it to the Dirichlet priors and Jelinek-Mercer smoothing method. We chose to use our smoothing method because the values of the smoothing factors λ_d and λ_c have a very intuitive meaning. Although we considered additional smoothing at the section level, we did not implement it, because the section elements could be nested in each other, so we would not have a constant number of smoothing factors.

The probabilities of the query term q_i given the element, document, or a collection are calculated in the following way:

$$P_{elem}(q_i|e) = \frac{tf(q_i|e)}{length(e)}, \tag{4}$$

$$P_{doc}(q_i|D) = \frac{tf(q_i|D)}{length(D)}, \tag{5}$$

$$P_{col}(q_i|C) = \frac{tf(q_i|C)}{length(C)}, \tag{6}$$

where $tf(q_i|e), tf(q_i|D), tf(q_i|C)$ are the term frequency of the query term q_i in the element, document, and collection, respectively; $length(e)$, $length(D)$, $length(C)$ are the length of the element, document, and collection, respectively, in terms of the number of words.

3.2   Ranking the Elements

In the ranking procedure, other factors may influence the scores for each element. Elements are first scored using the language model formula 1, and then filtered according to the structural constraints from the CAS query, if there are any. For example, if the CAS query specifies that the user wants to find a figure, then elements that contain the element figure in their XPath are promoted to the top of the rank. The promotion of these elements is done sequentially from top to bottom, so the order of relevance for these elements is preserved.

We also implemented the +/- keyword operators, meaning that keywords from the query marked with the plus operator must be contained in the returned element, and keywords marked with the minus operator must not be contained in the element. For performance reasons, this operation is integrated in the calculation of the probabilities in the language model, so elements that do not satisfy the constraints of these operators, i.e., those that do not contain keywords marked with the plus operator or do contain keywords marked with the minus operator, are automatically assigned a score of zero.

Finally, when all the retrieved elements are ranked, we have to eliminate overlapping elements so that the ranking procedure does not return duplicate content. This is done simply by iterating through the results from top to bottom and eliminating elements whose XPath is fully contained in any of the previous elements' XPath, or if any of the previous elements' XPath is fully contained in the XPath of the element currently being analyzed.

3.3   Index Database

The index database is the backbone of our search engine. In the database, we store the positions of words in every element from the document collection being

Table 1. Tag set of indexed elements

No.	Tag name
1	article
2	body
3	section
4	p
5	table
6	figure
7	image

indexed. Along with word indices, some other statistical data is also stored for use in language modelling. Each element is represented with its own language model, so some data must be stored separately for every element in the collection, e.g., term frequencies for every term that the element contains, and the element length. The data is stored for all overlapping elements in the document collection. The size of the index database is therefore dependent on the number of elements we want to index, i.e., the depth of the document structure we want to process. For example, it might not be very useful to index individual table cells as separate elements, so instead we can choose to index the entire table as one element and therefore reduce the size of the index database. We chose to index only elements that are most likely to be relevant, as shown in Table 1.

The document as a whole is also considered as an element. Other statistical data that is stored in the index database includes the term frequencies for the entire collection, number of elements containing each term, unique term count, total term count, and total element count.

Collection information is also stored in the index database. Information such as the structure of the elements being indexed, i.e., the parent child relations of the elements in the document, needs to be stored in order for the language model to perform the necessary smoothing operations at the document and collection level, and also to reconstruct the proper XPath for every retrieved element.

4 Ad Hoc Results

The evaluation in the INEX Ad Hoc Track was performed on the Wikipedia collection based on the English Wikipedia in early 2006 [2]. The collection contains 659,338 articles in XML format. Results for our runs are given in Table 2 and are sorted by the interpolated precision measure at 1% recall, i.e., iP[0.01], which is the official measure of the focused retrieval task in the INEX Ad Hoc Track. Other measures include interpolated precision at other early levels of recall, i.e., iP[0.00], iP[0.05], iP[0.10], and mean average interpolated precision over 101 standard levels of recall, i.e., MAiP. The best result for each measure is marked in boldface. The name of the run contains the type of query used, i.e., CO for content-only and CAS for content-and-structure query. It also contains the returned information unit, i.e., document or element, and the smoothing

Table 2. Official results for our runs

No.	Run	iP[0.00]	iP[0.01]	iP[0.05]	iP[0.10]	MAiP
1	co-document-lc6	0.6393	**0.5953**	**0.5094**	**0.4732**	**0.2566**
2	cas-element-ld5-lc4	0.6688	0.5539	0.4085	0.3275	0.1449
3	co-element-ld2-lc5	**0.6913**	0.5429	0.4038	0.2955	0.1008
4	co-element-ld2-lc1	0.6725	0.5251	0.3951	0.2987	0.0942
5	cas-element-ld2-lc5	0.6500	0.5207	0.3602	0.2629	0.1148
6	cas-element-ld1-lc6	0.6649	0.5064	0.3688	0.2638	0.1147

factors used, i.e., *ld* for document level and *lc* for collection level. Note that the smoothing factors are in the range from 0.0 to 1.0 with the restriction that $\lambda_d + \lambda_c < 1$, so for the run *cas-element-ld5-lc4* the smoothing factors are $\lambda_d = 0.5$ and $\lambda_c = 0.4$. Along with the focused retrieval ranking, article retrieval ranking was derived from the retrieval results and mean average precision, i.e., MAP, measure was used for the evaluation of article retrieval.

Immediately from the results, we can see that the retrieval of the whole document gives better performance at higher levels of recall than element retrieval, as can be seen from the precision-recall graph in Fig. 1. Only at very low levels of recall, i.e., iP[0.00], does element retrieval outperform document retrieval. Similar results, where the document retrieval outperforms element retrieval at iP[0.01] and at higher levels, were seen in previous years and in some of the systems in this year's Ad Hoc Track. This could, perhaps, be a consequence of the way in which we perform relevance assessments, where fully relevant documents are assigned to most of the topics. Another problem could be that the articles in the Wikipedia collection are very specific to their content, and the topics are usually not very specific. This leads to a situation where many articles are marked as fully relevant, and only a few have some specific relevant elements.

Smoothing factors also had a significant impact on the retrieval performance. As we mentioned previously, retrieving whole documents outperformed element retrieval at higher levels of recall, so it is reasonable to expect that higher smoothing at the document level would yield better results. This can be seen in our *cas-element-ld5-lc4* run in Fig. 1, where higher smoothing at the document level contributes to significantly better performance at higher levels of recall than other runs with lower smoothing at the document level, e.g., *cas-element-ld2-lc5* and *cas-element-ld1-lc6*. Liu et al. [4] also found that the document score greatly influenced the retrieval performance. They implemented a separate document and element index, and combined the document and element score.

The use of CAS query constraint filtering did improve retrieval performance overall, especially at midrange levels of recall. At low levels of recall the difference is not significant, and even at iP[0.00], the performance is slightly worse than the run using CO queries. Perhaps more complex processing of CAS queries could yield some improvement at low levels of recall, although most of the topics did not use the structural features of CAS queries.

Although we did not do a direct comparison on the influence of the +/- keyword operator and element priors on the retrieval performance, we did notice

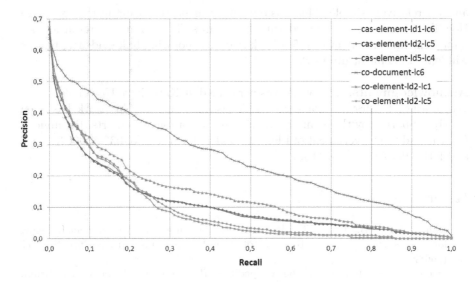

Fig. 1. Precision recall graph for our runs

during development that using both the operators and element priors did in fact improve performance slightly.

Overall, in the focused retrieval ranking, we achieved 29th place out of 61 runs, with an iP[0.01] measure of 0.5953, while the first place run achieved an iP[0.01] measure of 0.6896. In the document retrieval ranking, we achieved 9th place out of 76 runs, with an MAP measure of 0.3232, while the first place run achieved an MAP measure of 0.3629. This difference between our focused retrieval rank and article retrieval rank is to be expected, as our system performs better in ranking whole articles than ranking article elements.

5   Conclusion

We developed a search engine that exploits the structure of documents and implemented a simple ranking method that uses language modelling and smoothing at two levels: the document and the collection level. Retrieving whole documents performed better than element retrieval at higher levels of recall, which could perhaps be attributed to the nature of the topics. Element retrieval performed better than document retrieval only at the lowest level of recall, i.e., iP[0.00]. Filtering of elements' structural path to the CAS query constraints contributed to the improvement in retrieval performance, as well as the higher smoothing factor at the document level. We have also used element priors and implemented the +/- keyword operators, which we noticed tend to improve the retrieval performance, but we did not investigate their impact on performance in detail.

We developed our system from the ground up, putting an emphasis on simplicity, efficiency, and effectiveness. Language modelling proved to be very effective, and yet relatively simple. This was our first year participating in the INEX Ad Hoc Track, and we are pleased with the results. There is much room left for improvements, e.g., relevance feedback and incorporating link evidence, but we will leave that for future work.

Acknowledgments. This work was performed at Katholieke Universiteit Leuven, during the first author's stay as a visiting scholar. This work has been jointly supported by the Ministry of Science, Education and Sports, Republic of Croatia and the Government of Flanders under the grant No. 036-1300646- 1986 and KRO/009/06 (CADIAL).

References

1. Dalbelo Bašić, B., Tadić, M., Moens, M.-F.: Computer Aided Document Indexing for Accessing Legislation, Toegang tot de wet. Die Keure, Brugge, pp. 107–117 (2008)
2. Denoyer, L., Gallinari, P.: The wikipedia XML corpus. In: ACM SIGIR Forum, vol. 40, pp. 64–69. ACM Press, New York (2006)
3. Huang, F.: The role of shallow features in XML retrieval. In: INEX 2007 Workshop Proceedings, pp. 33–38 (2007)
4. Liu, J., Lin, H., Han, B.: Study on reranking XML retrieval elements based on combining strategy and topics categorization. In: INEX 2007 Workshop Proceedings, pp. 170–176 (2007)
5. Mijić, J., Dalbelo Bašić, B., Šnajder, J.: Building a search engine model with morphological normalization support. In: ITI 2008 Proceedings of the 30th International Conference on Information Technology Interfaces, pp. 619–624 (2008)
6. Ponte, J.M., Croft, W.B.: A language modeling approach to information retrieval. In: SIGIR 1998: Proceedings of the 21st Annual International ACM SIGIR Conference on Research and Development in Information Retrieval, pp. 275–281. ACM Press, New York (1998)
7. Porter, M.F.: An algorithm for suffix stripping. Program: electronic library and information systems 40(3), 211–218 (2006)
8. Šilić, A., Šarić, F., Dalbelo Bašić, B., Šnajder, J.: TMT: Object-oriented text classification library. In: ITI 2007 Proceedings of the 29th International Conference on Information Technology Interfaces, pp. 559–566 (2007)
9. Šnajder, J., Dalbelo Bašić, B., Tadić, M.: Automatic acquisition of inflectional lexica for morphological normalisation. Information Processing & Management 44(5), 1720–1731 (2008)
10. Wang, Q., Li, Q., Wang, S.: Preliminary work on XML retrieval. In: INEX 2007 Workshop Proceedings, pp. 70–76 (2007)

Indian Statistical Institute at INEX 2008 Adhoc Track

Sukomal Pal, Mandar Mitra, Debasis Ganguly, Samaresh Maiti,
Ayan Bandyopadhyay, Aparajita Sen, and Sukanya Mitra

Information Retrieval Lab, CVPR Unit
Indian Statistical Institute, Kolkata
India
{sukomal_r,mandar,samaresh_t,ayan_t,aparajita_t,sukanya_t}@isical.ac.in,
debasis@synopsys.com

Abstract. This paper describes the work that we did at Indian Statistical Institute towards XML retrieval for INEX 2008. Besides the Vector Space Model (VSM) that we have been using since INEX 2006, this year we implemented the Language Modeling (LM) approach in our text retrieval system (SMART) to retrieve XML elements against the INEX Adhoc queries. Like last year, we considered Content-Only (CO) queries and submitted three runs for the FOCUSED sub-task. Two runs are based on the Vector Space Model and one uses the Language Model. One of the VSM-based runs (*VSMfbElts0.4*) retrieves sub-document-level elements. Both the other runs (*VSMfb* and *LM-nofb-0.20*) retrieve elements only at the whole-document level. We applied blind feedback for both the VSM-based runs; no query expansion was used in the LM-based run. In general, the relative performance of our document-level runs is respectable (ranked 15/61 and 22/61 according to the official metric). Though our element retrieval run does reasonably (ranked 16/61 by *iP[0.01]*) according to the early-precision metrics, we think there is plenty of scope to improve our element retrieval strategy. Our immediate next task is therefore to focus on how to improve true element-level retrieval.

1 Introduction

Traditional Information Retrieval systems return whole documents in response to queries, but the challenge in XML retrieval is to return the most relevant parts of XML documents which meet the given information need. Since INEX 2007 [1], arbitrary passages are permitted as retrievable units, besides the usual XML elements. A retrieved passage consists of textual content either from within an element or spanning a range of elements. Since INEX 2007, the adhoc retrieval task has also been classified into three sub-tasks: a) the FOCUSED task which asks systems to return a ranked list of elements or passages to the user; b) the RELEVANT in CONTEXT task which asks systems to return relevant elements or passages grouped by article; and c) the BEST in CONTEXT task which expects systems to return articles along with one best entry point to the user.

S. Geva, J. Kamps, and A. Trotman (Eds.): INEX 2008, LNCS 5631, pp. 79–86, 2009.

Each of the three subtasks can be based on two different query variants: Content-Only(CO) and Content-And-Structure(CAS) queries. In the CO task, a user poses a query in free text and the retrieval system is supposed to return the most relevant elements/passages. A CAS query can provide explicit or implicit indications about what kind of element the user requires along with a textual query. Thus, a CAS query contains structural hints expressed in XPath [2] along with an *about()* predicate.

This year we submitted three adhoc focused runs, two using a Vector Space Model (VSM) based approach and one using a Language Modeling (LM) approach. VSM sees both the document and the query as bags of words, and uses their *tf-idf* based weight-vectors to measure the inner product *similarity* as a measure of closeness between the document and the query. The documents are retrieved and ranked in decreasing order of the similarity-value. In the LM approach, the probability of a document generating the query terms is taken as the measure of similarity between the document and the query.

We used a modified version of the SMART system for the experiments at INEX 2008. Of the two VSM-based runs, one retrieves elements at the whole-document level, while the other does sub-document-level element retrieval. For both these runs, we used blind feedback after the initial document retrieval. This year we also incorporated the LM approach within SMART. Our third run was based on this implementation and works at the document level. All three runs were for the *FOCUSED* sub-task of the Adhoc track, and considered CO queries only. In the following section, we describe our general approach for all these runs, and discuss results and further work in Section 4.

2   Approach

2.1   Indexing

We first shortlisted about thirty tags that correspond to elements containing useful information: $<p>$, $<ip1>$, $<it>$, $<st>$, $<fnm>$, $<snm>$, $<atl>$, $<ti>$, $<p1>$, $<h2a>$,$<h>$, $<wikipedialink>$, $<section>$, $<outsidelink>$, $<td>$, $<body>$, etc. Documents were parsed using the libxml2 parser, and only the textual portions included within the selected tags were used for indexing. Similarly, for the topics, we considered only the *title* and *description* fields for indexing, and discarded the *inex-topic*, *castitle* and *narrative* tags. No structural information from either the queries or the documents was used.

The extracted portions of the documents and queries were indexed using single terms and a controlled vocabulary (or pre-defined set) of statistical phrases following Salton's blueprint for automatic indexing [3]. Stopwords were removed in two stages. First, we removed frequently occurring common words (like *know, find, information, want, articles, looking, searching, return, documents, relevant, section, retrieve, related, concerning,* etc.) from the INEX topic-set. Next, words listed in the standard stop-word list included within SMART were removed from both documents and queries. Words were stemmed using a variation of the Lovins' stemmer implemented within SMART. Frequently occurring word bi-grams (loosely referred to as phrases) were also used as indexing units. We used the N-gram Statistics Package (NSP)[1] on the English Wikipedia text corpus and selected the 100,000 most frequent word bi-grams as the list of candidate phrases. Documents and queries were weighted using the *Lnu.ltn* [4] term-weighting formula. Based on our INEX 07 experiments, we used *slope* = 0.4 and *pivot* = 80, which yielded the best results for early precision. For each of 135 adhoc queries(544-678), we retrieved 1500 top-ranked XML documents or non-overlapping elements.

2.2   Document-Level Retrieval

We submitted two runs at the document level. One was based on the LM approach and the other on the VSM approach.

LM approach: (*LM-nofb-0.20*) We implemented the language modelling framework as described by Hiemstra [5] within the SMART system [6]. In this model, the similarity-score between a document D and a query $Q = (T_1, T_2, \ldots, T_n)$ of length n is given by

$$Sim(D, Q) = \log(\sum_t tf(t, d)) + \sum_{i=1}^{n} \log(1 + \frac{\lambda_i tf(t_i, d) \sum_t df(t)}{(1 - \lambda_i) df(t_i) \sum_t tf(t, d)}) \quad (1)$$

where $tf(t, d)$ denotes the term frequency of term t in document d, $df(t)$ denotes the document frequency of term t, and λ_i is the importance of term t_i. We used $\lambda_i = \lambda = 0.20$ for all i as this value yielded the best result over the INEX 2007 dataset. No feedback was used in this run.

VSM approach: For the *VSMfb* run, we used blind feedback to retrieve whole documents. We applied automatic query expansion following the steps given below for each query (for more details, please see [7]).

1. For each query, collect statistics about the co-occurrence of query terms within the set S of 1500 documents retrieved for the query by the baseline run. Let $df_S(t)$ be the number of documents in S that contain term t.
2. Consider the 50 top-ranked documents retrieved by the baseline run. Break each document into overlapping 100-word windows.
3. Let $\{t_l, \ldots, t_m\}$ be the set of query terms (ordered by increasing $df_S(t_i)$) present in a particular window. Calculate a similarity score Sim for the window using the following formula:

$$Sim = idf(t_1) + \sum_{i=2}^{m} idf(t_i) \times \min_{j=1}^{i-1}(1 - P(t_i|t_j))$$

where $P(t_i|t_j)$ is estimated based on the statistics collected in Step 1 and is given by

$$\frac{\# \text{ documents in } S \text{ containing words } t_i \text{ and } t_j}{\# \text{ documents in } S \text{ containing word } t_j}$$

---

[1] http://www.d.umn.edu/~tpederse/nsp.html

This formula is intended to reward windows that contain multiple matching query words. Also, while the first or "most rare" matching term contributes its full idf (inverse document frequency) to *Sim*, the contribution of any subsequent match is deprecated depending on how strongly this match was predicted by a previous match — if a matching term is highly correlated to a previous match, then the contribution of the new match is correspondingly down-weighted.

4. Calculate the maximum *Sim* value over all windows generated from a document. Assign to the document a new similarity equal to this maximum.
5. Rerank the top 50 documents based on the new similarity values.
6. Assuming the new set of top 20 documents to be relevant and all other documents to be non-relevant, use Rocchio relevance feedback to expand the query. The expansion parameters are given below:

$$\text{number of words} = 20$$
$$\text{number of phrases} = 5$$
$$\text{Rocchio } \alpha = 4$$
$$\text{Rocchio } \beta = 4$$
$$\text{Rocchio } \gamma = 2.$$

For each topic, 1500 documents were retrieved using the expanded query.

2.3 Element-Level Run

For element-level retrieval, we adopted a 2-pass strategy. In the first pass, we retrieved 1500 documents for each query using the method described in 2.1.

In the second pass, these documents were parsed using the libxml2 parser. All elements in these 1500 documents that contain text were identified, indexed and compared to the query. The elements were then ranked in decreasing order of similarity to the query. In order to avoid any overlap in the final list of retrieved elements, the nodes for a document are sorted in decreasing order of similarity, and all nodes that have an overlap with a higher-ranked node are eliminated.

3 Evaluation

Since INEX 2007, five metrics (viz. $iP[0.00]$, $iP[0.01]$, $iP[0.05]$, $iP[0.10]$ and $MAiP$) have been used to measure the retrieval performance of systems participating in the *FOCUSED* adhoc task. Among these $iP[0.01]$ was taken as the official measure to rank competing systems. As these measures are extensions of their counterparts in standard document retrieval, they are expected to behave similarly. In earlier work [8], we showed that early precision measures ($iP[0.00]$, $iP[0.01]$) are more error-prone and less stable to incomplete judgments, whereas $MAiP$ is the least vulnerable among these metrics. Our work considered evaluation with reduced query-sets (*query-sampling*) and reduced pool-sizes (*pool-sampling*), chosen at random.

The pool-sampling experiments reported in [8] have two major shortcomings. One, different systems get affected differently based on how the system in question retrieves the omitted elements, specifically at what position of the ranked list. This leads to non-uniform reductions in the score across systems. Two, all the retrievable units (here XML elements) irrespective of their ranks get the same probability of inclusion in (or exclusion from) the reduced sampled pool. Thus, this experiment does not necessarily indicate how system-rankings would differ if we gradually reduced the pool-depth. These experiments were, therefore, re-done by systematically re-creating the pool from the INEX 2008 adhoc focused submissions.

We considered all 76 submissions to create the pool, dynamically setting the pool-depth for each query in order to match the original poolsize for that query at INEX 2008. We did not get exactly matched numbers since some of the systems admissibly changed their submissions after pooling was done. We also surprisingly discovered that a substantial number of documents and their elements retrieved by various systems at early ranks did not figure in the pool, even though some of them on deeper investigation were found to be actually relevant. The fact was brought to the attention of the organizers. A corrected, more recent version of the relevance assessments as well as results were released.

Although the revision of the *qrels* and the result-set was a significant by-product of our pooling experiments, the actual experimental results are not at a reportable stage.

4   Results

The results as reported in the INEX08 website using relevance judgements for 70 topics are shown in Table 1. Table 2 shows the figures obtained if the evaluation is done at the document level, instead of at the element level.

Table 1. Results for the FOCUSED, CO task (element retrieval)

Run Id	iP@0.00	iP@0.01	iP@0.05	iP@0.10	MAiP	Official Rank
VSMfb	0.6363	0.6242	0.5509	0.5019	0.2735	22/61
VSMfbElts0.4	**0.7152**	0.6348	0.4805	0.4259	0.1538	16/61
LM-nofb-0.20	0.6854	**0.6364**	0.5565	0.5152	**0.2868**	**15/61**
Best Run (FOERStep)	0.7660	**0.6897**	0.5714	0.4908	0.2076	1/61

Table 2. Document-level evaluation for the FOCUSED, CO task

Run Id	MAP	Official Rank
VSMfb	0.3091	24/61
VSMfbElts0.4	**0.3195**	**14/61**
LM-nofb-0.20	0.2999	28/61
BEST (manualQE_indri03_focused)	0.3629	1/61

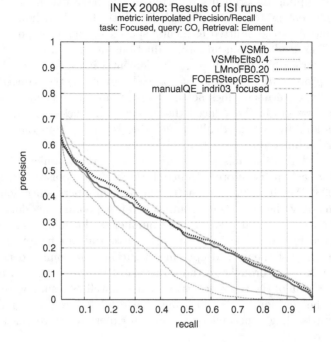

Fig. 1. Interpolated P - R graph for ISI runs

According to the official metric ($iP[0.01]$), our best run is *LM-nofb-0.20*, and considering precision at the first retrieval unit ($iP[0.00]$), our element run *VSMf-bElts0.4* is the best. However, these early-precision metrics are not very reliable [8,9]. The graphs in Figure 1 clearly show that our document-level runs (*VSMfb* and *LM-nofb-0.20*) are considerably better than the element-level run. This is reflected in their respective MAiP scores as well. The document-level runs achieve scores of 0.2735 and 0.2868 — the difference in these two scores was not found to be statistically significant ($p = 0.23$ for a two-tailed t-test). The element-level run achieves a score of only 0.1538, which was found to be significantly lower than the scores of the document-level runs on the basis of a t-test. More details about the relative performance of the two document-level runs and the element-level run are shown in Table 3.

Table 3. Comparison between document-level and element-level runs (AiP)

Run Id	Better than VSMfbElts0.4	Worse than VSMfbElts0.4
LM-nofb-0.20	72(69)	8(7)
VSMfb	60(58)	10(10)

Table 4. Comparison between document-level and element-level runs (iP[0.00] and ip[0.01])

Run Id	iP[0.00]		iP[0.01]	
	Better than VSMfbElts0.4	Worse than VSMfbElts0.4	Better than VSMfbElts0.4	Worse than VSMfbElts0.4
LM-nofb-0.20	28(22)	42(25)	35(28)	35(27)
VSMfb	28(20)	42(27)	36(29)	34(23)

The figures in the table show the number of queries for which a run performs better or worse than another run, as measured by AiP. The figures in parentheses correspond to the number of queries for which the relative performance difference is at least 5%.

A similar comparison on the basis of $iP[0.00]$ and $iP[0.01]$ is shown in Table 4. Since the elements retrieved by VSMfbElts are typically considerably shorter than full documents, the first relevant element retrieved by this run is likely to be more focused than the first relevant element (a full document, actually) retrieved by the other runs. Also, the irrelevant elements preceding the first relevant element are likely to be shorter. Thus, the element level run performs better in terms of $iP[0.00]$. At subsequent recall points, however, the performance of the element-level run drops. Thus, the differences in $iP[0.01]$ values between the various runs is marginal. This suggests that our ranking of sub-document level elements needs to be significantly improved.

5 Conclusion

This was our third year at INEX. Our main objective this year was to see the performance of the LM approach vis-a-vis VSM. Thus, we implemented LM within the SMART retrieval system. Its performance was encouraging as it fares the best among our runs. However the retrieval was at the document-level only. We need to extend this approach to element-level retrieval. Among the VSM runs, the document-level run is quite satisfactory. For the element-level run, however, there is plenty of room for improvement. We need to study the effect of document length normalization in-depth for XML collections in general and adopt a suitable strategy for the Wikipedia corpus. More generally, effective term-weighting schemes for different element-tags in the XML tree seems to be a promising field of enquiry. We hope these will be exciting exercises which we plan to do in the coming days.

References

1. INEX: Initiative for the Evaluation of XML Retrieval (2008),
   http://www.inex.otago.ac.nz
2. W3C: XPath-XML Path Language(XPath) Version 1.0,
   http://www.w3.org/TR/xpath

3. Salton, G.: A Blueprint for Automatic Indexing. ACM SIGIR Forum 16(2), 22–38 (1981)
4. Buckley, C., Singhal, A., Mitra, M.: Using Query Zoning and Correlation within SMART: TREC5. In: Voorhees, E., Harman, D. (eds.) Proc. Fifth Text Retrieval Conference (TREC-5), NIST Special Publication 500-238 (1997)
5. Hiemstra, D.: Using language models for information retrieval. PhD thesis, University of Twente (2001)
6. Ganguly, D.: Implementing a language modeling framework for information retrieval. Master's thesis, Indian Statistical Institute (2008)
7. Mitra, M., Singhal, A., Buckley, C.: Improving automatic query expansion. In: SIGIR 1998, Melbourne, Australia, pp. 206–214. ACM, New York (1998)
8. Pal, S., Mitra, M., Chakraborty, A.: Stability of inex 2007 evaluation measures. In: Proceedings of the Second International Workshop on Evaluating Information Access (EVIA), pp. 23–29 (2008),
http://research.nii.ac.jp/ntcir/workshop/OnlineProceedings7/
pdf/EVIA2008/06-EVIA2008-PalS.pdf
9. Fuhr, N., Kamps, J., Lalmas, M., Malik, S., Trotman, A.: Overview of the INEX 2007 Ad Hoc Track. In: Fuhr, N., Kamps, J., Lalmas, M., Trotman, A. (eds.) INEX 2007. LNCS, vol. 4862, pp. 1–23. Springer, Heidelberg (2008)

Using Collectionlinks and Documents as Context for INEX 2008

Delphine Verbyst[1] and Philippe Mulhem[2]

[1] LIG - Université Joseph Fourier, Grenoble, France
Delphine.Verbyst@imag.fr
[2] LIG - CNRS, Grenoble, France
Philippe.Mulhem@imag.fr

Abstract. We present in this paper the work of the Information Retrieval Modeling Group (MRIM) of the Computer Science Laboratory of
Grenoble (LIG) at the INEX 2008 Ad Hoc Track. We study here the use
of non structural relations between document elements (doxels) in conjunction with document/doxel structural relationships. The non structural links between doxels of the collection come from the *collectionlink*
doxels. We characterize the non structural relations with relative exhaustivity and specificity scores. Results of experiments on the test collection
are presented. Our best run is in the top 5 for iP[0.01] values for the
Focused Task.

1 Introduction

This paper describes the approach used by the MRIM/LIG research team for the
Ad Hoc Track of the INEX 2008 competition. Our goal here is to show that the
use of non structural links and the use of structural links lead to high quality
results for an information retrieval system on XML documents. We consider
that handling links between doxels in a "smart" way may help an information
retrieval system, not only to provide better results, but also to organize the
results in a way to overcome the usual simple list of documents. For INEX 2008
runs, we obtained very good results for low recall values (0.00 and 0.01).

First of all, we define one term: a *doxel* is any part of an XML document
between its opening and closing tag. We do not make any kind of difference
between a doxel describing the logical structure of the document (like a title
or a paragraph) or not, like anchors of links or words that are emphasized),
a relation between doxels may come from the structural composition of the
doxels, or from any other source. Assume that an xml document is "<A>This
is an <B>example</B> of <C>XML</C> document</A>". This document
contains 3 doxels: the first is delimited by the tag A, the second is delimited
by the tag B, and the third is delimited by the tag C. We also consider that
a compositional link relates A to B, and A to C. We will also depict B and
C as direct structural components of A. In the following, the non structural
relations between doxels will be referred to as the *non structural context* of the
doxels. Our assumption is that document parts are not only relevant because of

S. Geva, J. Kamps, and A. Trotman (Eds.): INEX 2008, LNCS 5631, pp. 87–96, 2009.

their content, but also because they are related to other document parts that answer the query. The document that contains a doxel d will be referred as the *structural context* of d. In some way, we revisit the *Cluster Hypothesis* of van Rijsbergen [13], by considering that the relevance value of each document is impacted by the relevance values of related documents.

In the proposal described here, we first build inter-relations between doxels, and then characterize these relations using relative exhaustivity and specificity at indexing time. We also build explicit relationships between the doxels and their container document. These two elements are used by the matching process.

The rest of this paper is organized as follows: after commenting in section 2 related works, we describe the non structural links that were used in our experiments in part 3, the structural links used in part 4, then the doxel space is described in detail in section 5, in which we propose a document model using the context. Section 6 introduces our *matching in context* process. Results of the INEX 2008 Ad Hoc track are presented in Section 7, where we present the five (three for the Focused task, and two for the Relevant in Context task) officially submitted runs by the LIG. We conclude in part 8.

2   Related Works

If we concentrate on approaches that retrieve doxels based on the documents structure, we can separate approaches that use the structure of documents at indexing time from those that use this structure during the query processing. Among the approaches that focus on the indexing of doxels, the work of Cui, Wen and Chua [2] propagates index terms of the document tree structure from the leaves to the root, by pruning the terms that appear in the content representation of the direct structural components of a doxel. The advantage of this approach is to reduce the size of index, since the index takes into account the transitivity of doxel composition. The difficulty of such work is to carry out a large number of pruning and still to maintain a good quality of results. From a more theoretical point of view, the work of Lalmas [6], based on Dempster-Shafer theory, indexes a doxel using the terms that index its components, proposing then a spreading of indexing terms. Among the proposals that process structured documents at query time the work of [4] manages a priori probability of doxels using their relative position and depth in the document. If we consider the history of the domain, Wilkinson's approach [16] falls into this category. Approaches using probabilistic networks [7] or [8] also use the structural relationships between doxels at query time. In the proposal described here, we make use of such structural links by combining the matching values of the whole documents and of the doxels. This approach has been proven to be effective in the context of INEX.

The use of non structural links, such as Web links or similarity links has been studied in the past. Well known algorithms such as Pagerank [1] or HITS [5] do not seamlessly integrate the links in the matching process. Savoy, in [11], showed that the use of non structural links may provide good results, without qualifying the strength of the inter-relations. In [12], Smucker and Allan show that similarity links may help navigation in the result space. Last year, for our first participation at INEX [14], our approach using non-structural links and a vector space model outperformed runs which did not make use of links. This year, we go further by refining the non structural relationships used, and by integrating during the matching phase the relevance status value (RSV) of the document that contains a relevant doxel.

3   Non Structural Context

The idea of considering non structural neighbours was proposed in [15], in order to facilitate the exploration of the result space by selecting the relevant doxels, and by indicating potential good neighbours to access from one doxel.

The INEX 2008 collection contains several links between doxels and documents, like *collectionlink, unknownlinks, languagelinks* and *outsidelinks* for instance. From these links, only the *collectionlink*s denote links inside the INEX 2009 collection, where the others link to extenal web pages for instance. In the following, we make only use of *collectionlink*, to avoid depending of oustide collection data. An example of a collectionlink (from the document 288041) is : *<collectionlink xmlns:xlink="http://www.w3.org/1999/xlink" xlink:type= "simple" xlink:href="10581.xml">*. For our needs, the most important attribute for such tag is *xlink:href*, that indicates the target of the link. In the original INEX 2009 collection, the targets of *collectionlink*s are only whole documents, and not documents part. As indicated earlier, we want to characterize in detail non structural relations between doxels, and not only from doxels to whole document doxels. We choose then to extend the original *collectionlink*s in a way to obtain relations from doxels to any doxel. To do that, we extend the initial collectionlink doxel c with a document target ct with the following processing :

- assume that c is a direct structural components of a doxel d of type Td,
- we compute the similarity s (using the cosine according to the vector space model) between d and all components of ct of type Td,
- we keep all the components above with the similarity value greater than a threshold T,
- we generate all of these components of ct in the context of c.
- we keep the initial link going from c to the document ct.

The reason why we do consider doxels composed of collectionlink is that very often the text of a collectionlink is very small and the matching with other doxels is unreliable.

The figure 1 describes graphically a simple example of such process. We see that a collectionlink (dotted line) goes from the doxel A (element of a document d1) to the document doxel d2. The doxel composed by A is B, and the type of the doxel B is t. The doxels B and C that compose d1 are also of type t. So, according to the steps described above, we check if the two links "A to C" and "A to D" are created or not (links with dashed lines in figure 1). Assume that the similarity s between B and C is 0.9 and the similarity between B and D is 0.4, and that the threshold T is equal to 0.8. Then, the link that goes from A to C is created and stored as a new non structural link, but the link from A to D is not created (it is crossed in figure 1).

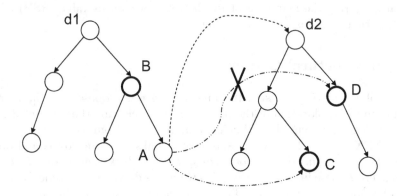

Fig. 1. Example of created non-structural links

Initially, there are 17 013 512 collectionlinks in the INEX 2008 collection. With the process described above we generated 115 million of non structural links.

4 Structural Links

We describe now the use of the structural links as they are used in our proposal. Assume that we build the transitive closure T_c of the compositional links (cf. introduction) between doxels, filtered to keep only the links with source a document doxel and target a non-document doxel. When there exists a link T_c in going from a document D to a doxel d, we generate a structural link going from the doxel d to the document D, expressed by a function Doc so that $Doc(d)=D$. Using this process, we generated 28 million structural links.

5 Doxel Space

5.1 Doxel Content

The representation of the content of doxel d_i is a vector generated from a usual vector space model using the full text content of the doxel: $d_i = (w_{i,1}, ..., w_{i,k})$. Such a representation has proved to give good results for structured document retrieval [3]. The weighting scheme retained is a simple $tf.idf$, with idf based on the whole corpus and with the following normalizations: the tf is normalized by the max of the tf of each doxel, and the idf is log-based, according to the document collection frequency. To avoid an unmanageable quantity of doxels, we kept only doxels having the following tags: article, p, collectionlink, title, section, item. The reason for using only these elements was because, except for the collectionlinks, we assume that the text content for these doxels is not too small. The overall number of doxels considered by us here is 29 291 417.

5.2 Characterizing Non Structural Doxel Context

To characterize the non structural relations between doxels, we propose to define relative exhaustivity and relative specificity. These features are inspired by the definitions of specificity and exhaustivity proposed at INEX 2005 [9], and are supposed to define precisely the nature of the link. Consider a non compositional relation from the doxel d_1 to the doxel d_2:

- The relative specificity of this relation, noted $Spe(d_1, d_2)$, denotes the extent to which d_2 focuses on the topics of d_1. For instance, if d_2 deals only with elements from d_1, then $Spe(d_1, d_2)$ should be close to 1.
- The relative exhaustivity of this relation, noted $Exh(d_1, d_2)$, denotes the extent to which d_2 deals with all the topics of d_1. For instance, if d_2 discusses all the elements of d_1, then $Exh(d_1, d_2)$ should be close to 1.

The values of these features are in $[0, 1]$. Generraky, these features behave in an opposite way: when $Spe(d_1, d_2)$ is high, then $Exh(d_1, d_2)$ is low, and vice versa. However this is not always true: when two doxels have the same texte content, then their relative exhaustivity and specificity should be equal to 1.

Relative specificity and relative exhaustivity between two doxels are extensions of the overlap function [10] of the index of d_1 and d_2: these values reflect the amount of overlap between the source and target of the relation. We define relative specificity and relative exhaustivity on the basis of the non normalized doxel vectors $w_{1,i}$ and $w_{2,i}$ (respectively for d_1 and d_2) as follows.

We estimate values of the exhaustivity and the specificity of d_1 and d_2, based on a vector where weights are $tf.idf$

$$Exh(d_1, d_2) = \frac{\sum_{i|w_{2,i} \neq 0} w_{1,i}^2}{\sum_i w_{1,i}^2} \tag{1}$$

$$Spe(d_1, d_2) = \frac{\sum_{i|w_{1,i} \neq 0} w_{2,i}^2}{\sum_i w_{2,i}^2} \tag{2}$$

These two values scores are in $[0, 1]$ if we assume that no doxel is indexed by a null vector.

6   Matching in Context

As we have characterized the doxel context, the matching process should return doxels relevant to the user's information needs regarding both content and structure aspects, and considering the context of each relevant doxel.

We define the matching function as a linear combination of a standard matching result without context, a matching result based on relative specificity and exhaustivity, and a matching coming from the rank of the documents according to the query processed. The retrieval status value $RSV(d,q)$ for a given doxel d and a given query q is thus given by:

$$RSV(d,q) = \alpha * RSV_{content}(d,q) + (1 - \alpha) * RSV_{context}(d,q) \qquad (3)$$

$$+revrank(RSV_{content}(Doc(d),q)),$$

where $\alpha \in [0,1]$ is experimentally fixed, $RSV_{content}(d,q)$ is the score without considering the set of neighbours $\mathcal{V}_d$ of d (i.e. cosine similarity),

$$RSV_{context}(d,q) = \sum_{d' \in \mathcal{V}_d} \frac{\beta * Exh(d,d') + (1 - \beta) * Spe(d,d')}{|\mathcal{V}_d|} RSV_{content}(d',q)$$

$$(4)$$

where $\beta \in [0,1]$ is used to focus on exhaustivity or specificity, and $revrank(RSV_{content}(Doc(d),q))$ the *reverse rank* of the matching result for each document of the collection: considering that the system returns N results, the reverse rank of the first document is N, the reverse rank of the second document is N-1, and so on.

According to what is done now, the part of the fomula above weighted by α and $1 - \alpha$ is in [0, 1], so adding the ranking of the documents induces a grouping by document of the doxels.

7   Experiments and Results

The INEX 2008 Adhoc track consists of three retrieval tasks: the Focused Task, the Relevant In Context Task, and the Best In Context Task. We submitted 3 runs for the Focused Task, and 2 runs for the Relevant In Context Task. For all these runs, we used only the *title* of the INEX 2008 queries as input for our system: we removed the words prefixed by a '-' character, and we did not consider the indicators for phrase search. The size of the vocabulary we used is 210 000.

First of all, we have evaluated our system with INEX 2007 collection to tune the α and β parameters and the number of non structural neighbours used. The best results were achieved with $\alpha = 0.5$ and $\beta = 0.0$, which means that the non structural context is as important as the context of the doxels, and that only the exhaustivity is considered. We considered 4 non structural neighbours, the ones with the higher similarity values according to part 3 . For the use of the structured context, we tested a ranking based on a Vector Space Model (similar to what was described earlier), abbreviated VSM, and a ranking based on a Language Model using a Dirichlet smoothing, abbreviated ML.

7.1   Focused Task

The INEX 2008 Focused Task is dedicated to find the most focused results that satisfy an information need, without returning "overlapping" elements. In our focused task, we experiment with two different rankings.

For the runs *LIG-ML-FOCRIC-4OUT-05-00* and *LIG-VSM-FOCRIC-4OUT-05-00*, as explained by the matching formula, the results are grouped by document, so results are somewhat similar to Relevant In Context (RIC) results (except for the ordering of the doxels in each document).

The last run, namely, FOC-POSTLM-4OUT-05-00 is a bit different in nature: we generated a binary value (1 for relevant document and 0 for non relevant document) for the document matching, and we filter the doxels belonging to relevant documents, without changing their matching value.

We present our results for the focused task in Table 1 showing precision values at given percentages of recall, and in Figure 2 showing the generalized precision/recall curve. These results show that runs based on the use of the Language Model outperform at recall values lower than 0.2 the VSM document based context. Between recall values of 0.3 and 0.5, the VSM gives better results than the LM. The post processing using language model document matching does not give good results. Each of these curves drops sharply, which leads to a low MAiP compared to other participants runs.

Table 1. Focused Task for LIG at INEX2008 Ad Hoc Track

Run	precision at 0.0 recall	precision at 0.01 recall	precision at 0.05 recall	precision at 0.10 recall
$LIG - ML - FOCRIC$ $MAiP = 0.1441$	0.7127	0.6678	0.5223	0.4229
$LIG - VSM - FOCRIC$ $MAiP = 0.1339$	0.5555	0.5187	0.4407	0.3762
$FOC - POSTLM$ $MAiP = 0.0958$	0.4756	0.4191	0.3741	0.3035

7.2   Relevant in Context Task

For the Relevant In Context Task, we take "default" focused results and re-ordered the first 1500 doxels such that results from the same document are clustered together. It considers the article as the most natural unit and scores the article with the score of its doxel having the highest RSV.

We submitted two runs :

- $LIG - VSM - RIC - 4OUT - 05 - 00$: a run similar to the *LIG-VSM-FOCRIC-4OUT-05-00*, except that the doxels are ordered in sequence of their apparition in each document. In this run, we set $\lambda = 0.5$ and $\beta = 0.0$ and four neighbours;

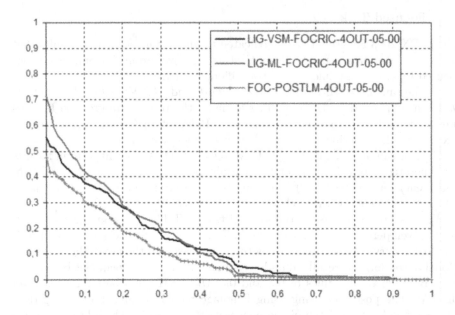

Fig. 2. Interpolated Precision/Recall - Focused Task LIG Ad Hoc

– $LIG - LM - RIC - 4OUT - 05 - 00$: a run similar to the *LIG-LM-FOCRIC-4OUT-05-00*, except that the doxels are ordered by their apparition in each document. In this run, we set $\lambda = 0.5$ and $\beta = 0.0$ and four neighbours.

For the relevant in context task, our results in terms of non-interpolated generalized precision at early ranks $gP[r], r \in \{5, 10, 25, 50\}$ and non-interpolated Mean Average Generalized Precision $MAgP$ are presented in Table 2, and the interpolated Recall/Precision curve is presented in Figure 3. In these results, we see that the use of language model document ranking for doxels retrieval always outperforms the use of vector space based (+55% for MAgP). Similarly to our Focused runs, our results for the Relevant in Context task drop sharply when recall values increase, but the precision is above 62% for a recall of 0, which means that our best approach gives accurate results for the first query results.

Table 2. Relevant In Context Task for INEX2007 Ad Hoc

Run	gP[5]	gP[10]	gP[25]	gP[50]
$LIG - VSM - RIC - 4OUT - 05 - 00$ $MAgP = 0.0961$	0.2444	0.2023	0.1756	0.1360
$LIG - LM - RIC - 4OUT - 05 - 00$ $MAgP = 0.1486$	0.3595	0.3069	0.2303	0.1708

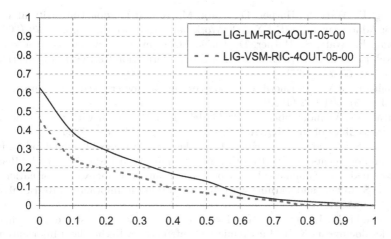

Fig. 3. Interpolated Recall/Precision - Relevant in Context Task LIG Ad Hoc

8   Summary and Conclusion

In the INEX 2008 Ad Hoc track, we integrated two contexts (the four most similar doxels in a collectionlink document target and the full document rank) with the RSV of doxels. We submitted runs implementing our proposals for the Focused and Relevant in Context Ad Hoc Tasks. For each of these tasks, we showed that combining content and context leads to good results, especially when considering full document language models. One explanation is that such models are well adapted for full documents, and using the context of doxel in a second step is a good approach. For our second participation to INEX, our best runs are ranked in the top 10 runs of participants systems at least in the Focused Task at iP[0.01]. However, we plan to improve our baseline to obtain better results in the following directions:

- For the best results obtained by our approach, we used a vector space model for the doxels and a language model for the full documents. We will focus in the future on proposing a more consistent approach that relies only on language models. We will then face the problem of modeling very short doxels with language models.
- We will also consider to redefine the non structural context of doxel, by studying other features that can be used to characterize the inter doxel relationships.

References

1. Brin, S., Page, L.: The anatomy of a large-scale hypertextual Web search engine. Computer Networks and ISDN Systems 30(1–7), 107–117 (1998)
2. Cui, H., Wen, J., Chua, T.: Hierarchical indexing and flexible element retrieval for structured documents. In: Sebastiani, F. (ed.) ECIR 2003. LNCS, vol. 2633, pp. 73–87. Springer, Heidelberg (2003)

3. Huang, F., Watt, S., Harper, D., Clark, M.: Robert Gordon University at INEX 2006: Adhoc Track. In: INEX 2006 Workshop Pre-Proceeding, pp. 70–79 (2006)
4. Huang, F., Watt, S., Harper, D., Clark, M.: Compact representation in xml retrieval. In: Comparative Evaluation of XML Information Retrieval Systems (INEX 2006), pp. 65–72 (2007)
5. Kleinberg, J.M.: Authoritative sources in a hyperlinked environment. J. ACM 46(5), 604–632 (1999)
6. Lalmas, M., Vannoorenberghe, P.: Modelling xml retrieval with belief functions. In: Proceedings of the CORIA 2004 conference, pp. 143–160 (2004)
7. Myaeng, S.-H., Jang, D.-H., Kim, M.-S., Zhoo, Z.-C.: A flexible model for retrieval of sgml documents. In: SIGIR 1998, pp. 138–145. ACM Press, New York (1998)
8. Piwowarski, B., Gallinari, P.: A bayesian framework for xml information retrieval: Searching and learning with the inex collection. Information Retrieval 8(4), 655–681 (1995)
9. Piwowarski, B., Lalmas, M.: Interface pour l'evaluation de systemes de recherche sur des documents XML. In: Premiere COnference en Recherche d'Information et Applications (CORIA 2004), Toulouse, France, March 2004, pp. 109–120. Hermes (2004)
10. Salton, G., McGill, M.J.: Introduction to Modern Information Retrieval, ch. 6, p. 203. McGraw-Hill, Inc., New York (1986)
11. Savoy, J.: An extended vector-processing scheme for searching information in hypertext systems. Inf. Process. Manage. 32(2), 155–170 (1996)
12. Smucker, M.D., Allan, J.: Using similarity links as shortcuts to relevant web pages. In: SIGIR 2007: Proceedings of the 30th annual international ACM SIGIR conference on Research and development in information retrieval, pp. 863–864. ACM Press, New York (2007)
13. van Rijsbergen, C.: Information retrieval, 2nd edn., ch. 3. Butterworths (1979)
14. Verbyst, D., Mulhem, P.: Lig at inex 2007 ad hoc track: Using collectionlinks as context. In: Fuhr, N., Kamps, J., Lalmas, M., Trotman, A. (eds.) INEX 2007. LNCS, vol. 4862, pp. 138–147. Springer, Heidelberg (2008)
15. Verbyst, D., Mulhem, P.: Doxels in context for retrieval: from structure to neighbours. In: SAC 2008: Proceedings of the 2008 ACM symposium on Applied computing, pp. 1122–1126. ACM Press, New York (2008)
16. Wilkinson, R.: Moeffective retrieval of structured documents. In: Proceedings of the 17th Annual International ACM-SIGIR Conference on Research and Development in Information Retrieval, pp. 311–317 (1998)

SPIRIX: A Peer-to-Peer Search Engine
for XML-Retrieval

Judith Winter and Oswald Drobnik

J.W. Goethe University, Institute for Informatics, Frankfurt, Germany
{winter,drobnik}@tm.informatik.uni-frankfurt.de

Abstract. At INEX 2008 we presented SPIRIX, a Peer-to-Peer search engine developed to investigate distributed XML-Retrieval. Such investigations have been neglected by INEX so far: while there is a variety of successful and effective XML-Retrieval approaches, all current solutions are centralized search engines. They do not consider distributed scenarios, where it is undesired or impossible to hold the whole collection on one single machine. Such scenarios include search in large-scale collections, where the load of computations and storage consumption is too high for one server. Other systems consist of different owners of heterogeneous collections willing to share their documents without giving up full control over their documents by uploading them on a central server. Currently, there are research solutions for distributed text-retrieval or multimedia-retrieval. With INEX and innovative techniques for exploiting XML-structure, it is now time to extend research to distributed XML-Retrieval. This paper reports on SPIRIX' performance at INEX'08.

Keywords: Distributed XML-Retrieval, INEX, Distributed Search, XML Information Retrieval, Efficiency, Peer-to-Peer.

1 Introduction

Many of the solutions presented at INEX are quite effective in terms of Information Retrieval (IR) quality. However, all current solutions are centralized search engines. This is sufficient for the current INEX test collection, consisting of nearly 660000 Wikipedia articles in XML format for the main ad-hoc track. In practice, collections such as those stored in digital libraries are much more voluminous and do not fit on one computer. Also, search should not be limited to one collection only, but a search engine should have access to a wide variety of collections offered, for instance, by public institutions such as universities or the private collections of users willing to share them. Thus, distributed XML-Retrieval systems should be investigated in order to search large-scale or scattered collections. Additionally, much more powerful search engines can be built by pooling computers together – the load of computations and storage consumption can then be distributed over all participating nodes.

However, this comes at the cost of traffic between those nodes. Hence, distributed search engines have to consider efficiency as well as effectiveness. In terms of XML-Retrieval, a distributed search engine has to ensure an appropriate distribution of both content and structural information. On the other hand, structure can be used to perform

S. Geva, J. Kamps, and A. Trotman (Eds.): INEX 2008, LNCS 5631, pp. 97–105, 2009.

the distributed search more effectively and efficiently, for example when selecting postings in the routing process.

To the best of our knowledge, none of the content-oriented XML-Retrieval solutions currently consider distributed aspects. In this paper, we present the first distributed XML-Retrieval system that performs content-oriented search for XML-documents. Our system is based on a structured P2P system.

P2P networks are emerging infrastructures for distributed computing, where peers – autonomous and equal nodes – are pooled together. Resources that are shared between the peers include hardware such as disk storage and CPU usage as well as information stored in files (e.g. documents). Unlike classical client/server systems, there is no central control, but a high degree of self-organization, such that P2P systems are very flexible, adapt to new situations (such as joining and leaving of nodes), and thus may scale to theoretically unlimited numbers of participating nodes.

A P2P network can be organized as a Structured Overlay Network where a logical structure is laid on top of a physical network, e.g. to allow for search of distributed resources by use of application-specific keys. Distributed hash tables (DHT) can be used to route messages to a peer storing an object specified by a unique identifier (ID), without knowing the peer's physical address. The data structure for the routing table, and thus the logical structure of the overlay network, depends on the DHT algorithm used. For example, the Chord protocol maintains a DHT ring and maps peers and keys to an identifier ring [9]. Commonly, DHT-based algorithms structure an overlay network such that efficient lookup of objects can be performed.

There are an increasing number of XML-documents among the growing amount of objects shared by P2P applications, especially since public and private institutions such as museums have started to share their Digital Libraries. There are a number of search engines for P2P networks, for example the DHT-based systems [2] and [7]. However, none of these approaches supports XML-Retrieval techniques. Schema-based P2P-networks [3], on the other hand, consider structure for the routing, but existing solutions (such as [1]) do not provide IR techniques for content-oriented search.

In this paper, we propose the first distributed XML-Retrieval system named SPIRX (Search Engine for P2P IR in XML-documents). It provides a series of features such as CAS queries, the capacity to weight different XML-elements differently, and element retrieval. On the one hand, this system takes advantage of XML-structure to make the search for XML-documents more precise, that is, structure is used to achieve a system that is more effective than current P2P-IR solutions. On the other hand, XML-structure can be involved in the routing to make the search process more efficient, e.g. by selecting posting list and postings based on evidence from both content and structure. To demonstrate the proposed approach as an XML-Retrieval system, we participated in INEX 2008. This paper describes motivation, research questions, technical details, submitted runs for the ad-hoc track, and preliminary evaluation results of our system.

2   A P2P Search Engine for XML-Documents

The following section describes SPIRX, a search engine for P2P Information Retrieval (IR) of XML-documents. First, we present the architecture and system design of SPIRIX, focusing in particular on the techniques for extracting and storing structured information while indexing. Second, we outline where and how structure can be used in the process of answering a given INEX query.

2.1 System Design – How XML-Structure Is Distributed While Indexing

The smallest object indexed by SPIRIX is what we denote by XTerm: a tuple consisting of a stemmed non-stopword term and its structure. A term's structure is denoted as the path from the document root element to the term element in the XML-document tree expressed with XPath but without element numbers. To reduce the variety of different structures, we apply methods such as mapping of syntactically different but semantically similar tags, as well as stopword removal and stemming of tags.

For each indexed collection, a temporary inverted index for all extracted XTerms is built locally, where the vocabulary consists of keys that are combinations of all XTerms with the same content while each XTerm maintains its own posting list (see figure 1). Each posting refers to a document, but its score is computed based on evidence from the document's weight (tf, idf), from its element weights, and from the peer assigned to the document (peerScore). More details about our impact ordering and the full formula are given in [10].

apple	\book\chapter	→	doc1(12.8), doc2(12.4)
	\article\p	→	doc2(25.3), doc3(12.7), doc4(10.7)

Fig. 1. Example of a key in the inverted index

In the same parsing phase, we build a temporary statistic index (document index and element index) by collecting the statistics for each document and for selected elements which are treated as independent documents. So far, only fixed tags (paragraphs and sections) are considered as potential relevant elements. We are currently extending the method to elements that have been found relevant in the past or were specially marked by the indexing user.

After creation of the temporary inverted index, we scan this index and build combinations of frequent keys to support multi term queries. These combinations are called highly discriminative keys (HDK) and are stored in the HDK index.

Finally, all temporary indexes are distributed over the network. For message transport, we integrated a P2P protocol named SpirixDHT that has been developed to support efficient transport of messages between peers when retrieving XML-documents. This protocol is based on Chord and has been adapted to the special requirements of XML-retrieval as described in [10]. That is, in a network of n peers, we can store and locate information identified by unique IDs with DHT techniques in $O(log(n))$ hops. The inverted index is distributed by storing the posting lists on different peers. As ID that is hashed to assign a posting list to its according peer, we use the posting list key's content. This guarantees that all posting lists referring to the same term are stored on the same peer, because in the routing they are most likely to be used together. The statistic index is distributed by hashing the unique ID of each document (which is calculated by the peer's IP-address, its port, and the local document name, or which is created by storing the exact number of global documents in the DHT). To ensure that element statistics are stored on the same peer as their root document, the hashed IDs of elements are chosen to be identical with those of their root documents.

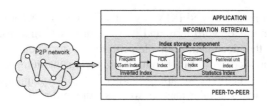

Fig. 2. Parts of the distributed global indexes are stored on each peer

By applying these distribution techniques, SPIRIX is based on global indexes (inverted index and statistics index) which are not stored centrally but distributed over all participating peers. If peers leave the network or new participants join, the indexes are automatically redistributed as the network organizes itself. For example, a new peer can take over some of the stored information for which its direct neighbour has been responsible. Figure 2 shows the different parts of the global indexes that are stored locally on each peer and that are managed by the index storage component as part of the IR complex on each peer.

2.2 System Use – Where XML-Structure Is Used While Querying

How is the distributed information used in the querying process, and above all, where is XML-structure applied to improve performance?

Retrieval includes routing and ranking, and consists of two steps: first, to locate the posting lists of all query terms, merge them and select promising postings. For this step, routing requests are sent over the P2P network. Second, locate the statistics for the selected postings, such that the referenced documents and elements can be ranked. For this step, ranking requests are sent. In contrast to common P2P search engines that perform routing and ranking solely based on content evidence, SPIRIX includes XML-structure in both steps.

Routing: Especially for multi term queries, where posting lists have to be sent between peers for merging, not all postings can be selected as this would significantly increase network traffic for large-scale collections with big posting lists. For the selection of posting lists, we compute the similarity between the structural hint of a CAS query and the structure of an XTerm's posting list using the functions described in [12]. Only those posting lists are considered where the structural similarity exceeds a threshold. Furthermore, only *top k* entries from the chosen posting lists are selected. For this purpose, the postings are ordered by impact based on evidence from document-, element-, and peer-level but also based on the structural similarity of the original posting list. Postings of XTerms that match the CAS hint closely thus get a higher impact factor than less similar ones.

Ranking: For each selected posting, a ranking request is sent to the peer that is assigned to the referenced document. The relevance is then computed using an extension of the vector space model: query, document, and elements are represented as vectors whereby each component contains the weight of an XTerm. That is, the weight of a term is split into individual weights for its different structures. Second, the weight itself is computed using structure: it is computed as the product of the weight of the XTerm's content and the structural similarity between XTerm and CAS query term. Third, the weight of the XTerm's content is computed with an adaptation of BM25E [8] such that different elements can be weighted differently, e.g. "titles" can get higher weights than "links". So far we have not been able to show improvement using this method so we set all weights to 1; however, we want to provide this feature in case we or others can use it to make future improvements to Wikipedia or maybe other, more semantically structured collections. Finally, all relevance computations are performed not only for the selected documents but also for their potential relevant element (whose statistics are stored on the same peer for better access) to achieve more focused and specific results. Similar to related work, e.g. [5], we assume a strong correlation between the relevance of an element and its parent document. Therefore, the parent document influences the ranking of elements: the score of an element is computed based on the stored statistics but smoothed by the score of its parent document.

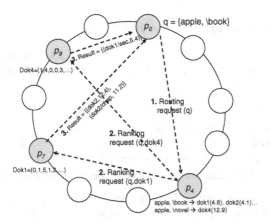

Fig. 3. Retrieval process for single term query $q = \{apple, \backslash book\}$ based on SpirixDHT

Figure 3 displays the process of answering a query q. First, q is routed to Peer p_4 which is assigned to the hash of *apple* and holds all posting lists for XTerms with content *apple*. On p_4, posting lists and postings are selected according to q by taking into account the postings' weight multiplied with the similarity to $\backslash book$. Postings *dok1* and *dok4* are selected, and routing requests are sent to peers p_7 and p_9 which are responsible for holding the statistics of these documents and their elements. Peer p_7 and peer p_9 both receive the query, calculate results and send these back to the querying peer p_0.

3 Participating in INEX 2008

3.1 INEX Tracks

2008 was the first year that the University of Frankfurt participated in INEX. We chose two tracks: the ad-hoc track and the efficiency track. Our participation in the efficiency track is described in [12]. We concentrate on the ad-hoc track.

3.2 INEX Tasks (Ad-Hoc Track)

Our main task was the focused task. To answer a topic for this task, 5000 documents are chosen from the posting list of each query term. Note that this is done on separate peers, that is, before the posting lists are merged. Their relevance plus the relevance of selected elements from these documents is then computed. Overlapping is filtered out by ordering the results by document ID and element offset. If two results have identical document IDs and one offset is the first part of the other one, only the result with the higher relevance is kept.

We submitted the same results – filtered by the corresponding rules – to the BestInContext task (BIC) and to the RelevantInContext task (RIC) as well to see how SPIRIX performs in these tasks. However, due to bugs in the BIC/RIC-Filters these runs were not successful and ended up with a precision of nearly 0%.

3.3 University of Frankfurt Runs (Ad-Hoc Track)

We participated in order to demonstrate SPIRIX as an XML-Retrieval system. Furthermore, we were interested in a comparison of different methods implemented in our system. These include:

- CO versus CAS: do structural hints help to improve ranking or routing?
- How do different ranking functions perform?
- How do different structural similarity functions perform (ranking and routing)?
- How does the amount of selected postings affect precision?
- Element versus document retrieval – where does SPIRIX perform better?

For the official runs, we decided to compare different ranking functions. For each run, 5000 postings were selected from the posting lists and the structural similarity function described in [11] was used. For each task, three runs were submitted for three different ranking functions: a tf*idf-Baseline, a variant of BM25E, and a variant of BM25E with weights>1 for the elements *article*, *body*, *section*, *paragraph*, *bold* and with weights<1 for the elements *collectionlink*, *languagelink*, *unknownlink*, *wikipedialink* and *template*.

3.4 Tuning the System – Balance between Effectiveness and Efficiency

Participating in the ad-hoc track, we aimed for high precision. Thus, for many parameters that influence the balance between effectiveness and efficiency, we chose values that aim at effectiveness including e.g. global statistics. For early termination (selection of postings), we decided for a compromise of 5000 postings. Regarding storage of structural information, we used methods to shorten the structure length which results in an index that is much smaller and easier to handle, but which also reduces precision by approximately 5%.

Global statistics: SPIRIX is based on a DHT which enables the collection and storage of global statistics. Usually, we estimate these statistics from the locally stored part of the distributed document index that contains randomly hashed samples from the collection. This estimation technique saves the messages necessary for distributing and accessing global statistics at the cost of reduced precision, depending on the estimations. Therefore, for the INEX runs, the exact global statistics were used.

Early termination (Selection of 5000 postings): Due to our system architecture, taking all postings from a posting list is not efficient as this leads to many ranking request messages. However, precision increases with the amount of selected postings (up until a collection specific point). Thus, the best 5000 documents were selected from each query term's posting list. Note that this is done on separate peers and thus without merging – we lose precision for multi term queries when good documents are on positions > 5000. For Wikipedia, 5000 postings are sufficient for most topics.

4   Evaluation

The performance of SPIRIX in the ad-hoc track (focused task) is shown in table 1. The official result for the submission in July 2008 shows a precision of 0,20 and rank 64 out of 76 participants for the document run (with 0,36 being the top achieved precision). For focused retrieval, the official result is 0,27. The achieved search quality establishes SPIRIX as an XML-Retrieval system but is rather low in comparison to other systems. This was due to bugs in the overlapping filters resulting in illegal duplicates such that only our baseline run – with a simple tf*idf-variant as ranking model – could be officially evaluated. After correcting the filtering bugs, we ran the evaluation ourselves and could report a precision of 0,52 at the INEX workshop in December 2008 (with 0,68 achieved at rank 1 by University of Waterloo).

Table 1. Performance at INEX2008, measured in iP[0.01]

	July 2008 (official result)	July 2008 (official result)	Dec. 2008 (presented at workshop)	Feb. 2009 (efficiency track)
Track/Task	Dokument Retrieval (AdHoc /Focused)	Focused Retrieval (AdHoc /Focused)	AdHoc/ Focused	AdHoc/ Focused
iP[0.01] SPIRIX	0,20	0,27	0,52	**0,679**
iP[0.01] Best system	0,36 (Lyon)	0,69 (Waterloo)	0,69 (Waterloo)	0,69 (Waterloo)
Rank	64 of 76	58 of 61	ca. 47 of 61	ca. 4 of 61

We can now report on a SPIRIX run of 0,679 precision for the focused task. This run was officially submitted (after the INEX workshop, though) to the efficiency track, in order to compare the 5 submitted P2P-based runs with an optimized run where all parameters were set to simulate a client/server search engine. The increase in precision was *not* achieved by adapting to the INEX 2008 topics but by improvements in the indexing and ranking process of SPIRIX and by keeping all posting lists on one peer to simulate a client/server-system.

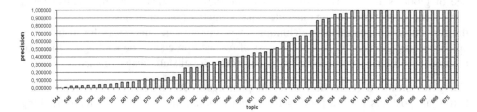

Fig. 4. iP[0.01] for each single topic

To analyze SPIRIX' performance in order to get hints for future improvements, we looked at the precision for each single topic as displayed in figure 4. The variety over the topics is as wide as possible: the precision is anything between 0 and 1. A third of the topics was answered successfully with a precision of more than 0,90. For 19 out of 70 topics, even a precision of 0,99 could be achieved. More than half of all topics could be answered with a precision of more than 0,45. Only for 15 topics was the achieved precision less than 0,10. These were mainly topics with many terms or difficult topics. For example, topic no. 677 asked for articles about terracotta figures showing horses. Only two documents in the whole Wikipedia collection have been assessed as relevant to this topic.

5 Conclusion

In this paper, we have proposed the first distributed search engine for XML-documents. SPIRIX is based on a structured P2P system and offers a whole variety of XML-Retrieval features such as taking advantage of CAS queries, weighting different elements differently, and element retrieval. We participated in INEX 2008 to demonstrate SPIRIX as an XML-Retrieval system with an IR quality comparable to centralized XML-Retrieval. University of Frankfurt participated in INEX 2008 for the first time, and the implementation of our first prototype was not ready until one week before the run submission deadline. Thus, most of our runs for the official submission failed due to technical errors. The only officially evaluated run was our baseline run in which we achieved a precision of 27% – high enough to claim that SPIRIX is indeed a XML-Retrieval system. However, after fixing the filters a re-evaluation with the new INEX 2008 tool could show a precision of 52%, which was reported at the INEX workshop in Dagstuhl in December 2008. Further improvements in SPIRIX's indexing and ranking process as well as simulating a client/server-architecture have now enabled us to report a precision of 67,9%, which is comparable to the search quality that the top-10 INEX systems achieved in 2008.

Acknowledgements. We would like to thank Andrew Trotman and Shlomo Geva for endless hours of discussion about the challenges and opportunities of XML-IR for P2P, for loads of critical remarks – always constructive –, and for sharing their view of XML-IR.

References

[1] Abiteboul, S., Manolescu, I., Polyzotis, N., Preda, N., Sun, C.: XML processing in DHT networks. In: IEEE 24th Internat Conference on Data Engineering (ICDE 2008) (2008)
[2] Bender, M., Michel, S., Weikum, G., Zimmer, C.: The MINERVA Project - Database Selection in the Context of P2P Search. In: BTW Conference 2005 (2005)
[3] Koloniari, G., Pitoura, E.: Peer-to-Peer Management of XML Data: Issues and Research Challenges. SIGMOD Rec. 34(2) (2005)
[4] Li, J., Loo, B., Hellerstein, J., Kaashoek, F., Karger, D., Morris, R.: On the Feasibility of Peer-to-Peer Web Indexing and Search. In: Proc. of the Second International Workshop on Peer-to-Peer Systems (2003)
[5] Mass, Y., Mandelbrod, M.: Component Ranking and Automatic Query Refinement for XML Retrieva. In: Fuhr, N., Lalmas, M., Malik, S., Szlávik, Z. (eds.) INEX 2004. LNCS, vol. 3493, pp. 73–84. Springer, Heidelberg (2005)
[6] Michel, S., Triantafillou, P., Weikum, G.: KLEE - A Framework for Distributed Top-k Query Algorithms. In: Proc. of 31st VLDB Conference, Trondheim, Norway (2005)
[7] Podnar, I., Rajman, M., Luu, T., Klemm, F., Aberer, K.: Scalable Peer-to-Peer Web Retrieval with Highly Discriminative Keys. In: Proc. of IEEE 23rd International Conference on Data Engineering (ICDE 2007), Istanbul, Turkey (2007)
[8] Robertson, S., Zaragoza, H., Taylor, M.: Simple BM25 extension to multiple weighted fields. In: Proc. of CIKM 2004. ACM Press, New York (2004)
[9] Stoica, I., Morris, R., Liben-Nowell, D., Karger, D., Kaashoek, F., Dabek, F., Balakrishnan, H.: Chord - A Scalable Peer-to-peer Lookup Protocol for Internet Applications. IEEE/ACM Transactions on Networking 11(1) (2003)
[10] Winter, J.: Routing of Structured Queries in Large-Scale Distributed Systems. In: LSDS_IR 2008 at ACM CIKM 2008, Napa Valley, California, USA (October 2008)
[11] Winter, J., Drobnik, O.: An Architecture for XML Information Retrieval in a Peer-to-Peer Environment. In: ACM PIKM2007 at CIKM 2007, Lisbon, Portugal (2007)
[12] Winter, J., Jeliazkov, N.: Recognition of Structural Similarity To Increase Performance. In: Preproceedings of the 7th International Workshop of the INitiative for the Evaluation of XML Retrieval (INEX 2008), Dagstuhl, Germany (2008)
[13] Yee, W.G., Nguyen, L.T., Jia, D., Frieder, O.: Efficient Query Routing by Improved Peer Description in P2P Networks. In: Proc. of ACM/ICST Infoscale (2008)
[14] Zhang, J., Suel, T.: Optimized Inverted List Assignment in Distributed Search Engine Architectures. In: Proc. of 21th IPDPS 2007, California, USA (2007)

Overview of the INEX 2008 Book Track

Gabriella Kazai[1], Antoine Doucet[2], and Monica Landoni[3]

[1] Microsoft Research Cambridge, United Kingdom
gabkaz@microsoft.com
[2] University of Caen, France
doucet@info.unicaen.fr
[3] University of Lugano
monica.landoni@unisi.ch

Abstract. This paper provides an overview of the INEX 2008 Book Track. Now in its second year, the track aimed at broadening its scope by investigating topics of interest in the fields of information retrieval, human computer interaction, digital libraries, and eBooks. The main topics of investigation were defined around challenges for supporting users in reading, searching, and navigating the full texts of digitized books. Based on these themes, four tasks were defined: 1) The Book Retrieval task aimed at comparing traditional and book-specific retrieval approaches, 2) the Page in Context task aimed at evaluating the value of focused retrieval approaches for searching books, 3) the Structure Extraction task aimed to test automatic techniques for deriving structure from OCR and layout information, and 4) the Active Reading task aimed to explore suitable user interfaces for eBooks enabling reading, annotation, review, and summary across multiple books. We report on the setup and results of each of these tasks.

1   Introduction

As a result of numerous mass-digitization projects [2], e.g., Million Book project[1], efforts of the Open Content Alliance[2], and the digitization work of Google[3], the full texts of digitized books are increasingly available on the Web and in digital libraries. The unprecedented scale of these efforts, the unique characteristics of the digitized material, as well as the unexplored possibilities of user interactions present exciting research challenges and opportunities, see e.g., [7].

Motivated by the need to foster research in this domain, the Book Track was launched in 2007 as part of the INEX initiative. The overall goal of the track is to promote inter-disciplinary research investigating techniques for supporting users in reading, searching, and navigating the full texts of digitized books and to provide a forum for the exchange of research ideas and contributions. In 2007, the track concentrated on identifying infrastructure issues, focusing

---

[1] http://www.ulib.org/
[2] www.opencontentalliance.org/
[3] http://books.google.com/

S. Geva, J. Kamps, and A. Trotman (Eds.): INEX 2008, LNCS 5631, pp. 106–123, 2009.
© Springer-Verlag Berlin Heidelberg 2009

on information retrieval (IR) tasks. In 2008, the aim was to look beyond and bring together researchers and practitioners in IR, digital libraries, human computer interaction, and eBooks to explore common challenges and opportunities around digitized book collections. Toward this goal, the track set up tasks to provide opportunities for investigating research questions around three broad topics:

- IR techniques for searching collections of digitized books,
- Users' interactions with eBooks and collections of digitized books,
- Mechanisms to increase accessibility to the contents of digitized books.

Based around these main themes, four specific tasks were defined

1. The Book Retrieval (BR) task, framed within the user task to build a reading list for a given topic, aimed at comparing traditional document retrieval methods with domain-specific techniques exploiting book-specific features, such as the back of book index or associated metadata, like library catalogue information,
2. The Page in Context (PiC) task aimed to test the value of applying focused retrieval approaches to books, where users expect to be pointed directly to relevant book parts,
3. The Structure Extraction (SE) task aimed to evaluate automatic techniques for deriving structure from OCR and layout information for building hyperlinked table of contents, and
4. The Active Reading task (ART) aimed to explore suitable user interfaces enabling reading, annotation, review, and summary across multiple books.

In this paper, we discuss the setup and results of each of these tasks. First, in Section 2, we give a brief summary of the participating organisations. In Section 3, we describe the corpus of books that forms the basis of the test collection. The following three sections detail the four tasks: Section 4 summarises the BR and PiC tasks, Section 5 reviews the SE task, and Section 6 discusses ART. We close in Section 7 with a summary and further plans.

2   Participating Organisations

A total of 54 organisations registered for the track (double from last year's 27), of which 15 took part actively throughout the year (up from 9 last year), see Tables 1 and 2. For active participants, the topics they created and assessed, and the runs they submitted are listed in Table 1. In total, 19 groups downloaded the book corpus, 11 groups contributed 40 search topics, 2 groups submitted runs to the Structure Extraction task, 4 to the Book Retrieval task, 2 to the Page in Context task, and 2 are currently participating in the Active Reading task. A total of 17 participants from 10 known[4] groups contributed to the relevance assessments.

---

[4] Three of the assessors did not provide an affiliation (topics assessed: 8, 60, 68).

Table 1. Active participants of the INEX 2008 Book Track, contributing topics, runs, and/or relevance assessments (BR = Book Retrieval, PiC = Page in Context, SE = Structure Extraction, ART = Active Reading Task)

ID	Organisation	Topics	Runs	Assessed topics
6	University of Amsterdam	51, 52, 65	3 BR, 7 PiC	8, 9, 21, 29, 51, 52, 57, 60
7	Oslo University College			12
14	University of California, Berkeley	66, 67	3 BR, ART	
17	University of Strathclyde			9, 21, 55
30	CSIR, Wuhan University	36, 38, 39, 42		
31	Faculties of Management and Information Technologies, Skopje	40, 46, 47, 48		
41	University of Caen	60, 61		31, 37, 60
43	Xerox Research Centre Europe		4 SE	
52	Kyungpook National University	44, 45, 49, 50	ART	1
54	Microsoft Research Cambridge	55, 56, 57, 58, 62, 63, 64, 70		1, 3, 5, 8, 21, 22, 27, 31, 36, 51, 53, 55, 56, 57, 62, 63, 64
56	JustSystems Corporation	53, 54, 59		53, 54
62	RMIT University	31, 37, 41, 43	10 BR	5, 8, 21, 27, 31, 36, 37, 39, 41, 57, 60, 64, 69
78	University of Waterloo	32, 33, 34, 35	2 BR, 6 PiC	12, 51, 53, 62
86	University of Lugano	68, 69		3, 15, 68, 70
125	Microsoft Development Center Serbia		3 SE	

3 The Book Corpus

The track builds on a collection of 50,239 digitized out-of-copyright books, provided by Microsoft Live Search and the Internet Archive. The corpus is made up of books of different genre, including history books, biographies, literary studies, religious texts and teachings, reference works, encyclopedias, essays, proceedings, novels, and poetry.

The OCR text of the books has been converted from the original DjVu format to an XML format referred to as BookML, developed by Microsoft Development Center Serbia. BookML provides additional structure information, including markup for table of contents entries. 50,099 of the books also come with an associated MAchine-Readable Cataloging (MARC) record, which contains publication (author, title, etc.) and classification information.

The basic XML structure of a typical book in BookML (ocrml.xml) is a sequence of pages containing nested structures of regions, sections, lines, and words ([coords] represents coordinate attributes, defining the position of a bounding rectangle for a region, line or word, or the width and height of a page):

```
<document>
 <page pageNumber=''I-N'' label=''PT_CHAPTER'' [coords] key=''0'' id=''0''>
  <region regionType=''Text'' [coords] key=''0'' id=''0''>
   <section label=SEC_BODY'' key=''408'' id=''0''>
    <line [coords] key=''0'' id=''0''>
     <word [coords] key=''0'' id=''0'' val=''Moby''/>
     <word [coords] key=''1'' id=''1'' val=''Dick''/>
    </line>
    <line [...]>
     <word [...] val=''Herman''/>
     <word [...] val=''Melville''/>
    </line>          [...]
   </section>        [...]
  </region>          [...]
 </page>             [...]
</document>
```

BookML provides a set of labels (as attributes) indicating structure information in the full text of a book and additional marker elements for more complex texts, such as a table of contents. For example, a label attribute may indicate the semantic unit that an XML element is likely to be a part of, e.g., a section may be part of a header (SEC_HEADER), a footer (SEC_FOOTER), the back of book

Table 2. Passive participants of the INEX 2008 Book Track

ID	Organisation	ID	Organisation
Passive participants (Corpus download only)			
4	University of Otago	42	University of Toronto
10	Max-Planck-Institut Informatik	116	University of the Aegean
Passive participants			
5	Queensland University of Technology	104	UCLV
8	University College London	107	University of Sci. and Tech. of China
9	University of Helsinki	112	Hitachi, Ltd.
15	University of Iowa	115	IIT
19	University of Ca Foscari di Venezia	117	Iran
21	MPP	118	M.Tech Student
27	University at Albany (also ID=76)	127	UNICAMP
29	Indian Statistical Institute	148	UEA
32	CUHK	158	George Mason University
39	University of New South Wales	160	Universite Jean Monnet
51	Suny-Albany	161	University of California, Santa Cruz
60	Saint Etienne University	164	Isfahan University
66	University of Rostock	165	Universidad de Oriente
88	Independent	166	Drexel University
91	Auckland University of Technology	171	Chinese University of Hong Kong
93	Wuhan Institute of Technology	174	Alexandria University
96	Cairo Microsoft Innovation Center	181	COLTEC
100	Seoul National University		

index (SEC_INDEX), the table of contents (SEC_TOC), or the body of the page (SEC_BODY), etc. A page may be labeled as a table of contents page (PT_TOC), an empty page (PT_EMPTY), a back of book index page (PT_INDEX), or as a chapter start page (PT_CHAPTER), etc. Marker elements provide detailed markup, e.g., for table of contents, indicating entry titles (TOC_TITLE), and page numbers (TOC_CH_PN), etc.

The full corpus, which totals around 400GB, was distributed on USB HDDs (at a cost of 70GBP). In addition, a reduced version (50GB, or 13GB compressed) was made available for download. The reduced version was generated by removing the word tags and propagating the values of the val attributes as text content into the parent (i.e., line) elements.

4   Information Retrieval Tasks

Focusing on IR challenges, two search tasks were investigated: 1) Book Retrieval (BR), in which users search for whole books in order to build a reading list on a given topic, and 2) Page in Context (PiC), in which users search for information in books on a given topic and expect to be pointed directly at relevant book parts. Both these tasks used the corpus of over 50,000 books described in Section 3, and the same set of test topics (see Section 4.3). This was motivated by the need to reduce the relevance assessment workload and to allow possible future comparisons across the two tasks.

A summary of the tasks, the test topics, the online relevance assessment system, the collected assessments, and the evaluation results are described in the following sections. Further details and the various DTDs, describing the syntax of submission runs, are available online in the track's Tasks and Submission Guidelines at http://www.inex.otago.ac.nz/tracks/books/taskresultspec.asp.

4.1   The Book Retrieval (BR) Task

This task was set up with the goal to compare book-specific IR techniques with standard IR methods for the retrieval of books, where (whole) books are returned to the user. The user scenario underlying this task is that of a user searching for books on a given topic with the intent to build a reading or reference list. The list may be for research purposes, or in preparation of lecture materials, or for entertainment, etc.

Participants of this task were invited to submit either single runs or pairs of runs. A total of 10 runs could be submitted. A single run could be the result of either generic (non-specific) or book-specific IR methods. A pair of runs had to contain both types, where the non-specific run served as a baseline which the book-specific run extended upon by exploiting book-specific features (e.g., back-of-book index, citation statistics, book reviews, etc.) or specifically tuned methods. One automatic run (i.e., using only the topic title part of a test topic for searching and without any human intervention) was compulsory. A run could contain, for each test topic, a maximum of 1000 books (identified by their 16 character long bookID[5]), ranked in order of estimated relevance.

A total of 18 runs were submitted by 4 groups (3 runs by University of Amsterdam (ID=6); 3 runs by University of California, Berkeley (ID=14); 10 runs by RMIT University (ID=62); and 2 runs by University of Waterloo (ID=78)), see Table 1.

4.2   The Page in Context (PiC) Task

The goal of this task was to investigate the application of focused retrieval approaches to a collection of digitized books. The task was thus similar to the INEX ad hoc track's Relevant in Context task, but using a significantly different collection while also allowing for the ranking of book parts within a book. The user scenario underlying this task was that of a user searching for information in a library of books on a given subject. The information sought may be 'hidden' in some books (i.e., it forms only a minor theme) while it may be the main focus of some other books. In either case, the user expects to be pointed directly to the relevant book parts. Following the focused retrieval paradigm, the task of a focused book search system is then to identify and rank (non-overlapping) book parts that contain relevant information and return these to the user, grouped by the books they occur in.

Participants could submit up to 10 runs, where one automatic and one manual run was compulsory. Each run could contain, for each topic, a maximum of 1000 books estimated relevant to the given topic, ordered by decreasing value of relevance. For each book, a ranked list of non-overlapping XML elements, passages, or book page results estimated relevant were to be listed in decreasing order of relevance. A minimum of one book part had to be returned for each book in the ranking. A submission could only contain one type of results, i.e., only XML elements or only passages; result types could not be mixed.

A total of 13 runs were submitted by 2 groups (7 runs by the University of Amsterdam (ID=6); and 6 runs by the University of Waterloo (ID=78)), see Table 1. All runs contained XML element results (i.e., no passage based submissions were received).

4.3   Test Topics

The test topics are representations of users' informational needs, i.e, the user is assumed to search for information on a given subject. As last year, all topics were limited to deal with content only aspects (i.e., no structural query conditions).

Participants were asked to create and submit topics for which at least 2 but no more than 20 relevant books were found using an online Book Search system (see Section 4.4).

---

[5] The bookID is the name of the directory that contains the book's OCR file, e.g., A1CD363253B0F403

```
<?xml version=''1.0'' encoding=''ISO-8859-1''?>
<!DOCTYPE inex_topic SYSTEM ''bs-topic.dtd''>
<inex_topic track=''book'' task=''book-retrieval/book-ad-hoc''
topic_id=''62''  ct_no=''2008-37''>
 <title> Attila the hun </title>
 <description> I want to learn about Attila the Hun's character, his way of
    living and leading his men, his conquests, and rule.
 </description>
 <narrative>
  <task> I was discussing with some friends about Attila the Hun. What I
     found interesting was the difference in our perceptions of Attila: As a
     great hospitable king vs. a fearsome barbarian. I want to find out more
     about Attila's character, his way of living as well as about his wars to
     better understand what he and his era of ruling represents to different
     nations.
  </task>
  <infneed> Any information on Attila's character, his treatment of others, his
     life, his family, his people's and enemies' view on him, his ambitions,
     battles, and in general information on his ruling is relevant, and so is any
     information that can shed light on how he is perceived by different nations.
     Poems that paint a picture of Attila, his court and his wars are also relevant.
  </infneed>
 </narrative>
</inex_topic>
```

Fig. 1. Example topic from the INEX 2008 Book Track test set

A total of 40 new topics (ID: 31-70) were contributed by 11 participating groups (see Table 1), following the topic format described below. These were then merged with the 30 topics created last year for the PiC task (ID: 1-30). An example topic is shown in Figure 1.

Topic Format. The topic format remained unchanged from 2007, each topic consisting of three parts, describing the same information need, but for different purposes and at different level of detail:

<title>: represents the search query that is to be used by systems for the automatic runs. It serves as a short summary of the user's information need.
<description>: is a natural language definition of the information need.
<narrative>: is a detailed and unambiguous explanation of the information need and a description of what makes a book part relevant or irrelevant. The narrative is taken as the only true and accurate interpretation of the user's need. It consists of the following two parts:
  <task>: a description of the user task for which information is sought, specifying the context, background and motivation for the information need.
  <infneed>: a detailed explanation of what information is sought and what is considered relevant or irrelevant.

4.4   Relevance Assessment System

The Book Search system (http://www.booksearch.org.uk), developed at Microsoft Research Cambridge, is an online web service that allows participants to search, browse, read, and annotate the books of the test corpus.

For the collection of relevance assessments, a game called the Book Explorers' Competition was designed and deployed, where assessors (as individuals or as members of teams) competed for prizes sponsored by Microsoft Research. The competition involved reading books and marking relevant content inside the books for which assessors were rewarded points. Assessors with the highest scores at the close of the competition were pronounced the winners. The game was modeled as a two-player game with competing roles: explorer vs. reviewer. An explorer's task was to judge the set of pooled pages as well as to locate and mark additional relevant content inside books. Reviewers then had the task of checking the quality of the explorers' work by providing their own relevance assessments for each page that has been judged by at least one explorer. During this process, the reviewers could see the relevance assessments of all the explorers who assessed a particular page. In addition to the passage level exploration, both explorers and reviewers were required, independently (information was not shared), to assign a degree of relevance to the book as a whole (on a scale from 0 to 5, with 5 designating the highest degree of relevance). For further details on the relevance assessment gathering process, please refer to [8].

Screenshots of the assessment system are shown in Figures 2 and 3. Figure 2 shows the list of books in the assessment pool to be judged for a given topic. The list was built by pooling all the submitted runs, i.e., both BR and PiC runs, using a round robin process and merging additional search results from the Book Search system itself. Selecting a book from the list, opened the Book Viewer window (see Figure 3). There, assessors could browse through the book and search inside it, or go through the pages listed in the Assessment tab, which were pooled from the submitted PiC runs. Assessors could highlight text fragments on a page by drawing a highlight-box over the page image. They could also mark a whole page or a range of pages as relevant/irrelevant. A detailed user manual and system description is available at http://www.booksearch.org. uk/BECRulesAndUserManual.pdf.

Two rounds of the Book Explorers' Competition were run. The first round (run in Dec 2008) lasted two weeks and resulted in three winners. One of them participated as an individual assessor and the other two formed a team. The second round (run in Jan 2009) spanned four weeks and yielded four winners. All four assessors belonged to the same team; one among them also achieving the highest individual score.

4.5   Collected Relevance Assessments

The collection of relevance assessments was frozen on the 25th of February 2009. The data collected includes the highlight-boxes drawn by assessors on a page, the binary relevance labels assigned by judges to a page, any notes and comments

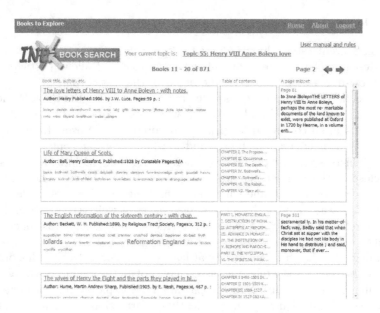

Fig. 2. Screenshot of the relevance assessment module of the Book Search system: List of books in the assessment pool for a selected topic

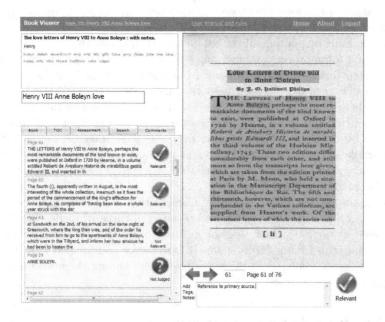

Fig. 3. Screenshot of the relevance assessment module of the Book Search system: Book Viewer window with Assessment tab listing pooled pages to judge

Table 3. Collected Relevance Assessments (25 February 2009)

Topic ID	Books				Pages		
	Total Judged	Relevant	Irrelevant	Skipped	Total Judged	Relevant	Irrelevant
25 topics - used in evaluation							
3	4	4	0	0	414	360	101
8	15	1	14	2	562	23	551
9	235	35	199	2	5991	285	5818
12	133	11	116	9	5660	48	5612
21	66	31	37	1	6026	1400	4696
22	30	12	18	0	956	244	712
27	35	21	14	1	365	101	274
31	18	9	17	0	129	46	97
36	9	7	2	0	1073	1043	30
37	15	7	11	0	120	34	99
39	25	7	18	0	358	27	331
41	14	9	5	0	370	276	94
51	135	15	107	14	1813	555	1270
52	41	23	18	0	1651	199	1456
53	1000	14	986	0	88	76	12
54	385	10	375	0	107	104	3
55	29	20	9	0	2108	397	1714
56	13	7	6	0	139	62	77
57	171	25	147	4	845	83	764
60	85	56	30	6	508	226	310
62	100	38	61	2	868	215	672
63	38	7	31	0	303	37	266
64	23	9	14	0	757	669	89
68	1	1	0	0	313	206	107
69	16	3	13	0	75	12	63
25	**2636**	**382**	**2248**	**41**	**31599**	**6728**	**25218**
Additional assessments - not used in evaluation							
1	999	0	999	0	55	0	54
5	33	0	33	2	495	145	495
15	1	0	1	0	0	0	0
29	5	0	4	1	421	0	421
70	0	0	0	0	495	7	488
5	**1038**	**0**	**1037**	**3**	**1466**	**154**	**1458**
29	**3674**	**382**	**3285**	**44**	**33120**	**6889**	**26724**

added for a page, and the relevance degree assigned to the books. In total, 3674 unique books and 33,120 unique pages were judged across 29 topics, and 1019 highlight boxes were drawn by 17 assessors. Table 3 provides a breakdown of the assessments per topic. For more details on the collected data, please refer to [8].

From the collected assessments, separate book-level and page-level assessment sets (qrels) were produced, where multiple relevance labels assigned by multiple assessors were averaged. For example, a book with assigned relevance degrees of 3 and 5 (by two assessors from the multi-grade scale of 0-5) yielded an averaged score of 4. Note that the score that appears in the qrels is this value multiplied by 10. The page-level qrel set is similarly the average of the binary scores (0-1) assigned by multiple assessors to a page, multiplied by 10. For example, a page with scores of $\{0, 1, 1, 1\}$ yielded 0.75*10. A weighted version of the qrel sets was also released to participants, where assessors' topic familiarity was taken into account: $w = \frac{\sum f \cdot r}{\sum f}$, where w is the weighted average, r is the relevance score given to a page or book by the assessor, and f is the assessor's familiarity with the topic (as provided by the assessor on a seven point scale, where 1 meant practically no knowledge about the topic and 7 represented an expert on the area). For example, if a book was rated as 3 by an assessor with familiarity of 6, and rated as 5 by an assessor with familiarity of 1, then the weighted score is $(3 \cdot 6 + 5 \cdot 1)/(6 + 1) = 3.28$.

Table 4. Evaluation results for the BR runs

ParticipantID+RunID	MAP	iP[0.00]	iP[0.10]	P5	P10	P20
14_BOOKSONLY	0.0837	0.3761	0.3135	0.192	**0.136**	0.082
14_MARCONLY	0.0386	0.2302	0.1421	0.088	0.056	0.046
14_MERGEMARCDOC	0.0549	0.3076	0.2528	0.144	0.088	0.064
54_BSS	0.0945	0.3715	0.2484	0.168	**0.136**	0.09
6_BST08_B_clean_trec	0.0899	**0.4051**	0.2801	0.176	0.132	0.096
6_BST08_B_square_times_sim100_top8_fw_trec	0.0714	0.2771	0.223	0.152	0.12	0.088
6_inex08_BST_book_sim100_top8_forward_trec	0.0085	0.1058	0.0406	0.032	0.02	0.01
62_RmitBookTitle	0.0747	0.2469	0.2195	0.128	0.104	0.094
62_RmitBookTitleBoolean	0.0747	0.2469	0.2195	0.128	0.104	0.094
62_RmitBookTitleInfneed	0.067	0.331	0.1999	0.136	0.1	0.086
62_RmitBookTitleInfneedManual	0.0682	0.2757	0.1868	0.112	0.108	0.088
62_RmitConPageMergeTitle	0.05	0.2414	0.2017	0.104	0.072	0.064
62_RmitConPageMergeTitleBoolean	0.05	0.2414	0.2017	0.104	0.072	0.064
62_RmitConPageMergeTitleInfneedManual	0.0544	0.2786	0.2126	0.128	0.084	0.058
62_RmitPageMergeTitle	0.0742	0.3022	0.2601	0.144	0.116	0.084
62_RmitPageMergeTitleBoolean	0.0741	0.3022	0.2601	0.144	0.116	0.084
62_RmitPageMergeTitleInfneedManual	**0.1056**	0.3671	**0.3456**	**0.216**	0.132	**0.098**
78_1	0.0193	0.117	0.0683	0.024	0.012	0.01
78_2	0.0214	0.1162	0.0678	0.024	0.012	0.008

4.6 Evaluation Measures and Results

Both IR tasks were evaluated using standard IR measures reported by trec_eval v8.1[6]. The ranking of books in both the BR and PiC runs was evaluated as traditional document retrieval, by comparing the ranked list of books returned by systems to the book-level qrel set. To do this, the runs were first converted to TREC format, during which some runs were truncated at rank 1000; *rank* values were derived based on the ordering of book results in a run; and the *rsv* was set to $1000 - rank$.

---

[6] http://trec.nist.gov/trec_eval/index.html

Table 5. Book-level evaluation results for the PiC runs

ParticipantID+RunID	MAP	iP[0.00]	iP[0.10]	P5	P10	P20
6_BST08_P_clean_trec	0.078	0.3359	0.2077	0.136	0.108	**0.096**
6_BST08_P_plus_B_trec	0.0761	0.3734	0.2028	0.136	**0.116**	0.078
6_BST08_P_plus_sim100_top8_fw_trec	0.0707	0.2794	0.1775	0.128	0.092	0.08
6_BST08_P_times_B_trec	0.0532	0.3179	0.1905	0.112	0.068	0.048
6_BST08_P_times_sim100_top8_fw_trec	0.0646	0.3408	0.1643	0.136	0.1	0.074
6_BST08_P_with_B_trec	**0.0785**	**0.3761**	**0.2189**	**0.152**	**0.116**	0.088
6_BST08_P_with_sim100_top8_fw_trec	0.053	0.2532	0.1645	0.128	0.096	0.062
78_3	0.0214	0.1162	0.0678	0.024	0.012	0.008
78_4	0.0513	0.278	0.2096	0.096	0.076	0.05
78_5	0.0214	0.1162	0.0678	0.024	0.012	0.008
78_6	0.0495	0.2744	0.205	0.096	0.076	0.048
78_7	0.0495	0.2744	0.205	0.096	0.076	0.048
78_8	0.0495	0.2744	0.205	0.096	0.076	0.048

Table 6. Page-level evaluation results for the PiC runs (precision, recall and the harmonic mean of precision and recall (F-measure))

ParticipantID+RunID	P	R	F
6_BST08_P_clean_trec	0.069	0.028	0.027
6_BST08_P_plus_B_trec	0.069	0.028	0.027
6_BST08_P_plus_sim100_top8_fw_trec	0.069	0.028	0.027
6_BST08_P_times_B_trec	0.069	0.028	0.027
6_BST08_P_times_sim100_top8_fw_trec	0.069	0.028	0.027
6_BST08_P_with_B_trec	0.064	0.027	0.025
6_BST08_P_with_sim100_top8_fw_trec	0.068	0.028	0.026
78_3	0.066	0.084	0.045
78_4	0.068	0.096	0.048
78_5	0.069	0.098	0.056
78_6	**0.070**	0.11	0.057
78_7	0.059	**0.14**	**0.065**
78_8	0.059	**0.14**	**0.065**

The ranking of book parts in the PiC task was evaluated at page-level for each book, treating each page as a document and comparing the ranked list of pages returned by systems to the page-level qrels for that book, and then averaging over the run (where additional relevant, but not retrieved books were given 0 scores). Note that retrieved XML elements that were at a finer granularity level than page elements were converted to page-level results to match the qrel set granularity.

Tables 4, 5, and 6 show the results for the BR, PiC book-level, and PiC page-level evaluations, respectively. In addition, Figure 4 shows the recall/precision curves for BR runs.

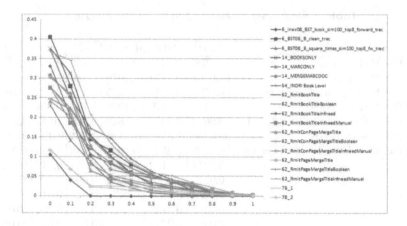

Fig. 4. Recall/precision curves for BR runs

We summarise below the main findings, but note that since the qrels vary greatly across topics, these should be treated more as preliminary observations.

For the BR task, the 4 submitting groups experimented with various techniques, e.g., using book content vs. MARC record information [9], ranking books by document score vs. best element score [5], or ranking books by the percentage of pages retrieved [12], as well as incorporating Wikipedia evidence [6]. The best performing run was a run submitted by RMIT (ID=62), ranking books by the percentage of pages retrieved using BM25 over a page level index (MAP=0.1056). The general conclusion, however, for the other 3 groups' experiments was that the simple book content based baseline performed better than any attempts to combine book-specific evidence to improve performance. This suggests that there is still plenty to be done in discovering suitable ranking strategies for books.

For the PiC task, the 2 submitting groups mostly experimented with ways of combining document and element level scoring methods [5,6]. The best performing runs, based on book-level scores, were submitted by the University of Amsterdam (ID=6), who found that while focused retrieval methods were able to locate relevant text within books, page level evidence was of limited use without the wider context of the whole book. The best page-level results were achieved by the University of Waterloo (ID=78), ranking book parts by element score and using no cutoff to limit the size of the ranked list (runs: 78_7 and 78_8).

5   The Structure Extraction (SE) Task

The goal of this task was to test and compare automatic techniques for extracting structure information from digitized books and building a hyperlinked table of contents (ToC). The task was motivated by the limitations of current digitization and OCR technologies that produce the full text of digitized books with only minimal structure markup: Pages and paragraphs are usually identified, but more sophisticated structures, such as chapters, sections, etc., are typically not recognised.

Participants of the task were provided a sample collection of 100 digitized books of different genre and styles in DjVu XML format. Unlike the BookML format of the main corpus, the DjVu files only contain markup for the basic structural units (e.g., page, paragraph, line, and word); no structure labels and markers are available. In addition to the DjVu XML files, participants were distributed the PDF of books or the set of JPEG image files (one per book page).

Participants could submit up to 10 runs, each containing the generated table of contents for the 100 books in the test set.

A total of 7 runs were submitted by 2 groups (3 runs by Microsoft Development Center Serbia (MDCS) (ID=125), and 4 runs by Xerox Research Centre Europe (XRCE) (ID=43)).

5.1   Evaluation Measures and Results

For the evaluation of the SE task, the ToCs generated by participants were compared to a manually built ground-truth, created by hired assessors, using a structure labeling tool built by Microsoft Development Center Serbia. The tool allowed assessors to attach labels to entries and parts of entries in the printed ToC of a book (using the PDF file as source).

Performance was evaluated using recall/precision like measures at different structural levels (i.e., different depths in the ToC). Precision was defined as the ratio of the total number of correctly recognized ToC entries and the total number of ToC entries; and recall as the ratio of the total number of correctly recognized ToC entries and the total number of ToC entries in the ground-truth. The F-measure was then calculated as the harmonic of mean of precision and recall. For further details on the evaluation measures, please see http://www.inex. otago.ac.nz/tracks/books/INEXBookTrackSEMeasures.pdf. The ground-truth and the evaluation tool can be downloaded from http://www.inex.otago. ac.nz/tracks/books/Results.asp#SE.

The evaluation results are given in Table 7. According to this, the best performance ($F = 53.47\%$) was obtained by the MDCS group (ID=125), who extracted

Table 7. Evaluation results for the SE task (complete ToC entries)

ParticipantID+RunID	F-measure
125_MDCS	**53.47%**
125_MDCS_NAMES_AND_TITLES	52.59%
125_MDCS_TITLES_ONLY	23.24%
43_HF_ToC_prg_Jaccard	10.27%
43_HF_ToC_prg_OCR	10.18%
43_HF_TPF_ToC_prg_Jaccard	10.10%
43_HF_ToC_lin_Jaccard	5.05%

ToCs by first recognizing the page(s) of a book that contained the printed ToC [10]. The XRCE group (ID=43) relied on title detection within the body of a book and achieved a score of $F = 10.27\%$ [3].

6   The Active Reading Task (ART)

The main aim of ART is to explore how hardware or software tools for reading eBooks can provide support to users engaged with a variety of reading related activities, such as fact finding, memory tasks, or learning. The goal of the investigation is to derive user requirements and consequently design recommendations for more usable tools to support active reading practices for eBooks. The task is motivated by the lack of common practices when it comes to conducting usability studies of e-reader tools. Current user studies focus on specific content and user groups and follow a variety of different procedures that make comparison, reflection, and better understanding of related problems difficult. ART is hoped to turn into an ideal arena for researchers involved in such efforts with the crucial opportunity to access a large selection of titles, representing different genres and appealing to a variety of potential users, as well as benefiting from established methodology and guidelines for organising effective evaluation experiments.

ART is based on the large evaluation experience of EBONI [11], and adopts its evaluation framework with the aim to guide participants in organising and running user studies whose results could then be compared.

The task is to run one or more user studies in order to test the usability of established products (e.g., Amazon's Kindle, iRex's Ilaid Reader and Sony's Readers models 550 and 700) or novel e-readers by following the provided EBONI-based procedure and focusing on INEX content. Participants may then gather and analyse results according to the EBONI approach and submit these for overall comparison and evaluation. The evaluation is task-oriented in nature. Participants are able to tailor their own evaluation experiments, inside the EBONI framework, according to resources available to them. In order to gather user feedback, participants can choose from a variety of methods, from low-effort online questionnaires to more time consuming one to one interviews, and think aloud sessions.

6.1   Task Setup

Participation requires access to one or more software/hardware e-readers (already on the market or in prototype version) that can be fed with a subset of the INEX book corpus (maximum 100 books), selected based on participants' needs and objectives. Participants are asked to involve a minimum sample of 15/20 users to complete 3-5 growing complexity tasks and fill in a customised version of the EBONI subjective questionnaire, usually taking no longer than half an hour in total, allowing to gather meaningful and comparable evidence. Additional user tasks and different methods for gathering feedback (e.g., video capture) may be added optionally. A crib sheet (see below) is provided to participants as a tool to define the user tasks to evaluate, providing a narrative describing the scenario(s) of use for the books in context, including factors affecting user performance, e.g., motivation, type of content, styles of reading, accessibility, location and personal preferences.

ART crib sheet. A task crib sheet is a rich description of a user task that forms the basis of a given user study based on a particular scenario in a given context. Thus, it aims to provide a detailed explanation of the context and motivation of the task, and all details that form the scenario of use:

- Objectives: A summary of the aims and objectives of the task from the users' point of view, i.e., what is it that users are trying to achieve in this task.
- Task: Description of the task.
- Motivation: Description of the reasons behind running the task.
- Context: Description of the context of the task in terms of time and resources available, emphasis and any other additional factors that are going to influence task performance.
- Background: Description of any background knowledge required to accomplish the task.
- Completion: Description of how to assess whether the task has been completed or not.
- Success: Description of whether the task has been completed successfully.

Participants are encouraged to integrate questionnaires with interviews and think aloud sessions when possible, and adapt questionnaires to fit into their own research objectives whilst keeping in the remit of the active reading task.

We also encourage direct collaboration with participants to help shape the tasks according to real/existing research needs. In fact one of the participants explained how English written material was not much use for their experiments as they were targeting Korean speaking users, so it was agreed that they would use their own book collection while still adopting the ART evaluation framework to ensure results were comparable at the end.

Our aim is to run a comparable but individualized set of studies, all contributing to elicit user and usability issues related to eBooks and e-reading.

Since ART is still ongoing, there is no data to be presented at this point.

7   Conclusions and Plans

The Book Track this year has attracted a lot of interest and has grown to double the number of participants from 2007. However, active participation remained a challenge for most due to the high initial set up costs (e.g., building infrastructure). Most tasks also require advance planning and preparations, e.g., for setting up a user study. This, combined with the late announcement and advertising of some of the tasks has limited active participation this year. In particular, we received expressions of interest for the Structure Extraction and the Active Reading tasks, but the deadlines prohibited most people from taking part. We aim to address this issue in INEX 2009 by raising awareness early on in the start of the INEX year and by ensuring continuity with the tasks established this year.

As a first step in this direction, we are proposing to run the Structure Extraction task both at INEX 2009 and at ICDAR 2009 (International Conference on Document Analysis and Recognition) with an increased set of 1,000 books.

Both the Book Retrieval and Page in Context tasks will be run again in 2009, albeit with some modifications. The BR task will be shaped around the user task of compiling a reading list for selected Wikipedia articles, while we aim to expand the PiC tasks to tree retrieval [1].

The greatest challenge in running these two tasks has been the collection of relevance assessments. Due to the huge effort required, we decided to depart from the traditional method of relevance assessment gathering (i.e., one judge per topic), and designed a system where multiple judges assess the same topic. Implemented as an online game, assessors contributed relevance labels for passages, pages, and whole books on the topics they were interested in and for any number of books on that topic. This way of collecting judgements is aimed to provide a more realistic expectation on the assessors, but it also comes with its own risks. Attracting a sufficiently large group of dedicated assessors is one of the risks, for example. To address this issue, we are currently looking at using Amazon's Mechanical Turk service, as well as investigating the possibility of opening up the Book Search system and allowing users to create their own topics and saving their searches and book annotations for these. Other risks include the question of the quality of the collected relevance data due to a mixture of expert and non-expert judges. Working toward a solution, we introduced a number of measures, such as requiring assessors to specify their familiarity with their selected topics, as well as allowing users to quality check each other's work. We aim to explore additional measures in our future work.

We also plan to re-run this year's Active Reading task in 2009. We found that the introduction of ART was a challenge for number of reasons:

- Because of its original approach to evaluation, which is quite far away from the classic TREC paradigm, and the relative difficulty in framing ART in a formal way, the task organisation has suffered delays that have affected the availability of participants to get fully involved in it;
- User studies are per se risky and unpredictable and the idea of running a number of those in parallel in order to compare and combine results added an extra layer of uncertainty to the task, somehow discouraging participants that were used to a more stochastic approach to evaluation;
- The formalisation of the procedure and protocols to be followed when running user studies was designed on purpose to be flexible and unconstructive in order to accommodate for participants' specific research needs. This flexibility, however, was interpreted by some as a lack in details that discouraged them from taking part.
- Opening up to different communities that were not yet involved in INEX required concentrated effort in order to advertise and raise awareness of

what INEX's aims and objectives and in particular what ART's goals were. Some of this effort was simply too late for some interested parties.

The organisation of ART has proved a valuable experience though that has given us the opportunity to explore different research perspective while focusing on some of the practical aspects of the task. We believe that the effort that has gone into setting up ART this year will be rewarded by a more successful task next year.

Acknowledgements

The Book Track in 2008 was supported by the Document Layout Team of Microsoft Development Center Serbia. The team contributed to the track by providing the BookML format and a tool to convert books from the original OCR DjVu files to BookML. They also contributed to the Structure Extraction task by helping us prepare the ground-truth data and by developing the evaluation tools.

References

1. Ali, M.S., Consens, M.P., Kazai, G., Lalmas, M.: Structural relevance: a common basis for the evaluation of structured document retrieval. In: CIKM 2008: Proceeding of the 17th ACM Conference on Information and Knowledge Management, pp. 1153–1162. ACM Press, New York (2008)
2. Coyle, K.: Mass digitization of books. Journal of Academic Librarianship 32(6), 641–645 (2006)
3. Déjean, H., Meunier, J.-L.: XRCE participation to the book structure task. In: Geva, et al. (eds.) [4]
4. Geva, S., Kamps, J., Trotman, A. (eds.): Advances in Focused Retrieval: 7th International Workshop of the Initiative for the Evaluation of XML Retrieval (INEX 2008). LNCS, vol. 5631. Springer, Heidelberg (2009)
5. Itakura, K., Clarke, C.: University of Waterloo at INEX 2008: Adhoc, book, and link-the-wiki tracks. In: Geva, et al. (eds.) [4]
6. Kamps, J., Koolen, M.: The impact of document level ranking on focused retrieval. In: Geva, et al. (eds.) [4]
7. Kantor, P., Kazai, G., Milic-Frayling, N., Wilkinson, R. (eds.): BooksOnline 2008: Proceeding of the 2008 ACM workshop on Research advances in large digital book repositories. ACM, New York (2008)
8. Kazai, G., Milic-Frayling, N., Costello, J.: Towards methods for the collective gathering and quality control of relevance assessments. In: SIGIR 2009: Proceedings of the 32nd Annual International ACM SIGIR Conference on Research and Development in Information Retrieval. ACM Press, New York (2009)
9. Larson, R.: Adhoc and book XML retrieval with Cheshire. In: Geva, et al. (eds.) [4]
10. Uzelac, A., Dresevic, B., Radakovic, B., Todic, N.: Book layout analysis: TOC structure extraction engine. In: Geva, et al. (eds.) [4]
11. Wilson, R., Landoni, M., Gibb, F.: The web experiments in electronic textbook design. Journal of Documentation 59(4), 454–477 (2003)
12. Wu, M., Scholer, F., Thom, J.A.: The impact of query length and document length on book search effectiveness. In: Geva, et al. (eds.) [4]

XRCE Participation to the Book Structure Task

Hervé Déjean and Jean-Luc Meunier

Xerox Research Centre Europe
6 Chemin de Maupertuis, F-38240 Meylan
`Firstname.lastname@xrce.xerox.com`

Abstract. We present here XRCE participation to the Structure Extraction task of the INEX Book track. After briefly explaining the method used for detecting table of contents and their corresponding entries in the book body, we will mainly discuss the evaluation and the main issues we faced, and eventually we will propose improvements for our method as well as for the evaluation framework/method.

1 Introduction

We present in this paper our participation to the Structure Extraction task of the INEX Book. Our objective was to assess a component, a table of contents detector, presented in [1], [2], with the minimal effort. By minimal effort, we mean: the initial input (segmentation and text) was taken almost as provided. Especially, no preprocessing was used to improve it. But more important, no specific tuning was done with regard to the collection. This fact will be highlighted in the Evaluation Section, since some books do not comply with our assumptions about table of contents.

The rest of the article is structured as follows: we will explain the processing done for the collection. Then the method for detecting the Table of contents is sketched. We will explain the postprocessing and the different parameters used in our runs. Eventually we will discuss the results of the evaluation.

2 Pre-processing

The first step simply consists in reformatting the XML INEX format into our internal format, mostly renaming tag names and adding some internal attributes (such as unique IDs for each tag). This was performed using XSLT technology, with some difficulty for the largest books.

A second step consists in detecting pages headers and footers, which often introduce noisy for our table of contents detector (see [1]).

A heuristics has been used for run 4 in order to improve the ToC page detection based on the detected page headers and footers.

3 The ToC Detector

The method is detailed in [1], [2] and in this section we will only sketch its outline. The design of this method has been guided by the interest in developing a generic

S. Geva, J. Kamps, and A. Trotman (Eds.): INEX 2008, LNCS 5631, pp. 124–131, 2009.
© Springer-Verlag Berlin Heidelberg 2009

method that uses very intrinsic and general properties of the object known as a table of contents. In view of the large variation in shape and content a ToC may display, we believe that a descriptive approach would be limited to a series of specific collections. Therefore, we instead chose a functional approach that relies on the functional properties that a ToC intrinsically respects. These properties are:

1. Contiguity: a ToC consists of a series of contiguous references to some other parts of the document itself;
2. Textual similarity: the reference itself and the part referred to share some level of textual similarity;
3. Ordering: the references and the referred parts appear in the same order in the document;
4. Optional elements: a ToC entry may include (a few) elements whose role is not to refer to any other part of the document, e.g. decorative text;
5. No self-reference: all references refer outside the contiguous list of references forming the ToC.

Our hypothesis is that those five properties are sufficient for the entire characterization of a ToC, independently of the document class and language. In the Evaluation and Discussion section, we will discuss the cases where theses hypotheses were not valid.

Three steps permit us to identify the area of the document containing the ToC text. Firstly, links are defined between each pair of text blocks in the whole document satisfying a textual similarity criterion. Each link includes a source text block and a target text block. The similarity measure we currently use is the ratio of words shared by the two blocks, considering spaces and punctuation as word separators. Whenever the ratio is above a predefined threshold, the similarity threshold, a pair of symmetric links is created. In practice, 0.5 is a good threshold value to tolerate textual variation between the ToC and the document body while avoiding too many noisy links. The computation of links is quadratic to the number of text blocks and takes most of the total computation time. However, searching for the ToC in the N first and last pages of the document leads to linear complexity without loss of generality.

Secondly, all possible ToC candidate areas are enumerated. A brute force approach works fine. It consists in testing each text block as a possible ToC start and extending this ToC candidate further in the document until it is no longer possible to comply with the five properties identified above. A ToC candidate is then a set of contiguous text blocks, from which it is possible to select one link per block so as to provide an ascending order for the target text blocks.

Thirdly, we employ a scoring function to rank the candidates. The highest ranked candidate table of contents is then selected for further processing. Currently, the scoring function is the sum of entry weights, where an entry weight is inversely proportional to the number of outgoing links. This entry weight characterizes the certainty of any of its associated links, under the assumption that the more links initiate from a given source text block, the less likely that any one of those links is a "true" link of a table of contents.

4 Post-processing

This step mainly transforms the output of the ToC detector into the INEX format. Our component marks up the ToC entry and the body heading. From this information, the required page number was extracted. For the required title, we selected the title of the ToC entry, which, as we will see in the Evaluation section, will impact the evaluation.

5 The Different Runs

Several runs were conducted with different values for the main parameters, in particular the processing can be performed either at the line or paragraph level and the similarity measure can be the one described above, called Jaccard, or a dynamic time warping alternative (DTW). So we performed the following runs:

1. Paragraph level, Jaccard similarity: The Jaccard similarity consists in computing the ratio of the cardinal of the intersection to the union of normalized words of two text blocks, i.e. the paragraphs in this run.
2. Paragraph level, DTW similarity: the DTW consists in finding the best alignment of words of two blocks of text, the similarity between two words being established from an edit distance. The DTW similarity is more robust to OCR errors than the Jaccard but is computationally more intensive, usually twice more.
3. Line level, Jaccard similarity: here the considered blocks of text are the lines.
4. Paragraph Level, Jaccard similarity, with an additional heuristic to determine the ToC position.

We encountered memory issues with some of the largest documents and had to break our batches in several parts. Eventually we are not able to report accurately on the processing time.

Applying our standard ToC method prevented us from computing the level of the ToC entries, so we voluntarily set it to 1 for all entries despite this is clearly wrong.

6 Evaluation and Discussion

Let us review the results of the various runs, with a particular look at the first one because the other runs share many identical issues with it.

6.1 Run 1: Paragraph Level, Jaccard Similarity

We reproduce below the result of the Inex metric.

Table 1. Inex results for the run 1

All books – run 1	Precision	Recall	F-Measure
Titles	25,98%	20,90%	22,13
Levels	14,43%	11,95%	12,45
Links	22,01%	18,16%	19,19
Complete entries	10,86%	9,30%	9,62
Entries disregarding depth	22,01%	18,16%	19,19

These results appear overall quite bad. Actually, because the title matching between the run and the ground truth is critical to the evaluation method, any error on the title induces an error for all other criteria. For instance, a link with a valid destination but incorrect title will count as an error and a miss. In addition the conditions for title to match were quite strict, tolerating 20% of the shortest string as maximum edit distance with an additional condition on the first and last 5 characters. It turned out that an additional or missing word such as 'Chapter' as the beginning of the title suffices to discard entirely the title.

Under those conditions, 22% of precision at the link level with 26% correct titles, means in fact that among the entries with a correct title, 85% of them had a valid link. To examine this phenomenon further, we computed another link measure that ignores the title. We compare two links by comparing the page number they point to, so we consider a run output as a sequence of page numbers and compute the edit distance between the run and groundtruth sequences, which gives us a precision & recall measure. In other words, the Inex measure views an output as a unordered set of entries identified by their title, while our proposed complementary measure views an output as an ordered sequence of entries identified by the page pointed by each entry. Our measure, which we shall call 'hyperlink' focuses more on document navigation needs, and the quality of the extracted titles can be measured in a second step. Our 'hyperlink' measure is given Table 2 below.

Table 2. 'Hyperlink' measure, which ignore title errors, for the run 1

All books	Precision	Recall	F-Measure
Hyperlinks (i.e. ignoring titles)	71%	40%	51

This result is more conform to what we generally observe although the recall is particularly low. The histogram below shows an interesting aspect, where books tend to go either well or bad but more rarely in the middle. This behavior can be exploited thanks to automated quality assurance methods.

Table 3. Histogram of the "hyperlink" F1 distribution, for the run 1

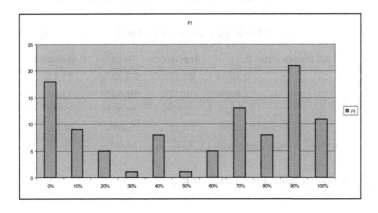

Now, looking in detail at the error on the first book, we found several main causes of errors and misses:

> Ground truth errors: we saw 7 wrong additional entries, which seems to have been automatically generated from the ToC of an advert page at the end of the book (see page 640).

> Title errors: our run does not include the 'Chapter XX" at the beginning of the title, discarding about 90% of the found entries.

> One error is caused by the ToC not conforming to the ordering property, our method rely upon.

> Combined OCR noise and segmentation issues, .e.g. the ToC contains 'Chapter XX – some title' while the same information is split over two paragraph in the document body. Combined with OCR errors, typically a 'B' instead of a 'E', the similarity threshold is not met.

Unfortunately, those problems combined with the importance of the title in the Inex measure lead to important measure variation, as exemplified below on the first book:

Table 4. Variable results depending on the measure, for book #0 in run 1

Book #0 – run 1	Precision	Recall	F-Measure
Inex (title + linkls)	5%	4%	4
Hyperlinks, INEX ground truth	92%	69%	79
Hyperlinks, **fixed** ground truth	92%	80%	86

With respect to the ToC level, we have observed that the distribution of entries for level 1, 2 and 3 was about 33%, 36% and 26% respectively. This stresses the importance of reconstructing the ToC hierarchy.

6.2 Run 2: Paragraph Level, DTW Similarity

The DTW similarity should better deal with OCR errors, but to our surprise the results are not better. It turned out that the SW extracted 1% more links (36):

Table 5. Inex results for the run 2

All books – run 2	Precision	Recall	F-Measure
Titles	24,38%	20,31%	21,45
Levels	13,15%	11,43%	11,90
Links	20,87%	18,02%	18,84
Complete entries	10,12%	9,26%	9,47
Entries disregarding depth	20,87%	18,02%	18,84

Our 'hyperlinks' measure shows a minor improvement:

Table 6. Hyperlink measure, which ignore title errors, for the run 2

All books	Precision	Recall	F-Measure
Hyperlinks (i.e. ignoring titles)	71%	41%	52

6.3 Run 3: Line Level, Jaccard Similarity

Working at line level does not make much sense with the Inex evaluation method.

Table 7. Inex results for the run 3

All books – run 3	Precision	Recall	F-Measure
Titles	16,53%	15,41%	15,46
Levels	7,44%	7,24%	7,13
Links	14,46%	13,47%	13,53
Complete entries	5,54%	5,42%	5,33
Entries disregarding depth	14,46%	13,47%	13,53

Our Hyperlink measure shows a loss in precision.

Table 8. Hyperlink measure, which ignore title errors, for the run 3

All books	Precision	Recall	F-Measure
Hyperlinks (i.e. ignoring titles)	61%	42%	50

6.4 Ground Truth Issues

The ground truth used for this evaluation suffers from several coherence issues:

> Should the links point to the title page of a chapter or to the actual start of its text? Compare for instance the documents #13 0050CA95E49A5E97 and #20 00A4141D9CC87E65.

> Should the label of the entry (chapter, section,...) or its number be part of the extracted title? This choice has an enormous impact on the whole evaluation because of the importance of the title in the measure design. The choice made for the groundtruth is not consistent across all books. In fact, given a book, the choice can be difficult since the body and ToC pages can differ on this matter.

> ToC entry segmentation for old-style ToC, shown below, the subentries should be extracted or not, independently of the presence of a page locator since the document body shows clearly that there are subsections.

6.5 Evaluation Method Issues and Suggestions

Given the previous observations, we suggest some improvement for measuring the quality of the results:

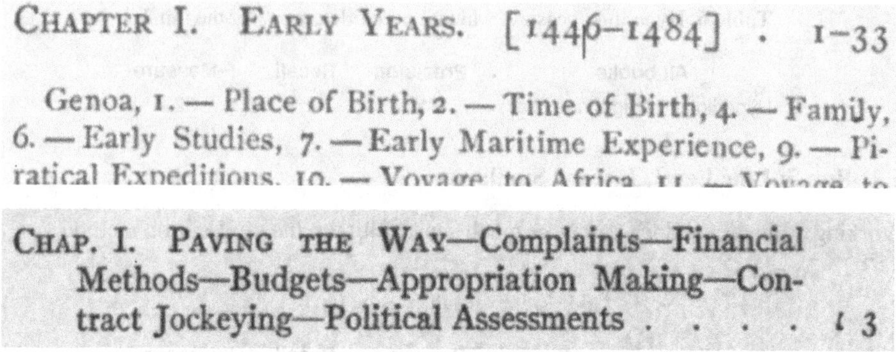

CHAPTER I.   EARLY YEARS.   [1446–1484]   .   1–33

Genoa, 1. — Place of Birth, 2. — Time of Birth, 4. — Family, 6. — Early Studies, 7. — Early Maritime Experience, 9. — Piratical Expeditions. 10. — Voyage to Africa 11 — Voyage to

CHAP. I.   PAVING THE WAY—Complaints—Financial Methods—Budgets—Appropriation Making—Contract Jockeying—Political Assessments 13

Fig. 1. Excerpt from book #3 (0008D0D781E665AD) and #52 (0E5E2F4BC9008492) showing a chapter title as well as the title of the subsections. In the first case the ground truth indicates the subsections but not in the second one (probably because of the absence of page number?).

> Results at the book level should be made available.
> Case normalization: the measure should be computed in a case-independent way. Indeed certain documents can have a title in uppercase in the ToC but in lowercase or capitalized form in the document body.
> In some applications, such as providing hyperlinks for navigation purpose, the quality of the links is more important than the exactitude of the title, provided it reads well and has some appropriate meaning. So we suggest measuring the link quality independently of the title quality. In fact, the latter should be measured as a complementary indication, e.g. computing an edit distance with the ground truth title.
> When the title is used as primary quality measure, a less strict title matching function should be used unless a sound and methodic way to determine uniformly the title has been designed.

Note: Some of those issues have been now corrected.

7 Conclusion

It is difficult to draw conclusions because of the issues found with the ground truth, and to less extent with the evaluation method. The numerical results of the evaluation remain disputable in our opinion. However, this evaluation confirms a known problem of our method: it can confuse a Table of Illustrations with a Table of Contents. This caused on this collection a loss of about 45% in precision and ~20% in recall. Similarly, the indexes caused a ~15% loss in recall. The challenge is now to turn this into an advantage by typing the found table.

We have found very interesting the corpus proposed for the INEX, composed of historical documents with a large variety of table of contents. Many of them were challenging because of the need for segmenting entries at a lower level than the paragraph level, as shown in figure 1.

Some ToC did not respect at all the properties we enforce, since some ToC entries were short sentence *summarizing* the section contents rather than reproducing some title present in the document body, e.g. book #3 0008D0D781E665AD.

It would have been very useful to have a human evaluation of the results, in order to give a perspective different than the one underlying the ground truth preparation or quality measure design.

We are grateful to the organizers and thank them for their work.

References

1. Déjean, H., Meunier, J.-L.: Structuring Documents according to their Table of Contents. In: Proceedings of the 2005 ACM symposium on Document engineering, pp. 2–9. ACM Press, New York (2005)
2. Déjean, H., Meunier, J.-L.: On Tables of Contents and how to recognize them. International Journal of Document Analysis and Recognition, doi:10.1007/s10032-009-0078-8

University of Waterloo at INEX 2008: Adhoc, Book, and Link-the-Wiki Tracks

Kelly Y. Itakura and Charles L.A. Clarke

University of Waterloo, Waterloo, ON N2L3G1, Canada
{yitakura,claclark}@cs.uwaterloo.ca

Abstract. In this paper, we describe University of Waterloo's approaches to the Adhoc, Book, and Link-the-Wiki tracks. For the Adhoc track, we submitted runs for all the tasks, the Focused, the Relevant-in-Context, and the Best-in-Context tasks. The results show that we ranked first among all participants for each task, by the simple scoring of elements using Okapi BM25. In the Book track, we participated in the Book retrieval and the Page-in-Context tasks, by using the approaches we used in the Adhoc track. We attribute our poor performance to lack of training. In the Link-the-Wiki track, we submitted runs for both File-to-File and Anchor-to-BEP tasks, using PageRank [1] algorithms on top of our previous year's algorithms that yielded high performance. The results indicate that our baseline approaches work best, although other approaches have rooms for improvement.

1 Introduction

In 2008, University of Waterloo participated in the Adhoc, the Book, and the Link-the-Wiki tracks. In the Adhoc track, we implemented both passage and element retrieval algorithms to compare the relation between the best passages and the best elements. This is in contrast to our 2007 runs [4] that compared the passage-*based* element retrieval algorithm against the simple element retrieval algorithm. We scored elements and passages using a *biased* BM25 and language modeling [6], in addition to Okapi BM25 [7] to see the effect of scoring functions in retrieval results.

In the Book track, we implemented our tried and true element retrieval algorithm with Okapi BM25 to retrieve best books and pages.

In the Link-the-Wiki track, we added PageRank algorithm [1] in addition to using anchor density [4,8] for the numbers of links to return for each topic.

This paper is organized as follows. In Section 2, we describe our approaches to the Adhoc track, and in Section 3, we describe our approaches to the Book track. In Section 4, we describe our approaches in the Link-the-Wiki track. We conclude this paper with directions for future work in Section 5.

2 Ad Hoc Track

In the Adhoc track, as in the past year, the basic retrieval algorithms used are element retrieval and passage retrieval.

S. Geva, J. Kamps, and A. Trotman (Eds.): INEX 2008, LNCS 5631, pp. 132–139, 2009.

The only difference between element retrieval and passage retrieval is the retrieval unit. In element retrieval, we only scored the following elements in corpus. These were the results of manual examination in [3],

```
<p>, <section>, <normallist>, <article>, <body>, <td>, <numberlist>,
<tr>, <table>, <definitionlist>, <th> ,<blockquote>, <div>, <li>,
<u>.
```

In passage retrieval, we scored passages of any word-lengths longer than 25 words, including elements.

To score an element or a passage, we converted each topic into a disjunction of query terms without negative query terms. We located positions of all query terms and XML tags using Wumpus [2]. We then used two versions of Okapi BM25 [7] to score passages and elements. The score of an element/passage P using Okapi BM25 is defined as follows.

$$s(P) \equiv \sum_{t \in Q} W_t \frac{f_{P,t}(k+1)}{f_{P,t} + k(1 - b + b\frac{pl_P}{avgdl})} \ , \tag{1}$$

where Q is a set of query terms, W_t is an IDF value of the term t in the collection, $f_{P,t}$ is the sum of term frequencies in a passage P, pl_P is a passage length of P, and $avgdl$ is an average document length in Wikipedia collection to act as a length normalization factor.

Using a biased Okapi BM25, the first α words' term frequency is multiplied by β. The reason behind this is that because in Best-in-Context task it appears that the closer the best entry point is to the beginning of an article, the more relevant it is judged, we thought that the closer the terms are to the beginning of an element the more relevant they are.

We tuned parameters using INEX2007 Adhoc track evaluation scripts distributed via email by the organizers. Our tuning approach was such that the sum of all relevance scores, ip[0.00], ip[0.01], ip[0.05], ip[0.10], and MAiP are maximized. However, by looking at the training results, the choice of the parameters did not seem much different if we had chosen the official metrics, ip[0.01] for the Focused task, and MAiP for Relevant-in-Context and Best-in-Context task.

2.1 Focused Task

In the focused task, after scoring all elements and passages, we eliminated overlaps and returned the top 1500 elements and passages. There are four runs we submitted. For element retrieval, we submitted three runs using Okapi BM25 (FOER) with parameters $k = 4$ and $b = 0.8$, the biased Okapi BM25 (FOER-Step) with $\alpha = 10$, $\beta = 2$, $k = 3$, and $b = 0.8$. For passage retrieval, we submitted a run using Okapi BM25 with $k = 4$ and $b = 1.2$.

Our biased Okapi BM25 approach ranked first amongst 19 participants and the 61 runs. Table 1 shows the results of our individual runs.

Although our biased Okapi BM25 approach performed better than the simple BM25, it did not give substantial improvement. However, it may give an insight into efficient scoring by possibly only scoring the first words of an element/passage and ignoring the rest. This would not only reduce the length of elements/passages to score, but also the number of elements/passages to score because many of them overlap at the beginning. A look into the scores of all elements/passages that start at the same character is necessary.

The passage run was disqualified because of overlaps. While training with correct overlap elimination, the evaluation scores of the passage run was somewhat lower than that of BM25 run. Re-evaluation of the correct unofficial run confirms this. However, the overall impression is that Okapi BM25 is the best scoring function and that users prefer element results to passage results. One question left is, would users prefer ranges of elements over single elements or passages?

Table 1. Results of Waterloo's Runs in the Adhoc Track Focused Task

Run (run ID)	Rank	ip[0.01]
Biased Okapi (FOERStep)	1	0.68965708
Okapi BM25 (FOER)	2	0.68800831
Passage Retrieval (unofficial)	(42)	0.554286743

2.2   Relevant-in-Context Task

In the Relevant-in-Context task, the results of the top 1500 elements/passages using Okapi BM25 in the Focused task was grouped in two different ways. The first way (RICArt) is to rank the articles according to the score of the articles themselves with parameters $k = 2$ and $b = 0.8$, the second way (RICBest) is to rank the articles according to the scores of the highest scoring elements/passages the article contains with $k = 2$ and $b = 0.8$.

Our run with best element score ranked first amongst 11 participants and their 40 runs. Table 2 shows the results of our individual runs.

Because there was no substantial difference between ranking articles by the article scores and ranking articles by their best element scores, it may be as effective to fetch the articles with the highest articles score first, and then run element retrieval on the retrieved set.

Table 2. Results of Waterloo's Runs in the Adhoc Track Relevant-in-Context Task

Run (run ID)	Rank	ip[0.01]
Best Element Score (RICBest)	1	0.22779217
Article Score (RICArt)	2	0.22705669

2.3   Best-in-Context Task

In the Best-in-Context task, we recorded the scores of the best element/passage in each article for biased and simple Okapi BM25, and returned the top 1500 elements/passages. The parameters used for element retrieval with Okapi BM25 (BICER) are $k = 1.2$ and $b = 0.4$, for element retrieval with biased Okapi BM25 (BICERStep) are $\alpha = 10$, $\beta = 2$, $k = 0.6$, and $b = 0.4$, and for passage retrieval using Okapi BM25 (BICPRPlus) are $k = 1.4$ and $b = 0.8$.

Our simple Okapi BM25 based run scored the first amongst 13 participants and their 35 runs. Table 3 shows the results of our individual runs.

The performance of all the approaches are similar during the training phase and the results are fully what we had expected. As in the Focused task, the biased Okapi function did similarly well to a simple Okapi BM25, implying a possible efficiency improvement. The results of the passage run is not impressive as expected. This fact is quite alarming given that the only difference between the two approaches is the unit of scoring; the highest scoring passage must be quite far apart from the highest scoring element. This explains why our passage-*based* element retrieval run of INEX 2007 was not as effective as the simple element retrieval run.

Table 3. Results of Waterloo's Runs in the Adhoc Track Best-in-Context Task

Run (run ID)	Rank	ip[0.01]
Okapi BM25 (BICER)	1	0.2237906
Biased Okapi BM25 (BICERStep)	3	0.21267588
Passage Retrieval (BICPRPlus)	24	0.12535625

3   Book Track

In the Book track, we employed the element retrieval algorithm with Okapi BM25 as described in Section 2. The only difference is the unit of scoring, which are document, page, region, and section. Since we had no training data available from previous years, we ran our algorithms with arbitrary parameters.

3.1   Book Search Task

In the Book search task, Run 78-1 was obtained by ranking books according to the document scores, and Run 78-2 was obtained by ranking books according to their best element scores. The results in Table 4 indicate that both runs perform similarly. Because in the relevant-in-context task in the adhoc track, article scoring performed similarly to best-element scoring, and article scoring is more efficient, we think article-scoring is preferable in tasks that require us to rank articles/books. Our runs performed very poorly compared to other institutions

Table 4. Results of Waterloo's Runs in the Book Track Book Retrieval Task

Run (run ID)	Rank (out of 19)	MAP
by best element scores (78-2)	17	0.0214
by book scores (78-1)	18	0.0193

and we think it is due to the lack of training. Because training takes time, in the coming years, we would like to use language modeling to score elements.

3.2   Page-in-Context Task

All runs for the Page-in-Context task differed in how the books were ranked. Within books, elements are scored by element scores. Run 78-3 ordered books by their best element scores, Run 78-4 ordered books by the books' score with manual query expansion. The manual query expansion was done by observing the query phrases and adding any extra query phrases that may be helpful to disambiguate the queries. For example, if the query phrase is just "mouse", but the description of the user's information need suggests that it pertains to the animal mouse, as opposed to the computer mouse, the expanded query phrase would be "animal mouse". Query expansion normally increases precision, but it also increases the processing time. Since one of the goal of book search is to create an efficient search system, our assumption is that applying query expansion only on the whole book score maybe reasonable. Other effort to shorten search processing time was to make a distributed search. The problem with this approach was that in order to create a merged list of top scoring elements precisely, for each distributed node, it was necessary to compute and store all the element scores, which was costly. To see how much of top elements for each node could be cut-off without affecting the results, we made the above two runs with various cut-off values, including no cut-off.

Table 5 shows the results of our runs. There was only one other participant, so we cannot compare our performance against others well, but we think we performed poorly and need training. Between the two approaches, it seems that screening out books by the book score not only performs better than by their best element scores, but more efficient. However, the better performance may be

Table 5. Results of Waterloo's Runs in the Book Track Page-in-Context Task

Run (run ID)	Rank (out of 13)	MAP
by book scores+1000 cut-off (78-4)	8	0.0513
by best book scores+3000 cut-off (78-6)	9	0.0495
by best element scores+no cut-off (78-7)	9	0.0495
by best book scores+no cut-off (78-8)	9	0.0495
by best element scores+1000 cut-off (78-3)	12	0.0214
by best element scores+3000 cut-off (78-5)	12	0.0214

due to manual query expansion. Nevertheless, manual query expansion to score all the possible passages/elements in the book collection seems very inefficient, so along with the experience from the adhoc relevant-in-context, and the book retrieval tasks, we think a good heuristic is to screen documents/books by their scores first.

As for how many top elements to compute/store in each distributed node, we can only say that we only need to score and store a small fraction of all possible elements.

4   Link the Wiki Track

Following the previous year's successful run, we decided to extend our basic approaches for both incoming and outgoing links by incorporating PageRank [1]. Our File-to-File runs mirror our Anchor-to-BEP runs by employing the same algorithms, but abstracting out to the article level. Therefore, in this section, we describe our approaches to Anchor-to-BEP runs.

4.1   Outgoing Links

As in the last year, the basic ingredient to computing outgoing links is the following ratio, γ.

$$\gamma = \frac{\sharp \text{ of files that has a link from anchor } a \text{ to a file } d}{\sharp \text{ of files in which } a \text{ appears at least once}}$$

Because this year, we are allowed to specify multiple destinations for a given anchor phrase, for each anchor phrase that appear in the corpus, we computed γ for the most frequent destination, but kept up to four other destinations on the list. We sorted all anchor phrases by γ, and then for each topic file, we looked for the locations of the anchor phrase. Once the anchor phrases are located, we listed the destinations in the order of frequency in the corpus. We specified the best entry point as the beginning of the article. We call this baseline run (a2a#1 for Anchor-to-BEP f2f#2 for File-to-File).

In the second, anchor-density-based run (a2a#2,f2f#2), we computed the maximum number of anchor phrases that a topic file can contain using the size of the topic file. As in [4], we define the *anchor density* δ as

$$\delta = \frac{\text{\# number of anchor strings in the file}}{\text{size of the file in bytes}}.$$

We computed that anchor density is linear and that there are 3.584 anchor per KB of a document in the corpus and set the number of anchor phrases in the topic files accordingly.

In the third, PageRank-based run (a2a#3,f2f#3), instead of ordering the anchor phrases by the γ values, we ordered them by the PageRank value. For this run, there is no cut-off values as in the second run.

Official results in Table 6 and Table 7 indicate that both for File-to-File and Anchor-to-BEP runs, our first runs without an anchor density performed

Table 6. Results of Waterloo's File-to-File Runs in LTW Outgoing Tasks

Run (run ID)	Rank (out of 21)	MAP
baseline (f2f#1)	3	0.33453
anchor density (f2f#2)	7	0.29203
PageRank (f2f#3)	10	0.20532

Table 7. Results of Waterloo's Anchor-to-BEP Runs LTW Outgoing Tasks

Run	Rank (out of 28)	MAP
baseline (a2a#1)	5	0.4071
PageRank (a2a#3)	10	0.3835
anchor density (a2a#2)	11	0.3835

best among our submissions. For File-to-File task, the run with anchor density performed much better than the PageRank run, and for Anchor-to-BEP task, the two runs performed similarly. Compared to other participants, our File-to-File run ranked the third, and Anchor-to-BEP run the second.

Our anchor density approach did not perform as well as the baseline run because the official metrics is mean average precision.

4.2   Incoming Links

The first run of the incoming links (a2a#1, f2f#1) is done exactly the same as in Waterloo's last year's run. For each topic file, we created a query that consists of the topic title, and looked for the files that contains the title. We did not differentiate between the articles that contain the query term, but we simply picked the first 250 articles in the corpus. The best entry point to the topic file was set to the beginning of the article. We call this baseline.

For the second, element-retrieval-based run (a2a#2,f2f#2), we used the same set of query terms, but applied the element retrieval algorithm in the Adhoc track to rank the article that contains the query terms according to its highest element score.

For the third, PageRank-based run (a2a#3,f2f#3), we took the result of our third outgoing run (a2a#3,f2f#3) to compute a topic oriented PageRank [5] and reranked all articles containing the query term by these values.

Currently the official evaluation is only available at the file-to-file level as in Table 8. Although our PageRank-based run performed closed to the QUT's

Table 8. Results of Waterloo's File-to-File Runs in LTW Incoming Tasks

Run	Rank (out of 24)	MAP
PageRank (f2f#3)	2	0.55633
baseline (f2f#1)	3	0.55403
element retrieval (f2f#2)	5	0.53501

best performing run, it is a little disappointing that our baseline run performed equally as well. The element-retrieval-based run may be improved if we train on previous year's data set.

5   Conclusions and Future Work

This year, we extended our previous year's best performing algorithms to improve the performance. Unfortunately, our simple algorithms from the previous years not only did the best amongst all our runs in the Adhoc and Link-the-Wiki tracks, but also did best amongst all the runs in all the tasks in the Adhoc track. This may indicate the uselessness of the XML structure. On the other hand, since it seems that the passage runs do not perform as well as our element runs, marking up elements do seem useful. Moreover, the effect of specifying ranges of elements as opposed to the current approaches of choosing the single elements or passages is a new area to investigate.

A very interesting area of future research is the effect of positioning within a document to the relevance. Maybe the poor performance of passage retrieval in the Best-in-Context task is because the highest scoring passage within a document is located further down in the document than the second highest scoring passage that starts close to the beginning of the document. Therefore combining the score of a passage with the positional/structured information seems promising.

References

1. Brin, S., Page, L.: The anatomy of a large-scale hypertextual web search engine. Computer Networks and ISDN Systems, Proceedings of the Seventh International World Wide Web 30(1-7), 107–117 (1998)
2. Büttcher, S.: The Wumpus Search Engine (2007), http://www.wumpus-search.org
3. Clarke, C.L.A.: Controlling Overlap in Content-oriented XML retrieval. In: SIGIR 2005: Proceedings of the 28th annual international ACM SIGIR conference on Research and development in information retrieval, pp. 314–321. ACM Press, New York (2005)
4. Itakura, K.Y., Clarke, C.L.A.: University of Waterloo at INEX2007: Adhoc and link-the-wiki tracks. In: Fuhr, N., Kamps, J., Lalmas, M., Trotman, A. (eds.) INEX 2007. LNCS, vol. 4862, pp. 417–425. Springer, Heidelberg (2008)
5. Page, L., Brin, S., Motwani, R., Winograd, T.: The PageRank citation ranking: Bringing order to the Web. Technical Report 1999-66, Stanford InfoLab (November 1999)
6. Ponte, J.M., Croft, W.B.: A language modeling approach to information retrieval. In: SIGIR 1998: Proceedings of the 21st annual international ACM SIGIR conference on Research and development in information retrieval, pp. 275–281. ACM Press, New York (1998)
7. Robertson, S., Walker, S., Beaulieu, M.: Okapi at TREC-7: Automatic ad hoc, filtering, vlc and interactive track. In: 7th Text REtrieval Conference (1998)
8. Zhang, J., Kamps, J.: Link detection in XML documents: What about repeated links. In: Proceedings of the SIGIR 2008 Workshop on Focused Retrieval, pp. 59–66 (2009)

The Impact of Document Level Ranking on Focused Retrieval

Jaap Kamps[1,2] and Marijn Koolen[1]

[1] Archives and Information Studies, Faculty of Humanities, University of Amsterdam
[2] ISLA, Faculty of Science, University of Amsterdam

Abstract. Document retrieval techniques have proven to be competitive methods in the evaluation of focused retrieval. Although focused approaches such as XML element retrieval and passage retrieval allow for locating the relevant text within a document, using the larger context of the whole document often leads to superior document level ranking. In this paper we investigate the impact of using the document retrieval ranking in two collections used in the INEX 2008 Ad hoc and Book Tracks; the relatively short documents of the Wikipedia collection and the much longer books in the Book Track collection. We experiment with several methods of combining document and element retrieval approaches. Our findings are that 1) we can get the best of both worlds and improve upon both individual retrieval strategies by retaining the document ranking of the document retrieval approach and replacing the documents by the retrieved elements of the element retrieval approach, and 2) using document level ranking has a positive impact on focused retrieval in Wikipedia, but has more impact on the much longer books in the Book Track collection.

1 Introduction

In this paper we investigate the impact of document ranking for focused retrieval by comparing standard document retrieval systems to element retrieval approaches. In the evaluation of focused retrieval as studied in INEX, document retrieval techniques have proven to be competitive methods when compared with sub-document level retrieval techniques[3]. Although focused approaches such as XML element retrieval and passage retrieval allow for locating the relevant text within a document, using the larger context of the whole document often leads to better document ranking [7]. Our aim is to investigate the relative effectiveness of both approaches and experiment with combining the two approaches to get the best of both worlds. That is, we want to exploit the better document ranking performance of a document retrieval strateties and the higher within-document precision of an element retrieval strategy. To study the impact of using the document retrieval ranking we perform our experiments on the two collections used in the INEX 2008 Ad hoc and Book Tracks; the relatively short documents of the Wikipedia collection and the much longer books in the Book Track collection.

The paper is structured as follows. First, in Section 2, we report the results for the Ad Hoc Track. Then Section 3 presents our retrieval approach in the Book Track. Finally, in Section 4, we discuss our findings and draw some conclusions.

S. Geva, J. Kamps, and A. Trotman (Eds.): INEX 2008, LNCS 5631, pp. 140–151, 2009.

2 Ad Hoc Track

For the INEX 2008 Ad Hoc Track we investigate several methods of combining article retrieval and element retrieval approaches. We will first describe our indexing approach, then the run combination methods we adopted, the retrieval framework, and finally per task, we present and discuss our results.

The document collection for the Ad hoc track is based on the English Wikipedia [14]. The collection has been converted from the wiki-syntax to an XML format [1]. The XML collection has more than 650,000 documents and over 50,000,000 elements using 1,241 different tag names. However, of these, 779 tags occur only once, and only 120 of them occur more than 10 times in the entire collection. On average, documents have almost 80 elements, with an average depth of 4.82.

2.1 Retrieval Model and Indexing

Our retrieval system is based on the Lucene engine with a number of home-grown extensions [5, 9]. For the Ad Hoc Track, we use a language model where the score for a element e given a query q is calculated as:

$$P(e|q) = P(e) \cdot P(q|e) \tag{1}$$

where $P(q|e)$ can be viewed as a query generation process—what is the chance that the query is derived from this element—and $P(e)$ an element prior that provides an elegant way to incorporate query independent evidence [4].

We estimate $P(q|e)$ using Jelinek-Mercer smoothing against the whole collection, i.e., for a collection D, element e and query q:

$$P(q|e) = \prod_{t \in q} \left((1 - \lambda) \cdot P(t|D) + \lambda \cdot P(t|e) \right), \tag{2}$$

where $P(t|e) = \frac{\text{freq}(t,e)}{|e|}$ and $P(t|D) = \frac{\text{freq}(t,D)}{\sum_{e' \in D} |e|}$.

Finally, we assign a prior probability to an element e relative to its length in the following manner:

$$P(e) = \frac{|e|^\beta}{\sum_e |e|^\beta}, \tag{3}$$

where $|e|$ is the size of an element e. The β parameter introduces a length bias which is proportional to the element length with $\beta = 1$ (the default setting). For a more thorough description of our retrieval approach we refer to [12]. For comprehensive experiments on the earlier INEX data, see [10].

Our indexing approach is based on our earlier work [2, 6].

- *Element index*: Our main index contains all retrievable elements, where we index all textual content of the element including the textual content of their descendants. This results in the "traditional" overlapping element index in the same way as we have done in the previous years [11].
- *Article index*: We also build an index containing all full-text articles (i.e., all wiki-pages) as is standard in IR.

For all indexes, stop-words were removed, but no morphological normalization such as stemming was applied. Queries are processed similar to the documents, we use either the CO query or the CAS query, and remove query operators (if present) from the CO query and the about-functions in the CAS query.

2.2 Combining Article and Element Retrieval

Our experiments with combining runs all use the same two base runs:

- *Article*: a run using the Article index; and
- *Element*: a run using the element index.

Both runs use default parameters for the language model ($\lambda = 0.15, \beta = 1.0$). As shown by Kamps et al. [7], article retrieval leads to a better document ranking, whereas element retrieval fares better at retrieving relevant text within documents. For the Ad hoc Focused task, where the retrieved elements of different documents maybe interleaved in the ranking, we would expect that element retrieval achieves high early precision, while document retrieval, given that it will return whole documents which are often not relevant in their entirety, will have lower early precision. On the other hand, we expect that a document retrieval approach will have relatively little difficulty identifying long articles that have large a fraction of text highlighted as relevant, and therefore return them in the top ranks. The first few returned documents will thus contain a relatively large fraction of all the highlighted text with good within-document precision, resulting in a fairly slow drop in precision across the first recall percentages. For the Relevant in Context task, where retrieved elements have to be grouped by document, and introducing a document ranking score and a within-document retrieval score, we expect the document retrieval approach to rank the documents better and with a perfect within-document recall (due to it retrieving all text in the document) have a reasonable within-document score. With element retrieval, we expect the within-document precision to be better than that of the document retrieval approach, but it will have less recall and a worse document ranking. We therefore assume that a combined approach, using the document ranking of an article level run with the within document element ranking of an element level run, outperforms both runs on the "in context" tasks.

We experiment with three methods of combining the article and element results.

1. *ArtRank*: Retain the article ranking, replacing each article by its elements retrieved in the element run. If no elements are retrieved, use the full article.
2. *Multiplication*: Multiply element score with article score of the article it belongs to. If an element's corresponding article is not retrieved in the top 1,000 results of the article run, use only the element score.
3. *CombSUM*: Normalise retrieval scores (by dividing by highest score in the results list) and add the article score to each element score (if article is not in top 1,000 results for that topic, only element score is used). Thus elements get a boost if the full article is retrieved in the top 1,000 results of the article run.

Our Focused and Relevant in Context submissions are all based on the following base "Thorough" runs:

Table 1. Results for the Ad Hoc Track Focused Task (runs in emphatic are not official submissions)

Run	iP[0.00]	iP[0.01]	iP[0.05]	iP[0.10]	MAiP
Article	0.5712	0.5635	0.5189	0.4522	**0.2308**
Element	**0.6627**	0.5535	0.4586	0.4062	0.1710
ArtRank	0.6320	**0.6025**	0.5054	0.4569	0.1991
CombSUM	0.6556	0.5901	0.4983	0.4553	0.1989
Multiplication	0.6508	0.5614	0.4547	0.4117	0.1815
Element CAS	0.6196	0.5607	0.4941	0.4396	0.2000
ArtRank CAS	0.6096	0.5891	**0.5361**	**0.4629**	0.2140
CombSUM CAS	0.6038	0.5811	0.5158	0.4506	0.2044
Multiplication CAS	0.6077	0.5855	0.5328	0.4601	0.2126

- ArtRank: submitted as inex08_art_B1_loc_in_100_and_el_B1_T
- CombSUM: submittedinex08_art_B1_loc_in_100_comb_sum_el_B1_T
- Multiplication: inex08_art_B1_loc_in_100_x_el_B1_T

We also made CAS versions of these Thorough runs, using the same filtering method as last year [2]. That is, we pool all the target elements of all topics in the 2008 topic set, and filter all runs by removing any element type that is not in this pool of target elements. Our official runs for all three tasks are based on these Thorough runs. Because of the lengthy names of the runs, and to increase clarity and consistency of presentation, we denote the official runs by the methods used, instead of the official run names we used for submission.

2.3 Focused Task

To ensure the Focused run has no overlap, it is post-processed by a straightforward list-based removal strategy. We traverse the list top-down, and simply remove any element that is an ancestor or descendant of an element seen earlier in the list. For example, if the first result from an article is the article itself, we will not include any further element from this article. In the case of the CAS runs, we first apply the CAS filter and then remove overlap. Doing this the other way around, we would first remove possibly relevant target elements if some overlapping non-target elements receive a higher score.

Table 1 shows the results for the Focused Task. Somewhat surprisingly, the Article run outperforms the Element run on the official Focused measure iP[0.01], although the Element run fares much better at the earliest precision level iP[0.00]. Thus, already after 1% recall, document retrieval has a higher precision than element retrieval. A possible explanation is that, given the encyclopedic nature of the collection, for many of the Ad hoc topics there will be a Wikipedia entry that is almost entirely relevant and form more than 1% of the total relevant text. As mentioned earlier, it seems plausible that a document retrieval approach finds these pages relatively easy, and 1% recall is often achieved with the first one or two results, thus with reasonable precision. Both CombSUM and Multiplication attain higher scores for iP[0.00] than ArtRank, but the latter keeps higher precision at further recall levels. The Multiplication method loses much more precision than the other two methods. Compared to the baseline runs Article

Table 2. Results for the Ad Hoc Track Relevant in Context Task (runs in emphatic are not official submissions)

Run	gP[5]	gP[10]	gP[25]	gP[50]	MAgP
Article	0.3376	0.2807	0.2107	0.1605	0.1634
Element	0.2784	0.2407	0.1879	0.1471	0.1484
ArtRank	0.3406	0.2820	0.2120	0.1627	0.1692
CombSUM	0.3281	0.2693	0.2099	0.1615	0.1665
Multiplication	0.3295	0.2827	0.2136	0.1654	0.1695
Element CAS	0.3378	0.2837	**0.2236**	0.1719	0.1703
ArtRank CAS	0.3437	0.2897	0.2207	0.1712	0.1734
CombSUM CAS	0.3481	**0.2991**	0.2200	**0.1726**	**0.1752**
Multiplication CAS	**0.3482**	0.2888	0.2198	0.1724	0.1748

and Element, the combination methods ArtRank and CombSUM lead to substantial improvements at iP[0.01], where the Multiplication method performs slightly worse than the Article run. However, the standard Article run clearly outperforms all other runs when looking at overall precision.

Looking at the CAS runs, we see that the differences are small, with ArtRank leading to the highest iP[0.01] and MAiP scores. The CAS filtering method leads to improvements in overall precision—all MAiP scores go up compared to the non CAS variants—but has a negative effect for early precision as both iP[0.00] and iP[0.01] scores go down, except for the Multiplication run, where the iP[0.01] score goes up. Also, the CAS version of the Multiplication run does improve upon the Article run for precision up to 10% recall.

2.4   Relevant in Context Task

For the Relevant in Context task, we use the Focused runs and cluster all elements belonging to the same article together, and order the article clusters by the highest scoring element. Table 2 shows the results for the Relevant in Context Task. The Article run is better than the Element across the ranking, which is to be expected, given the results reported in [7]. It has a superior article ranking compared to the Element run, and as we saw in the previous section, it even outperformed the Element run on the official measure for the Focused task. However, this time, the combination methods ArtRank and Multiplication do better than the Article run on all reported measures, except for the Multiplication run on gP[5]. Since they use the same article ranking as the Article run, the higher precision scores of the ArtRank and Multiplication show that the elements retrieved in the Element run can improve the precision of the Article run. The CombSUM method, while not far behind, fails to improve upon the Article run on early precision levels (cutoffs 5, 10, and 25). Through the weighted combination of article and element scores, its article ranking is somewhat different from the article ranking of the Article run (and the ArtRank and Multiplication runs).

The CAS filtering method leads to further improvements. The Element CAS run outperforms the standard Article run, and the combination methods show higher precision scores than their non CAS counterparts at all rank cutoffs. This time, the CombSUM method benefits most from the CAS filter. Whereas it was well behind on performance

Table 3. Results for the Ad Hoc Track Best in Context Task (runs in emphatic are not official submissions)

Run	gP[5]	gP[10]	gP[25]	gP[50]	MAgP
Element	0.2372	0.2213	0.1778	0.1384	0.1394
Article	**0.3447**	**0.2870**	**0.2203**	**0.1681**	**0.1693**
Article offset 190	0.2462	0.2042	0.1581	0.1204	0.1228
ArtRank	0.2954	0.2495	0.1849	0.1456	0.1580
CombSUM	0.2720	0.2255	0.1872	0.1487	0.1560
Multiplication	0.2782	0.2399	0.1866	0.1496	0.1577
Element CAS	0.2758	0.2410	0.1929	0.1517	0.1487
ArtRank CAS	0.3101	0.2616	0.1952	0.1539	0.1587
CombSUM CAS	0.3081	0.2547	0.1942	0.1532	0.1581
Multiplication CAS	0.3098	0.2595	0.1944	0.1545	0.1596

compared to the other two combination methods, its CAS version has the highest scores for gP[10], gP[50] and MAgP. Perhaps surprisingly, the Element CAS run is even on par with the combined runs. For the Focused task, the Element CAS run scored well below the combined runs at later rank cutoffs, but when grouped by article, the differences at the later cutoff levels are very small. In fact, the Element CAS run has the highest score at gP[25]. The CAS filter could have an effect on the document ranking of the Element run.

2.5 Best in Context Task

The aim of the Best in Context task is to return a single result per article, which gives best access to the relevant elements. We experimented with three methods of selecting the best entry point:

- *Highest Scoring Element*: the highest scoring element (HSE) returned for each article. We use this on the ArtRank combined run;
- *offset 0*: the start of each returned article; and
- *offset 190*: the median distance from the start of the article of the best entry points in the 2007 assessments.

Table 3 shows the results for the Best in Context Task.

The Article run is far superior to the Element run for the Best in Context Task, at all rank cutoffs and in MAgP. In fact, the Article run outperforms all combined runs and CAS runs. The combined ArtRank run does better than the pure article run with BEPs at offset 190. Note that both these two runs have the same article ranking as the standard Article run. The highest scoring element is thus a better estimation of the BEP than the median BEP offset over a large number of topics. However, using the start of the element clearly outperforms both other runs. Of the three run combination methods, ArtRank gets better scores at early precision levels (cutoffs 5 and 10), but is overtaken by the Multiplication method at further cutoff levels. All three combinations do outperform the Element run and the article run with fixed offset of 190.

The CAS runs again improve upon their non CAS variants, showing that our filtering method is robust over tasks, retrieval approaches and combination methods. As for the non CAS variants, ArtRank gives the best early precision, but the Multiplication gets better precision at later cutoff levels.

The combination methods consistently improve upon the Element retrieval approach, but are far behind the standard Article run. This means that our focused retrieval techniques fail to improve upon an article retrieval approach when it comes to selecting the best point to start reading a document. A closer look at the distribution of BEPs might explain the big difference between the standard Article run and the other runs. The median BEP offset for the 2008 topics is 14 and 49% of all BEPs is at the first character. This shows that choosing the start of the article will in most cases result in a much better document score than any offset further in the document.

2.6 Findings

To sum up, the combination methods seem to be effective in improving early precision. For the official Focused measure, iP[0.01], they lead to improvements over both the Article run and the Element run. The ArtRank method gives the best results for the official measure. Although the Element run scores slightly better at iP[0.00], the combination methods show a good trade off between the good overall precision of the Article run and the good early precision of the Element run. Combining them with the CAS filter improves their overall precision but hurts early precision.

For the Relevant in Context task, all three methods improve upon the Article and Element runs for MAgP. The ArtRank method shows improvement across all cutoff levels. The Multiplication method leads to the highest MAgP scores of the three methods. The CAS filter further improves their effectiveness, although the differences are small for the ArtRank method. Here, the combined runs show the best of both worlds: the good article ranking of the Article run and the more precise retrieval of relevant text within the article of the Element run.

In the Best in Context task, of the three combination methods ArtRank scores better on early precision, while the other two methods do better at later cutoff levels. However, no focused retrieval method comes close to the effectiveness of the pure Article run. With most of the BEPs at, or *very* close to, the start of the article, there seems to be little need for focused access methods for the Wikipedia collection. This result might be explained by the nature of the collection. The Wikipedia collection contains many short articles, where the entire article easily fits on a computer screen, and are all focused on very specific topics. If any text in such a short article is relevant, it usually makes sense to start reading at the beginning of the article.

Finally, the CAS filtering method shows to be robust over all tasks and focused retrieval methods used here, leading to consistent and substantial improvements upon the non CAS filtered variants.

3   Book Track

For the Book Track we investigate the effectiveness of using book level evidence for page level retrieval, and experiment with using Wikipedia as a rich resource for topical descriptions of the knowledge found in books, to mediate between user queries and books in the INEX Book Track collection. We use Indri [13] for our retrieval experiments, with default settings for all parameters. We made one index for both book and page level, using the Krovetz stemmer, no stopword removal, and created two base runs, one at the *book* level and one at the *page* level. The INEX Book Track collection contains 50,239 out-of-copyright books. The books have on average 321 pages and just over 100,000 words. An average page has 323 words. An important difference with the Wikipedia collection, apart from document length, is the difference in structural information in the form of XML markup. In the Wikipedia articles, the markup is based on the layout, containing markup for sections, paragraphs, tables, lists, figures, etc. The books contain only minimal markup, based on the individually scanned pages. That is, there is no layer of elements about sections or chapters in between the page level and book level. Although there is information about the start of chapters and sections in the attributes of `<marker>` elements, they provide no information about where these chapters and sections end. To make use of this information for retrieval, it would require either substantial changes to our indexing approach or a pre-processing step to adjust the XML markup by introducing actual chapter and section elements.

Before we analyse the impact of book level ranking on the retrieval of individual pages, we will discuss the various book level runs we submitted for the Book Retrieval Task.

3.1 Book Retrieval Task

Koolen et al. [8] have used Wikipedia as an intermediary between search queries and books in the INEX Book collection. They experimented with using the link distance between so called *query* pages—Wikipedia pages with titles exactly matching the queries—and *book* pages—each book in the collection is associated with one or more Wikipedia pages based on document similarity—as external evidence to improve retrieval performance. We adopt this approach with the aim to investigate its effectiveness on queries that have no exact matching Wikipedia page.

We obtained the *query* pages by sending each query to the online version of Wikipedia and choosing the first returned result. If the query exactly matches a Wikipedia page, Wikipedia automatically returns that page. Otherwise, Wikipedia returns a results list, and we pick the top result. The idea is that most search topics have a dedicated page on Wikipedia. With the 70 topics of the 2008 collection, we found dedicated Wikipedia pages for 23 queries (38.6%). The *book* pages are obtained by taking the top 100 $tf.idf$ terms of each book (w.r.t. the whole collection) as a query to an Indri index of all Wikipedia pages.[1] Next, we computed the link distance between *query* pages and *book* pages by applying a random walk model on the Wikipedia link graph to obtain a measure of closeness between these pages. Books associated with Wikipedia pages closer in the link graph to the *query* page have a higher probability of being relevant [8]. We then combine these closeness scores with the retrieval scores from an Indri run.

The probability of going from node j at step s from the *query* node to node k is computed as:

$$P_{s+1|s}(k|j) = P_{s|s-1}(j) * \frac{l_{jk}}{l_j} \tag{4}$$

---

[1] This is based on the Wikipedia dump of 12 March, 2008.

Table 4. Results for the Book Retrieval Task (the Closeness ordered run is not an official submission

Run	MAP	P(0.0)	P(0.1)	P5	P10
$Book$	0.0899	0.4051	0.2801	0.1760	0.1320
$Book^2 * Closeness$	0.0714	0.2771	0.2230	0.1520	0.1200
$Closeness$	0.0085	0.1058	0.0406	0.0320	0.0200
$Closeness\ ordered$	0.0302	0.2163	0.0978	0.0960	0.0600

where l_{jk} is the number of links from node j to node k, l_j is the total number of links from node j and $P_{s|s-1}(j)$ is the probability of being at node j after step s. Experimentally, using the INEX 2007 Book Track data, we found that the best closeness scores for the books are obtained by simply adding the closeness scores of the top 8 Wikipedia pages retrieved for that book.

We submitted the following runs:

- $Book$: a baseline book level Indri run (submitted as 6_BST08_B_clean_trec)
- $Closeness$: a run using only the closeness scores (submitted as 6_inex08_BST_book_sim100_top8_forward_trec)
- $Book^2 * Closeness$: a combination of the baseline Indri and the closeness scores, computed as $Indri(q,b)^2 * closeness(q,b)$ for a book b and topic q (submitted as 6_BST08_B_square_times_sim100_top8_fw_trec)

Table 4 shows the results for our submitted runs based on the first release of the relevance judgements, containing judgements for 25 topics. The number of judgements per topic varies greatly. Some topics have only one or two judged books, while others have hundreds of judged books. We have to be careful in drawing conclusions from these results. The standard run performs best on all measures. The official run based on closeness scores alone performs very poorly, based on a simple error. Only 1,000 results per topic were allowed to be submitted. In generating a run from the closeness scores, the first 1,000 scores for each topic were used. However, the closeness scores were not ordered, the first 1,000 were not the highest scores. Therefore, we add the results of on unofficial run – $Closeness\ ordered$ – based on the 1,000 highest closeness scores per topic. Although still well below the baseline run, it is clearly much better than the erroneous official run. As the baseline run is clearly the best of these runs, and the Page in Context runs submitted to the INEX 2008 Book Track are derived from this baseline, we will only use this book level run in the following section.

3.2 Page in Context

As in the Ad Hoc Track (Section 2), we experiment with methods of re-ranking the *page* level runs using the ranking of a *book* level run. Because Indri scores are always negative (the log of a probability, i.e. ranging from $-\infty$ to 0), combining scores can lead to unwanted effects (page score + book score is lower than page score alone). We therefore transform all scores back to probabilities by taking the exponents of the scores.

We experimented with the following three methods.

1. *CombSum*: add exponents of page score and book score (if the book is not retrieved, use only page score. Submitted as 6_BST08.P_plus_B.xml).
2. *Multiplication*: multiply exponents of page and book scores (if book is not retrieved, discard page. Submitted as 6_BST08.P_times_B.xml).
3. *BookRank*: retain the book ranking, replacing each book by its pages retrieved in the *Page* run. If no pages are retrieved, use the whole book.

The official evaluation measures and results of the Page in Context are not yet released, but the relevance judgements are available. To allow a direct comparison of our methods on the Ad hoc and Book Tracks, we evaluated our Page in Context runs using the Focused and Relevant in Context measures of the Ad hoc track.

We transformed the Book Track assessments into FOL format in the following way. First, we computed the number of pages of each book and the length of the actual text, and the average page length. Giving all pages in a book this same average length, we then compute the page offsets of judged pages and retrieved pages by multiplying the average page length by $1-$ the page number. That is, a book with 500 pages and 1 million characters has an average page length of 2,000 characters. Thus, page 54 in that book has length 2,000 and starts at offset $(54 - 1) * 2,000 = 106,000$. For the Focused task, we rank the individual pages on their scores, without grouping them per book. For the Relevant in Context evaluation, which requires results from the same document to be grouped, we use the officially submitted Page in Context runs, where the book ranking is based on the highest scoring pages of each book.

The results of the Page in Context runs evaluated using the Ad hoc measures for the Focused task (see Section 2.3) are shown in Table 5. We see that for overall precision, the *Book* run has low precision scores compared to the Book Retrieval Task, because it is penalised for retrieving the whole books instead of only the relevant pages. However, the more focused page level runs have even lower precision scores (except for the earliest precision score of the *BookRank* run). This somewhat surprising result can be explained by the fact that the original *Page* run contains only a very small portion – 141 out of 6,477 – of the relevant pages. The early precision is comparable to that of the *Book* run, but rapidly drops. The reason for the very low precision scores after 5% recall is that our runs contain up to 1,000 retrieved pages, which is not enough for most topics to reach even 5% recall.

Among the page level runs, the *BookRank* run clearly outperform the standard *Page* run and *CombSUM*, showing that the book level ranking helps. Boosting pages from highly ranked books leads to substantial improvements in precision across the ranking. Apart from that, retaining the whole book when no individual pages for that have been retrieved has a big impact on recall. Especially further down the results list, the *Book* run finds relevant books that are not found by the *Page* run. The *BookRank* run also improves upon the *Book* run at iP[0.00], showing that focused methods can indeed locate the relevant text within books. The big difference between the *Page* and *BookRank* runs, as well as the low precision of the *Page* run by itself show that page level evidence is of limited use without the wider context of the whole book.

The results of the Page in Context runs evaluated using the Ad hoc measures for the Relevant in Context task (see Section 2.4) are shown in Table 6. We see a similar pattern as with the Focused Task results. The *Book* run receives low scores because

Table 5. Results for the Book Track Page in Context Task (using Focused measures)

Run	iP[0.00]	iP[0.01]	iP[0.05]	iP[0.10]	MAiP
Book	0.1690	**0.1690**	**0.0999**	**0.0957**	**0.0393**
Page	0.1559	0.1002	0.0030	0.0017	0.0037
BookRank	**0.2650**	0.1618	0.0838	0.0838	0.0280
CombSUM	0.1666	0.1045	0.0095	0.0083	0.0054
Multiplication	0.0349	0.0247	0.0035	0.0030	0.0015

Table 6. Results for the Book Track Page in Context Task (using Relevant in Context measures)

Run	gP[5]	gP[10]	gP[25]	gP[50]	MAgP
Book	0.0567	0.0309	**0.0147**	**0.0087**	0.0254
Page	0.0242	0.0164	0.0098	0.0058	0.0088
BookRank	**0.0581**	**0.0315**	0.0147	0.0082	**0.0273**
CombSUM	0.0231	0.0158	0.0090	0.0064	0.0102
Multiplication	0.0061	0.0031	0.0027	0.0015	0.0047

it retrieves a lot of irrelevant text. Focused techniques should be able to achieve much better precision. Again, only the *BookRank* run can compete with the *Book* run, and improves upon it with early and overall precision. The fact that the *BookRank* run has lower precision than the *Book* run further down the ranking shows that at these lower ranks, the whole books do better than the individually retrieved pages of these books. Although this might partly be caused by the low number of pages retrieved, the low precision scores for the Focused evaluation show that the content of individual pages is not very effective for locating the relevant information in books containing hundreds of pages.

4   Discussion and Conclusions

For the *Ad Hoc Track*, we investigated the effectiveness of combining article and element retrieval methods and found that the ArtRank method, where the article run determines the article ranking, and the element run determines which part(s) of the text is returned, gives the best results for the Focused Task. For the Relevant in Context Task, the Multiplication method is slightly better than ArtRank and CombSUM, but for the CAS runs, where we filter on a pool of target elements based on the entire topic set, the CombSUM method gives the best performance overall. The combination methods are not effective for the Best in Context Task. The standard article retrieval run is far superior to any focused retrieval run. With many short articles in the collection, all focused on very specific topics, it makes sense to start reading at the beginning of the article, making it hard for focused retrieval techniques to improve upon traditional document retrieval. The CAS pool filtering method is effective for all three tasks as well, showing consistent improvement upon the non CAS variants for all measures.

For the *Book Track*, we experimented with the same run combination methods as in the Ad Hoc Track. As for the Ad hoc Track using the Wikipedia collection, we see that for the Book Track, a document retrieval approach is a non-trivial baseline.

However, for the long documents in the Book Track collection, where an individual pages forms only a small part in a much wider context, the impact of the document level ranking on focused retrieval techniques is much bigger than for the short documents in the Wikipedia collection. Using only page level evidence, the precision is very low, indicating that the content of individual pages seems not very effective in locating all the relevant text spread over multiple pages in a book. By using the ranking of the book level run, and replacing the whole content of a book only when individual pages of that book are retrieved, the combination can improve upon standard document level retrieval.

Acknowledgments. Jaap Kamps was supported by the Netherlands Organization for Scientific Research (NWO, grants # 612.066.513, 639.072.601, and 640.001.501). Marijn Koolen was supported by NWO under grant # 640.001.501. # 639.072.601.

References

[1] Denoyer, L., Gallinari, P.: The Wikipedia XML Corpus. SIGIR Forum 40, 64–69 (2006)

[2] Fachry, K.N., Kamps, J., Koolen, M., Zhang, J.: Using and detecting links in wikipedia. In: Fuhr, N., Kamps, J., Lalmas, M., Trotman, A. (eds.) INEX 2007. LNCS, vol. 4862, pp. 388–403. Springer, Heidelberg (2008)

[3] Fuhr, N., Kamps, J., Lalmas, M., Malik, S., Trotman, A.: Overview of the INEX 2007 ad hoc track. In: Fuhr, N., Kamps, J., Lalmas, M., Trotman, A. (eds.) INEX 2007. LNCS, vol. 4862, pp. 1–23. Springer, Heidelberg (2008)

[4] Hiemstra, D.: Using Language Models for Information Retrieval. PhD thesis, Center for Telematics and Information Technology, University of Twente (2001)

[5] ILPS: The ILPS extension of the Lucene search engine (2008), http://ilps.science.uva.nl/Resources/

[6] Kamps, J., Koolen, M., Sigurbjörnsson, B.: Filtering and clustering XML retrieval results. In: Fuhr, N., Lalmas, M., Trotman, A. (eds.) INEX 2006. LNCS, vol. 4518, pp. 121–136. Springer, Heidelberg (2007)

[7] Kamps, J., Koolen, M., Lalmas, M.: Locating relevant text within XML documents. In: Proceedings SIGIR 2008, pp. 847–849. ACM Press, New York (2008)

[8] Koolen, M., Kazai, G., Craswell, N.: Wikipedia Pages as Entry Points for Book Search. In: Proceedings of the Second ACM International Conference on Web Search and Data Mining (WSDM 2009). ACM Press, New York (2009)

[9] Lucene: The Lucene search engine (2008), http://lucene.apache.org/

[10] Sigurbjörnsson, B.: Focused Information Access using XML Element Retrieval. SIKS dissertation series 2006-28, University of Amsterdam (2006)

[11] Sigurbjörnsson, B., Kamps, J., de Rijke, M.: An Element-Based Approach to XML Retrieval. In: INEX 2003 Workshop Proceedings, pp. 19–26 (2004)

[12] Sigurbjörnsson, B., Kamps, J., de Rijke, M.: Mixture models, overlap, and structural hints in XML element retrieval. In: Fuhr, N., Lalmas, M., Malik, S., Szlávik, Z. (eds.) INEX 2004. LNCS, vol. 3493, pp. 196–210. Springer, Heidelberg (2005)

[13] Strohman, T., Metzler, D., Turtle, H., Croft, W.B.: Indri: a language-model based search engine for complex queries. In: Proceedings of the International Conference on Intelligent Analysis (2005)

[14] Wikipedia: The free encyclopedia (2008), http://en.wikipedia.org/

Adhoc and Book XML Retrieval with Cheshire

Ray R. Larson

School of Information
University of California, Berkeley
Berkeley, California, USA, 94720-4600
`ray@ischool.berkeley.edu`

Abstract. For this year's INEX UC Berkeley focused on the Book track and also submitted two runs for the Adhoc Focused Element search task and one for the Best in Context task. For all of these runs we used the TREC2 logistic regression probabilistic model. For the Adhoc Element runs and Best in Context runs we used the "pivot" score merging method to combine paragraph-level searches with scores for document-level searches.

1 Introduction

In this paper we will first discuss the algorithms and fusion operators used in our official INEX 2008 Book Track and Adhoc Focused and Best in Context track runs. Then we will look at how these algorithms and operators were used in the various submissions for these tracks, and finally we will discuss problems in implementation, and directions for future research.

2 The Retrieval Algorithms and Fusion Operators

This section largely duplicates earlier INEX papers in describing the probabilistic retrieval algorithms used for both the Adhoc and Book track in INEX this year. Although These are the same algorithms that we have used in previous years for INEX and in other evaluations (such as CLEF), including a blind relevance feedback method used in combination with the TREC2 algorithm, we are repeating the formal description here instead of refering to those earlier papers alone. In addition we will again discuss the methods used to combine the results of searches of different XML components in the collections. The algorithms and combination methods are implemented as part of the Cheshire II XML/SGML search engine [10,8,7] which also supports a number of other algorithms for distributed search and operators for merging result lists from ranked or Boolean sub-queries.

2.1 TREC2 Logistic Regression Algorithm

Once again the principle algorithm used for our INEX runs is based on the *Logistic Regression* (LR) algorithm originally developed at Berkeley by Cooper, et al.

S. Geva, J. Kamps, and A. Trotman (Eds.): INEX 2008, LNCS 5631, pp. 152–163, 2009.

[5]. The version that we used for Adhoc tasks was the Cheshire II implementation of the "TREC2" [4,3] that provided good Thorough retrieval performance in the INEX 2005 evaluation [10]. As originally formulated, the LR model of probabilistic IR attempts to estimate the probability of relevance for each document based on a set of statistics about a document collection and a set of queries in combination with a set of weighting coefficients for those statistics. The statistics to be used and the values of the coefficients are obtained from regression analysis of a sample of a collection (or similar test collection) for some set of queries where relevance and non-relevance has been determined. More formally, given a particular query and a particular document in a collection $P(R \mid Q, D)$ is calculated and the documents or components are presented to the user ranked in order of decreasing values of that probability. To avoid invalid probability values, the usual calculation of $P(R \mid Q, D)$ uses the "log odds" of relevance given a set of S statistics derived from the query and database, such that:

$$\log O(R|C, Q) = log \frac{p(R|C, Q)}{1 - p(R|C, Q)} = log \frac{p(R|C, Q)}{p(\overline{R}|C, Q)}$$

$$= c_0 + c_1 * \frac{1}{\sqrt{|Q_c|} + 1} \sum_{i=1}^{|Q_c|} \frac{qtf_i}{ql + 35}$$

$$+ c_2 * \frac{1}{\sqrt{|Q_c|} + 1} \sum_{i=1}^{|Q_c|} \log \frac{tf_i}{cl + 80}$$

$$- c_3 * \frac{1}{\sqrt{|Q_c|} + 1} \sum_{i=1}^{|Q_c|} \log \frac{ctf_i}{N_t}$$

$$+ c_4 * |Q_c|$$

where C denotes a document component and Q a query, R is a relevance variable, and

$p(R|C, Q)$ is the probability that document component C is relevant to query Q,

$p(\overline{R}|C, Q)$ the probability that document component C is not relevant to query Q, (which is 1.0 - $p(R|C, Q)$)

$|Q_c|$ is the number of matching terms between a document component and a query,

qtf_i is the within-query frequency of the ith matching term,

tf_i is the within-document frequency of the ith matching term,

ctf_i is the occurrence frequency in a collection of the ith matching term,

ql is query length (i.e., number of terms in a query like $|Q|$ for non-feedback situations),

cl is component length (i.e., number of terms in a component), and

N_t is collection length (i.e., number of terms in a test collection).

c_k are the k coefficients obtained though the regression analysis.

Assuming that stopwords are removed during index creation, then ql, cl, and N_t are the query length, document length, and collection length, respectively. If the query terms are re-weighted (in feedback, for example), then qtf_i is no longer the original term frequency, but the new weight, and ql is the sum of the new weight values for the query terms. Note that, unlike the document and collection lengths, query length is the relative frequency without first taking the log over the matching terms.

The coefficients were determined by fitting the logistic regression model specified in $\log O(R|C, Q)$ to TREC training data using a statistical software package. The coefficients, c_k, used for our official runs are the same as those described by Chen[1]. These were: $c_0 = -3.51$, $c_1 = 37.4$, $c_2 = 0.330$, $c_3 = 0.1937$ and $c_4 = 0.0929$. Further details on the TREC2 version of the Logistic Regression algorithm may be found in Cooper et al. [4].

2.2 Blind Relevance Feedback

It is well known that blind (also called pseudo) relevance feedback can substantially improve retrieval effectiveness in tasks such as TREC and CLEF. (See for example the papers of the groups who participated in the Ad Hoc tasks in TREC-7 (Voorhees and Harman 1998)[12] and TREC-8 (Voorhees and Harman 1999)[13].)

Blind relevance feedback is typically performed in two stages. First, an initial search using the original queries is performed, after which a number of terms are selected from the top-ranked documents (which are presumed to be relevant). The selected terms are weighted and then merged with the initial query to formulate a new query. Finally the reweighted and expanded query is run against the same collection to produce a final ranked list of documents. It was a simple extension to adapt these document-level algorithms to document components for INEX.

The TREC2 algorithm has been been combined with a blind feedback method developed by Aitao Chen for cross-language retrieval in CLEF. Chen[2] presents a technique for incorporating blind relevance feedback into the logistic regression-based document ranking framework. Several factors are important in using blind relevance feedback. These are: determining the number of top ranked documents that will be presumed relevant and from which new terms will be extracted, how to rank the selected terms and determining the number of terms that should be selected, how to assign weights to the selected terms. Many techniques have been used for deciding the number of terms to be selected, the number of top-ranked documents from which to extract terms, and ranking the terms. Harman [6] provides a survey of relevance feedback techniques that have been used.

Obviously there are important choices to be made regarding the number of top-ranked documents to consider, and the number of terms to extract from those documents. For this year, having no truly comparable prior data to guide us, we chose to use the top 10 terms from 10 top-ranked documents. The terms were chosen by extracting the document vectors for each of the 10 and computing the Robertson and Sparck Jones term relevance weight for each document. This weight is based on a contingency table where the counts of 4 different conditions for combinations of (assumed) relevance and whether or not the term is, or is not in a document. Table 1 shows this contingency table.

Table 1. Contingency table for term relevance weighting

	Relevant	Not Relevant	
In doc	R_t	$N_t - R_t$	N_t
Not in doc	$R - R_t$	$N - N_t - R + R_t$	$N - N_t$
	R	$N - R$	N

The relevance weight is calculated using the assumption that the first 10 documents are relevant and all others are not. For each term in these documents the following weight is calculated:

$$w_t = log\frac{\frac{R_t}{R-R_t}}{\frac{N_t-R_t}{N-N_t-R+R_t}} \tag{1}$$

The 10 terms (including those that appeared in the original query) with the highest w_t are selected and added to the original query terms. For the terms not in the original query, the new "term frequency" (qtf_i in main LR equation above) is set to 0.5. Terms that were in the original query, but are not in the top 10 terms are left with their original qtf_i. For terms in the top 10 and in the original query the new qtf_i is set to 1.5 times the original qtf_i for the query. The new query is then processed using the same TREC2 LR algorithm as shown above and the ranked results returned as the response for that topic.

2.3 Result Combination Operators

As we have also reported previously, the Cheshire II system used in this evaluation provides a number of operators to combine the intermediate results of a search from different components or indexes. With these operators we have available an entire spectrum of combination methods ranging from strict Boolean operations to fuzzy Boolean and normalized score combinations for probabilistic and Boolean results. These operators are the means available for performing fusion operations between the results for different retrieval algorithms and the search results from different components of a document.

For the Adhoc Focused and Best In Context runs we used a merge/reweighting operator based on the "Pivot" method described by Mass and Mandelbrod[11] to combine the results for each type of document component considered. In our case the new probability of relevance for a component is a weighted combination of the initial estimate probability of relevance for the component and the probability of relevance for the entire article for the same query terms. Formally this is:

$$P(R \mid Q, C_{new}) = (X * P(R \mid Q, C_{comp})) + ((1 - X) * P(R \mid Q, C_{art})) \tag{2}$$

Where X is a pivot value between 0 and 1, and $P(R \mid Q, C_{new})$, $P(R \mid Q, C_{comp})$ and $P(R \mid Q, C_{art})$ are the new weight, the original component weight, and article weight for a given query. Although we found that a pivot value of 0.54 was most effective for INEX04 data and measures, we adopted the "neutral" pivot value of 0.4 for all of our 2008 adhoc runs, given the uncertainties of how this approach would fare with the new database.

3   Database and Indexing Issues

We used the latest version of the Wikipedia database for this year's Adhoc runs, and created a number of indexes similar to those described in previous INEX papers[9].

Table 2. Wikipedia Article-Level Indexes for INEX 2008

Name	Description	Contents	Vector?
docno	doc ID number	//name@id	No
names	Article Title	//name	Yes
topic	Entire Article	//article	Yes
topicshort	Selected Content	//fm/tig/atl //abs //kwd //st	Yes
xtnames	Template names	//template@name	No
figure	Figures	//figure	No
table	Tables	//table	No
caption	Image Captions	//caption	Yes
alltitles	All Titles	//title	Yes
links	Link Anchors	//collectionlink //weblink //wikipedialink	No

Table 2 lists the document-level (/article) indexes created for the INEX database and the document elements from which the contents of those indexes were extracted.

Table 3. Wikipedia Components for INEX 2006

Name	Description	Contents
COMPONENT_SECTION	Sections	//section
COMPONENT_PARAS	Paragraphs	//p | //blockquote | //indentation1| //indentation2|//indentation3
COMPONENT_FIG	Figures	//figure

As noted above the Cheshire system permits parts of the document subtree to be treated as separate documents with their own separate indexes. Tables 3 & 4 describe the XML components created for INEX and the component-level indexes that were created for them.

Table 3 shows the components and the path used to define them. The first, refered to as COMPONENT_SECTION, is a component that consists of each identified section in all of the documents, permitting each individual section of a article to be retrieved separately. Similarly, each of the COMPONENT_PARAS and COMPONENT_FIG components, respectively, treat each paragraph (with all of the alternative paragraph elements shown in Table 3), and figure (<figure> ... </figure>) as individual documents that can be retrieved separately from the entire document.

Table 4. Wikipedia Component Indexes for INEX 2006

Component or Index Name	Description	Contents	Vector?
COMPONENT_SECTION			
sec_title	Section Title	//section/title	Yes
sec_words	Section Words	* (all)	Yes
COMPONENT_PARAS			
para_words	Paragraph Words	* (all)	Yes
COMPONENT_FIG			
fig_caption	Figure Caption	//figure/caption	No

Table 4 describes the XML component indexes created for the components described in Table 3. These indexes make the individual sections (such as COM-PONENT_SECTION) of the INEX documents retrievable by their titles, or by any terms occurring in the section. These are also proximity indexes, so phrase searching is supported within the indexes. Individual paragraphs (COM-PONENT_PARAS) are searchable by any of the terms in the paragraph, also with proximity searching. Individual figures (COMPONENT_FIG) are indexed by their captions.

Few of these indexes and components were used during Berkeley's simple runs of the 2006 INEX Adhoc topics. The two official submitted Adhoc runs and scripts used in INEX are described in the next section.

We decided to try the same methods on the Book Track data this year, but we did not use multiple elements or components, since the goal of the main Books Adhoc task was to retrieval entire books and not elements. We did, however create the same indexes for the Books and MARC data that we created last year as shown in Table 5, for the books themselves we used a single index of the entire document content. We did not use the Entry Vocabulary Indexes used in last year's Book track runs.

The indexes used in the MARC data are shown in Table 5. Note that the tags represented in the "Contents" column of the table are from Cheshire's MARC to

Table 5. MARC Indexes for INEX Book Track 2008

Name	Description	Contents	Vector?
names	All Personal and Corporate names	//FLD[1670]00, //FLD[1678]10, //FLD[1670]11	No
pauthor	Personal Author Names	//FLD[170]00	No
title	Book Titles	//FLD130, //FLD245, //FLD240, //FLD730, //FLD740, //FLD440, //FLD490, //FLD830	No
subject	All Subject Headings	//FLD6..	No
topic	Topical Elements	//FLD6.., //FLD245, //FLD240, //FLD4.., //FLD8.., //FLD130, //FLD730, //FLD740, //FLD500, //FLD501, //FLD502 //FLD505, //FLD520, //FLD590	Yes
lcclass	Library of Congress Classification	//FLD050, //FLD950	No
doctype	Material Type Code	//USMARC@MATERIAL	No
localnum	ID Number	//FLD001	No
ISBN	ISBN	//FLD020	No
publisher	Publisher	//FLD260/b	No
place	Place of Publication	//FLD260/a	No
date	Date of Publication	//FLD008	No
lang	Language of Publication	//FLD008	No

XML conversion, and are represented as regular expressions (i.e., square brackets indicate a choice of a single character).

3.1   Indexing the Books XML Database

Because the structure of the Books database was derived from the OCR of the original paper books, it is primarily focused on the page organization and layout and not on the more common structuring elements such as "chapters" or "sections". Because this emphasis on page layout goes all the way down to the individual word and its position on the page, there is a very large amount of markup for page with content. For this year's original version of the Books database, there are actually NO text nodes in the entire XML tree, the words actually present on a page are represented as attributes of an empty word tag in the XML. The entire document in XML form is typically multiple megabytes in size. A separate version of the Books database was made available that converted these empty tags back into text nodes for each line in the scanned text. This provided a significant reduction in the size of database, and made indexing much simpler. The primary index created for the full books was the "topic" index containing the entire book content.

We also created page-level "documents" as we did last year. As noted above the Cheshire system permits parts of the document subtree to be treated as

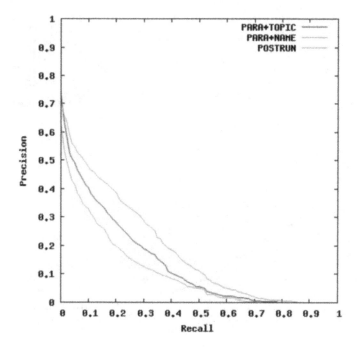

Fig. 1. Berkeley Adhoc Element Retrieval Results

separate documents with their own separate indexes. Thus, paragraph-level components were extracted from the page-sized documents. Because unique object
(page) level indentifiers are included in each object, and these identifiers are
simple extensions of the document (book) level identifier, we were able to use
the page-level identifier to determine where in a given book-level document a
particular page or paragraph occurs, and generate an appropriate XPath for it.

Indexes were created to allow searching of full page contents, and component
indexes for the full content of each of individual paragraphs on a page. Because
of the physical layout based structure used by the Books collection, paragraphs
split across pages are marked up (and therefore indexed) as two paragraphs.
Indexes were also created to permit searching by object id, allowing search for
specific individual pages, or ranges of pages.

We encountered a number of system problems dealing with the Books database this year, since the numbers unique terms exceeded the capacity of the
integers used to store them in the indexes. For this year, at least, moving to
unsigned integers has provided a temporary fix for the problem but we will need
to rethink how statistical summary information is handled in the future – perhaps moving to long integers, or even floating point numbers and evaluating
the tradeoffs between precision in the statistics and index size (since moving to
Longs could double index size).

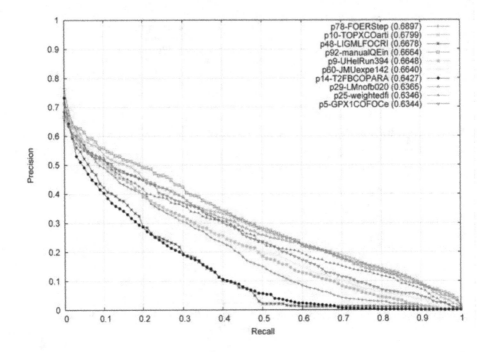

Fig. 2. Top 10 (by group) Adhoc Retrieval Runs

4   INEX 2008 Adhoc Track Runs

We submitted three runs this year to the Adhoc "Focused" track, two for the CO
Element search tasks, and one for the "Best in context" task. Figure 1 shows the
precision/recall curves for the two Element search runs. The better performing
run used a fusion of paragraph with full document (topic) search and had an
iP at 0.01 of 0.6395. This was ranked number twelve out of the sixty-one runs
submitted, and since many of the top-ranked runs came from the same groups,
the run appeared in the "top ten" graph on the official site (reproduced as Figure
2). As Figure 2 shows, the run was fairly strong for precision at low recall levels,
but overall showed a lack of recall placing it much lower than the majority of
the top ten submissions at higher recall levels. We had intended this run to use
the blind feedback mechanism described above, but fail to specify it correctly.
As a result the run used only the TREC2 logistic regression algorithm and the
weighted merging described above, but with no blind feedback.

Our second "Focused" run used a fusion of paragraphs with searches on the
"name" element of the Wikipedia documents. This run, also shown in Figure
1 had an iP at 0.01 of 0.5410, and ranked forty-third out of sixty-one runs - a
much poorer showing. This run also was intended to use the blind feedback, but
did not.

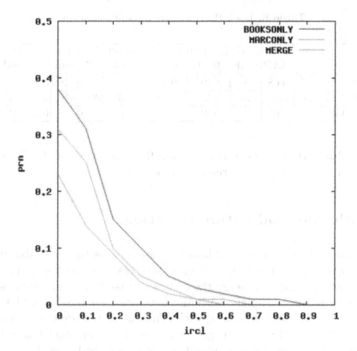

Fig. 3. Book Retrieval Task Runs (ircl_prn0.00-1.00

As an unofficial experiment after we discovered the lack of blind feedback in these runs, we ran the same script as for our first official run (using paragraphs and the topic index) but correctly specified the use of blind feedback in addition to the TREC2 Logistic Regression algorithm. This unofficial run obtained a iP at 0.01 of 0.6586, which would have ranked tenth if it had been submitted. This run is shown in Figure 1 as "POSTRUN" and indicates how the lack of blind feedback in our official submissions adversely impacted the results.

Our Best in context run basically took the raw results data from the Adhoc focused run using paragraphs and topics, and selected the top-ranked paragraphs for each document as the "best in context", all other elements were eliminated from the run. The results clearly show that this is not a very effective strategy, since our results of a MAgP of 0.0542 was ranked thirty-second out of thirty-five runs submitted overall.

5   INEX 2008 Book Track Runs

We submitted three runs for the Book Search task of the Books track, one using MARC data only, one using full Book contents only, and a third performing a merge of the MARC and Book data. The results of these runs are compared in Figure 3. Other measures of effectiveness for these runs (including their relative ranks across all runs in the track) are shown in Table 6.

Table 6. Book Retrieval Task result measures

Name	MAP	MAP rank	ircl_prn 0.00	ircl_prn 0.10	ircl_prn rank	P5	P5 rank	P10	P10 rank
BOOKSONLY	0.08	4	0.38	0.31	2	0.19	2	0.14	1
MARCONLY	0.04	16	0.23	0.14	16	0.09	16	0.06	16
MERGE	0.05	12	0.31	0.25	6	0.14	6	0.09	12

We are also participating in the "Active Reading" task of the Books track which is still underway with no result to report yet.

6   Conclusions and Future Directions

For all of the Adhoc (focused and best in context) runs that we submitted this year, only paragraphs were used as the retrieved elements, and we did not (as in previous years) attempt to merge the results of searches on multiple elements. This helps to partially explain both the low recall for the Focused task and the low score of the Best in context task. The failure to correctly specify blind feedback in the runs also had a negative impact. However, since this is the first year when we have managed to submit any runs for the Focused task without them being disqualified for overlap, we can consider it a significant improvement. We note for the future that double-checking scripts before running them, and submitting the results is very good idea.

References

1. Chen, A.: Multilingual information retrieval using english and chinese queries. In: Peters, C., Braschler, M., Gonzalo, J., Kluck, M. (eds.) CLEF 2001. LNCS, vol. 2406, pp. 44–58. Springer, Heidelberg (2002)
2. Chen, A.: Cross-Language Retrieval Experiments. In: Peters, C., Braschler, M., Gonzalo, J. (eds.) CLEF 2002. LNCS, vol. 2785, pp. 28–48. Springer, Heidelberg (2003)
3. Chen, A., Gey, F.C.: Multilingual information retrieval using machine translation, relevance feedback and decompounding. Information Retrieval 7, 149–182 (2004)
4. Cooper, W.S., Chen, A., Gey, F.C.: Full Text Retrieval based on Probabilistic Equations with Coefficients fitted by Logistic Regression. In: Text REtrieval Conference (TREC-2), pp. 57–66 (1994)
5. Cooper, W.S., Gey, F.C., Dabney, D.P.: Probabilistic retrieval based on staged logistic regression. In: 15th Annual International ACM SIGIR Conference on Research and Development in Information Retrieval, Copenhagen, Denmark, June 21-24, pp. 198–210. ACM, New York (1992)
6. Harman, D.: Relevance feedback and other query modification techniques. In: Frakes, W., Baeza-Yates, R. (eds.) Information Retrieval: Data Structures & Algorithms, pp. 241–263. Prentice Hall, Englewood Cliffs (1992)

7. Larson, R.R.: A logistic regression approach to distributed IR. In: SIGIR 2002: Proceedings of the 25th Annual International ACM SIGIR Conference on Research and Development in Information Retrieval, Tampere, Finland, August 11-15, pp. 399–400. ACM Press, New York (2002)

8. Larson, R.R.: A fusion approach to XML structured document retrieval. Information Retrieval 8, 601–629 (2005)

9. Larson, R.R.: Probabilistic retrieval approaches for thorough and heterogeneous xml retrieval. In: Fuhr, N., Lalmas, M., Trotman, A. (eds.) INEX 2006. LNCS, vol. 4518, pp. 318–330. Springer, Heidelberg (2007)

10. Larson, R.R.: Probabilistic retrieval, component fusion and blind feedback for XML retrieval. In: Fuhr, N., Lalmas, M., Malik, S., Kazai, G. (eds.) INEX 2005. LNCS, vol. 3977, pp. 225–239. Springer, Heidelberg (2006)

11. Mass, Y., Mandelbrod, M.: Component ranking and automatic query refinement for xml retrieval. In: Fuhr, N., Lalmas, M., Malik, S., Szlávik, Z. (eds.) INEX 2004. LNCS, vol. 3493, pp. 73–84. Springer, Heidelberg (2005)

12. Voorhees, E., Harman, D. (eds.): The Seventh Text Retrieval Conference (TREC-7). NIST (1998)

13. Voorhees, E., Harman, D. (eds.): The Eighth Text Retrieval Conference (TREC-8). NIST (1999)

Book Layout Analysis:
TOC Structure Extraction Engine

Bodin Dresevic, Aleksandar Uzelac, Bogdan Radakovic, and Nikola Todic

Microsoft Development Center Serbia,
Makedonska 30, 11000 Belgrade, Serbia
{bodind,aleksandar.uzelac,bogdan.radakovic,nikola.todic}@microsoft.com
http://www.microsoft.com/scg/mdcs

Abstract. Scanned then OCRed documents usually lack detailed layout
and structural information. We present a book specific layout analysis
system used to extract TOC structure information from the scanned and
OCRed books. This system was used for navigation purposes by the live
books search project. We provide labeling scheme for the TOC sections
of the books, high level overview for the book layout analysis system, as
well as TOC Structure Extraction Engine. In the end we present accuracy
measurements of this system on a representative test set.

Keywords: book layout analysis, TOC, information extraction, TOC
navigation, ocrml, bookml.

1 Introduction

Book layout analysis as described in this paper is a process of extracting structural information from scanned and OCRed books. The main purpose of this
work was to enable navigation experience for the live book search project, a
clickable TOC experience, in mid 2007. More precisely, the task was to isolate
TOC pages from the rest of the book, detect TOC entries and locate the target
page where TOC entries are pointing to. In this paper we shall focus on the
second part, TOC structure extraction.

There are two aspects of the problem; the first one is related to acquir-
ing/extracting information from raw OCR (words, lines and bounding boxes),
i.e. performing in depth analysis of TOC pages. On any TOC page there are several types of information which we shall roughly divide in two: TOC entries and
other. Each TOC entry is a single smallest group of words with the same title
target somewhere in the book (usually specified by the page number at the end
of the entry). Everything other than TOC entries we shall ignore, implying other
stuff does not yield significant information to be treated. With this defined —
the TOC Structure Extraction Engine is responsible for providing following information about TOC entries:

- Separating TOC entries from other less important content of the page
- Separating TOC entries among themselves

S. Geva, J. Kamps, and A. Trotman (Eds.): INEX 2008, LNCS 5631, pp. 164–171, 2009.
© Springer-Verlag Berlin Heidelberg 2009

- Establishing TOC entry target page
- Establishing TOC entry relative significance
- Determining TOC entry internal structure

The second aspect of the problem is related to the presentation of extracted information. For presentation purposes we have devised a labeling scheme which supports all possible scenarios for TOC pages.

2 Labeling Scheme

At this point we shall introduce several new ideas regarding the actual labeling of TOC pages. We have previously mentioned TOC entries in an informal way and now we shall define a TOC entry in a more rigorous manner. A TOC entry is a single referee to a target (part, chapter section, first line, first couple of words in some line...) somewhere in the book. The target cannot begin in the middle of a line (we emphasize the case of a single word in the middle of the target line — this is not considered to be TOC entry, but rather an index entry). In the same manner we shall disregard entries with a number of targets (again, we would consider this entry to be an index rather than TOC).

To illustrate the point we have prepared an example, taken from Charles Francis Adams 1890 book "Richard Henry Dana", published by Houghton Boston. We took an excerpt from the table of contents of that book. In Fig. 1 below we have presented three TOC entries. All three have the same structure; they all consist of a chapter number (IV, V, and VI) at the beginning of the line, then there is a title ("Round the World", and so on), a separator (a number of dots in each line), and finally a page number for each entry (178, 248, and 282).

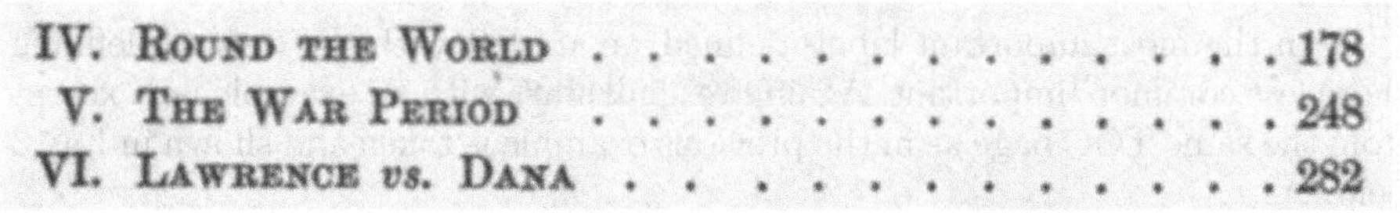
IV. Round the World 178
V. The War Period 248
VI. Lawrence vs. Dana 282

Fig. 1. Example of three TOC entries

To ensure that each entry is in fact a TOC entry it is not sufficient to consider a TOC page alone, target pages need be considered as well. As an illustration we shall observe page 178 and its content (shown in Fig. 2 below). After inspection it is clear that the first entry from Fig. 1 indeed targets page 178, i.e. there are both chapter number (the same as indicated in TOC entry) and chapter title (again, the same as indicated in TOC entry). We observe that the chapter number is of somewhat different structure than in the TOC entry on TOC page (namely, there is keyword "CHAPTER" in front of the actual chapter number, and the chapter name appears in the line preceding the actual title line), nevertheless the target page is correct. Furthermore, both the chapter number and the chapter title are spatially separated from the rest of the text on that page (we shall refer to the rest of the page as a body), the font, although the same type and size, it is in all capital letters, which is yet another significant feature to further ensure that our target page is in fact correct.

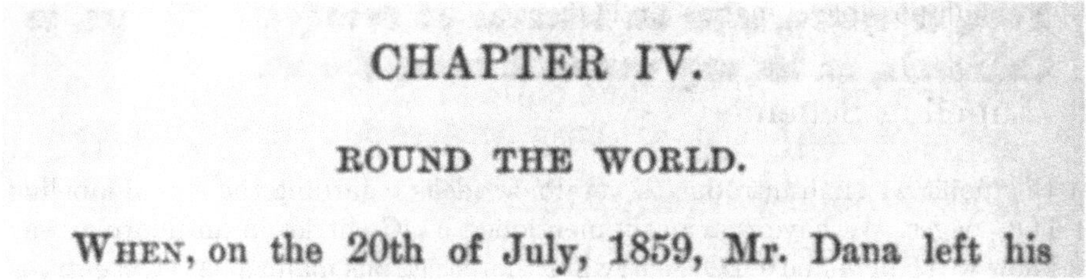
CHAPTER IV.

ROUND THE WORLD.

WHEN, on the 20th of July, 1859, Mr. Dana left his

Fig. 2. Example of TOC entry target page

Given a simple example (such as the previous one), one can induce a minimal set of labels needed for labeling TOC entries. The most important label is clearly a title one, indicating a number of words that represent a TOC entry title. In our labeling scheme we refer to this label as a TOC CHAPTER TITLE. Next in line of importance is the page number, providing information about the target page for each entry. In our labeling scheme we refer to this label as a TOC CHAPTER PAGE NUMBER. Then we have TOC CHAPTER NAME label, which in the example above refers to a chapter number IV; in general TOC chapter name label could consist of both keyword (like "CHAPTER" or "SECTION") and a number. At last there is TOC CHAPTER SEPARATOR label, which is self explanatory.

With the most important labels defined we are left with the duty of defining those less common/important. Again, we shall start with an example, an excerpt from the same TOC page as in the previous example is taken and shown in Fig. 3 below.

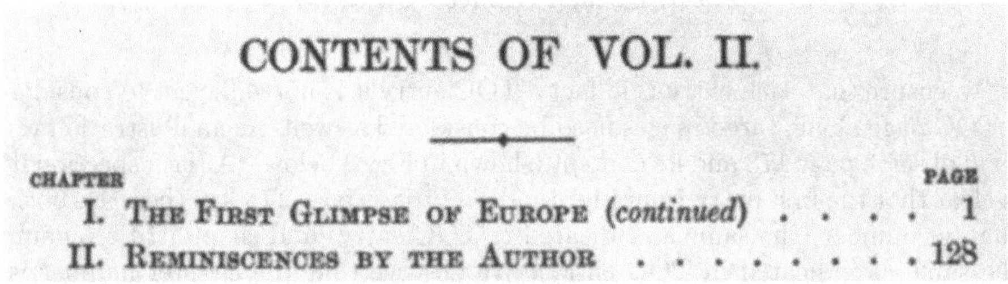
CONTENTS OF VOL. II.

CHAPTER PAGE

I. THE FIRST GLIMPSE OF EUROPE (*continued*) 1

II. REMINISCENCES BY THE AUTHOR 128

Fig. 3. Example of TOC TITLE and NONE labels

Simple visual inspection is sufficient to ensure "CONTENTS OF VOL. II." should never be identified as a TOC entry, but rather as the title of the given TOC page. That is why we label this (and similar) group of words with a TOC TITLE label.

Another quick inspection is enough to rule out "CHAPTER" and "PAGE" words in the line preceding the first TOC entry; these two are surely not a TOC entry (neither of the two is targeting any title later in the book). To a viewer it is clear that the first word ("CHAPTER") stands for the horizontal position of chapter numbers, while the second word ("PAGE") stands for the horizontal position of page numbers for each TOC entry below. These two words are not of particular importance, especially compared to TOC entry labels, and that is why we introduce another label for negligible text, TOC NONE label. Because neither of the two words in question are actually linking to any target page, these words are not relevant for navigation nor for indexing purposes — that is how we validate our labeling decision. Furthermore, any other text that cannot possibly link to a target page is labeled the same way. A good example would be a page with severe errors in OCR, e.g. some text is recognized where no text exist.

Finally, there is one last type of less-important words that could be found on TOC pages. Observe the example in Fig. 4 (last three entries), where the author of the given title is specified in the same line with a chapter title, separator and page number. This example is taken from "Representative one-act plays" by Barrett Harper Clark, published by Brown Boston Little in 1921. It makes sense to distinguish a title from an author name; yet again, instead of an author name there could be any other text, and so far we observed several patterns, a year of publishing, a short summary of a chapter, rating, etc. All such text we label using TOC CHAPTER ADDITION label. We may decide to further fine sub-class the addition label in the future, and use several labels instead.

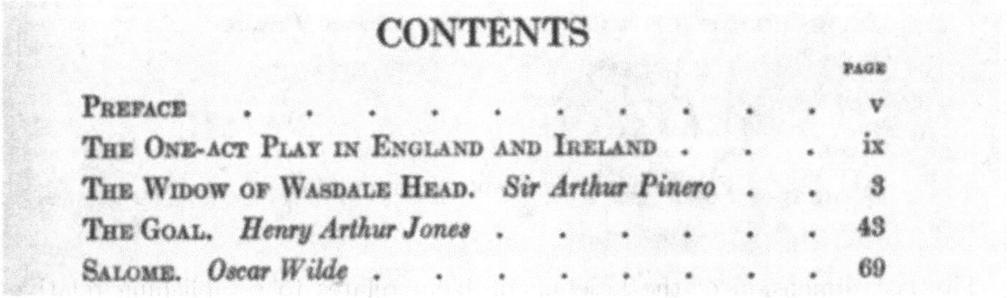
CONTENTS

PAGE

PREFACE v

THE ONE-ACT PLAY IN ENGLAND AND IRELAND ix

THE WIDOW OF WASDALE HEAD. *Sir Arthur Pinero* . . 3

THE GOAL. *Henry Arthur Jones* 43

SALOME. *Oscar Wilde* 69

Fig. 4. Example of TOC CHAPTER ADDITION labels

So far we have discussed labels for each word on a TOC page, i.e. each word is labeled using one of the proposed labels. As a short summary we shall list them all here once more:

- TOC NONE label, for negligible text
- TOC TITLE label
- TOC entry labels:
  - TOC CHAPTER NAME label
  - TOC CHAPTER TITLE label
  - TOC CHAPTER SEPARATOR label
  - TOC CHAPTER PAGE NUMBER label
  - TOC CHAPTER ADDITION label

At this point we shall introduce another dimension to the labeling problem. It is clear that labeling of individual words is not sufficient. For instance by looking at Fig. 4 one can observe that the first two entries are somewhat different than the others; the later entries are referencing one-act plays (as indicated by the book title), and the first two are giving preface and historical background for the plays in the book. On further inspection it can be observed that the first two entries have roman numerals for chapter page numbers, while the remaining TOC entries have arabic digits. This same pattern is observed to be common for most of the books we had access to. That is why we decided to introduce an additional label/attribute for each TOC entry, where an entry is either of introductory or of regular type (with introductory entries being ones with roman numerals).

The third dimension of labeling problem is related to defining a TOC entry. There are cases of TOC pages where entries are in more than one line, e.g. the second entry in Fig. 5 is in two lines (there are three entries in Fig. 5). In that and similar cases (where entries are in two or more lines) it is necessary to introduce yet another label for each word of each TOC entry, specifying whether any given word is a continuation of previous entry. If any given word is not a continuation of previous, then it is the beginning of a new TOC entry.

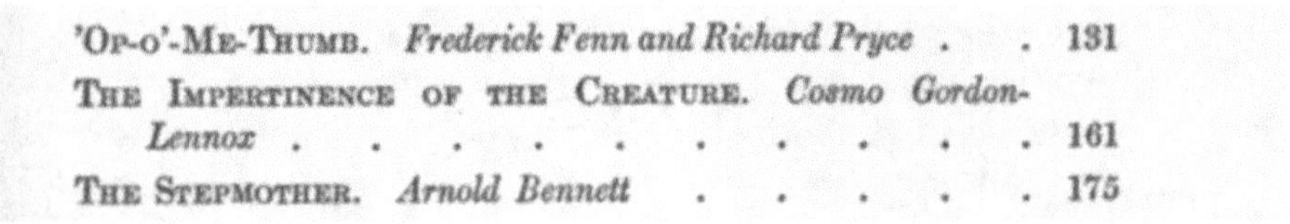

'OP-O'-ME-THUMB.   *Frederick Fenn and Richard Pryce* .   . 131

THE IMPERTINENCE OF THE CREATURE. *Cosmo Gordon-*
Lennox 161

THE STEPMOTHER. *Arnold Bennett*   175

Fig. 5. Example of entry continuation

The last dimension of the labeling problem relates to establishing relative significance of each TOC entry on the TOC page. Relative significance of the entry is represented in the form of a logical depth level. Observe the structure presented in Fig. 6 (excerpt taken from "Managing and Maintaining a Microsoft Windows Server 2003 Environment for an MCSE Certified on Windows 2000" by Orin Thomas, published by Microsoft Press, 2005), there are five TOC entries presented in three structural/hierarchical groups, with a different relative significance assigned to each.

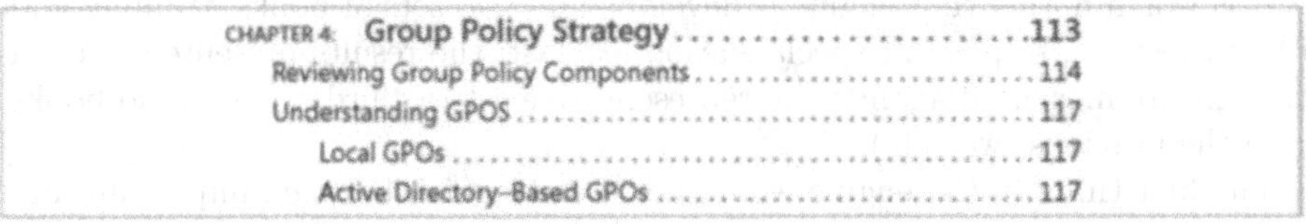
CHAPTER 4 **Group Policy Strategy** 113
Reviewing Group Policy Components 114
Understanding GPOS 117
Local GPOs 117
Active Directory–Based GPOs 117

Fig. 6. Example of depth level

The entry in the first line is at highest hierarchical level, which is in this case indicated with keyword "CHAPTER", blue font color, bold characters, and higher font size. Not to be forgotten, horizontal level of indentation is also a strong feature, especially for the following entries. The second and third TOC entries in Fig. 6 are one step below the level of the first one, resulting in level 2 in the hierarchy. At last, TOC entries four and five are at level 3 in the hierarchy.

We are now ready to summarize proposed labels for TOC page, four dimensions of the labeling problem are (categorized in hierarchical order):

- Hierarchical level of an entry
  - entry type (introductory vs. regular entries)
  - entry depth level
- Hierarchical level of a word
  - beginning vs. continuation of a word within an entry
  - word labels (name, title, separator, page number, addition, negligible text, toc page title)

3 Book Layout Engine

As previously stated, we consider raw OCR information as the input. This consists of words (strings), lines (list of words in each line) and bounding boxes of each word. Provided with this information we aim to detect and process TOC pages. There are a few prerequisites for the TOC Structure Extraction Engine, each of them listed below.

The first step in this process is to perform page classification and detect TOC pages. The next step would be detection of the page numbers, e.g. assigning each physical page of the book with a unique logical page number. At last, each TOC page is processed to detect the scope of the TOC section. These three steps are sufficient for the TOC Structure Extraction Engine to perform.

4 TOC Structure Extraction Engine

In the introduction we have specified the responsibilities of the TOC Structure Extraction Engine. Later on while discussing the labeling scheme we have specified the means of exposing this information. Here we shall discuss the engine itself.

It is worth noting that engine is developed on two sets of books, training and a blind test set, to prevent specialization. At last, the results presented later in the text are measured against the representative set (a third set, with no books from the previous two sets).

The first thing in the engine would be to distinguish between important and negligible portions of the TOC section. As specified by the labeling scheme, there are several possible cases of negligible text, the title of the TOC page, false positive OCR errors (usually words instead of the pictures), and random text. The engine is detecting each based on the pattern occurrences in the training set.

While it is feasible to separate entries among themselves without additional information, we have chosen to detect chapter names and chapter page numbers first, and only then proceed to the spatial entry detection. Again, this part of the engine is based on the pattern occurrences in the training set. As one possible implementation we can suggest Conditional Random Fields[1], which is a good framework for labeling sequential data.

Once we have entries we can proceed to the linking, where a fuzzy search technique is used to locate the entry title on the target page. It is worth noting that detection of page numbers (assigning each physical page of the book with a unique logical page number) comes handy here, because it provides a significant hint of where to look for the target title. Parameters of the fuzzy search are based on the pattern occurrences in the training set.

At last, relative significance of the entries is obtained after clustering, where each cluster represent a single level of significance. Parameters and features of the clustering are based on the training set.

5   Representative Set

In order to measure the accuracy of the TOC engine a fairly large test set needs to be selected due to the various layouts applied to books throughout history. Also, topics and publishers typically have specific templates for the book and TOC layout. Since storing large quantities of book data and performing tests on it is a costly process, a representative set has been created.

The representative set is a subset (200 books) of a much larger test set (180,000 books). It has better feature distribution (compared to the entire book corpus) than 93% of the random sets of the same size and gives a good picture of how the engine will perform in real-time situations.

6   Results

In this section we shall present some results on the representative set (blind set). Each of the four major parts of the TOC Structure Extraction are measured for precision and recall.

We shall start with the entry internal structure (word label) engine; chapter names are detected with 98.56% precision and 98.63% recall; chapter page numbers are detected with 99.97% precision and 99.07% recall; not surprising joint chapter titles and additions (as a single measure) are detected with 99.36% precision and 99.80% recall. The total number of human labeled chapter page number words is 9,512, while the total number of chapter name words is 7,206. The total number of chapter title and addition words is 52,571.

Spatial entry detection (we only consider entries which are entirely correct in a spatial sense — all words in an entry must be correct for the entry to be considered spatially correct) is with 92.91% precision and 95.45% recall. Total number of human labeled entries is 9535.

Linking detection is with 91.29% precision and 93.78% recall (98.25% conditional to entry being spatially correct).

Depth level 1 is detected with 80.51% precision and 84.58% recall (90.91% conditional to entry being spatially correct).

Reference

1. Ye, M., Viola, P.: Learning to Parse Hierarchical Lists and Outlines Using Conditional Random Fields. In: Proceedings of the Ninth international Workshop on Frontiers in Handwriting Recognition, pp. 154–159. IEEE Computer Society, Washington (2004)

The Impact of Query Length and Document Length on Book Search Effectiveness

Mingfang Wu, Falk Scholer, and James A. Thom

RMIT University, Melbourne, Australia
{mingfang.wu,falk.scholer,james.thom}@rmit.edu.au

Abstract. This paper describes the RMIT group's participation in the book retrieval task of the INEX booktrack in 2008. Our results suggest that for book retrieval task, using a page-based index and ranking books based on the number of pages retrieved may be more effective than directly indexing and ranking whole books.

1 Introduction

This paper describes the participation of the RMIT group in the Initiative for the Evaluation of XML retrieval (INEX) book search track in 2008, specifically the book retrieval task. Book search is an important track at INEX – books are generally much larger documents than the scientific articles or the wikipedia pages that have been used in the main ad hoc track at INEX. The book corpus as provided by Microsoft Live Book Search and the Internet Archive contains 50 239 digitized out-of-copyright books. The contents of books are marked up in an XML format called BookML [1]. The size of this XML marked up corpus is about 420 Gb. With book retrieval, structure is likely to play a much more important role than in retrieval from collections of shorter documents.

This is the first year of RMIT's participation in the book search track at INEX, and we explore the effectiveness of book retrieval by experimenting with different parameters, namely: the length of queries and the length of documents being indexed and retrieved. We begin by describing our approach in the next section, which is followed by our results, and then the conclusion.

2 Our Approach to Book Retrieval Task

The book retrieval task investigates on a typical scenario in which a user searches for relevant books on a given topic with the intent to build a reading or reference list. With this task, we attempted to explore the following research questions.

1. What is a suitable length for a query? Does adding more query terms improve retrieval effectiveness?
2. What is the most suitable index and search granularity? Should we treat the whole book or a just a section or page of a book as a searchable document unit?

S. Geva, J. Kamps, and A. Trotman (Eds.): INEX 2008, LNCS 5631, pp. 172–178, 2009.
© Springer-Verlag Berlin Heidelberg 2009

To answer these research questions, we experimented with various query construction strategies, two different units of searchable document, as well as the combinations of query length and searchable document unit. We undertook 12 runs in total, of which 10 official runs were submitted into the pool for evaluation.

2.1   Query Construction

The test collection has a total of 70 test topics that were contributed by participating groups in 2007 and 2008. Participating groups also contributed to the relevance assessment of these topics. At the time of writing, there are 25 (out of 70) topics assessed (or partially assessed), with at least one relevant book being identified. Thus our evaluation is based only on the completed assessments for these 25 topics (this is the version of the book track assessments known as v250209 from 25th February 2009).

```
<inex_topic track="book" task="book-retrieval/book-ad-hoc"
                    topic_id="41"  ct_no="2008-13">
<title>
     Major religions of the world
</title>
<description>
     I am interested to learn about the origin of the world's major
     religions, as well as their commonalities and differences.
</description>
<narrative>
  <task>
     Having met people from different cultural and religious
     background, I am keen to learn more about their religions
     in order to understand them better.
  </task>
  <infneed>
     A concise book that has direct comparison of the world's
     popular religions would be an ideal pick. If this book can
     not be found, multiple books about the origin and believes
     of each religion should also be acceptable.
  </infneed>
</narrative>
</inex_topic>
```

Fig. 1. A sample topic

Topics from the book track collection contain three components: title, description and narrative. The narrative component is further decomposed into two sub-components: task and information need. An example of such a topic is shown in Figure 1. As we can see, the title, description and information need components provide specificity of a search context in an increasing order: the title mostly represents a query that a user would type into a search box, the description provides more details of what information is required, while the task and the information need depict the context of the potential use to be made of the searched books, and the criteria on which books should be counted as being relevant or irrelevant.

Generally, long queries describe information needs more specifically than those short ones [3]. It has also been reported that long and short queries perform differently [2]. We set out to investigate, given a book collection where the length of a book is much longer than other types of documents such as a newswire article or a web page from TREC test collection, whether long queries would perform better than short queries. So we explored the following four approaches to constructing queries from different components of a topic.

Title: Use all words from the topic title element.

Title+Infneed: Use all words from the topic title and the topic information need element.

TitleBoolean: Boolean operator "AND" is inserted between query words as in **Title**.

Title+InfneedManual: Use all words as in **Title**, and add some manually selected words from the information need element.

Consider the sample topic shown in Figure 1. Applying our four different approaches results in the following queries (after stopping and stemming):

Title: major religion world

TitleBoolean: major AND religion AND world

Title+Infneed: major religion world concis book direct comparison world popular religion ideal pick book found multipl book origin believ religion accept

Title+InfneedManual: major religion world direct comparison world popular religion

The average query length for the set of 25 assessed topics is 2.6 terms for the set of **Title** (or **TitleBoolean**) queries, 25.3 terms for the set of **Title+Infneed** queries, and 13.4 terms for the set of **Title+InfneedManual** queries.

2.2   Index Construction and Runs

We used the Zettair search engine[1] for indexing and searching in all of our submitted runs. After preprocessing into documents but before indexing, we removed off all XML tags, leaving with a corpus of about 30GB for indexing. It took about 2 hours elapsed time to create an index on a shared 4 CPU (2.80GHz) Intel Pentium running Linux. For retrieval, we applied the BM25 similarity function [4] for document weighting and ranking (with $k1 = 1.2$ and $b = 0.75$). During indexing and searching, words from a stoplist were removed and the Porter stemmer was applied. We created separate indexes based on book-level evidence and page-level evidence.

---

[1] http://www.seg.rmit.edu.au/zettair/

Ranking Based on Book-level Evidence

The book-level index treated each book as a document. We sent the four sets of queries (**Title, TitleBoolean, Title+Infneed**, and **Title+InfneedManual**) to the search engine using this index, generating four book-level runs.

RmitBookTitle
RmitBookTitleBoolean
RmitBookTitleInfneed
RmitBookTitleInfneedManual

Ranking Based on Page-level Evidence

On average, a book from the collection contains 36 455 words. Sometimes, a topic is only mentioned in passing in a book, and may not be the main theme of the book. In order to promote those books dedicated primarily to the topic, we require a topic be mentioned in most parts of a book. We therefore experimented to break a book down into pages by using the "page" tag and constructed a corpus in which each page was treated as a document.

There are a total of 16 975 283 pages (documents) in this page corpus. Both book and page collections have 1 831 505 097 indexable terms of which 23 804 740 are distinct terms. The average document length is 589 370.9 bytes per book and 1 793.6 per page, and the average number of terms in each document is 102 041 per book and 302 per page.

We used the following two methods to estimate a book's relevance to the topic.

1. In the first page-level evidence ranking method, we first retrieve the top 3 000 pages and then rank books according to percentage of pages retrieved per book.
2. The second page-level evidence ranking method is similar to the first one but ranks books based on the maximum number of continuous pages retrieved from the book as a percentage of the total number of pages in the book.

Combing these two page-level evidence ranking methods and our four query types, we had another eight runs as follows:

RmitPageMergeTitle: Query terms are the same as in **RmitBookTitle**, books are ranked according to the method 1;

RmitConPageMergeTitle: Query terms are the same as in **RmitBookTitle**, books are ranked according to the method 2;

RmitPageMergeTitleBoolean: Queries are the same as in **RmitBookTitleBoolean**, books are ranked according to the method 1;

RmiConPageMergeTitleBoolean: Queries are the same as in **RmitBookTitleBoolean**, books are ranked according to the method 2;

RmitPageMergeTitleInfneed: Query terms are the same as in **RmitBookTitleInfneed**, books are ranked according to the method 1;

RmitConPageMergeTitleInfneed: Query terms are the same as in **Rmit-BookTitleInfneed**, books are ranked according to the method 2.
RmitPageMergeTitleManual: Query terms are the same as in **RmitBook-TitleInfneedManual**, books are ranked according to the method 1;
RmitConPageMergeTitleManual: Query terms are the same as in **Rmit-BookTitleInfneedManual**, books are ranked according to the method 2.

3   Results

Table 1 shows performance of twelve runs as measured in precision at 5 (P@5), 10 (P@10) and 20 (P@20) books retrieved, average interpolated precision averages at 0.00 and 0.10 recall, and MAP (mean average precision). In what follows, we make observations about both query type and length, and document type and length, based on the measure P@5.

Table 1. Experimental result for the book search task (the additional runs in italics were not included in the pool of submitted runs)

					incl_prn	incl_prn
Run ID	P@5	P@10	P@20	MAP	0.00	0.10
RmitBookTitle	0.128	0.104	0.094	0.075	0.247	0.220
RmitBookTitleBoolean	0.128	0.104	0.049	0.075	0.247	0.220
RmitBookTitleInfneed	0.136	0.100	0.086	0.067	0.331	0.200
RmitBookTitleInfneedManual	0.112	0.108	0.088	0.068	0.276	0.187
RmitPageMergeTitle	0.144	0.116	0.084	0.074	0.302	0.260
RmitPageMergeTitleBoolean	0.144	0.116	0.084	0.074	0.302	0.260
RmitPageMergeTitleInfneed	0.168	0.108	0.090	0.079	0.358	0.291
RmitPageMergeTitleInfneedManual	0.216	0.132	0.098	0.106	0.367	0.346
RmitConPageMergeTitle	0.104	0.072	0.064	0.050	0.241	0.202
RmitConPageMergeTitleBoolean	0.104	0.072	0.064	0.050	0.241	0.202
RmitConPageMergeTitleInfneed	0.104	0.072	0.046	0.039	0.224	0.130
RmitConPageMergeTitleInfneedManual	0.128	0.084	0.058	0.054	0.279	0.213

Query Type and Length

We observe the following trends regarding queries.

- The three runs with Boolean queries (RmitBookTitleBoolean, RmitPage-MergeTitleBoolean and RmitConPageMergeTitleBoolean) have almost the same performance as their corresponding runs without Boolean operators (RmitBookTitle, RmitPageMergeTitle and RmitConPageMergeTitle). This might be because the topic title is typically short (average of 2.6 terms), indeed 7 out of the 25 topics have only one query term.
- When a whole book is treated as a document, including the information need in the queries (RmitBookTitleInfneed) has a small improvement of 6.3% over just using the topic title as the queries (RmitBookTitle). However, manually adding terms from the information need (RmitBookTitleInfneedManual) is worse than the plain title queries.

- When a page of a book is treated as a document, the two runs with manually added terms from the information need (RmitPageMergeTitleInfneed-Manual and RmitConPageMergetitleInfneedManual) performed better than their corresponding runs with the title only queries (RmitPageMergeTitle and RmitConPageMergeTitle) (by 50% and 23.0% for ranking method 1 and 2 respectively), and the corresponding runs with queries including the information need as well as the title (RmitPageMergeTitleInfneed and Rmit-ConPageMergeTitle) (by 28.6% and 23.0% for ranking method 1 and 2 respectively).
- The average length of the queries that added all the terms from the information need to the title (Title+Infneed) is almost double the length of the queries where the added terms were manually selected from the information need (Title+InfneedManual). This might be an explanation for why the runs with the Title+Infneed queries worked better for documents of book length, while the Title+InfneedManual queries worked better for documents of page length.

Document Type and Length

We observe the following trends regarding document ranking.

- The first page-level evidence ranking method (run IDs starting with Rmit-Page) has the best performance regardless of query type. The four runs from this method improved over counterpart book runs by 12.5%, 12.5%. 23.5% and 92.8% respectively. In particular, the run where terms from the information need were manually added to title (RmitPageMergeTitleInfneed-Manual), improved the performance over the base run (RmitBookTitle) by 68.8%. Incidentally, this run also performed the best amongst all submitted runs across all participating groups in terms of the measures MAP, P@5, P@20 and incl_prn0.10.
- The second page ranking method (run IDs starting with RmitConPage) gives worse performance for almost every query type, except the run RmitCon-PageMergeTitleInfneedMnaual, which is better than its corresponding book run RmitBookTitleInfneedManual by 14.3%.

4   Concluding Remarks

This paper has reported the results of the RMIT group's participation in the INEX book search track in 2008. We explored the impact of variation of query length and document size on the effectiveness of the book retrieval task. Based on the current relevance assessments, the evaluation shows that treating a page of a book as a searchable unit and ranking books based on the percentage of pages retrieved performs better than indexing and retrieving whole book as a search unit. The search performance can be further improved by adding additional query words that describe information need.

This work provides a baseline for further experiments in structured information retrieval, in particular developing new approaches to book retrieval and in exploring other tasks such page-in-context.

References

1. Kazai, G., Doucet, A., Landoni, M.: Overview of the INEX 2008 book track. In: INEX 2008 Workshop Proceedings (2008)
2. Kumaran, G., Allan, J.: A case for shorter queries, and helping users create them. In: Proceedings of Human Language Technologies: The Annual Conference of the North American Chapter of the Association for Computational Linguistics, pp. 220–227 (2006)
3. Phan, N., Bailey, P., Wilkinson, R.: Understand the relationship of information need specificity to search query length. In: Proceedings of SIGIR 2007, pp. 709–710 (2007)
4. Robertson, S., Walker, S., Hancock-Beaulieu, M.M., Gatford, M.: Okapi at TREC-3. In: Proceedings of TREC-3, pp. 109–126 (1994), trec.nist.gov/pubs/

Overview of the INEX 2008 Efficiency Track

Martin Theobald[1] and Ralf Schenkel[1,2]

[1] Max Planck Institute for Informatics, Saarbrücken, Germany
[2] Saarland University, Saarbrücken, Germany

Abstract. This paper presents an overview of the Efficiency Track that was newly introduced to INEX in 2008. The new INEX Efficiency Track is intended to provide a common forum for the evaluation of both the effectiveness and efficiency of XML ranked retrieval approaches on *real data* and *real queries*. As opposed to the purely synthetic XMark or XBench benchmark settings that are still prevalent in efficiency-oriented XML retrieval tasks, the Efficiency Track continues the INEX tradition using a rich pool of manually assessed relevance judgments for measuring retrieval effectiveness. Thus, one of the main goals is to attract more groups from the DB community to INEX, being able to study effectiveness/efficiency trade-offs in XML ranked retrieval for a broad audience from both the DB and IR communities. The Efficiency Track significantly extends the Ad-Hoc Track by systematically investigating different types of queries and retrieval scenarios, such as classic ad-hoc search, high-dimensional query expansion settings, and queries with a deeply nested structure (with all topics being available in both the NEXI-style CO and CAS formulations, as well as in their XPath 2.0 Full-Text counterparts).

1 General Setting

1.1 Test Collection

Just like most INEX tracks, the Efficiency Track uses the 2007 version of the INEX-Wikipedia collection [2] (without images), an XML version of English Wikipedia articles initially introduced for INEX 2006 and slightly revised in 2007. The collection is available for download from the INEX website `http://www.inex.otago.ac.nz/` for registered participants, or directly from `http://www-connex.lip6.fr/~denoyer/wikipediaXML/`. Although this 4.38 GB XML-ified Wikipedia collection is not particularly large from a DB point-of-view, it has a rather irregular structure with many deeply nested paths, which will be particularly challenging for traditional DB-style approaches, e.g., using path summaries. There is no DTD available for INEX-Wikipedia.

1.2 Topic Types

One of the main goals to distinguish the Efficiency Track from traditional Ad-Hoc retrieval is to cover a broader range of query types than the typical NEXI-style CO or CAS queries, which are mostly using either none or only very little structural information and only a few keywords over the target element of the query. Thus, two natural extensions are to extend Ad-Hoc queries with high-dimensional query expansions

S. Geva, J. Kamps, and A. Trotman (Eds.): INEX 2008, LNCS 5631, pp. 179–191, 2009.
© Springer-Verlag Berlin Heidelberg 2009

and/or to increase the amount of structural query conditions without sacrificing the IR aspects in processing these queries (with topic `description` and `narrative` fields providing hints for the human assessors or allowing for more semi-automatic query expansion settings, see Figure 1). The Efficiency Track focuses on the following types of queries (also coined "topics" in good IR tradition), each representing different retrieval challenges:

- **Type (A) Topics:** 540 topics (no. 289–828) are taken over from previous Ad-hoc Track settings used in 2006–2008, which constitute the major bulk of topics used also for the Efficiency Track. These topics represent classic, Ad-Hoc-style, focused passage or element retrieval (similar to the INEX Ad-Hoc Focused subtask 2006–2008, see for example [3]), with a combination of NEXI CO and CAS queries. Topic ids are taken over from the Ad-Hoc track as well, thus allowing us to reuse assessments from the Ad-Hoc Track for free.
- **Type (B) Topics:** 21 topics (no. 829–849) are derived from interactive, feedback-based query expansion runs, kindly provided by the Royal School Of Library And Information Science, Denmark, investigated in the context of the INEX Interactive Track 2006 [5,6]. These CO topics are intended to simulate high-dimensional query expansion settings with up to 112 keywords (topic no. 844), which cannot be evaluated in a conjunctive manner and are expected to pose a major challenge to any kind of search engine. Respective expansion runs have been submitted by RSLIS also to the 2006 Ad-Hoc track, such that relevant results are expected to have made it into the relevance pools of INEX 2006 Ad-Hoc track assessments as well. An additional `adhocid` attribute marks the original Ad-Hoc id of the topic that it has been derived from, such that—at least incomplete—assessments are available for these type (B) topics.
- **Type (C) Topics:** 7 new topics (no. 850–856) have been developed and submitted by Efficiency Track participants in 2008. These topics represent high-dimensional, structure-oriented retrieval settings over a DB-style set of CAS queries, with deeply nested structure but only a few keyword conditions. Assessments were originally intended to get accomplished by Efficiency Track participants as well, but were then skipped due to the low amount of newly proposed type (C) topics and the low respective impact on overall result effectiveness as compared to the more than 500 Ad-Hoc topics that already come readily assessed. The evaluation of runtimes however remains very interesting over this structure-enhanced set of type (C) topics.

1.3 Topic Format and Assessments

Just like the original NEXI queries, type (A) queries have some full-text predicates such as phrases (marked by quotes ""), mandatory keywords (+), and keyword negations (-). Although participants were encouraged to use these full-text hints, they are not mandatory just like in other INEX tracks. Because of their high-dimensional nature, most type (B) and (C) queries require IR-style, non-conjunctive (aka. "andish") query evaluations that can either preselect the most significant query conditions or dynamically relax both the structure- and content-related conditions at query processing time. The reuse of type (A) and (B) lead to 308 topics for which assessments from the INEX 2006–2008

Ad-hoc Tracks are available. An additional conversion to the new 2008 version of the INEX-Eval tool and the (passage-based) assessments format was needed to incorporate the 2008 assessment files (QRels) and has meanwhile been made available online for download from the track homepage[1].

```
<topic id="856" type="C">
<co_title>
  State Parks Geology Geography +Canyon
</co_title>
<cas_title>\
  //article//body[about(.//section//p, State Park) and
                  about(.//section//title, Geology) and
                  about(.//section//title, Geography)]
                //figure[about(.//caption, +Canyon)]
</cas_title>
<xpath_title>
  //article//body[.//section//p ftcontains "State Park" and
                  .//section//title ftcontains "Geology" and
                  .//section//title ftcontains "Geography"]
                //figure[.//caption ftcontains "Canyon"]
</xpath_title>
  <description>
    I'm looking for state parks with sections describing
    their geology and/or geography, preferably with a figure of
    a canyon as target element.
  </description>
  <narrative>
    State park pages often follow the common pattern of having
    sections entitled with "Geology" or "Geography". I'm
    particularly interested in those pages with a figure of a
    canyon, e.g., the Grand Canyon.
  </narrative>
</topic>
```

Fig. 1. Example type (C) topic (no. 856)

All topic titles are provided in the NEXI syntax (in both their CO and CAS formulations) and (new for the Efficiency Track) in their corresponding XPath 2.0 Full-Text specification. XPath 2.0 queries were automatically generated from the respective NEXI CAS titles, while the CAS title itself was taken over from the CO title and wrapped into a pseudo target element of the form //*[about(...)] whenever there was no actual CAS title available.

A "Call for Structure-Enhanced Queries" for the new type (C) queries was issued to all registered INEX participants in early May 2008. The final set of 568 Efficiency Track topics was released in early July 2008, with the intention to keep a relatively tight time window between the release of the topics and the run submission deadline to prevent people from overtuning to particular topics.

1.4 Sub-tasks

The Efficiency Track particularly encourages the use of top-k style query engines. The result submission format includes options for marking runs as top-15, top-150, and top-1500 (the latter corresponding to the traditional Ad-hoc submission format), using

---

[1] http://www.inex.otago.ac.nz/tracks/efficiency/efficiency.asp

either a Focused (i.e., non-overlapping), Thorough (incl. overlap), or Article retrieval mode (see below). Automatic runs may use either the title field, including the NEXI CO, CAS, or XPATH titles, additional keywords from the narrative or description fields, as well as automatic query expansions if desired. At least one automatic and sequential run with topics being processed one-by-one is mandatory for each participating group. Participants are invited to submit as many runs in different retrieval modes as possible.

- **Article:** This is the mode that corresponds to a classical search engine setting to the largest extent. All documents may be considered either in their plain text or XML version. Moreover, queries can be used in both their CO and CAS (incl. XPath 2.0 Full-Text) formulation. In the Article-Only/CO combination this setting resembles a classical IR setting with entire documents as retrieval units and plain keywords as queries. Article-only runs are always free of overlapping results.
- **Thorough:** The Thorough mode represents the original element-level retrieval mode used in INEX 2003-2005. Here, any element correctly identified as relevant to the query will contribute to the recall of the run. This setting intentionally allows overlapping elements to be returned, since removing overlap may mean a substantial burden for different systems. We thus re-conceal the Thorough setting used in previous INEX years with respect to efficiency aspects, such that actual query processing runtimes can be clearly distinguished from the runtime needed to remove overlapping results (which typically is a costly post-processing step).
- **Focused:** Focused (i.e., overlap-free) element- and/or passage-level retrieval typically is favorable from a user point-of-view and therefore replaced the Thorough retrieval as primary retrieval mode in the Ad-hoc Track in 2006. Here, the reported runtimes should include the time needed to remove overlap, which may give rise to interesting comparisons between systems following both Thorough and Focused retrieval strategies.

2   Run Submissions

The submission format for all Efficiency Track retrieval modes is defined by the following DTD, depicted in Figure 2. The following paragraph provides a brief explanation of the DTD fields:

- Each *run* submission *must* contain the following information:
  - participant-id - the INEX participant id
  - run-id - your run id
  - task - either focused, thorough, or article
  - query - either automatic or manual mode (at least one automatic mode using exactly one of the title fields is required; in manual mode any form of manual query expansion is allowed)
  - sequential - queries being processed sequentially or in parallel (independent of whether distribution is used)

```
<!ELEMENT efficiency-submission (topic-fields,
  general_description,
   ranking_description,
   indexing_description,
   caching_description,
   topic+) >
<!ATTLIST efficiency-submission
 participant-id CDATA #REQUIRED
 run-id         CDATA #REQUIRED
 task           (article|thorough|focused) #REQUIRED
 query          (automatic|manual) #REQUIRED
 sequential     (yes|no) #REQUIRED
 no_cpu         CDATA #IMPLIED
 ram            CDATA #IMPLIED
 no_nodes       CDATA #IMPLIED
 hardware_cost  CDATA #IMPLIED
 hardware_year  CDATA #IMPLIED
    topk          (15|150|1500) #IMPLIED >
<!ELEMENT topic-fields EMPTY>
<!ATTLIST topic-fields
 co_title       (yes|no) #REQUIRED
 cas_title      (yes|no) #REQUIRED
 xpath_title    (yes|no) #REQUIRED
 text_predicates(yes|no) #REQUIRED
 description    (yes|no) #REQUIRED
 narrative      (yes|no) #REQUIRED >
<!ELEMENT general_description   (#PCDATA)>
<!ELEMENT ranking_description   (#PCDATA)>
<!ELEMENT indexing_description  (#PCDATA)>
<!ELEMENT caching_description   (#PCDATA)>
<!ELEMENT topic (result*)>
<!ATTLIST topic
 topic-id       CDATA #REQUIRED
 total_time_ms  CDATA #REQUIRED
 cpu_time_ms    CDATA #IMPLIED
 io_time_ms     CDATA #IMPLIED >
<!ELEMENT result (file, path, rank, rsv?) >
<!ELEMENT file   (#PCDATA)>
<!ELEMENT path   (#PCDATA)>
<!ELEMENT rank   (#PCDATA)>
<!ELEMENT rsv    (#PCDATA)>
```

Fig. 2. DTD for Efficiency Track run submissions

– Furthermore, each *run* submission *should* contain some basic system and retrieval statistics:

- no_cpu - the number of CPUs (cores) in the system (sum over all nodes for a distributed system)
- ram - the amount of RAM in the system in GB (sum over all nodes for a distributed system)
- no_nodes - the number of nodes in a cluster (only for a distributed system)
- hardware_cost - estimated hardware cost (optional)
- hardware_year - date of purchase of the hardware (optional)
- topk - top-k run or not (if it is a top-k run, there may be at most k elements per topic returned)

– Each *run* submission *should* also contain the following brief system descriptions (keywords), if available:

- general_description - a general system and run description

- `ranking_description` - the ranking strategies used
- `indexing_description` - the indexing structures used
- `caching_description` - the caching hierarchies used

- Each *topic* element in a run submission *must* contain the following elements:

  - `topic_id` - the id of the topic
  - `total_time_ms` - the total processing time in milliseconds: this should include the time for parsing and processing the query but does not have to consider the extraction of resulting file names or element paths (needed to create the above format for the run submission)

- Furthermore, each *topic* element of a run submission *should* contain the following elements:

  - `cpu_time_ms` - the CPU time spent on processing the query in milliseconds
  - `io_time_ms` - the total I/O time spent on physical disk accesses in milliseconds

Providing CPU and I/O times is optional for each topic. Also, it is sufficient to provide a list of matching elements along with their `path` locators (as canonical XPath expressions—see, again, the Ad-hoc Track settings [3]). Providing relevance score values `rsv` was also optional. In `article` retrieval mode, all results' element paths had to be `/article[1]`. Moreover, the many different types of (optional) description fields are supposed to encourage participants to provide detailed descriptions along with their run submissions.

Particularly interesting for the Efficiency Track submissions is the `runtime` field, of course. This can optionally be split into `cpu_time` and `io_time`, the latter two of which had not been used by any of the participants, though. We therefore focus on actual wallclock running times as efficiency measure for our 2008 setting. Top-k runs with less than 1,500 ranks have only been submitted by University of Frankfurt and Max-Planck-Institut Informatik. Distribution has only been used by the University of Frankfurt, using a cluster with 8 nodes.

3   Metrics

To assess the quality of the retrieved results, the Efficiency Track applies the same metrics as used in the Ad-Hoc track. Runs in `Focused` or `Article` mode were evaluated with the interpolated precision metric [4], using the evaluation toolkit from INEX 2008; the assessments for the topics from 2006 and 2007 have been converted to the new Qrel-based format. Runs in `Thorough` mode were evaluated with the precision-recall metric as implemented in `inex_eval` [1] after converting the Qrels from 2008 to the old XML-based assessment format. In the future, the track will probably use metrics implemented in EvalJ[2].

---

[2] http://evalj.sourceforge.net/

4 Participants

An overall amount of 20 runs were submitted by 5 participating groups. The following paragraphs provide short system descriptions submitted by the participants.

Max-Planck-Institut Informatik [Part.ID 10]. For the INEX Efficiency Track 2008, we were just on time to finish and (for the first time) evaluate our brand-new TopX 2.0 prototype. Complementing our long-running effort on efficient top-k query processing on top of a relational back-end, we now switched to a compressed object-oriented storage for text-centric XML data with direct access to customized inverted files, along with a complete reimplementation of the engine in C++. Core of the new engine is a multiple-nested block-index structure that seamlessly integrates top-k-style sorted access to large blocks stored as inverted files on disk with in-memory merge-joins for efficient score aggregations.

University of Frankfurt [Part.ID 16]. University of Frankfurt has developed Spirix, a Peer-to-Peer (P2P) search engine for Information Retrieval of XML-documents. The underlying P2P protocol is based on a Distributed Hash Table (DHT). Due to the distributed architecture of the system, efficiency aspects have to be considered in order to minimize bandwidth consumption and communication overhead. Spirix is a top-k search engine aiming at efficient selection of posting lists and postings by considering structural information, e.g. taking advantage of CAS queries. As collections in P2P systems are usually quite heterogeneous, no underlying schema is assumed but schema-mapping methods are of interest to detect structural similarity. The runs submitted to the INEX efficiency track compare different structure similarity functions which are then used to improve efficiency of routing and ranking.

University of Toronto [Part.ID 42]. Task: Thorough. The query is sequential and automatic. System description: one virtual CPU running over an AMD Opteron Processor 250 at 2.39 GHz. Note the virtual machine runs Red Hat 4 which runs GSX (of which the virtual machines have Windows XP Professional 2002). There is 2,048 MB of RAM topk = 1500. Only the CAS title is used. Text predicates are not used. Description of the run: structural relevance and/or relaxed heuristics with boosting of terms based on summary plus multiple index selection. Ranking description: Lucene's default scoring is boosted with values obtained from the MCH probabilities for the summary and boost parameters in use. Post-processing for the removal of overlap relies on the summary and boost parameters in use as well. Indexing description: top-level article, section, and paragraph element-level indexes are built then combined depending on the tags in the CAS title of the topic. Caching description: Lucene's internal caching which is reset between topics. The summary and MCH parameters are reloaded before each topic.

University of Twente & CWI [Part.ID 53]. The efficiency submission from CWI and University of Twente was testing the performance of the current PF/Tijah XML retrieval system. PF/Tijah does not use top-k query processing but profits from highly tuned containment join and scoring algorithms that are implemented as physical operators in the MonetDB database back-end and have direct low-level access to the data. With the 4 submissions we also evaluated two different query semantics: the first interpreting an about predicate as a boolean filter, the other regarding it only as scoring operator. While the first interpretation shows a higher performance dealing with smaller intermediary results, we expected the second to yield a higher retrieval quality.

JustSystems Corporation [Part.ID 56]. JustSystems Corporation has developed an IR-based XML search engine. The system has an inverted-file index and a relative inverted-path (RIP) list. For CO queries, the system computes RSV scores with TF-IDF-based node weights in a bottom-up manner. For CAS queries, the system computes RSV scores with TF-IDF-based node weights in a bottom-up manner, which are also constrained by manually-rewritten XPath expressions taken from the CAS titles. A distinctive feature of the system is that RSV scores are computed over fragmented nodes from bottom-up at the time of retrieving, thus the system makes its index size comparatively small. Additionally, the new system is built in Java 1.5, which is also able to eliminate undesignated XPaths at the time of retrieving.

5   Results

Table 1 summarizes all run parameters as they were delivered in the runs' headers. Table 2 summarizes all effectiveness (iP, MAiP) and efficiency results (avg.&sum of wallclock runtimes in milliseconds) for the respective number of topics processed (#topics). Tables 3–5 summarize the results by topic type for all Focused runs (with effectiveness results only being available for type (A) and (B) topics). Figure 3 depicts detailed interpolated precision plots for all Focused and (the only) Article-only run(s); while Figure 4 depicts classic precision-recall plots for the Thorough runs. Figures 5–6 finally depict the respective interpolated precision plots split by type (A) and (B) topics (type (C) plots are skipped due to the lack of assessments).

We received an overall amount of 21 runs submitted by 5 different groups. According to the run descriptions submitted by the participants, systems varied from classic IR engines with XML-specific ranking capabilities to highly specialized XQuery engines with full-text extensions. As for efficiency, average running times per topic varied from 91 ms to 17.19 seconds over the entire batch of 568 topics, from 19 ms to 4.72 seconds over the 540 type (A) topics, from 845 ms to 14.58 seconds over the 21 type (B) topics, and from 41 ms to 18.19 seconds over the 7 type (C) topics, respectively. Similarly to the Ad-Hoc Track results, article-only runs generally yielded very good efficiency results, as they clearly constitute an easier retrieval mode, however also at a comparable effectiveness level. Overall effectiveness results were generally comparable to the Ad-hoc Track (albeit using different topics), with the best runs achieving a MAiP value of 0.19 and interpolated (early) precision values of 0.67 at 1% recall (iP[0.01]) and 0.49 at 10% recall (iP[0.10]), respectively. Up to now, none of the systems made use of the XPath-FT-based topic format, which leads to the conclusion that so far only systems traditionally used in INEX were also used for the Efficiency Track.

Table 1. Run parameters as taken from the submission headers

Part.ID	Run ID	Task	#CPU	RAM	#Nodes	Hardw.Cost	Year	Top-k	Cache	Seq.	Aut.	Title Fields
10	TOPX2-Eff08-CAS-15-Focused-W	Foc.	4	16	1	8,000 Eur	2005	15	OS+TopX	Yes	Yes	CAS
10	TOPX2-Eff08-CAS-15-Thorough-W	Tho.	4	16	1	8,000 Eur	2005	15	OS+TopX	Yes	Yes	CAS
10	TOPX2-Eff08-CAS-150-Focused-	Foc.	4	16	1	8,000 Eur	2005	150	OS+TopX	Yes	Yes	CAS
10	TOPX2-Eff08-CAS-1500-Focused-W	Foc.	4	16	1	8,000 Eur	2005	1500	OS+TopX	Yes	Yes	CAS
10	TOPX2-Eff08-CO-15-Focused-W	Foc.	4	16	1	8,000 Eur	2005	15	OS+TopX	Yes	Yes	CO
10	TOPX2-Eff08-CO-15-Thorough-W	Tho.	4	16	1	8,000 Eur	2005	15	OS+TopX	Yes	Yes	CO
10	TOPX2-Eff08-CO-150-Focused-W	Foc.	4	16	1	8,000 Eur	2005	150	OS+TopX	Yes	Yes	CO
10	TOPX2-Eff08-CO-1500-Focused-W	Foc.	4	16	1	8,000 Eur	2005	1500	OS+TopX	Yes	Yes	CO
16	SPIRIX-ARCH	Foc.	8	8	8	n/a	2008	150	n/a	Yes	Yes	CAS
16	SPIRIX-CSRU	Foc.	8	8	8	n/a	2008	1500	n/a	Yes	Yes	CAS
16	SPIRIX-FINE	Foc.	8	8	8	n/a	2008	150	n/a	Yes	yes	CAS
16	SPIRIX-NOSIM	Foc.	8	8	8	n/a	2008	150	n/a	Yes	Yes	CAS
16	SPIRIX-PATHSIM	Foc.	8	8	8	n/a	2008	150	n/a	Yes	Yes	CAS
16	SPIRIX-STRI	Foc.	8	8	8	n/a	2008	150	n/a	Yes	Yes	CAS
42	B2U0_full-depth-heur	Foc.	1	2	1	n/a	2008	1500	Lucene	Yes	Yes	CAS
42	B2U0_full-depth-sr	Tho.	1	2	1	n/a	n/a	1500	Lucene	Yes	Yes	CAS
53	pftijah_article_strict	Art.	1	8	1	1,000 Eur	2008	1500	DBMS	Yes	Yes	CO
53	pftijah_asp_strict	Tho.	1	8	1	1,000 Eur	2008	1500	DBMS	Yes	Yes	CAS
53	pftijah_asp_vague	Tho.	1	8	1	1,000 Eur	2008	1500	DBMS	Yes	Yes	CAS
53	pftijah_star_strict	Tho.	1	8	1	1,000 Eur	2008	1500	DBMS	Yes	Yes	CAS
56	VSM_RIP	Foc.	1	2	1	1,500 USD	2004	1500	None	Yes	Yes	CO+CAS

Table 2. Effectiveness/efficiency summary of all runs

Part.ID	Run ID	iP[0.00]	iP[0.01]	iP[0.05]	iP[0.10]	MAiP	AVG MS.	SUM MS.	#Topics
Focused									
10	TOPX2-Eff08-CAS-15-Focused-W	0.4587	0.3878	0.2592	0.1918	0.0662	90.99	51,499	566
10	TOPX2-Eff08-CAS-150-Focused-W	0.4747	0.4282	0.3494	0.2915	0.1094	112.32	63,574	566
10	TOPX2-Eff08-CAS-1500-Focused-W	0.4824	0.4360	0.3572	0.3103	0.1241	253.42	143,436	566
10	TOPX2-Eff08-CO-15-Focused-W	0.4751	0.4123	0.2793	0.1971	0.0726	49.79	28,180	566
10	TOPX2-Eff08-CO-150-Focused-W	0.4955	0.4520	0.3674	0.3114	0.1225	85.96	48,653	566
10	TOPX2-Eff08-CO-1500-Focused-W	0.4994	0.4560	0.3749	0.3298	0.1409	239.73	135,688	566
16	SPIRIX-ARCH	0.4953	0.4950	0.4544	0.3892	0.1601	100.97	28,779	70
16	SPIRIX-CSRU	0.7134	0.6787	0.5648	0.4915	0.1890	4,723.80	1,346,284	70
16	SPIRIX-FINE	0.4888	0.4882	0.4528	0.3898	0.1628	101.78	29,010	70
16	SPIRIX-NOSIM	0.4943	0.4854	0.4443	0.3940	0.1651	103.23	29,421	70
16	SPIRIX-PATHSIM	0.4997	0.4957	0.4550	0.3885	0.1588	105.30	30,013	70
16	SPIRIX-STRI	0.4821	0.4821	0.4260	0.3942	0.1573	100.48	28,637	33
42	B2U0_full-depth-heur	0.4388	0.3964	0.3344	0.3013	0.1357	2,994.00	1,679,634	561
56	VSM_RIP	0.4836	0.4058	0.3077	0.2553	0.0895	4,807.55	2,730,687	568
Article									
53	pftijah_article_strict	0.4599	0.4272	0.3689	0.3346	0.1839	701.98	398,722	568
Thorough		P@0.01	P@0.05	P@0.10	MAP			
10	TOPX2-Eff08-CAS-15-Thorough-W		0.1811	0.0288	0.0069	0.0053	89.31	50,549	566
10	TOPX2-Eff08-CO-15-Thorough-W		0.1890	0.0357	0.0084	0.0065	70.91	40,133	566
42	B2U0_full-depth-sr		0.2196	0.0541	0.0077	0.0080	3,519.59	1,974,492	561
53	pftijah_asp_strict		0.2674	0.1008	0.0294	0.0136	2,306.08	1,309,854	568
53	pftijah_asp_vague		0.2653	0.1120	0.0357	0.0141	8,213.05	4,665,010	568
53	pftijah_star_strict		0.2415	0.1029	0.0471	0.0169	17,186.03	9,761,663	568

Table 3. Summary over all 540 type (A) topics (Focused runs only)

Part.ID	Run ID	MAiP	AVG MS.	SUM MS.	#Topics
10	TOPX2-Eff08-CO-15-Focused-W	0.0712	18.88	10,157	538
10	TOPX2-Eff08-CO-150-Focused-W	0.1234	49.12	26,427	538
10	TOPX2-Eff08-CO-1500-Focused-W	0.1430	191.27	102,903	538
10	TOPX2-Eff08-CAS-15-Focused-W	0.0643	48.84	26,276	538
10	TOPX2-Eff08-CAS-150-Focused-W	0.1094	61.25	32,953	538
10	TOPX2-Eff08-CAS-1500-Focused-W	0.1249	165.53	89,055	538
16	SPIRIX-ARCH	0.1601	100.97	28,779	70
16	SPIRIX-CSRU	0.1890	4,723.80	1,346,284	70
16	SPIRIX-FINE	0.1628	101.78	29,010	70
16	SPIRIX-NOSIM	0.1651	103.23	29,421	70
16	SPIRIX-PATHSIM	0.1588	105.30	30,013	70
16	SPIRIX-STRI	0.1573	100.48	28,637	33
42	B2U0_full-depth-heur	0.1373	2,716.45	1,450,584	534
53	pftijah_article_strict	0.1884	604.51	326,438	540
56	VSM_RIP	0.0936	4,253.85	2,297,077	540

Table 4. Summary over all 21 type (B) topics (Focused runs only)

Part.ID	Run ID	MAiP	AVG MS.	SUM MS.	#Topics
10	TOPX2-Eff08-CO-15-Focused-W	0.0915	844.67	17,738	21
10	TOPX2-Eff08-CO-150-Focused-W	0.1094	1038.90	21,817	21
10	TOPX2-Eff08-CO-1500-Focused-W	0.1125	1468.67	30,842	21
10	TOPX2-Eff08-CAS-15-Focused-W	0.0915	1044.71	21,939	21
10	TOPX2-Eff08-CAS-150-Focused-W	0.1096	1074.66	22,568	21
10	TOPX2-Eff08-CAS-1500-Focused-W	0.1124	1479.33	31,066	21
42	B2U0_full-depth-heur	0.1143	8,052.14	169,095	21
53	pftijah_article_strict	0.1224	3,212.52	67,463	21
56	VSM_RIP	0.0329	14,583.33	306,250	21

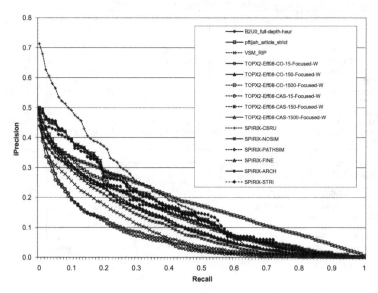

interpolated precision - focused - all topics

Fig. 3. Interpolated precision plots for all Focused and Article runs

Table 5. Summary over all 7 type (C) topics (Focused runs only)

Part.ID	Run ID	MAiP	AVG MS.	SUM MS.	#Topics
10	TOPX2-Eff08-CO-15-Focused-W	n/a	41.00	287	7
10	TOPX2-Eff08-CO-150-Focused-W	n/a	58.86	412	7
10	TOPX2-Eff08-CO-1500-Focused-W	n/a	277.57	1,943	7
10	TOPX2-Eff08-CAS-15-Focused-W	n/a	469.42	3,286	7
10	TOPX2-Eff08-CAS-150-Focused-W	n/a	1150.14	8,051	7
10	TOPX2-Eff08-CAS-1500-Focused-W	n/a	3330.71	23,315	7
42	B2U0_full-depth-heur	n/a	14,629.86	102,409	7
53	pftijah_article_strict	n/a	688.71	4,821	7
56	VSM_RIP	n/a	18,194.29	127,360	7

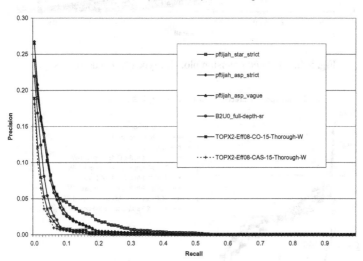

Fig. 4. Precision-recall plots for all Thorough runs

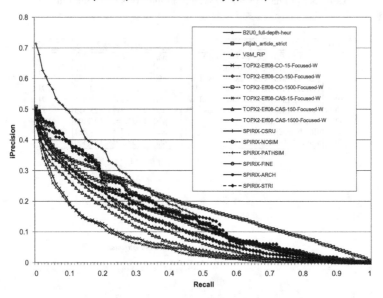

Fig. 5. Interpolated precision plots for type (A) Focused runs

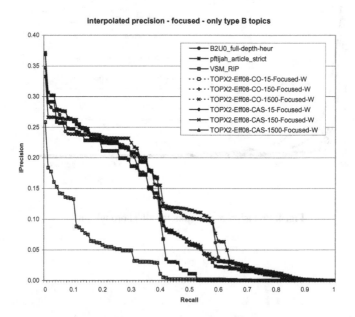

Fig. 6. Interpolated precision plots for type (B) Focused runs

6 Conclusions

This paper gave an overview of the INEX 2008 Efficiency Track. We intend to continue and expand this track in upcoming INEX years, thus hoping to increase the general visibility of this project and to attract more people from the DB&IR fields to efficient XML-IR settings. We also aim to establish the Efficiency Track, along with its large body of IR-style topics and readily available assessments, as a reference benchmark for more realistic XML-IR experiments outside the INEX community. One step towards this direction was to introduce queries in the more common XPath 2.0 Full-Text syntax.

References

1. Overview of the INitiative for the Evaluation of XML retrieval (INEX). In: Fuhr, N., Gövert, N., Kazai, G., Lalmas, M. (eds.) INitiative for the Evaluation of XML Retrieval (INEX). Proceedings of the First INEX Workshop, Dagstuhl, Germany, December 8–11 (2002); ERCIM Workshop Proceedings, Sophia Antipolis, France, ERCIM (March 2003), http://www.ercim.org/publication/ws-proceedings/INEX2002.pdf
2. Denoyer, L., Gallinari, P.: The Wikipedia XML Corpus. In: SIGIR Forum (2006)
3. Fuhr, N., Kamps, J., Lalmas, M., Malik, S., Trotman, A.: Overview of the INEX 2007 Ad Hoc Track. In: Fuhr, N., Kamps, J., Lalmas, M., Trotman, A. (eds.) INEX 2007. LNCS, vol. 4862, pp. 1–23. Springer, Heidelberg (2008)
4. Kamps, J., Pehcevski, J., Kazai, G., Lalmas, M., Robertson, S.: INEX 2007 Evaluation Measures. In: Fuhr, N., Kamps, J., Lalmas, M., Trotman, A. (eds.) INEX 2007. LNCS, vol. 4862, pp. 24–33. Springer, Heidelberg (2008)
5. Malik, S., Larsen, B., Tombros, A.: Report on the INEX 2005 Interactive Track. SIGIR Forum 41(1), 67–74 (2006)
6. Malik, S., Tombros, A., Larsen, B.: The Interactive Track at INEX 2006. In: INEX, pp. 387–399 (2006)

Exploiting User Navigation to Improve Focused Retrieval

M.S. Ali, Mariano P. Consens, Bassam Helou, and Shahan Khatchadourian

University of Toronto
{sali,consens,bassam,shahan}@cs.toronto.edu

Abstract. A common approach for developing XML element retrieval systems is to adapt text retrieval systems to retrieve elements from documents. Two key challenges in this approach are to effectively score structural queries and to control overlap in the output across different search tasks. In this paper, we continue our research into the use of navigation models for element scoring as a way to represent the user's preferences for the structure of retrieved elements. Our goal is to improve search systems using structural scoring by boosting the score of desirable elements and to post-process results to control XML overlap. This year we participated in the Ad-hoc Focused, Efficiency, and Entity Ranking Tracks, where we focused our attention primarily on the effectiveness of small navigation models. Our experiments involved three modifications to our previous work; (i) using separate summaries for boosting and post-processing, (ii) introducing summaries that are generated from user study data, and (iii) confining our results to using small models. Our results suggest that smaller models can be effective but more work needs to be done to understand the cases where different navigation models may be appropriate.

1   Introduction

At INEX 2008, the University of Toronto investigated the effectiveness of using XML summaries [8] in structural scoring for XML retrieval. An XML summary is a graph-based model that is found by partitioning elements in the collection. By weighting the graph, it represents a navigation model of users traversing elements in their search for relevant information to satisfy their information need. Our use of navigation models was originally developed for use with the performance evaluation measure structural relevance (SR) [5,2]. SR is a measure of the expected relevance value of an element in a ranked list, given the probability of whether the user will see the element one or more times while seeking relevant information in the higher-ranked results [5]. SR has been shown to be a stable measure that effectively evaluates element, passage, document and tree retrieval systems [2]. Its effectiveness has been validated using navigation models based on either collection statistics or user assessments.

Our search engine uses the Lucene text retrieval system as its basis. Our main adaptation of it is that we index the collection based on XML elements

S. Geva, J. Kamps, and A. Trotman (Eds.): INEX 2008, LNCS 5631, pp. 192–206, 2009.

as documents. Specifically, for INEX 2008, we considered indexes for `article`, `section` and `p` elements. This allowed us to consider the scoring of elements in much the same way as documents would be scored in classical text retrieval. To structurally score elements, we employed boosts corresponding to the label path of the candidate element. In focused retrieval, a significant problem of using Lucene in this way is that it does not prevent overlap (which is known to degrade the quality of results) and so a post-processor is used to control overlap. A key feature of our approach is that both score boosting and post-processing use navigation models in structural scoring.

Our approach to structural scoring involved: (i) using separate and independent navigation models for boosting and post-processing, (ii) introducing navigation models that are generated from user study data, and (iii) focusing on the effectiveness of very small models. In this paper, we show how navigation models have been integrated into structural scoring by using concepts from structural relevance. In particular, we show experimentally how post-processing can be used to not only control overlap, but also to improve the system effectiveness. We do this by presenting three different approaches to post-processing: (a) INEX overlap control where the lowest ranking element in a pair of overlapping elements is removed from the results, (b) Navigation overlap control where the element which was most highly weighted using a pre-selected navigation model is removed from the results, and (c) Ranked list control where the set of elements from the same document that had the highest structural relevance [2] would be included in the system output.

Existing approaches to XML retrieval have relied on rote return structures and ad-hoc tuning parameters for structural scoring of elements. A naive approach assumes that XML documents are structured as articles, and so only logical elements such as articles, sections and paragraphs are returned in the search results. Another approach is to allow users to specify structure, such as using NEXI which is a notation for expressing XML queries that includes structural constraints and hints [15]. NEXI can be used in conjunction with XPATH to retrieve strict XML structural paths according to what the user specifies in the query. Other approaches to structural retrieval, like XRANK [9] or Clarke's Re-ranking Algorithm [7], use element weighting schemes to iteratively score and re-rank results to improve the final system output. In this work, we rely on a probabilistic model of navigation developed for the evaluation measure structural relevance. Other models exist, most notably the model that underlies the PRUM evaluation measure [12].

We first investigated the effectiveness of using XML summaries as a way to introduce structural scoring into XML retrieval at INEX 2007 in the Thorough Ad Hoc Track in element retrieval [3]. Our initial approach allowed complex modeling of user navigation using large models that were derived solely from collection statistics in XML summaries (such as element length, element depth and element label paths). In this work, we greatly extend this work, and, in particular, focus on the effectiveness of using smaller models, and weighting schemes that are based on user assessments.

The paper is structured as follows. In Section 2, we review the preliminary concepts of the information seeking behaviour of users and structural relevance. In Section 3, we present the navigation models that we used in INEX 2008. In Section 4, we present the XML retrieval system that was used in this work. In Section 5, we present our results for INEX 2008. Finally, in Section 6 we discuss our findings and future work.

2   Preliminary Concepts

In structural scoring in XML retrieval, the key challenge is to differentiate among candidate elements those that meet the user's preference for elements whose structure supports how they fulfill their information need or those that minimize the redundancy that a user will experience while seeking for relevant information from search results. The goal of a focused system is to retrieve non-overlapping elements (or passages) that contain relevant information. For INEX 2008, only focused results, which contain no overlapping elements, are considered. We show how the more general concept of redundancy in structural relevance can be used as an effective way to address problems with overlap in XML retrieval [10] by providing a means to structurally score XML elements based on user preferences and XML element structural characteristics.

In this work, the redundancy between elements is measured using structural relevance and found using different navigation models of how users experience redundancy while browsing for relevant information in retrieved documents. In Section 2.1, the information seeking behaviour of users fulfilling an information need is presented. This is followed by Section 2.2 where structural relevance and its underlying probabilistic model of redundancy based on user navigation is presented.

2.1   Information Seeking Behaviour

In element retrieval, we assume that users *consult* the system output going from one rank to the next, *visiting* the element in context [2]. The user stops consulting the output when their information need has been satisfied. A visited element is considered to have been *seen* by the user. After seeing an element, the user may visit additional elements by *browsing* out of the element into the rest of its parent XML document. This process of visiting, browsing and seeing content (in elements) within documents to seek relevant information is called *user navigation*. During this process the user may encounter content in already seen elements, and, thus, redundant content. If the user tolerates redundant content, then the relevant content that is seen a number of times remains relevant to the user. If the user does not tolerate redundant content, then relevant content is considered non-relevant if already seen. In this work, we assume that the user does not tolerate (redundancy) seeing content more than once. In structural relevance, overlap is a special-case of redundancy where overlapping elements are on the same XML branch in the same document instance [5].

2.2   Structural Relevance

Structural relevance (SR) [2] is a measure of the relevance of the results given that the user may find some of the results in the output redundant. SR depends on the user's browsing history, which is the set of elements that a user has viewed, and a model of how the user navigates to obtain relevant information in documents. We calculate SR as the expected relevance value of a ranked list given that the user does *not* find results in the output redundant (seen more than once) given their browsing history:

$$SR(R) = \sum_{i=1}^{k} rel(e_i) \cdot (1 - \pi_{(e_i)}^{m(e_i)}) \tag{1}$$

where the system output $R = \{e_1, e_2, \ldots, e_k\}$ is a ranked list of k elements, the browsing history $m(e_i)$ is the number of elements in R that are higher-ranked than e_i and are from the same document as e_i, $\pi_{(e_i)}$ is the probability that element e_i will be navigated to from a higher-ranked element from the same document in R, and $rel(e_i) \in [0, 1]$ is the relevance value of element e_i where $rel(e) > 0$ if e is relevant, and not relevant otherwise, $rel(e) = 0$.

User navigation to element e, $\pi_{(e)}$, is the probability that the user will see the element e each time he or she visits the document that contains element e. The probability $\pi_{(e)}$ is a steady-state probability. We calculate $\pi_{(e)}$ using a weighted graph of the elements in the collection. Each node in the graph corresponds to a partition of the elements in the collection s.t. every element in the collection is in a partition. Weights are then ascribed to the edges (or paths) in the graph. We denote the weighted graph as a Navigation Graph. The partitioning scheme and weighting scheme together are called the Navigation Model. To determine the steady-state probability, first, the weights on the outgoing edges of each node are normalized by dividing the weight of each edge of a node by the sum of all of the weights of the outgoing edges of the node. These normalized values are put into matrix format with the rows and columns corresponding to the normalized weight between nodes in navigation graph. We call this a transition matrix. The transition matrix is iteratively multiplied with itself until all rows are equal [13] (pp. 200–213). The equal rows contain the steady-state probabilities $\pi_{(e)}$ which are used in this work to calculate SR as shown in Equation 1.

3   Proposed Navigation Models for INEX 2008

Navigation models represent how users navigate between elements while seeking relevant information from structured search results. In structural scoring, we use these models to both boost candidate retrieval elements and for post-processing ranked lists to control overlap. In this section, we provide the background to understand what is entailed in a navigation model and the navigation models considered in INEX 2008.

XML structural summaries (referred to as summaries) provide a way to represent navigation models based on the structure of documents in the collection. We

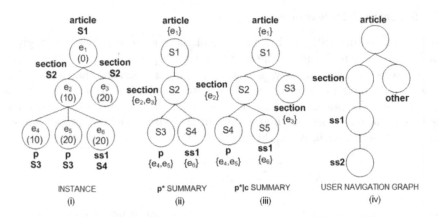

Fig. 1. Examples of (i) Document instance, (ii) p^* summary, (iii) $p^*|c$ summary, and (iv) user navigation graph

refer to these models as summary navigation models. A structural summary is a graph partition which divides the original graph into *blocks* so that each block's *extent* contains the elements having a similar neighbourhood (paths, subtrees, etc.); a structural summary then represents the relationships between sets of document elements with a similar neighbourhood.

The summaries were built using the AxPRE framework [8] which captures all existing structural summaries available in the literature by using an *axis path regular expression* language to define the neighbourhood by which to group similar document elements. For example, a p^* summary partitions XML elements based on their incoming label paths, since p^* is the axis path regular expression describing paths of parent (p) axis traversals to the root element, and groups elements having the same label path to the root into the extent of the same block. Similarly, a $p^*|c$ summary is the axis path regular expression describing paths of parent (p) with a single child (c) axis traversals, thus grouping elements having the same label path to the root *and the same set of children element labels*. Figure 1 shows an example Wikipedia article instance in (i), its p^* summary in (ii), and its $p^*|c$ summary in (iii). For instance, in Figure 1(ii) block S2 contains document elements e_2 and e_3 in its extent.

In our summary navigation models, edges between blocks are bi-directionally weighted based on collection statistics. The different weighting statistics explored in our submission include content length, extent size, and element depth. Content length weighting uses the sum of character lengths of each block's extent elements. Extent size is the number of elements in a block's extent. Finally, depth weights are the same as content length but damped (divided) by the path depth of the block (easily determined by counting the number of labels in a block's label path to the root).

Using the methodology for finding $\pi_{(e)}$ described in the previous section, and 2343 randomly selected Wikipedia articles summarized using a p^* summary and whose partitions were mapped to the user navigation graph shown in Figure 1(iv), we get Table 2C which shows summary navigation models for Wikipedia based on path, content and depth weights.

Navigation models can also be generated from user assessments. We refer to these models as user navigation models. In [6], assessments from the INEX 2006 Interactive Track user study were used to produce user navigation models. The INEX 2006 user study consisted of 83 participants for 12 assessment topics with user activity recorded for 818 documents from the INEX 2006 Wikipedia collection [11]. Figure 1(iv) shows the five types of XML elements that participants visited in the 2006 user study; namely, ARTICLE, SEC, SS1, SS2, and OTHER. These correspond to elements whose label paths are the root */article* (ARTICLE), a section path */article/body/section* (SEC), a subsection path SEC/*section* (SS1), a sub-subsection path SS1/*section* (SS2), and all other elements' paths (OTHER). We call Figure 1(iv) the user navigation graph.

Table 1. Number of visits (mean time spent)

Source	Destination ARTICLE	SEC	SS1	SS2	OTHER
ARTICLE	0 (0)	138 (100.4)	18 (48.7)	1 (22)	2 (76)
SEC	278 (57.0)	372 (14.7)	41 (11.3)	0 (0)	0 (0)
SS1	46 (13.1)	50 (10.2)	50 (9.52)	0 (0)	1 (48)
SS2	4 (12.3)	2 (264.5)	13 (5.3)	0 (0)	0 (0)
OTHER	7 (27.7)	0 (0)	1 (4)	0 (0)	4 (26)

Table 1 tabulates the observed visits and mean time spent in visits for element assessments by participants in the INEX 2006 user study. For instance, participants visited SS2 elements and then navigated to element ARTICLE 4 times. The mean time spent in SS2 before navigating to element ARTICLE was on average 12.3 seconds. This led to an overall time, which we refer to as an episode, of 12.3 x 4 = 49.2 seconds. The most visited element was SEC, and the largest mean time spent occurred in navigations to SEC elements from ARTICLE. These assessments of user navigation can be used to weight the paths in the user navigation graph in Figure 1(iv). The resultant navigation probabilities $p(e; f)$ for the user navigation model, based on normalizing the number of visits in Table 1, are shown in Table 2A. Similarly, we can generate user navigation models (based on the same user navigation graph) for the observed time-spent and episodes. The user navigation models developed in [6] are shown in Table 2B.

Additionally, we investigated the use of trivial navigation models that were composed of two nodes; a main node that would contain one type of element and the other node which would include all other elements in the collection. The main node has a high steady-state probability (we used 0.999) and the other node would have a correspondingly small steady-state probability (0.001). We proposed two trivial models; namely the article model and the ss2 model. These have the effect of an exclusion filter in that the elements in the main node will be less preferred in the results than other nodes. The proposed trivial navigation models are shown in Figure 2. For INEX 2008, five navigation models

Table 2. (A) Example transition matrix, (B) User models, & (C) Summary models

A. Normalized Weights for Visits

	Destination				
Source	ARTICLE	SEC	SS1	SS2	OTHER
ARTICLE	0.0	0.87	0.11	0.01	0.01
SEC	0.40	0.54	0.06	0.0	0.0
SS1	0.31	0.34	0.0	0.0	0.01
SS2	0.21	0.11	0.68	0.0	0.0
OTHER	0.58	0.0	0.08	0.0	0.33

B. User Navigation Models

	ARTICLE	SEC	SS1	SS2	OTHER
Visit	0.281	0.606	0.105	0.002	0.006
Episode	0.410	0.531	0.050	0.001	0.009
Time spent	0.318	0.209	0.129	0.028	0.317

C. Summary Navigation Models

	ARTICLE	SEC	SS1	SS2	OTHER
Path	0.361	0.537	0.087	0.014	0.001
Content	0.103	0.434	0.089	0.013	0.361
Depth	0.309	0.435	0.067	0.008	0.181

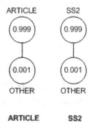

Fig. 2. Trivial navigation models used in INEX 2008

were considered: summary navigation models path and depth (Table 2C), user navigation model visit (Table 2B), and, the article and ss2 trivial navigation models (Figure 2).

In this section, we have presented how different navigation models are generated, and presented the 5 navigation models that we used in INEX 2008. In the next section, we present a description of our search system, how we implemented boosts, and a description of the different post-processors that we used in INEX 2008.

4   System Description

This section provides details on how we implemented our search engine, which is based on Apache Lucene.

4.1   Lucene

The p^* structural summary of the Wikipedia XML collection, originally generated using code from DescribeX [4], consisted of 55486 summary nodes (with
aliasing on tags containing the substrings `link`, `emph`, `template`, `list`, `item`,
or `indentation`). The extents in the structural summary were then mapped to
the navigation graph shown in Figure 1(iv) to produce the summary navigation
models. As the collection was summarized, modified Apache Lucene [1] code was
used to index the tokens. The posting list also included character offsets. Tokens
not excluded from the stop word filter had punctuation symbols removed and
the tokens maintained their case. The structural summary was generated at the
same time as each document was indexed, and the payload information for each
token occurrence included the summary partition in which the token appears.

To accommodate the structural hints in the INEX topics, separate indexes
were built for each tag identified by the structural hint present within the set
of INEX topics which included `article`, `section`, and `p`. For example, building
an index for the p tag would index the first p element and its children, including
nested p elements, until its respective closing tag. Thus, a file with multiple non-
overlapping indexed elements will create multiple documents within the index,
and these non-overlapping elements are easily identified since the index stores the
character offsets as previously mentioned. This results in having element-level
documents which allows the calculation of idf scores for terms within elements.
The index sizes (which includes term frequencies, file paths and the payload
information) were 6.07GB for tag `article`, 4.84GB for tag `section`, and 4.63GB
for tag `p`.

Lucene's query parser was adapted to accept Top-X [14] NEXI queries with
structural hints. The queries were encoded using boolean operators to represent
tokens that were mandatory, optional, or to be excluded. Double quotes indicating adjacent tokens were removed since token positions were not indexed. Prior
to running a query, the query was examined for any structural hints and the
required indexes were searched as a single merged index using Lucene's regular
application interface. If no structural hints were identified, the complete set of
element indexes were used in the search.

In the Content Only (CO) sub-task, queries are specified using keywords and
content-related conditions. Structure can be included in the query as hints to
reduce the number of returned elements. CO queries with structural hints are
called Content Only + Structure (CO+S) queries. For CO+S queries, it is left
to the discretion of the search engine to interpret it as either strict or vague.
Our system used the element labels in structural hints to include or exclude
specific search indexes while processing the query. Queries were not explicitly
interpreted as either strict or vague. The score of elements was composed of
two main factors; (i) content relevance, and (ii) a score boost based on the label
path of the smallest element (Section 4.2) that enclosed the content. The highest
scoring elements from Lucene were then post-processed (Section 4.3) to ensure
that the final results returned were focused.

4.2   Boosting Strategies

The collection was indexed at the element-level for `article`, `section`, and `p`. In our experiments, we included runs with score boosting per term occurrence and using the average of the term scores as a modifier to Lucene's original document score. The boost used was the stationary probability $\pi_{(e)}$ of the partition in the summary of the element in which the term occurs. The baseline payload score per occurrence was set to 1 and the boosted term score was the baseline plus the stationary probability. The scenarios reported in this paper include runs with either no boosting, boosted using the path summary navigation model (Table 2B), or boosted using the visit user navigation model (Table 2C).

4.3   Post-processing Algorithms

The purpose of post-processing is to control overlap and produce focused runs from Lucene's results. We present three different approaches to post-processing. The first approach, called INEX Overlap Control and shown in Figure 3, removes all parent-child overlap from the ranked list by comparing each pair of elements e_i and e_j in R, and removing the lower-ranked element if they are on the same branch.

The second approach, called Navigation Overlap Control and shown in Figure 4, involves removing parent-child overlap where overlapped elements were ranked closely to one another (in the results reported here, the window size was set to 10 rank positions). Instead of removing the lowest ranked overlapped element, the element with the highest steady-state probability was removed from the ranked list. This is akin to removing the element most likely to be visited by an element within the same document using any path.

The third approach, called Ranked List Overlap Control and shown in Figure 5, was developed in INEX 2007 and involves computing SR for scenarios where redundant elements (i.e., from the same document) are systematically removed or reordered in the ranked list until the highest scoring scenario is found. We assume that all redundant elements are relevant and place the restriction that a document cannot be excluded from the ranked list by removing all of its elements that were present in the original result list.

So, the post-processing of overlap used either; (i) a simple heuristic to remove the lowest ranked elements that were overlapped; (ii) a more complex heuristic to remove the most redundant overlapped elements; or (iii) an algorithm to find the most structurally relevant non-overlapped set of elements. To determine redundancy and structural relevance in post-processing, we used four navigation models; namely, a trivial navigation model based on article elements being redundant, a second trivial navigation model based on sub-subsection elements being redundant, a depth summary navigation model, and a path user navigation model.

Algorithm. *INEX Overlap Control*
Input: Ranked list $R = e_1, e_2, \ldots$ of k elements.
Output: Ranked list with overlapped elements removed R^*
1: let m be the length of R^*, initialize as $m = 1$
2: **for** $i = 1$ to k **do**
3:    $skip = false$
4:    **for** $j = 1$ to $i - 1$ **do**
5:       **if** e_i and e_j are on same branch **then**
6:          $skip = true$
7:       **end if**
8:    **end for**
9:    **if** $skip = false$ **then**
10:       $R^*[m] = e_i$
11:       $m = m + 1$
12:    **end if**
13: **end for**

Fig. 3. Remove lowest ranked overlapped elements from a ranked list

Algorithm. *Navigation Overlap Control*
Input: Ranked list $R = e_1, e_2, \ldots$ of k elements.
Input: Summary graph S with navigation $\pi_{(e)}$ for element e.
Output: Ranked list with overlapped elements removed R^*
1: let $window$ be the minimum distance between competing elements
2: let m be the length of R^*, initialize as $m = 1$
3: **for** $i = 1$ to k **do**
4:    $skip = false$
5:    **for** $j = 1$ to k **do**
6:       **if** e_i and e_j are on same branch **then**
7:          **if** $|i - j| > 10$ OR $\pi e_i > \pi_{e_j}$ **then**
8:             $skip = true$
9:          **end if**
10:       **end if**
11:    **end for**
12:    **if** $skip = false$ **then**
13:       $R^*[m] = e_i$
14:       $m = m + 1$
15:    **end if**
16: **end for**

Fig. 4. Remove the least isolated overlapped elements

Algorithm. *Ranked List Overlap Control*
Input: Ranked list $R = e_1, e_2, \ldots$ of k elements.
Input: Summary graph S
Output: Ranked list R' in Ω
Output: Ranked list in Ω with highest SRP R^{sr}
Output: Ranked list with overlapped elements removed R^*
1: let n be number of overlapped elements in R
2: let Ω be the scenarios of R
3: **for** $R' \in \Omega$ **do**
4:    **if** $SR(R')/k > SR(R^*)/k$ AND **then**
5:       $R^{sr} = R'$
6:    **end if**
7: **end for**
8: $R^* =$ INEX List Control(R^{sr})

Fig. 5. Find the highest scoring scenario

5   Results INEX 2008

In this section, we present our results from our participation in INEX 2008 in the Ad-Hoc Focused Track (Section 5.1), the Efficiency Track (Section 5.2), and the Entity Ranking Track (Section 5.3).

5.1   Ad-Hoc Focused Element Retrieval Content-Only Sub-Task

We submitted results for the Ad-hoc Focused Track using element retrieval in the Content-Only Sub-Task. We submitted 3 runs: B2U0_visit-heur, B2U0_tiny-path-sr, and B2U0_tiny-path-heur. Runs B2U0_visit-heur and B2U0_tiny-path-heur were boosted using the navigation models visit (Table 2B) and path (Table 2C), respectively. Both of these runs were post-processed using INEX overlap control. They showed similar performance. Unfortunately, the B2U0_tiny-path-sr run (which was boosted and post-processed using the path summary navigation model in Table 2C), was not admissible in the focused task due to overlap in the run (because of a syntax error in the ranked list overlap control post-processor code). Our results reported here show an extended set of runs that are indicative of the effectiveness of our proposed approach to structural scoring, and we rename the runs B2U0_visit-heur and B2U0_tiny-path-heur to PATH INEX NONE and VISIT INEX NONE, respectively in Table 3 and Figure 6.

The runs reported here are top-100 results for the Wikipedia collection across 235 topics in INEX 2008 evaluated using the official INEX measures (inex_eval) MAiP and (inex_eval) interpolated precision across interpolated recall points. The purpose of these runs was to investigate empirically whether there existed predictable combinations of boosting and post-processing that would result in more effective systems.

In Table 3, we show the MAiP evaluations for all tested configurations. The configuration for each run consisted of the type of navigation model used to boost results (Boost), the approach used to post-process the run to remove overlap (Overlap), and the navigation model used by the approach for removing overlap (Navigation). For instance, our best run was first boosted using the visit user navigation model, and then post-processed with the depth summary navigation model. The NAV runs (navigation overlap control using the algorithm shown in Figure 4) did not perform well (a maximum MAiP of 0.102 with boosting using the visit user navigation model was observed). In INEX 2007 [3], we observed that post-processing with the depth summary navigation model improved the effectiveness of systems. This was a full summary of the collection with a navigation graph that consisted of 55486 nodes, as opposed to the smaller model of only 5 nodes used this year. Moreover, we note that regardless of the boost, the best overall post-processor (by MAiP) was the depth summary navigation model. Additionally, we observed that, for each boost (NONE, PATH, VISIT), the relative performance of the post-processor configurations (Overlap-Navigation pairs in Table 3) was consistent, and was (listed from best configuration to worst) SR-DEPTH ≻ INEX-NONE ≻ SR-SS2 ≻ NAV-DEPTH ≻ SR-VISIT ≻ SR-ARTICLE.

Table 3. Mean-average interpolated precision using HiXEval for INEX 2008 Focused Runs (k=100, 235 topics)

Boost	Overlap	Navigation	MAiP
NONE	INEX	NONE	0.111
NONE	NAV	DEPTH	0.0924
NONE	SR	ARTICLE	0.0685
NONE	SR	DEPTH	0.130
NONE	SR	SS2	0.10172
NONE	SR	VISIT	0.0817
PATH	INEX	NONE	0.115
PATH	NAV	DEPTH	0.107
PATH	SR	ARTICLE	0.0745
PATH	SR	DEPTH	0.139
PATH	SR	SS2	0.106
PATH	SR	VISIT	0.080
VISIT	INEX	NONE	0.123
VISIT	NAV	DEPTH	0.102
VISIT	SR	ARTICLE	0.0723
VISIT	SR	DEPTH	**0.145**
VISIT	SR	SS2	0.116
VISIT	SR	VISIT	0.089

In Figure 6, we show the interpolated I-R curves for the six best runs reported in Table 3. The runs were within +/-0.05 precision of each other across recall points. Using MAiP, significant differences in overall performance were observed; the depth summary navigation model and INEX overlap control consistently performed better than the other configurations. From these results (and observations from INEX 2007 using large models), it seems that system effectiveness can be improved by using separate summaries for boosting and post-processing. Moreover, we observed similar performance within the small models (specifically that the depth model out-performed other models) as in large models, suggesting that the size of the model is not as important as the type of model used. Finally, these preliminary results suggest that boosting is effective, and that boosting with a user navigation model is more effective than using summary navigation models. It remains to be seen whether these observations can be generalized across search tasks.

5.2   Efficiency Track

The Efficiency Track is a forum for the evaluation of both the effectiveness and efficiency of XML ranked retrieval approaches. The approach taken at INEX 2008 is to time sequential query runs. Three query categories were provided: ad-hoc-style (category A), high-dimensional content retrieval (category B), and high-dimensional content retrieval (category C). Even though categories B and C involved many structural hints, the queries were interpreted in a relaxed

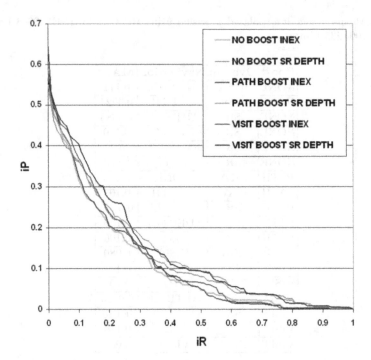

Fig. 6. Interpolated precision-recall in Focused Task using summary boosts, INEX overlap heuristics, and optimal structural relevance ($k = 100$, 285 topics)

manner, meaning that a result missing tags or keywords is still a valid result. Each query's execution time includes two main components. First, the time required to interpret the query, retrieve candidate elements from one or more indexes, and return back a posting list with thorough results. Second, the time to post-process the results using SR to output focused results.

Figure 7 shows a histogram of the query execution times for all 568 queries returning the top 1500 result using the SR-DEPTH boost and post-processor combination as it was the best navigation model and overlap post-processor from the different combinations tested in Figure 3. Just over half of the queries took under 2s to execute with a median query time of 1703ms. A sizable proportion (about 16%) of the queries took more than 5 seconds to execute and includes 35 queries which needed more than 10s to complete.

By each query type, our times were as follows:

(A) 540 Queries. Ad-Hoc-style, average query time is 3166 ms with times varying from 15 to 26063 ms.

(B) 21 Queries. High-dimensional content retrieval, average query time is 8460 ms, with times varying from 188 to 61141 ms.

(C) 7 Queries. High-dimensional structure retrieval. average query time is 12090 ms, with times varying from 2948 to 35328 ms.

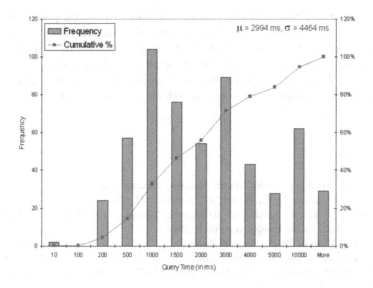

Fig. 7. Histogram of query times in the INEX 2008 Efficiency Track

Our system was run on Windows XP in a virtual machine using VMWare GSX on a Sun Fire V20xz Server cluster running on Red Hat Enterprise Linux. The VM was configured the 2048 MB of RAM and one virtual 2.39 GHz CPU running over an AMD Opteron Processor 250. Our performance is affected to some extent by virtualization. Nevertheless, we prefer to conduct our experiments in a virtual environment for convenience reasons (e.g., highly simplified reproducibility of experiments).

5.3   Entity Ranking

The Entity Ranking Track uses the Wikipedia data, where systems may exploit the category metadata associated with entities in entity retrieval. For example, consider a category "Dutch politicians". The relevant entities are assumed to be labelled with this category or other closely related category in the categorization hierarchy, e.g. "politicians". Our participation involved developing 6 new topics and conducting assessments. We did not submit runs this year and leave this as an area of future work for applying the methods presented in this paper.

6   Conclusions

In this work, we have shown how navigation models can be used effectively in both element score boosting and in the post-processing of overlap. The models can be derived either from collection statistics or user assessments. The most significant observation in this work is that small models based on either assessments or collection statistics can be used. Our results in INEX 2007 suggested that the depth summary navigation model was a good model to use for post-processing. In this study, our results have corroborated with the observations in INEX 2007, but, importantly we have shown that smaller models can be used. Small models are more easily interpreted and more efficient to use in our computations. In the future, we hope to generalize this methodology for structural scoring as a way to test and compare how different search engines handle different structural constraints and hints.

References

1. Apache Lucene Java (2008), http://lucene.apache.org
2. Ali, M.S., Consens, M.P., Kazai, G., Lalmas, M.: Structural relevance: a common basis for the evaluation of structured document retrieval. In: CIKM 2008, pp. 1153–1162. ACM Press, New York (2008)
3. Ali, M.S., Consens, M.P., Khatchadourian, S.: XML retrieval by improving structural relevance measures obtained from summary models. In: Fuhr, N., Kamps, J., Lalmas, M., Trotman, A. (eds.) INEX 2007. LNCS, vol. 4862, pp. 34–48. Springer, Heidelberg (2008)
4. Ali, M.S., Consens, M.P., Khatchadourian, S., Rizzolo, F.: DescribeX: Interacting with AxPRE Summaries. In: ICDE 2008, pp. 1540–1543. IEEE Computer Society Press, Los Alamitos (2008)
5. Ali, M.S., Consens, M.P., Lalmas, M.: Structural Relevance in XML Retrieval Evaluation. In: SIGIR 2007 Workshop on Focused Retrieval, pp. 1–8 (2007)
6. Ali, M.S., Consens, M.P., Larsen, B.: Representing user navigation in XML retrieval with structural summaries. In: ECIR 2009 (in press, 2009)
7. Clarke, C.: Controlling overlap in content-oriented XML retrieval. In: SIGIR 2005, pp. 314–321. ACM Press, New York (2005)
8. Consens, M.P., Rizzolo, F., Vaisman, A.A.: AxPRE Summaries: Exploring the (Semi-)Structure of XML Web Collections. In: ICDE 2008, pp. 1519–1521 (2008)
9. Guo, L., Shao, F., Botev, C., Shanmugasundaram, J.: XRANK: Ranked keyword search over xml documents. In: SIGMOD 2003. ACM Press, New York (2003)
10. Kazai, G., Lalmas, M., de Vries, A.P.: The overlap problem in content-oriented xml retrieval evaluation. In: SIGIR 2004, pp. 72–79. ACM Press, New York (2004)
11. Malik, S., Tombros, A., Larsen, B.: The Interactive Track at INEX 2006. In: Fuhr, N., Lalmas, M., Trotman, A. (eds.) INEX 2006. LNCS, vol. 4518, pp. 387–399. Springer, Heidelberg (2007)
12. Piwowarski, B., Gallinari, P., Dupret, G.: Precision recall with user modeling (PRUM): Application to structured information retrieval. ACM Trans. Inf. Syst. 25(1), 1 (2007)
13. Ross, S.M.: Introduction to Probability Models, 8th edn. Academic Press, New York (2003)
14. Theobald, M., Schenkel, R., Weikum, G.: An efficient and versatile query engine for TopX search. In: Proc. VLDB Conf., pp. 625–636 (2005)
15. Trotman, A., Sigurbjörnsson, B.: Narrowed Extended XPath I (NEXI). In: Fuhr, N., Lalmas, M., Malik, S., Szlávik, Z. (eds.) INEX 2004. LNCS, vol. 3493, pp. 16–40. Springer, Heidelberg (2005)

Efficient XML and Entity Retrieval
with PF/Tijah:
CWI and University of Twente at INEX'08

Henning Rode[1], Djoerd Hiemstra[2], Arjen de Vries[1], and Pavel Serdyukov[2]

[1]CWI Amsterdam, The Netherlands
[2]CTIT, University of Twente, The Netherlands

Abstract. The paper describes the submissions of CWI and University of Twente to the efficiency and entity ranking track of INEX 2008. With the INEX participation, we demonstrate and evaluate the functionality of our open source XML retrieval system PF/Tijah.

1   Introduction

PF/Tijah is a research prototype created by the University of Twente and CWI Amsterdam with the goal to create a flexible environment for setting up search systems. By integrating the PathFinder (PF) XQuery system [1] with the Tijah XML information retrieval system [2] it combines database and information retrieval technology. The PF/Tijah system is part of the open source release of MonetDB/XQuery developed in cooperation with CWI Amsterdam and the University of Tübingen.

PF/Tijah is first of all a system for *structured retrieval* on XML data. Compared to other open source retrieval systems it comes with a number or unique features [3]:

- It can execute any NEXI query without limits to a predefined set of tags. Using the same index, it can easily produce a "focused", "thorough", or "article" ranking, depending only on the specified query and retrieval options.
- The applied retrieval model, score propagation and combination operators are set at query time, which makes PF/Tijah an ideal experimental platform.
- PF/Tijah embeds NEXI queries as functions in the XQuery language. This way the system supports ad hoc result presentation by means of its query language. The efficiency task submission described in the following section demonstrates this feature. The declared function INEXPath for instance computes a string that matches the desired INEX submission format.
- PF/Tijah supports text search combined with traditional database querying, including for instance joins on values. The entity ranking experiments described in this article intensively exploit this feature.

With this year's INEX experiments, we try to demonstrate the mentioned features of the system. All experiments were carried out with the least possible

S. Geva, J. Kamps, and A. Trotman (Eds.): INEX 2008, LNCS 5631, pp. 207–217, 2009.
© Springer-Verlag Berlin Heidelberg 2009

pre- and post-processing outside PF/Tijah. Section 2 shows with the application of the system to the INEX efficiency track, how a wide range of different NEXI queries can be executed efficiently. Section 3 demonstrates how combined database and retrieval queries provide a direct solution to specialized tasks like entity ranking.

2   Efficiency

The INEX efficiency task combines retrieval quality and performance. In order to test the performance on a wide range of different queries, the task uses a query set of 568 structured queries combined from other tasks and collected over the last several years. The queries vary with respect to the number of terms contained in them and their structural requirements. A subset for instance represents typical relevance feedback queries containing a considerable higher number of query terms.

The retrieval efficiency of PF/Tijah was improved in the last year with respect to several aspects, which we wanted to test by our submissions. The index structure, containment joins, and score computation had been changed [4] to improve the execution of simple query patterns such as

```
//tag[about(., term query)]
```

PF/Tijah creates a full-text index on top of Pathfinder's pre/post encoding of XML files [5]. Instead of assigning a pre-order value to complete text-nodes as done by the Pathfinder, the Tijah full-text index enumerates each single term. Both the Pathfinder encoding and the separate full-text index are held in database tables. An "inverted" table is created by clustering the table (pre-order values) on tag- and term ID.

PF/Tijah does not use top-k query processing strategies. Neither tag-term pairs nor scores are precalculated or indexed in order avoid redundancy on the one hand, and to allow at query time the application of arbitrary ranking functions on the other hand. The applied ranking function is specified in PF/Tijah for each single query. Furthermore, PF/Tijah's containment join operator relies on input sorted in document order. Node sequences sorted on score order as they are typically accessed in the top-k query processing framework do not match this requirement. PF/Tijah does not implement any caching strategy itself. However, the underlying database system tries to make use of the operating system's caching functionalities.

2.1   Submissions

We submitted four runs, one "article" ranking and three "thorough" element rankings. Since PF/Tijah does not support top-k query processing, all submitted runs return the complete set of the 1500 highest ranked elements for each query. The applied ranking function for all submissions follows the language modeling framework for retrieval. The so-called NLLR, normalized logarithmic likelihood ratio, compares the within-element and within-query distribution for each query term. The ranking aggregates single term scores on the level of scored elements. Query terms marked by a leading '-' to indicate that they should *not* occur in relevant elements were removed from the queries, since PF/Tijah currently does not support this feature. For the same reason, phrases were treated as normal query terms only.

For repeatability we report here the complete XQuery that was used to produce the ranking in PF/Tijah. The XQuery below was generated for Topic 856. The individual queries only substitute the inside NEXI string accordingly. The costly function call producing the required INEX path string was omitted when running time measurements, since it does not reflect the retrieval performance itself:

```
declare function INEXPath($n as node()) as xs:string
{
    let $paths :=
        for $a in $n/ancestor-or-self::*
        where local-name($a) ne "docs"
        return if (local-name($a) eq "article")
            then concat(local-name($a),"[1]")
            else concat(local-name($a),"[",
                string(1 + count($a/preceding-sibling::*
                [local-name() eq local-name($a)])),"]")
    return string-join($paths, "/")
};

let $opt := <TijahOptions returnNumber="1500" ir-model="NLLR"
                prior="NO_PRIOR" txtmodel_returnall="FALSE"/>
let $nexi := "//article//body[about(.//section//p, State Park) and
                about(.//section//title, Geology) and
                about(.//section//title, Geography)]
                //figure[about(.//caption, Canyon)]"
return <topic id="856"> {
    for $res at $rank in tijah:queryall($nexi, $opt)
    return <result><file> {
        concat("",$res/ancestor-or-self::article/name/@id)}</file>
        <path>{INEXPath($res)}</path>
        <rank>{$rank}</rank></result> }
</topic>
```

For the *article* ranking we automatically created NEXI queries by substitution of the placeholder ?CO-TITLE? below with the content-only (CO) field of the query topic:

```
//article[about(., ?CO-TITLE?)]
```

The run should show how our XML retrieval system performs when used as a standard document retrieval system.

In contrast, the "thorough" element rankings use the content-and-structure (CAS) field of each query topic. The first "thorough" run, called *star-strict*, executes the unmodified CAS query as provided in the query topic. The final two runs perform a slight modification. Since the new PF/Tijah system is tuned towards queries starting with a tag-name selection rather than searching in all element nodes, we translated queries starting with the pattern

```
//*[about(., terms)]...
```

to

```
//(article|section|p)[about., terms)]...
```

The runs based on this modification are called *asp-strict* and *asp-vague*. The distinction between these is explained below.

Thinking in terms of XPath, the base of the NEXI language, the scoring predicates [about(., terms)] are first of all evaluated to a boolean value, causing those elements to pass that satisfy the query. If the predicates are translated to a scoring operator in the algebra tree, that only assigns scores to all elements, the predicate becomes obsolete as a filter and the side effect of the predicate evaluation, the score assignment, has become the primary aim. This is clearly not the only possible query interpretation. We can require that an element has to reach a certain score threshold in order to satisfy the predicate condition. The least strict setting of such a threshold would be to filter out all zero scored element. In other words, the about function would assign a true value to all elements that contain at least one of the query terms. For a query of the form

```
//article[about(., xml)]//p[about(.,ir)]
```

strict semantics will pass only those articles that match the keywords of the first about, whereas *vague* semantics also considers results of paragraphs about "ir" that are not occurring within articles about "xml". The two submitted runs, *asp-strict* and *asp-vague*, compare the different query interpretation with respect to retrieval quality and performance.

2.2   Results

The test system used for all time measurements in this article was an INTEL Core2 Quad machine running on 2.4 Ghz with 8 GB main memory. The necessary index structures could hence be held in memory, but not in the considerably smaller CPU caches. Queries were executed sequentially. For time measurements, we omitted the generation of the INEXPath as mentioned above and stored only node identifiers instead. We measured the full execution of the query, including the query compilation phase.

Table 1 shows an overview on the execution times of the different runs. The article ranking is obviously faster on average than the three other runs evaluating

Table 1. Execution time overview in sec

run	avg time	sum time	min time	max time
article	0.702	399	0.327	11.814
star-strict	17.186	9762	0.324	330.495
asp-strict	2.306	1310	0.324	52.388
asp-vague	8.213	4665	0.444	1235.572

the CAS query. Since some CAS queries in the query set issue a simple fielded search, it is not surprising that the minimal execution time stays almost the same for all runs. Looking at the average and maximal execution time for a single query, we observe, however, huge differences. Most of the time differences can be attributed to queries that contain the pattern //* in the NEXI path. If a posting list of a certain tagname is fetched from the inverted index, the system guarantees the pre-order sortedness of the list, which is required for the subsequent containment evaluation. Fetching the entire inverted index, however, will not return a pre-order sorted element list, and therefore requires a resorting of the entire node set. The difference becomes apparent when comparing the execution times of the two runs *star-strict* and *asp-strict*. Even the expensive substitute pattern selecting all `article`, `section`, and `p` nodes shows still a better performance.

Evidently, the application of *strict* query semantics yield a better query performance. The average total execution is around four times faster than in the case of a *vague* interpretation. The early filtering on intermediary result sets especially helps on highly structured queries. Consequently, we observe similar minimal execution times but clear differences when looking at the highest times measured for evaluating a single query. The differences of the two query interpretations needs to be studied as well in terms of retrieval quality.

Table 2. Retrieval quality presented in official measures

run	MAiP	iP[0.10]	iP[0.5]	iP[0.01]
article	0.1839	0.3346	0.3689	0.4272
	MAP	P@0.10	P@0.5	P@0.01
star-strict	0.0169	0.0471	0.1029	0.2415
asp-strict	0.0136	0.0294	0.1008	0.2674
asp-vague	0.0141	0.0357	0.1120	0.2653

Table 2 reports the official measurements used in the efficiency track, which differ for "article" and "thorough" run submissions. Therefore, we can only compare our three "thorough" runs. The substitution of //*-queries sacrifices recall but not early precision. The two *asp* runs even yield a slightly higher precision on top of the ranked list. Comparing the *strict* and *vague* semantics we observe as expected a better retrieval quality when applying the vague "andish" interpretation. The differences, however, stay again small when looking at the top of the retrieved list.

3   Entity Ranking

The INEX entity ranking task searches for entities rather than articles or elements with respect to a given topic. With entities we mean here unique instances of a given type, such as "Hamburg" and "München" being an instance of type "German cities". For a given query topic such as "hanseatic league" and target entity type "German cities" a good entity retrieval system should return "Hamburg", but not "München" since it is off topic, or "Novgorod" since it is not a German city.

The target type is given as a Wikipedia category in the INEX task. Furthermore, each retrieved entity needs to have its own article in the Wikipedia collection. Obviously, this decision is only suitable for entity ranking within an encyclopedia, where we can assume that most mentioned entities in fact have their own entry. In consequence, a baseline ranking is achieved by a straightforward article ranking on the Wikipedia corpus combined with an appropriate category filtering mechanism.

The INEX task further provides a few relevant example entities for each query topic. The given entities can be used as relevance feedback to improve the initial text query or to redefine the set of target categories. Another application for the example entities comes with the list completion task. This task asks to derive appropriate target categories automatically from the given relevant entities.

Our main aim for this year's track participation was to express entity ranking queries completely in the XQuery language. Hence, we wanted to show that PF/Tijah is "out of the box" able to express and evaluate complex entity ranking queries with a high retrieval quality. One preprocessing step, however, turned out to be unavoidable. The INEX wikipedia corpus comes without category tagging in the provided XML format. Instead, the categorization of all articles is provided by separate plain text files. In order to unify all given information, we integrated the category tagging in the XML corpus itself as shown in the following example:

```
<article><name id="13467">Hamburg</name>
    <body>....</body>
    <category id="5654">cities in germany</category>
    <category id="52414">port cities</category>
</article>
```

In addition to the title keywords, target categories, and relevant entities provided with each search topic, we generated for each search topic an additional list of relevant derived categories. Namely those categories assigned to the specified relevant entities. The derived relevant categories are used as mentioned above for refinement of the target categories as well as for the list completion task:

```
for $topic in doc("topics.xml")//inex_topic
let $relevant_entities := $topic//entity/@id
return collection("wikipedia")//
        article[name/@id =
$relevant_entities]//category/@id
```

3.1   Submissions

We submitted six runs, four runs for the entity ranking task, and two list completion submissions. The submissions can also be divided into three runs based solely on a direct article ranking, and three other runs using also the scores of adjacent articles in the link graph.

We start by describing the direct article rankings. The ranking and category filtering is performed by a single XQuery, which is shown below. The fields fields ?QID?, ?QTERMS?, ?CATS?, ?DERIVEDCATS? were substituted according to the given query topic:

```
(: part1 - retrieval :)
let $query_num := "?QID?"
let $q_terms := tijah:tokenize("?QTERMS?")
let $opt := <TijahOptions ir-model="LMS" returnNumber="1000"
               collection-lambda="0.5"/>
let $nexi := concat("//article[about(.,", $q_terms, ")]")
let $tijah_id := tijah:queryall-id($nexi, $opt)
let $nodes := tijah:nodes($tijah_id)

(: part2 - determine target categories :)
let $targetcats := distinct-values(((?CATS?), (?DERIVEDCATS?)))

(: part3 - filtering and output generation :)
for $a at $rank in $nodes
let $score := if ($a//category/@id = $targetcats)
                then tijah:score($tijah_id, $a)
                else tijah:score($tijah_id, $a) * 0.0000001
order by $score descending
return string-join((string($query_num), "Q0", concat("WP",$a/name/@id),
        string($rank), string($score), "ER_TEC"), " ")
```

The presented XQuery ranks in the first part all articles of the Wikipedia collection according to the topic of the query. We applied here a standard language modeling retrieval model with the smoothing factor set to $\lambda = 0.5$. Moreover, the result set was limited to the top 1000 retrieved articles.

The second part determines the target categories. Whereas our first run *ER_TC* uses only the categories provided with the query topic, the second run *ER_TEC* refines the target category set by uniting the given and derived categories as shown in the query. The list completion *LC_TE*, on the other hand, uses only the derived but not the given categories.

The final part performs the actual filtering and required TREC-style output generation. Notice that the applied filtering in fact only performs a reordering and does not remove articles from the ranked list. Last year's experiments had clearly shown that the reordering comes with a higher recall compared to the filtering technique.

The other three runs *ER_TC_idg*, *ER_TEC_idg*, *LC_TE_idg* exploit the retrieval scores of adjacent nodes and follow otherwise a symmetrical experiment schema with respect to the used target categories. The underlying idea behind the exploitation of link structure is adopted from other entity ranking tasks such as expert finding, where we typically find a number of topical relevant documents that mention relevant entities, but entities do not have a textual description themselves. A sample cutout of such a graph is visualized in Figure 1. The edges here symbolize containment of entities within documents. Entities are then ranked by a propagation of scores from adjacent documents.

Although entity ranking on the Wikipedia corpus is different since entities are represented by their own articles and have a text description themselves, it still often occurs the articles outside the target category carry valuable information for the entity ranking. Recall the above given example query searching for German cities in the hanseatic league. We will find Wikipedia entries about the history of the hanseatic league listing and linking to all major participating cities. While such article remains outside the target category, the links to relevant city pages are of high value for the ranking. Especially, when a city's description itself does not reach far enough into history. We developed last year a ranking method matching this condition [6]. The personalized weighted indegree measure tries to combine the article ranking itself $w(e|q)$ with the ranking of other Wikipedia entries $w(e'|q)$ linking entity e:

$$PwIDG(e) = \mu w(e|q) + (1 - \mu) \sum_{e' \in \Gamma(e)} w(e'|q) \tag{1}$$

A corresponding indegree score computation can be expressed as well in XQuery. The below shown query part substitutes the score computation in the previous entity ranking example and sets the parameter μ to 0.85:

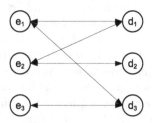

Fig. 1. Part of a link graph containing entities e_i and other documents d_j

```
for $a at $rank in $nodes
let $in_score := sum(
      for $l in $nodes//collectionlink[@*:href =
          concat($a/name/@id, ".xml")]
      let $source_article := exactly-one($l/ancestor::article)
      return tijah:score($tijah_id, $source_article)
)
let $score := if ($a//category/@id = $targetcats)
      then 0.85 * tijah:score($tijah_id, $a) + 0.15 * $in_score
      else (0.85 * tijah:score($tijah_id, $a) + 0.15 * $in_score)
          * 0.0000001
order by $score descending
return string-join((string($query_num), "Q0", concat("WP",$a/name/@id),
      string($rank), string($score), "1_cirquid_ER_TEC_idg"), " ")
```

Notice that each link between two entities is counted separately here. We tested before a version of the query that establishes only one link between two entities e_1 and e_2 even if e_1 links e_2 multiple times. Initial tests on last years data indicated, however, a higher retrieval quality for the above presented query.

3.2   Training

We trained the parameter μ on the data of last year's entity ranking task. For the chosen relevance propagation method a setting of $\mu = 0.85$ showed the best performance with respect to precision on top of the retrieved list as well as for mean average precision. The training results are presented in Table 3.

Table 3. Retrieval quality depending on the setting of μ

μ	0.8	0.85	0.9	0.95
MAP	0.3373	**0.3413**	0.3405	0.3349
P10	0.3739	**0.3783**	0.3717	0.3630

3.3   Results

The official evaluation measure for the INEX XER track is called "xinfAP" [7]. It makes use of stratified sampling for estimating average precision. Table 4 gives an overview of the results for all our submitted runs. For the entity ranking task, it shows that extending the target category set by derived categories from the provided feedback entities considerably improves the results. This observation was expected. We showed in last year's INEX experiments that extending the target category set by child categories improves retrieval as well. Both results can be explained by the fact that articles in Wikipedia that are assigned to a certain category are not necessarily assigned to the parent category as well. Hence, extending the target category set by similar categories always improves

Table 4. Retrieval quality presented in official measures

	ER_TC	ER_TEC	LC_TE
xinfAP	0.235	0.277	0.272
	ER_TC_idg	ER_TEC_idg	LC_TE_idg
xinfAP	0.274	**0.326**	**0.274**

recall. Applying the indegree method results in another clear improvement for the entity ranking task, however we did not achieve the same positive effect for list completion.

4   Conclusions

We demonstrated with this article the flexibility and effectiveness of the chosen approach to integrate the retrieval language NEXI with the database query language XQuery. The PF/Tijah system allows to express a wide range of INEX experiments without changes to the system itself. Often time consuming pre- and post-processing of data is not necessary or reduced to simple string substitutions of query terms for each given query.

Although PF/Tijah does not apply top-k query processing techniques, it shows a good performance on a wide range of NEXI queries. Future developments should address the currently bad supported retrieval on the entire node set, issued by //*-queries.

The INEX entity ranking task demonstrates how standard retrieval functions can be applied to non-standard retrieval tasks with the help of score propagation expressed on the XQuery level. A combined DB/IR system as PF/Tijah can demonstrate here its full advantage.

References

1. Boncz, P., Grust, T., van Keulen, M., Manegold, S., Rittinger, J., Teubner, J.: MonetDBXQuery: a fast XQuery processor powered by a relational engine. In: SIGMOD 2006: Proceedings of the 2006 ACM SIGMOD international conference on Management of data, pp. 479–490. ACM, New York (2006)
2. List, J., Mihajlovic, V., Ramírez, G., de Vries, A., Hiemstra, D., Blok, H.E.: Tijah: Embracing Information Retrieval methods in XML databases. Information Retrieval Journal 8(4), 547–570 (2005)
3. Hiemstra, D., Rode, H., van Os, R., Flokstra, J.: PFTijah: text search in an XML database system. In: Proceedings of the 2nd International Workshop on Open Source Information Retrieval (OSIR), Seattle, WA, USA, Ecole Nationale Supérieure des Mines de Saint-Etienne, pp. 12–17 (2006)
4. Rode, H.: From Document to Entity Retrieval. PhD thesis, University of Twente, CTIT (2008)
5. Grust, T., van Keulen, M., Teubner, J.: Accelerating XPath evaluation in any RDBMS. ACM Trans. Database Syst. 29, 91–131 (2004)

6. Rode, H., Serdyukov, P., Hiemstra, D.: Combining Document- and Paragraph-Based Entity Ranking. In: Proceedings of the 31th Annual International ACM SIGIR Conference on Research and Development in Information Retrieval (SIGIR 2008), pp. 851–852 (2008)
7. Yilmaz, E., Kanoulas, E., Aslam, J.A.: A simple and efficient sampling method for estimating ap and ndcg. In: SIGIR 2008: Proceedings of the 31st annual international ACM SIGIR conference on Research and development in information retrieval, pp. 603–610. ACM Press, New York (2008)

Pseudo Relevance Feedback Using Fast XML Retrieval

Hiroki Tanioka

Innovative Technology R&D, JustSystems Corporation,
108-4 Hiraishi-Wakamatsu Kawauchi-cho Tokushima-shi Tokushima, Japan
hiroki.tanioka@justsystems.com

Abstract. This paper reports the result of experimentation of our approach using the vector space model for retrieving large-scale XML data. The purposes of the experiments are to improve retrieval precision on the INitiative for the Evaluation of XML Retrieval (INEX) 2008 Adhoc Track, and to compare the retrieval time of our system to other systems on the INEX 2008 Efficiency Track. For the INEX 2007 Adhoc Track, we developed a system using a relative inverted-path (RIP) list and a Bottom-UP approach. The system achieved reasonable retrieval time for XML data. However the system has a room for improvement in terms of retrieval precision. So for INEX 2008, the system uses CAS titles and Pseudo Relevance Feedback (PRF) to improve retrieval precision.

1 Introduction

There are two approaches for XML information retrieval (IR): one based on database models, the other based on information retrieval models. Our system is based on the vector space model[3] from information retrieval.

Our system uses keywords (multi-word terms, single words) as the query and separates XML[1] documents into two parts: content information (the keywords) and structural information. XML nodes correspond to retrieval units, and nodes that include query terms can be quickly retrieved using an inverted-file list. For very large XML documents, all XML nodes are indexed to each term directly included in the node itself, but not the node's children or more distantly related nodes. During the retrieval phase, the score of a retrieved node is calculated by merging the scores from its descendant nodes. To merge scores while identifying parent-child relationships, the system employs a relative inverted-path list (*RIP* list)[6] that uses nested labels with offsets to save the structural information.

For INEX 2008, our experiments target both CO and CAS titles. The system accepts CO titles, which are terms enclosed in <title> tags. Furthermore, the system can accept CAS titles as a constrained condition, which are XPath[2] representations enclosed in <castitle> tags. Additionally, for improving retrieval precision, the system adopts Pseudo Relevance Feedback (PRF)[5]. The rest of this article is divided into three sections. In section 2, we describe the IR model for XML documents. In section 3, we describe experimental results. And in section 4, we discuss results and future work.

S. Geva, J. Kamps, and A. Trotman (Eds.): INEX 2008, LNCS 5631, pp. 218–223, 2009.
© Springer-Verlag Berlin Heidelberg 2009

2   XML Information Retrieval

2.1   TF-IDF Scoring

Our system uses a TF-IDF scoring function for retrieval. TF-IDF is additive, therefore a node score can be easily calculated by merging the scores of its descendant nodes. The TF-IDF score L_j of the jth node is composed of the term frequency tf_i of the ith term in the query, the number of nodes f_i including the ith term, and the number of all the nodes n in the XML collection.

$$L_j = \sum_{i=1}^{t} tf_i \cdot \log(\frac{n}{f_i}) \tag{1}$$

However, if the node score is the sum of the scores of its descendants, there is the problem that the root node always has the highest score in the document. Therefore, the score R_j of the jth node is composed of the number T_j of terms contained in the jth node, the score L_k of the kth descendant of the jth node, and the number T_k of terms contained in the kth node.

$$T_j = \sum_{k\ children of\ j} T_k \tag{2}$$

$$R_j = \sum_{k\ children of\ j} D \cdot L_k \tag{3}$$

$$R'_j = \frac{R_j}{T_j} \tag{4}$$

where $D(= 0.75)$ is a decaying constant. And the TF-IDF score R_j is normalized by the number of terms T_j,

Then let α be the set of terms included in the query and β_j be the set of terms included in the jth node. The conjunction, $\gamma_j = \alpha \cap \beta_j$, is the set of query terms included in the jth node. For every node,

$$\delta_j = \bigcup_{k\ children of\ j} \gamma_k \tag{5}$$

$$S_j = \frac{Q}{q} \cdot count(\delta_j) \tag{6}$$

where $Q(= 30)$ is a constant number, and q is the number of terms in the query. S_j is one of the heuristic scores we call a leveling score. If a node contains all terms in the query, the leveling score is the highest.

$$RSV_j = R'_j + S_j \tag{7}$$

After that, the score RSV_j of jth node is composed of the TF-IDF score R'_j and the leveling score S_j. Then the retrieved results are chosen from the node list, which is sorted in descending order of RSV scores. All the parameters are determined using a few topics of the Focused task of the INEX 2007 Adhoc Track[6].

2.2 Simplified XML Database

Our system uses either CO titles or CAS titles for XML retrieval. For CAS titles, the system separates a CAS title into some parts which consist of a combination of terms with XPaths in advance of retrieving. If a retrieved node including the terms is matched with the XPath, the retrieved node's score is multiplied by 1.1. For example, a CAS title is separeted as follows,

> *A CAS title*:
> <castitle>
> //article[about(., philosophy)]//section[about(., meaning of life)]
> </castitle>

> *Terms*: philosophy meaning of life
> *XPaths*: //article//section

The system treats an XML tag included in an XPath like a term, the following tuple is appended into the postings file. And an XPath can be instantly cross-checked with the simplified XML database which uses a postings file and a RIP list.

$TF(=1)$ means a term frequency which is a constant number. The RIP list is expanded for recording a tag ID to each node ID, and the RIP list also preserves a distance between a node and its parent node.

$$\text{Tag-ID: } \{\text{Node-ID}, TF\}$$

$$\text{Node-ID: } \{\text{Distance, Tag-ID}\}$$

For checking $XPath_A$, a node N_0 which is equal to the tail node <title> can be retrieved in the postings file. Then, the parent node <figure> of the node <title> is compared with the parent node N_1 of the retrieved node N_0 using the RIP list. If the compared node N_1 means <figure>, the parent node <sec> of the node <figure> is compared with the parent node N_2 of the node N_1.

$$XPath_A: \text{/sec/figure/title}$$

$$XPath_{ret}: /N_2/N_1/N_0$$

If all the nodes in $XPath_A$ are matched the nodes N_0, N_1 and N_2 in a row, $XPath_{ret}$ ($/N_2/N_1/N_0$) corresponds to $XPath_A$ (/sec/figure/title). The simplified XML database accepts an asterisk operator (*) which means the repetition of an arbitrary XML tag.

For checking $XPath_B$, the simplified XML database just retrieves the node <title> in the postings file, because the parent node of <title> is unconcerned. For checking $XPath_C$, all the retrieved nodes in the postings file are targeted for the tail node (*). Inevitably, the simplified XML database retrieves <figure> with the node (*) skipped.

$$XPath_B: \text{/*/title}$$

$$XPath_C: \text{/figure/*}$$

Therefore, the simplified XML database can accept inquiries regarding XPath using both the postings list and the RIP list with a Bottom-UP approach.

Table 1. Some results on the Focused Task in the Adhoc Track

Run ID	Condition	iP[0.00]	iP[0.01]	iP[0.05]	iP[0.10]	MAiP
RIP_01	(official) VSM	0.7162	0.5225	0.3872	0.2897	0.0992
RIP_02	(official) VSM/CAS	**0.7233**	0.5549	0.4275	0.3259	0.1153
RIP_03	(official) VSM/PRF	0.6289	0.5241	0.3854	0.3107	0.1044
RIP_04	(unofficial) VSM/CAS/PRF	0.6378	**0.5585**	**0.4490**	**0.3584**	**0.1179**

*Metric: iP[x] means interpolated Precision / Recall (invalid submissions are discarded.), as interpolated precision at x recall.

2.3 Pseudo Relevance Feedback

The retrieving method with Pseudo Relevance Feedback (PRF) is reported as a powerful yet simple approach in RIA Workshop[5]. When we extend our system with a PRF method to improve retrieval precision, the system operates as follows,

1. Retrieving XML nodes with original query terms.
2. Extracting terms in the top 10 retrieved XML nodes.
3. Choosing the top 20 ranked terms in the extracted terms.
4. Retrieving XML nodes with both original query terms and the chosen terms.

The PRF method chooses the top 20 ranked terms in the TF-IDF score as additional query terms in the top 10 retrieved XML nodes.

3 Experimental Results

To index the INEX 2008 Adhoc Track document collection, the system first parses all the structures of each XML document with an XML parser and then parses all the text nodes of each XML document with an English parser[1]. The size of the index containing both content information and structure information is about 8.32 GB. Thereafter, the system uses the same index in every experiment.

3.1 INEX 2008 Adhoc Track

In INEX 2008 Adhoc Track, our experiments target both CO and CAS titles. The system accepts CO titles, which are terms enclosed in <title> tags. The system can accept CAS titles with XPath as a constrained condition, which are XML tags enclosed in <castile> tags.

There are the Focused task, the Relevant in Context task, and the Best in Context task in the INEX 2008 Adhoc Track. All the tasks are the same as in the INEX 2007 Adhoc Track. Hence the system parameters are tuned for the Focused task on a few topics from INEX 2007.

---

[1] The English parser of a morphological analyzer uses the Hidden Markov Model and the bigram made by JustSystems Corporation. Also the English parser stems terms and removes some terms which consist of two characters or smaller.

Table 2. Run parameters in the Efficiency Track

Run ID	#CPU	RAM GB	#Nodes	Year	Top-k	Cache
TOPX2-Eff08-CO-1500-Focused-W	4	16	1	2005	1,500	OS+TopX
003-Uni Frankfurt,Architect-S	8	16	8	n/a	1,500	n/a
B2U0 full-depth-heur	1	2	1	n/a	1,500	Lucene
VSM RIP	1	2	1	2008	1,500	None

∗VSM RIP works on a machine which has 2.10GHz CPU, 2GHz RAM and 150GB SATA HDD in Java 1.6.0_12. Each run is chosen as the best MAiP in each affiliate.

Table 3. Some results on the Focused Task in the Efficiency Track

Run ID	iP[0.00]	iP[0.01]	iP[0.05]	iP[0.10]	MAiP	AVG MS..
TOPX2-Eff08-CO-1500-Focused-W	0.4994	0.4560	0.3749	0.3298	0.1409	239.73
003-Uni Frankfurt,Architect-Sim	0.2070	0.1960	0.1812	0.1669	0.0768	326,098.75
B2U0 full-depth-heur	0.4388	0.3964	0.3344	0.3013	0.1357	2,994.00
VSM RIP	0.4836	0.4058	0.3077	0.2553	0.0895	3,167.86
VSM RIP A	-	-	-	-	0.0936	2,751.27
VSM RIP B	-	-	-	-	0.0329	10,905.61
VSM RIP C	-	-	-	-	n/a	12,091.42

∗Metric: iP[x] means interpolated Precision / Recall (invalid submissions are discarded.), as interpolated precision at x recall.

The system is installed on the PC which has 2.1GHz CPU, 2GB RAM, and 150GB SATA HDD, and the system is implemented in Java 1.6.0_12. The time it takes to parse and load the 659,388 files on the PC is about 8.17 hours excluding file-copying time. The database size is about 3.18 GB on HDD.

Table 1 shows revised results for the Focused task using the main evaluation measure for INEX 2008, because submitted scores included erroneous results. RIP_02, which uses the CAS titles, was the best of our runs, ranking 41th in the official ranking with a iP[0.01] score of 0.5535 (revised score of 0.5549). Then, RIP_04 was unofficially the best score as MAiP. As RIP_01 is a baseline run, RIP_02 and RIP_04 both use the CAS titles instead of the CO titles and runs RIP_03 and RIP_04 use PRF. RIP_04 is an unofficial run, which achieves the highest MAiP score.

Also, RIP_03 scores better than RIP_01 at MAiP but lower at iP[0.00]. So the additional query terms boosted up the overall precision but introduced some noise as well, resulting in lower early precision.

3.2 INEX 2008 Efficiency Track

In INEX 2008 Efficiency Track, our experiment uses Type A Topics, Type B Topics and Type C Topics. Table 2 shows the PC environment. Table 3 shows some results[4] on the Focused Task in the Efficiency Track. VSM RIP means results regarding a total of three topic types.

Even though we ran our system on a single processor, it achieved reasonable retrieval time for type A topics while retaining good early precision.

4 Conclusions

According to the experimental results, the runs using the CAS titles achieve higher precision than the runs using the CO titles, indicating that the described approach can effectively use structural hints to improve performance. The runs using PRF improve precision at 1% recall and overall precision, showing that element-based relevance feedback can augment the query while keeping focus on the topic.

According to the experimental results in the Efficiency Track, VSM RIP did relatively well in precision and retrieval time. If the system uses the PRF method, it is important that the system has high-speed performance in order to adjust parameters. Even the precision of the system is expected to be further improved with a combination of CAS titles and the PRF method, a bunch of problems still remain. These problems are regarding not only a term weighting as a vector space model issue but also a noise rejection as an issue for using structural hints.

Acknowledgement

This paper was inspired by the advice of Dr. David A. Evans at JustSystems Evans Research, Inc. The author would like to thank David and other JSERI members who commented on the problem of experiment. This scoring method was simplified based on comments of Dr. Minoru Maruyama at Shinshu University. The author would like to express his appreciation for Dr. Maruyama's brilliant comments.

References

1. Extensible Markup Language (XML) 1.1, 2nd ed., http://www.w3.org/TR/xml11/
2. XML Path Language (XPath) Version 1.0., http://www.w3.org/TR/xpath
3. Salton, G., Wong, A., Yang, C.S.: A vector space model for automatic indexing. Communications of the ACM 18, 613–620 (1975)
4. Kamps, J., Pehcevski, J., Kazai, G., Lalmas, M., Robertson, S.: INEX 2007 Evaluation Measures. In: Fuhr, N., Kamps, J., Lalmas, M., Trotman, A. (eds.) INEX 2007. LNCS, vol. 4862, pp. 24–33. Springer, Heidelberg (2008)
5. Montgomery, J., Luo, S.L., Callan, J., Evans, D.A.: Effect of varying number of documents in blind feedback: analysis of the 2003 NRRC RIA workshop "bf_numdocs" experiment suite. In: SIGIR 2004: Proceedings of the 27th annual international ACM SIGIR conference on Research and development in information retrieval, pp. 476–477. ACM, New York (2004)
6. Tanioka, H.: A Fast Retrieval Algorithm for Large-Scale XML Data. In: Fuhr, N., Kamps, J., Lalmas, M., Trotman, A. (eds.) INEX 2007. LNCS, vol. 4862, pp. 129–137. Springer, Heidelberg (2008)

TopX 2.0 at the INEX 2008 Efficiency Track

A (Very) Fast Object-Store for Top-k-Style XML Full-Text Search

Martin Theobald[1], Mohammed AbuJarour[3], and Ralf Schenkel[1,2]

[1] Max Planck Institute for Informatics, Saarbrücken, Germany
[2] Saarland University, Saarbrücken, Germany
[3] Hasso Plattner Institute, Potsdam, Germany

Abstract. For the INEX Efficiency Track 2008, we were just on time to finish and evaluate our brand-new TopX 2.0 prototype. Complementing our long-
running effort on efficient top-k query processing on top of a relational back-end,
we now switched to a compressed object-oriented storage for text-centric XML
data with direct access to customized inverted files, along with a complete reimplementation of the engine in C++. Our INEX 2008 experiments demonstrate efficiency gains of up to a factor of 30 compared to the previous Java/JDBC-based
TopX 1.0 implementation over a relational back-end. TopX 2.0 achieves overall runtimes of less than 51 seconds for the entire batch of 568 Efficiency Track
topics in their content-and-structure (CAS) version and less than 29 seconds for
the content-only (CO) version, respectively, using a top-15, focused (i.e., non-
overlapping) retrieval mode—an average of merely 89 ms per CAS query and 49
ms per CO query.

1 Introduction

TopX is a native IR-engine for semistructured data with IR-style, non-conjunctive (aka.
"andish") query evaluations, which is particular challenging for efficient XPath-like
query evaluations because of the huge intermediate set of candidate result elements, i.e.,
when any XML element matching any of the query conditions may be a valid result. In
this "andish" retrieval mode, the result ranking is solely driven by score aggregations,
while the query processor needs to combine both content-related and structural aspects
of so-called content-and-structure (CAS) queries into a single score per result element.
Here, high scores for some query dimensions may compensate weak (even missing)
query matches at other dimensions. Thus, the query processor may dynamically relax
query conditions if too few matches would be found in a conjunctive manner, whereas
the ranking allows for the best (i.e., top-k) matches be cut-off if too many results would
be found otherwise. Queries are more difficult to evaluate in "andish" mode than typical
DB-style conjunctive queries, as we can no longer use conjunctive merge-joins (corresponding to intersections of index list objects), but we need to find efficient ways of
merging index lists in a non-conjunctive manner (corresponding to unions of index list
objects, or so-called "outer-joins" in DB terminology). Top-k-style evaluation strategies are crucial in these XML-IR settings, not only for pruning index list accesses (i.e.,
physical I/O's) but also for pruning intermediate candidate objects that need to be managed dynamically (i.e., in main memory) at query processing time. Pruning also the

S. Geva, J. Kamps, and A. Trotman (Eds.): INEX 2008, LNCS 5631, pp. 224–236, 2009.

latter in-memory data structures is particularly beneficial for good runtimes if CPU-time becomes a dominating factor, e.g., when XPath evaluations are costly, or when many index lists already come from cache.

TopX 2.0 combines query processing techniques from our original TopX 1.0 prototype [10,9] and carries over ideas for block-organized inverted index structures from our IO-Top-k algorithm [1] to the XML case. The result is a novel, object-oriented storage (albeit our index objects have a very regular structure) for text-oriented XML data, with sequential, stream-like access to all index objects. Just like the original engine, TopX 2.0 also supports more sophisticated cost models for sequential and random access scheduling along the lines of [5,6,9]. In the following we focus on a description of our new index structure and its relation to the element-specific scoring model we propose, while the non-conjunctive XPath query processor remains largely unchanged compared to TopX 1.0 (merely using merge-joins over entire index blocks instead of per-document hash-joins). Moreover, the TopX core engine has been refurbished in the form of a completely reimplemented prototype in C++, with carefully designed data-structures and our own cache management for inverted lists, also trying to keep the caching capabilities of modern CPU's in mind for finding appropriate data structures.

Core of the new engine is a multiple-nested block-index structure that seamlessly integrates top-k-style sorted access to large blocks stored as inverted files on disk with in-memory merge-joins for efficient score aggregations. The main challenge in designing this new index structure was to marry no less than three different paradigms in search engine design: 1) sorting blocks in descending order of the maximum element score they contain for threshold-based candidate pruning and top-k-style early termination; 2) sorting elements within each block by their id to support efficient in-memory merge-joins; and 3) encoding both structural and content-related information into a single, unified index structure.

2 Scoring Model

We refer the reader to [9] for a thorough discussion of the scoring model, while we merely aim to briefly review the most important concepts here. Our XML-specific version of Okapi BM25 has proven to be effective (and therefore remained unchanged, compare to, e.g., [2]) during four years of INEX Ad-hoc Track participations. It allows for the exact precomputation of fine-grained scores for the major building blocks of a CAS query—tag-term pairs in our case—with specific scores for each element-type/term combination. Computing individual weights for tag-term pairs introduces a certain factor of redundancy compared to a plain per-article/term scoring model, as each term occurrence is recursively propagated "upwards" the document tree, and its frequency is then aggregated with more occurrences of the same term (thus treating each XML element in the classic IR notion of a document). This volitional redundancy factor for materializing the scoring model (formally defined below) roughly corresponds to the average depth of a text node in the collection, while we aim to compensate the storage and I/O overhead through index compression, a novel feature for TopX 2.0. In the following, we distinguish between *content scores*, i.e., scores for the tag-term pairs of a query, and *structural scores*, i.e., scores assigned to additional navigational tags as they occur in longer path conditions or branching path queries.

2.1   Content Scores

For content scores, we make use of element-specific statistics that view the *full-content* of each XML element (i.e., the concatenation of all its descending text nodes in the entire subtree) as a bag of words:

1) the *full-content term frequency*, $ftf(t, n)$, of term t in node n, which is the number of occurrences of t in the full-content of n;
2) the *tag frequency*, N_A, of tag A, which is the number of nodes with tag A in the entire corpus;
3) the *element frequency*, $ef_A(t)$, of term t with regard to tag A, which is the number of nodes with tag A that contain t in their full-contents in the entire corpus.

The score of a tag-term pair of an element e with tag A with respect to a content condition of the form `//T[about(., t)]` (where `T` either matches A or is the tag wildcard operator `*`) is then computed by the following BM25-inspired formula:

$$score(e, \texttt{T[about(., t)]}) = \frac{(k_1 + 1)\, ftf(t, e)}{K + ftf(t, n)} \cdot \log\left(\frac{N_A - ef_A(t) + 0.5}{ef_A(t) + 0.5}\right)$$

$$\text{with} \quad K = k_1\left((1 - b) + b\frac{\sum_{t'} ftf(t', e)}{avg\{\sum_{t'} ftf(t', e') \mid e' \text{ with tag } A\}}\right)$$

We used the default values of $k_1 = 1.25$ and $b = 0.75$ as Okapi-specific tuning parameters (see also [3] for tuning Okapi BM25 on INEX). Note that our notion of tag-term pairs enforces a strict evaluation of the query conditions, i.e., only those elements whose tag matches the target element of the query are returned.

For a content condition with multiple terms, the score of an element satisfying the tag constraint is computed as the sum of the element's content scores for the corresponding content conditions, i.e.:

$$score(e, \texttt{//T[about(., } t_1 \ldots t_m \texttt{)]}) = \sum_{i=1}^{m} score(e, \texttt{//T[about(., } t_i \texttt{)]})$$

Note that content-only (CO) queries have not been in the primary focus when defining this scoring function, rather than keyword conditions as sub-conditions of structural queries. CO queries are therefore evaluated with the tag wildcard operator "*" which matches any tag. The score for a "*"-term pair then is the same as the score for the original tag-term pair. Hence an XML document yields as many "*"-tag matches as there a distinct tags that also match the term condition, possibly each with a different score.

2.2   Structural Scores

Given a query with structural and content conditions, we transitively expand all structural query dependencies. For example, in the query `//A//B//C[about(., t)]` an element with tag `C` has to be a descendant of both `A` and `B` elements (where branching path expressions can be expressed analogously). This process yields a *directed acyclic query graph* with tag-term conditions as leaves, tag conditions as inner nodes, and all transitively expanded descendant relations as edges. Our structural scoring model then counts the number of navigational (i.e., tag-only) conditions that are completely satisfied by the structure of a candidate element and assigns a small and constant score mass c for every such tag condition that is matched. This structural score mass is then aggregated with the content scores, again using summation. In our INEX 2008 setup, we have set $c = 0.01$, whereas content scores are normalized to $[0, 1]$. That is, we emphasized the relative weights of structural conditions much less than in previous INEX years, where we used a structural score mass of $c = 1.0$.

3 Index Structures

Just like in the original TopX 1.0 prototype, our key for efficient query evaluation is a combined inverted index for XML full-text search that combines content-related and structural information of tag-term pairs, as well as a smaller structure-only index that lets us evaluate additional tag conditions of a CAS query. These two index structures directly capture the scoring model described in Section 2. For both index structures, XML elements (matching a given tag-term pair or an individual tag, respectively) are grouped into so-called *element blocks* of elements that share the same tag, term, and document id. Novel for TopX 2.0 is a second level of nesting, where element blocks are in turn grouped into larger, so-called *document blocks*, with element blocks being sorted in ascending order of their document id within each document block, which allows for efficient m-way merge-joins between document blocks when evaluating multi-dimensional queries. Top-k-style index pruning is further supported by sorting the document blocks in descending order of their maximum element-block score. By default, TopX uses a pre-/post-order/level-based labeling scheme [7] to capture the structural information, which proved to be superior to, for-example, Dataguide-based techniques over heterogeneous collections such as INEX-Wikipedia and/or typical NEXI-style CAS queries that heavily make use of the descendant axis. Note that we may skip the *level* information when using the NEXI-style descendant axis only, and we thus focus on $(pre, post, score)$ triplets in the following. In summary, TopX maintains two separate inverted files:

- the *content index* that stores, for each tag-term pair, all elements with that tag that contain the term, including their BM25-based relevance score and their pre- and post labels.
- the *structure index* that stores, for each tag, all elements with that tag, including their pre- and post label.

3.1 Previous Relational Encoding

TopX uses tuples of the format *(tag, term, docid, pre, post, score, maxscore)* as basic schema for the content index, and $(docid, tag, pre, post)$ for the structure index, respectively. Note that $maxscore$ is the maximum $score$ per document id $docid$, element id pre, tag, and $term$, which is needed to enforce a respective grouping of tuples into the element block structure in a relational schema with sorted access in descending order of $maxscore$ [10]. To illustrate the redundancy that arises from a relational encoding of this multi-attribute schema, we quickly review the previous DB schema used in TopX 1.0 that is highly redundant in two orthogonal ways: first, because of the materialization of the scoring model discussed above, with a different score for each tag and term; and second, because of the frequent repetition of attributes (used as keys) required for a relational encoding. Here, a denormalized schema is necessary to implement efficient sorted (i.e., sequential) access to element blocks stored in descending order of their *maxscore* in a database table. In typical DBMS's, B^+-trees are the method of choice for such an index structure with precomputed blocks and sorted access to leaf nodes via linked lists (pointers to other leaf blocks). Note that some DBMS's support the use of so-called Index-Only-Tables (IOT's) to store the entire table directly as a B^+-tree. Assuming a simple 32-bit (i.e., 4-byte) based encoding of integers for the *docid*, *tag*,

Content Index

Doc ID	Tag	Term	Pre	Post	Score	Max Score
2	sec	xml	2	15	0.9	0.9
2	sec	xml	10	8	0.5	0.9
17	title	xml	5	3	0.5	0.5
1	par	xml	6	4	0.7	0.7
...	...	...	...	...	...	...

(4+4+4+4+4+4+4) bytes X 567,262,445 *tag-term pairs*

16 GB

Structure Index

Doc ID	Tag	Pre	Post
1	article	1	245
2	article	1	123
3	sec	2	15
3	sec	10	8
...	...	...	...

(4+4+4+4) bytes X 52,561,559 *tags*

0.85 GB

Fig. 1. Relational encoding and rough space consumption for TopX 1.0 on INEX-Wikipedia

term, *pre*, *post* attributes and 32-bit floating-point numbers for *score* and *maxscore*, we get a rough space consumption of 16 GB for the more than 560 million tag-term pairs and 0.85 GB for the more than 50 million tags we extract from INEX-Wikipedia, respectively (see also Section 5). This corresponds to an index blowup of a factor of about 4 compared to the 4.38 GB of XML sources for this collection.

Thus, a better approach is to keep the volitional redundancy that arises from our fine-grained scoring model but to encode this into a customized block structure that avoids the redundancy that is due to a relational encoding. These blocks are stored directly on disk in the spirit of a more compact object-oriented storage—however with very regular object structure. Our goal is to compress the index to a sensible extent, thus to save CPU time by keeping the fully precomputed scoring model over tag-term-pairs (with element-specific scores of terms), but to use moderate compression techniques to keep the index size at least similar to the overall size of the original XML data and thus to also spare I/O costs.

3.2   Content Index

TopX 2.0 maintains a single, binary inverted-file for content constraints, coined *content index*, whose structure is depicted in Figure 2. Here, the content index consists of individual index lists for two example tag-term pairs //sec[about(.,''XML'')] and //title[about(.,''XML'')], along with their respective file offsets. These are further subdivided into histogram, document and element blocks.

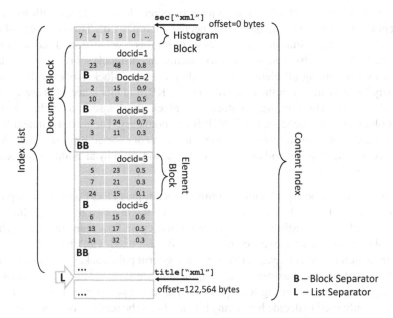

Fig. 2. Inverted and nested block structure of the content index for TopX 2.0

Index Lists. Index lists are the largest units in our index structure. For the structural index, an index list contains all tag-term pairs of elements contained in the collection (whereas for the content index, an index list contains all the elements' tags that occur in the collection). Sequential and random access to these inverted files is implemented by two auxiliary dictionaries each (see below). The physical end of an index list is marked by a special list separator byte L to prevent the stream decoder from jumping into the next list when an index list has been entirely scanned.

Element Blocks. Each element block consists of a header, which contains the document's id (*docid*), and one or more entries of (*pre, post, score*) triplets, where each triplet corresponds to one element within that document that matches the tag-term pair. For efficient evaluation of queries with structural constraints, each entry in an element block consists not only of the element's id (which corresponds to its *pre* value) and *score*, but also encodes the *post* attribute, i.e., the entire information to locate and score an element within the XML document. Element block sizes are data-dependent, i.e., an element block contains as many (*pre, post, score*) triplets as there are XML elements in the document that match the term condition. The physical end of an element block is marked by another reserved separator byte B that indicates the beginning of the next document block.

Document Blocks. As opposed to element blocks, the size of a document block is a configurable parameter of the system and merely needs to be chosen larger than the size of the largest element block. The sequence of document blocks for a tag-term pair is constructed by first computing, for each element block, the maximal score of any element in that block. Within a document block, element blocks are then resorted by document id, to support efficient merge joins of other element blocks with the same document id (as opposed to the hash-joins needed in TopX 1.0). This sequence of element blocks is grouped into document blocks of up this fixed size, and the document blocks are then sorted by descending maximum element-block score, i.e., the maximum $maxscore$ among all element blocks they contain. Block sizes can be chosen generously large, typically in the order of 256–512KB, and block accesses are counted as a single I/O operation on disk. A document block ends as soon as adding another element block would exceed such a 256KB block boundary. The next document block then again contains element blocks with similar $maxscore$ sorted by descending document id $docid$. The physical end of document block is marked by an additional (second) separator byte B.

Histograms. To support probabilistic pruning techniques [10,9] and more sophisticated cost models for random access scheduling [1], the inverted file can include histograms that allow us to estimate, for a given tag-term pair, how many documents have a maximal score below or above a certain value. We use fixed-width histograms with a configurable number of h buckets. For a given tag-term pair, bucket i stores the number of documents whose maximal score is in the interval $[1 - \frac{i-1}{h}; 1 - \frac{i}{h}[$. bounds can be derived from h and i) for tag-term pairs with at least two document blocks, as histograms are only used to decide how many blocks should be read beyond the first block. For these tag-term pairs, the dictionary entry points to the beginning of the histogram instead of the first document block.

Auxiliary Access Structures. TopX maintains an additional, hash-based and persistent dictionary [8] to store the offset, for each tag-term pair, in the above binary inverted file, pointing to where the first document block in the inverted list for that tag-term pair starts for sorted access (SA). The dictionary itself is addressed via random access only, similarly to an extensible hash-index in a DBMS, using a 64-bit hash of the tag-term condition as key and fetching the offset from the structural index as 64-bit value. This dictionary is usually small enough to be largely cached in main memory, allowing for a very efficient access to the file offsets. A second such file-based dictionary can optionally be maintained to find the right offset for random accesses (RA) to document blocks, using 64-bit hashes of tag-term pair plus document id as key in this case.

3.3 Structure Index

TopX maintains a similar inverted file to store, for a given tag, all elements with that tag name, along with their pre and $post$ labels. The overall structure of this file, coined *structure index*, is similar to that of the inverted file for content constraints—a sequence of document blocks for each tag, which in turn consist of a sequence of element blocks. Each element block consists of one pre-/post-order entry for each element in the document matching the tag. Note that there is no score entry needed, as scores for structural constraints are constant and can therefore be dropped from the index. No histogram headers are needed in this index structure for the same reason. In contrast to the content index, element blocks are simply stored in ascending order of document id.

For the sorted access (SA) entry points, a dictionary stores, for each tag, the offset to the first document block in the inverted file for this tag. Analogously to the content index, another optional dictionary can be used to also support random access (RA) to this structural index. This stores, for each tag and document id, the offset to the document block that contains the document's element block, if such a block exists.

3.4 CPU-Friendly Compression

Switching from a relational encoding to a more object-oriented storage already reduces the index size down to about 8 GB—less than half the size needed for a relation schema but still almost twice as much as for the XML source files. Fast decompression speed and low CPU overhead is more important than a maximum-possible compression ratio for our setting. Moreover, with the large number of different attributes we need to encode into our block index, we do not assume a specific distribution of numbers per block, e.g., with most numbers being small, which rules out Huffman or Unary codes. We also intentionally avoid more costly compression schemes like PFor-Delta (compare to [11]) that need more than one scan over the compressed incoming byte stream (not to mention dictionary-based techniques like GZip and the a-like). We thus employ different (simple) compression techniques for different parts of the index based on variations of delta and variable-length encodings requiring only linear scans over the incoming byte stream, thus touching each byte fetched from disk only exactly once.

Within each document block, there is a sequence of element blocks which are ordered by document id. Here, we exploit the order and store, instead of the document id of an element block, the delta to the document id of the previous block. This value is compressed with a variable-byte compression scheme. Moreover, within each element block, we first sort the entries by their *pre* value. We can now also use delta encoding for the *pre* values to save storage space for smaller numbers. However, many element blocks contain just a single entry, so delta encoding alone is not sufficient. For most short documents, values of *pre* and *post* attributes are fairly small (i.e., they can be stored with one or two bytes), but there may still be some values which require three bytes. We therefore encode these values with a variable number of bytes and use an explicit length indicator byte to signal the number of bytes used for each entry. Scores on the other hand are 32-bit floating point numbers between 0 and 1, but storing them in this format would be far too precise for our needs. Instead, we designate a fixed number of bytes to store such a score and store, instead of a floating point value s, the corresponding integer value $\lfloor s \cdot 2^8 \rfloor$ (if we use a single byte), or $\lfloor s \cdot 2^{16} \rfloor$ if we use two bytes. These integer values are then stored with a variable number of bytes, where the actual number of bytes needed is again stored in the encoding byte of the entry. To further reduce space consumption, we write each $(pre, post, score)$ triplet within an element block into a single bit code sequence, allowing us to use even less than one byte per attribute if the numbers are small. Using a fixed precision of only 6 bits for the *score* attribute and up to 13 bits for the *pre* and *post* attributes, we only need a single length-indicator byte per triplet. If *pre* and *post* exceed $2^{13} = 8,196$, we can switch to two bytes per length indicator.

4   Caching

TopX 2.0 also comes with its own cache management, as opposed to the previous TopX 1.0 that could only rely on the underlying system's and DBMS's general caching strategies. Index lists (either top-k-style pruned index list prefixes, or even entire lists) are decoded, and the in-memory data structures are now explicitly kept as cache for subsequent queries. If TopX 2.0 is compiled and run in a 64-bit environment, it may be configured to use generous amounts of memory for the cache. For our Efficiency Track experiments, we set the maximum cache size to up to 2,048 content and structure-related index lists, which led to maximum memory consumption of up to about 4 GB of main memory. For the caching itself, we currently employ a simple least-frequently-used (LFU) pruning strategy, which keeps the frequently used index lists in memory, such that for example the entire set of structural index lists used in the benchmark topics such as `article` or `section` quickly come completely from cache after a few queries are issued.

5   Experiments

5.1   Index Construction and Size

Indexing the INEX Wikipedia corpus [4] consisting of 659,388 XML documents with TopX yields about 567 million tag-term pairs and about 52 million tags (number of indexed elements) as depicted in Table 3. The average depth of an element is 6.72 which is a good indicator for the redundancy factor our scoring model involves. A comparison of storage sizes for the indexes is also depicted in Figure 3, while the uncompressed size of the corpus is 4.38 GB. We used no stemming for faster and more precise top-k runs. Stopwords were removed for the index construction.

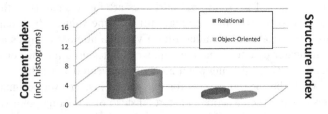

Fig. 3. Comparison of storage space needed for a relational encoding (TopX 1.0) and a more object-oriented, compressed encoding (TopX 2.0), in GB

All experiments in this paper were conducted on a quad-core AMD Opteron 2.6 GHz with 16 GB main memory, running 64-bit Windows Server. Caching was set to up to 2,048 index lists which resulted in a maximum main memory consumption of about 4 GB during the benchmark execution. All runtimes were measured over a hot system cache (i.e., after an initial execution of all queries), and then with varying usages of the TopX 2.0 internal cache (with a C suffix of runs ids indicating that the internal cache was cleared after each query, and with W indicating that the cache was kept

Table 1. Statistics for the content and structure indexes

	Content Index	Structure Index
# index objects	567,262,445 (overall tag-term pairs)	52,561,559 (overall tags)
# index lists	20,810,942 (distinct tag-term pairs)	1,107 (distinct tags)
# document blocks	20,815,884	2,323
# element blocks	456,466,649	8,999,193
index size (relational, uncompressed)	16 GB	0.85 GB
index size (object-oriented, uncompressed)	8.6 GB	0.43 GB
index size (object-oriented, compressed)	3.47 GB	0.23 GB
size of dictionary (SA)	0.55 GB	176 KB
size of dictionary (RA, optional)	2.24 GB	0.27 GB

and managed in a LFU manner). Each benchmark run however started with an empty internal cache and warm disk cache. The size of the RA dictionary is listed here only for completeness, as random-access scheduling has no longer been used for the TopX 2.0 experiments. Note that TopX 1.0 heavily relied on random access scheduling to accelerate candidate pruning because of the relatively bad sequential throughput when using a relational back-end, as compared to the relatively good random-access caching capabilities of most DBMS's, which makes random access scheduling more attractive over a relational back-end. With plain disk-based storage and large document block sizes of 256 KB, we were able to spare random accesses altogether.

Index construction still is a rather costly process, taking about 20 hours over INEX-Wikipedia to fully materialize the above index structure from the XML source collection—due to our explicit pre-computations but to the benefit of very good query response times. We achieve a compression factor of more than 2 compared to an uncompressed storage, which helps us keep the index size in the same order as the original data. Altogether, we obtain a factor of 4 less storage space compared to an uncompressed relational encoding.

5.2 Summary of Runs

Table 2 summarizes all Efficiency Track runs, using various combinations of Focused vs. Thorough, CO vs. CAS, as well as top-k points of $k = 15$, 150, $1,500$, the latter being the original result size demanded by the Ad-Hoc track.

Table 2. Effectiveness vs. efficiency summary of all TopX 2.0 runs

Run ID	iP[0.00]	iP[0.01]	iP[0.05]	iP[0.10]	MAiP	AVG MS.	SUM MS.	#Topics
Focused								
TOPX2-Eff08-CO-15-Focused-W	0.4751	0.4123	0.2793	0.1971	0.0726	49.79	28,180	566
TOPX2-Eff08-CO-150-Focused-W	0.4955	0.4520	0.3674	0.3114	0.1225	85.96	48,653	566
TOPX2-Eff08-CO-1500-Focused-W	0.4994	0.4560	0.3749	0.3298	0.1409	239.73	135,688	566
TOPX2-Eff08-CAS-15-Focused-W	0.4587	0.3878	0.2592	0.1918	0.0662	90.99	51,499	566
TOPX2-Eff08-CAS-150-Focused-W	0.4747	0.4282	0.3494	0.2915	0.1094	112.32	63,574	566
TOPX2-Eff08-CAS-1500-Focused-W	0.4824	0.4360	0.3572	0.3103	0.1241	253.42	143,436	566
Thorough								
TOPX2-Eff07-CO-15-Thorough-W	n/a	n/a	n/a	n/a	n/a	70.91	40,133	566
TOPX2-Eff07-CAS-15-Thorough-W	n/a	n/a	n/a	n/a	n/a	89.31	50,549	566
Focused (cold internal cache)								
TOPX2-Eff07-CO-15-Focused-C	0.4729	0.4155	0.2795	0.1979	0.0723	51.65	29,234	566
TOPX2-Eff07-CAS-15-Focused-C	0.4554	0.3853	0.2583	0.1905	0.0655	96.22	54,461	566

The iP and MAiP effectiveness measures reflect the 308 Ad-Hoc topics from INEX 2006–2008 for which assessments were readily available. Only 566 out of 568 topics were processed due to a rewriting problem that led to empty results for two of the topics. We generally observe a very good early precision at the lower recall points, which is an excellent behavior for a top-k engine. Compared to overall results from the Ad-Hoc Track, we however achieve lower recall at the top-1,500 compared to the best participants (also due to not using stemming and strictly evaluating the target element of CAS queries). TopX 2.0 however shows an excellent runtime behavior of merely 49.70 ms. average runtime per CO and 90.99 ms. average runtime per CAS query. Also, starting each query with a cold internal cache (but warm system cache) instead of a warm internal cache consistently shows about 10 percent decrease in runtime performance for both the CO and CAS modes.

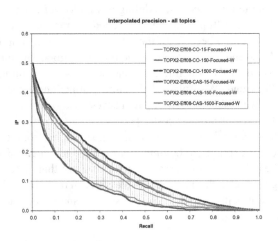

Fig. 4. Interpolated precision plots of all TopX 2.0 Efficiency Track runs

5.3   Efficiency Runs by Topic Type

Tables 3–5 depict the TopX 2.0 performance summaries grouped by topic type. We see a very good runtime result of only 19 ms average processing time for CO topics and 49 ms on average per CAS query for the type (A) (i.e., classic Ad-Hoc topics) and top-15 runs. Also, CO retrieval does not only seem to be more efficient but also more effective than a respective CAS mode, with a maximum MAiP value of 0.14 for CO compared to 0.12 for CAS (returning the top-1,500 elements in Focused mode). As expected, there is a serious runtime increase for type (B) topics, with up to 112 query dimensions, but only comparably little additional overhead for larger values of k, as runtimes are dominated by merging the huge amount of index lists here (in fact most top-k approaches seem to degenerate for high-dimensional queries). Type (C) topics were very fast for the small value of $k = 15$ but then showed a high overhead for the larger values of k, i.e., when returning the top-150 and top-1,500. As no assessments were available for the 7 type (C) (structure-enhanced) topics, the respective effectiveness fields are left blank.

Table 3. Summary of all TopX 2.0 Focused runs over 538 type (A) topics

Run ID	MAiP	AVG MS.	SUM MS.	#Topics
TOPX2-Eff08-CO-15-Focused-W	0.0712	18.88	10,157	538
TOPX2-Eff08-CO-150-Focused-W	0.1234	49.12	26,427	538
TOPX2-Eff08-CO-1500-Focused-W	0.1430	191.27	102,903	538
TOPX2-Eff08-CAS-15-Focused-W	0.0643	48.84	26,276	538
TOPX2-Eff08-CAS-150-Focused-W	0.1094	61.25	32,953	538
TOPX2-Eff08-CAS-1500-Focused-W	0.1249	165.53	89,055	538

Table 4. Summary of all TopX 2.0 Focused runs over 21 type (B) topics

Run ID	MAiP	AVG MS.	SUM MS.	#Topics
TOPX2-Eff08-CO-15-Focused-W	0.0915	844.67	17,738	21
TOPX2-Eff08-CO-150-Focused-W	0.1094	1038.90	21,817	21
TOPX2-Eff08-CO-1500-Focused-W	0.1125	1468.67	30,842	21
TOPX2-Eff08-CAS-15-Focused-W	0.0915	1044.71	21,939	21
TOPX2-Eff08-CAS-150-Focused-W	0.1096	1074.66	22,568	21
TOPX2-Eff08-CAS-1500-Focused-W	0.1124	1479.33	31,066	21

Table 5. Summary of all TopX 2.0 Focused runs over 7 type (C) topics

Run ID	MAiP	AVG MS.	SUM MS.	#Topics
TOPX2-Eff08-CO-15-Focused-W	n/a	41.00	287	7
TOPX2-Eff08-CO-150-Focused-W	n/a	58.86	412	7
TOPX2-Eff08-CO-1500-Focused-W	n/a	277.57	1,943	7
TOPX2-Eff08-CAS-15-Focused-W	n/a	469.42	3,286	7
TOPX2-Eff08-CAS-150-Focused-W	n/a	1150.14	8,051	7
TOPX2-Eff08-CAS-1500-Focused-W	n/a	3330.71	23,315	7

6   Conclusions and Future Work

This paper introduces our new index structure for the TopX 2.0 prototype and its initial evaluation on the INEX 2008 Efficiency Track. TopX 2.0 demonstrates a very good allround performance, with an excellent runtime for keyword-oriented CO and typical CAS queries and still good runtimes for very high-dimensional content (type B) and strucural (type C) query expansions. Overall we believe that our experiments demonstrate the best runtimes reported in INEX so far, while we are able to show that this performance does not have to be at the cost of retrieval effectiveness. The scoring model and non-conjunctive query evaluation algorithms remained unchanged compared to TopX 1.0, which also managed to achieve ranks 3 and 4 in the Focused Task of the 2008 Ad-hoc Track.

References

1. Bast, H., Majumdar, D., Theobald, M., Schenkel, R., Weikum, G.: IO-Top-k: Index-optimized top-k query processing. In: VLDB, pp. 475–486 (2006)
2. Broschart, A., Schenkel, R., Theobald, M., Weikum, G.: TopX @ INEX 2007. In: Fuhr, N., Kamps, J., Lalmas, M., Trotman, A. (eds.) INEX 2007. LNCS, vol. 4862, pp. 49–56. Springer, Heidelberg (2008)

3. Clarke, C.L.A.: Controlling overlap in content-oriented XML retrieval. In: Baeza-Yates, R.A., Ziviani, N., Marchionini, G., Moffat, A., Tait, J. (eds.) SIGIR, pp. 314–321. ACM Press, New York (2005)
4. Denoyer, L., Gallinari, P.: The Wikipedia XML Corpus. In: SIGIR Forum (2006)
5. Fagin, R., Lotem, A., Naor, M.: Optimal aggregation algorithms for middleware. In: PODS. ACM Press, New York (2001)
6. Fagin, R., Lotem, A., Naor, M.: Optimal aggregation algorithms for middleware. J. Comput. Syst. Sci. 66(4), 614–656 (2003)
7. Grust, T.: Accelerating XPath location steps. In: Franklin, M.J., Moon, B., Ailamaki, A. (eds.) SIGMOD Conference, pp. 109–120. ACM Press, New York (2002)
8. Helmer, S., Neumann, T., Moerkotte, G.: A robust scheme for multilevel extendible hashing. In: Computer and Information Sciences - 18th International Symposium (ISCIS), pp. 220–227 (2003)
9. Theobald, M., Bast, H., Majumdar, D., Schenkel, R., Weikum, G.: TopX: efficient and versatile top-k query processing for semistructured data. VLDB J. 17(1), 81–115 (2008)
10. Theobald, M., Schenkel, R., Weikum, G.: An efficient and versatile query engine for TopX search. In: Böhm, K., Jensen, C.S., Haas, L.M., Kersten, M.L., Larson, P.-Å., Ooi, B.C. (eds.) VLDB, pp. 625–636. ACM Press, New York (2005)
11. Zhang, J., Long, X., Suel, T.: Performance of compressed inverted list caching in search engines. In: WWW '08: Proceeding of the 17th international conference on World Wide Web, pp. 387–396. ACM Press, New York (2008)

Aiming for Efficiency by Detecting Structural Similarity

Judith Winter, Nikolay Jeliazkov, and Gerold Kühne

J.W. Goethe University, Department of Computer Science, Frankfurt, Germany
{winter,kuehne}@tm.informatik.uni-frankfurt.de,
nikolay.jeliazkov@gmail.com

Abstract. When applying XML-Retrieval in a distributed setting, efficiency issues have to be considered, e.g. reducing the network traffic involved in answering a given query. The new Efficiency Track of INEX gave us the opportunity to explore the possibility of improving both effectiveness and efficiency by exploiting structural similarity. We ran some of the track's highly structured queries on our top-k search engine to analyze the impact of various structural similarity functions. We applied those functions first to the ranking and based on that to the query routing process. Our results indicate that detection of structural similarity can be used in order to reduce the amount of messages sent between distributed nodes and thus lead to more efficiency of the search.

Keywords: XML Retrieval, Structural Similarity, Distributed Search, INEX.

1 Introduction and Motivation

While most systems participating in INEX aim at effectiveness in terms of precision and recall, we focus our research on distributed IR solutions where good performance includes both effectiveness and efficiency. Therefore, we took a great interest in the new Efficiency Track, especially after the cancellation of the heterogeneous track. Our motivation to participate in it was based on several practical and theoretical considerations. First of all, our system is based on a peer-to-peer (P2P) network where network traffic between peers has to be considered. We cannot send long posting lists but have to prune them which corresponds to the particular encouragement of top-k style search engines in the Efficiency Track. Secondly, our system exploits structural user hints to improve both effectiveness in the ranking process and efficiency in the routing process. Contrary to INEX's ad-hoc topics, which consist of many CO and poorly structured CAS queries, the high-dimensional structured type-C queries of the efficiency track offer an excellent opportunity to test more sophisticated structural similarity functions. Thirdly, many ad-hoc participants tune their systems for the Wikipedia collection whereas most P2P systems face heterogeneous collections based on different schemas and varying in content and structure. Hence, a more database-oriented view of INEX would be in our interest, e.g. a discussion on the application of schema-mapping methods in XML-retrieval. Finally, efficiency issues such as reducing the amount of messages between peers are of major concern in P2P-IR. The opportunity to discuss these challenges with other participants interested in efficiency issues is highly appreciated, e.g. to analyze how XML structure can help.

S. Geva, J. Kamps, and A. Trotman (Eds.): INEX 2008, LNCS 5631, pp. 237–242, 2009.

In this paper, different structural similarity functions are compared in order to take advantage of richly structured CAS queries. Our runs were performed with the top-k search engine Spirix [6] that is based on a P2P network. We first evaluate the use of CAS queries in the ranking of XML documents and then apply our results on the routing process of Spirix to improve its effectiveness and efficiency.

2 Measuring Structural Similarity

Detecting similarity between structures is a current area of research. Existing solutions propose different strategies and functions to calculate the similarity between the given structural conditions in a query and the structures found in a collection. For example, [1] proposes a novel XML scoring method that accounts for both structure and content while considering query relaxations. According to [2], scoring strategies for structure can be divided into four groups depending on the thoroughness they achieve in analyzing the similarity: *perfect match*, *partial match*, *fuzzy match*, and *baseline (flat)*. In the first case, only exact matching structures are considered, thus only retrieving relevant documents with the search terms found in the exact XML context as specified by the user. In the case of the *partial match* strategies, one of the compared structures has to be a sub-sequence of the other and the overlapping ratio is measured. The *fuzzy match* type of functions takes into account gaps or wrong sequences in the different tags. The *baseline* strategy ignores the specified XML structures, thus resulting in a conventional IR technique and losing the possible benefits of XML structured documents.

We analyzed several functions as representatives of the mentioned types of strategies. A formula which determines whether the query or the found structure is a subsequence of the other one and then measures the overlapping ratio between them (*partial match*) is proposed in [2]. The group of the *fuzzy match* type of similarity functions performs deeper analysis of the considered structures and noticeably expands the flexibility and the possibility of a precise ranking of all found structures for a search term according to their similarity to the query structure. In order to achieve this, [5] handles the structures as sequences of tags and uses a strategy of counting the costs for transforming one structure into another. It is based on the *Levenstein Edit Distance* [4], a method used to compute the similarity between two strings by counting the operations *delete*, *insert* and *replace* needed to transform one string into another. This method allows similar measuring of the difference between XML structures by considering every tag as a single character.

Another approach for the similarity analysis of XML structures within the scope of *fuzzy* type strategies is the definition of a number of factors describing specific properties of the compared structures and combining them in a single function. Five such factors are proposed in [3], divided in two major groups - *semantic* and *structural*. The first group consists of the factors *semantic completeness (SmCm)*, measured by the ratio of found query tags to the total amount of tags in the query structure, and *semantic correctness (SmCr)*, measured as a function of all semantic similarities between the tags in the query and the target structures. Within the *structural* group, three factors are distinguished in [3]. *Structural completeness (StCm)* represents the overall coverage of the query XML tree by the target tree - how many of the wanted hierarchical parent-child relationships between the tags are satisfied by an analogous pair in the found structure. The *structural correctness (StCr)* is computed as the complement of the amount of found but reversed hierarchical pairs in respect to all found pairs. The *structural cohesion (StCh)* represents the deviation of the found XML structure from the query and is calculated by the complement of the ratio between the non-relevant tags and the total amount of tags in the target.

3 Using Structural Similarity for XML-Retrieval

Can detecting structural similarity help to improve IR performance? We analyzed several different types of functions, applied appropriate enhancements and evaluated them with the INEX measures by using the Wikipedia document collection and a selection of topics with structural conditions.

We developed the following formula based on [2] as a representative of the *partial match* type of strategies:

$$Sim_1(s_q, s_{ru}) = \begin{cases} \left(\dfrac{1 + |s_q|}{1 + |s_{ru}|}\right)^{\alpha}, & \text{if } s_q \text{ is sub - sequence of } s_{ru} \\ \beta \cdot Sim_1(s_{ru}, s_q), & \text{if } s_{ru} \text{ is sub - sequence of } s_q \\ 0, & \text{else} \end{cases} \qquad \textbf{(ArchSim)}$$

s_q represents the structural condition in the query and s_{ru} stands for the structure of the search term found in the collection. Both parameters α and β allow for finer tuning of the calculated similarity value.

We also implemented a tag dictionary that contains values for the similarity between known tags. For example, *<author>* and *<writer>* are rather similar and a precise similarity value can be assigned to these tags. We are considering the possibility of giving the user the opportunity to define such similarities himself. As a representative of the class of functions based on cost calculation for transforming the query structure into the target one, we used the method proposed in [5]. We implemented and evaluated this method with a suitable normalization and, as above, an enhancement with a tag dictionary was also applied **(PathSim)**.

The five similarity factors from [3] capture different similarity aspects between two XML structures. We used the arithmetic mean to compute the *SmCr* and measured the similarities between tags with methods based on [4]. We used these factors to construct combined similarity functions. In order to compare these functions, we built a small but heterogeneous document collection with search terms occurring in many different XML contexts, resulting in a number of structures to be compared and ranked. Several functions were tested in the process with a number of parameters. We achieved the best ranking results with the following formula:

$$Sim_3 = \alpha \cdot \left(\frac{\beta.SmCm + (2 - \beta).SmCr}{2}\right) + (1 - \alpha) \cdot \left(\frac{\delta.StCm + \gamma.StCr + (3 - \delta - \gamma).StCh}{3}\right) \quad \textbf{(FineSim)}$$

All similarity factors were normalized and parameter boundaries were set such that the resulting single similarity value remains within the interval [0,1]. Parameter α provides an opportunity to shift the weight of the similarity between the two classes of factors – *semantic* and *structural*. The other parameters β, δ and γ can be used for further fine-tuning. After evaluation, we chose the value $\alpha = 0.7$. All other factors were set to 1. We designed an appropriate index in order to support these strategies in the context of P2P networks. For each term, a separate posting list for each of its structures is stored. This allows efficient selecting of postings according to structural similarity between hints in CAS topics and the stored structures of a specific term.

4 Evaluation

All similarity functions were implemented as a part of our P2P-based search engine Spirix [6]. The hardware used was a Linux system with 8x2,5GHz Intel Xeon CPU and 16GB of RAM. Of all Efficiency Track topics, we chose 80 richly structured topics from the INEX 2007 ad-hoc track, which are part of the type-A topics in 2008, and all type-C topics as those are high-dimensional structured queries. However, the type-C topics were not assessed and have therefore not been officially evaluated. We later submitted additional runs with all 70 assessed topics from the ad-hoc track of 2008, which are also part of the type-A topics of the efficiency track.

Our runs aimed at comparing the three proposed structural similarity functions with a baseline run, where no structure was used, as well as with a run where only perfect matches were considered. We first compared the functions when using them for ranking, i.e. aimed at improved effectiveness by using structural hints. Figure 1 shows the evaluation of the INEX 2007 topics, Table 2 displays the results using INEX 2008 topics (focused task). Secondly, we applied the best performing similarity function to improve efficiency during routing and retrieval.

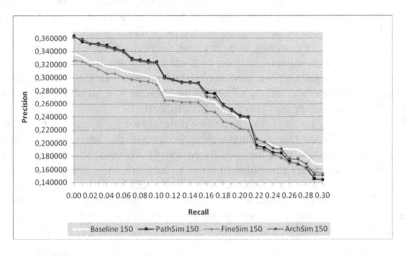

Fig. 1. iP at recall <0.30, for the compared similarity functions (INEX'07 topics), 150 Postings

The *strict/perfect match* case resulted in very low retrieval quality, as the most relevant documents were disqualified in the ranking process due to slight structural differences. Thus, this strict handling of the structural hints specified by the user can only be advantageous in database-oriented approaches and is of no use in the context of IR. At recall-levels up to iP[0.22], an increase in interpolated precision can be observed for the similarity functions *PathSim* and *ArchSim*, compared to the baseline. At iP[0.01], for example, the baseline run achieves only 0,332 while the *PathSim* run achieve 0,355 precision, which is an increase of 7% precision (+2,3 absolute precision). For higher recall-levels (above iP[0.30]), this advantage disappears and the baseline strategy shows a better interpolated precision than any other case. Nevertheless, most users are interested in early precision only – for which the use of similarity functions can lead to an improvement.

Table 2. Early Precision for the compared similarity functions (INEX'08 topics) and C-/S-Run

	Baseline (noSim)	PathSim	FineSim	ArchSim	Strict	C-/S-Run
iP[0.00]	0,494295	0,503136	0,489814	0,495336	0,017178	0,713448
iP[0.01]	0,485375	0,499124	0,489141	0,494979	0,017178	**0,678677**
iP[0.05]	0,436580	0,446879	0,445057	0,445081	0,014737	0,564768
iP[0.10]	0,394112	0,389201	0,388817	0,388549	0,009871	0,491477

For the INEX 2008 topics, we did not select structured topics only but chose all 70 assessed topics. A significant improvement of 2,83% (+1,37 absolute precision) by using the compared structural similarity functions is shown for the official INEX measure, recall level iP[0.01]. However, this measure is known to be rather unstable and the improvement decreases for higher recall level. From iP[0.09] on, the baseline without using structure performs best.

Also shown in Table 2 is the performance of a tuned run for the INEX 2008 topics. This run simulates a client/server environment by using only one node, by selecting 500 postings (which in other experiments we showed to be sufficient for this kind of setting), by retrieving elements and by using BM25 (default parameters k, b). This run performed quite well, with a interpolated precision of 0,6787 at iP[0.01]. Only 3 systems performed better at the focused task of the ad-hoc track.

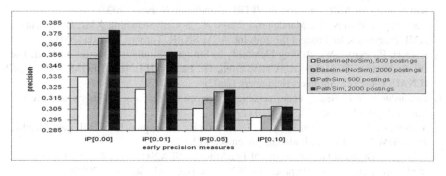

Fig. 2. Using structure to improve routing and reduce the size of posting lists

Based on the results above, we applied the best performing structure function *PathSim* on the routing process (this run was not submitted to the Efficiency Track but is based on the results in Figure 1; interpolated precisions are slightly different due to changes in the ranking algorithm). The postings (500 respective 2000 postings) were selected using BM25E for the baseline run. In the better performing *PathSim* run, the postings were selected by an impact factor proportional to the product of a BM25E-based weight and the calculated structural similarity. Figure 2 displays how calculating structural similarity can improve efficiency by helping to select the adequate postings. For example, at iP[0.01] we achieve a better interpolated precision (0,3580) by using *PathSim*, even if we select only 500 postings instead of 2000 for the *baseline*. Thus, we can save more than 1500 postings.

5 Discussion

In this paper, our participation in the new Efficiency Track of INEX 2008 was discussed. We have welcomed the possibility of using higher dimensioned structural hints, like the type-C topics, in order to retrieve more relevant and precise results, especially in the context of a convergence between IR-based and database-oriented approaches, where additional information such as XML structure can be exploited to achieve good performance. The newly started Efficiency Track offers new possibilities to continue the research in this field.

We have extended and evaluated various structural similarity functions. For the selected structured topics, an improvement of performance could be achieved during ranking and routing. This evaluation is an indication that structural hints can help indeed. In order to confirm the generality of this observation, we will extend our research further. For this purpose, we will need more highly-structured topics which should be more carefully constructed than the current ones. Additionally, we think that the Wikipedia collection lacks the variety of needed rich semantic structures.

References

[1] Amer-Yahia, S., Koudas, N., Marian, A., Srivastava, D., Toman, D.: Structure and Content Scoring for XML. In: Proc. of VLDB, Trondheim, Norway (2005)
[2] Carmel, D., Maarek, Y., Mandelbrod, Mass, Y.: Soffer: Searching XML Documents via XML Fragments. In: Proc. of the 26th Int. ACM SIGIR, Toronto, Canada (2003)
[3] Ciaccia, P., Penzo, W.: Adding Flexibility to Structure Similarity Queries on XML Data. In: Andreasen, T., Motro, A., Christiansen, H., Larsen, H.L. (eds.) FQAS 2002. LNCS (LNAI), vol. 2522. Springer, Heidelberg (2002)
[4] Levenshtein, V.I.: Binary codes capable of correcting deletions, insertions, and reversals. Soviet Physics Doklady 10, 707–710 (1966)
[5] Vinson, A., Heuser, C., Da Silva, A., De Moura, E.: An Approach to XML Path Matching. In: WIDM 2007, Lisboa, Portugal, November 9 (2007)
[6] Winter, J., Drobnik, O.: University of Frankfurt at INEX2008 – An Approach For Distributed XML-Retrieval. In: Preproc. of INEX 2008, Dagstuhl, Germany (2008)

Overview of the INEX 2008 Entity Ranking Track

Gianluca Demartini[1], Arjen P. de Vries[2], Tereza Iofciu[1], and Jianhan Zhu[3]

[1] L3S Research Center
Leibniz Universität Hannover
Appelstrasse 9a D-30167 Hannover, Germany
{demartini,iofciu}@L3S.de
[2] CWI & Delft University of Technology
The Netherlands
arjen@acm.org
[3] University College London
Adastral Park Campus
Ipswich, IP5 3RE, UK
jianhan.zhu@ucl.ac.uk

Abstract. In many contexts a search engine user would prefer to retrieve entities instead of just documents. Example queries include "Italian nobel prize winners", "Formula 1 drivers that won the Monaco Grand Prix", or "German spoken Swiss cantons". The XML Entity Ranking (XER) track at INEX creates a discussion forum aimed at standardizing evaluation procedures for entity retrieval. This paper describes the XER tasks and the evaluation procedure used at the XER track in 2008, focusing specifically on the sampled pooling strategy applied first this year. We conclude with a brief discussion of the predominant participant approaches and their effectiveness.

1 Introduction

Many user tasks would be simplified if search engines would support typed search, and return entities instead of just web pages. In 2007, INEX has started the XML Entity Ranking track (INEX-XER) to provide a forum where researchers may compare and evaluate techniques for engines that return lists of entities. In entity ranking and entity list completion, the goal is to evaluate how well systems can rank entities in response to a query; the set of entities to be ranked is assumed to be loosely defined by a generic category, given in the query itself, or by some example entities. The 2008 track continues to run the entity ranking (ER) and list completion (LC) tasks. In addition, we setup an entity relation search (ERS) pilot task investigating how well systems could establish correct relations between entities. For evaluation purpose we adopted a stratified sampling strategy for creating the assessment pools, using xinfAP as the official evaluation metric [9].

S. Geva, J. Kamps, and A. Trotman (Eds.): INEX 2008, LNCS 5631, pp. 243–252, 2009.

The remainder of the paper is organized as follows. In Section 2 we present details about the collection used in the track and the three different search tasks. Next, in Section 3 we describe some experiments with respect to possible sampling strategies to be used for creating the assessment pool. In Section 4 we summarize the evaluation results. Finally, in Section 5, we conclude the paper.

2   INEX-XER Setup

The goal of the XER track at INEX is to evaluate systems built for returning entities instead of documents. Entity retrieval can be characterized as 'typed search'. In the specific case of the INEX XER track, categories assigned to Wikipedia articles are used to define the *entity type* of the results to be retrieved. Topics are composed of a set of keywords, the entity type(s), and, for the LC task, a set of relevant entity examples.

Given this setup, expert finding [5] can be viewed as an instance of entity ranking, where the entity type is fixed to "people", and the query describes the desired expertise. The analogy is however not perfect: in entity retrieval the keywords in the query do not necessarily indicate expertise, but may capture other types of people finding (including, e.g., general searches for politicians or actors).

2.1   Data

Since data set and main tasks have not been modified, we only give a short summary while referring to the XER 2007 overview paper for more information [1]. The track uses the Wikipedia XML collection, where we exploit the category metadata about the pages to define the entity types. Participants are challenged to exploit fully Wikipedia's rich text, structure and link information.

Entities are assumed to correspond loosely to those Wikipedia pages that are labeled with the given category (or perhaps a sub-category of the given category). Retrieval models for entity ranking should handle the situation that the category assignments to Wikipedia pages are not always consistent and far from complete. The human assessor is of course not constrained by the category assignments made in the corpus when making his or her relevance assessments!

2.2   Tasks

XML Entity Ranking (XER) and List Completion (LC) concern information needs represented as triples of type `<query, category, entity>`. The `category` (that is entity type), specifies the type of objects to be retrieved. The `query` is a free text description that attempts to capture the information need. The `entity` attribute specifies a set of example instances of the given entity type. ER runs are given as input the `query` and `category` attributes, where LC runs are based on `query` and `entity`. In both cases, the system should return the relevant entities, which are represented by the their corresponding Wikipedia pages.

2.3   Topics

Participants from eleven institutions have created a small number of (partial) entity lists with corresponding topic text. Candidate entities correspond to the names of articles that loosely belong to categories (for example may be sub-category) in the Wikipedia XML corpus. As a general guideline, the topic title should be type explanatory, i.e., a human assessor should be able to understand from the title what type of entities should be retrieved. Some topics have been extended for the ERS pilot task, as we will detail below.

2.4   The 2008 Test Collection

The test collection created during INEX XER 2008 consists of 35 topics and their assessments in an adapted trec_eval format (adding strata information) for the xinfAP evaluation script. Topics 101-149 are genuine XER topics, in that the participants created these topics specifically for the track, and (almost all) topics have been assessed by the original topic authors. From the originally proposed topics, topics with less than 7 relevant entities and topics with more than 74 relevant entities have been excluded from the test collection (because they would be unstable or incomplete, respectively). Three more topics were dropped, one on request of the topic assessor and two due to unfinished assessments, resulting in a final XER 2008 test collection consisting of 35 topics with assessments.

2.5   Relation Search Pilot (ERS)

The motivation of the relation search (ERS) pilot task is that searchers may want to know details about previously retrieved entities, and, specifically, their relations with other entities. An example relation search seeks museums in the Netherlands exhibiting Van Goghs artworks, and the cities where these museums are located. A system needs to first find a number of relevant museums, and then establish correct correspondence between each museum and a city. The ERS task could help explore connections between information retrieval and related fields like information extraction, social network analysis, natural language processing, the semantic web, and question answering.

We divide ERS into an entity ranking phase followed by a relation search phase. ERS information needs are represented as tuples of type `<query, category, entity, relation-query, target-category, target-entity>`. The first three attributes have already been defined for the ER and LC tasks. The `relation-query` gives a free text description of the relation between an entity and a target entity. The `target-category` specifies the type of the target entity, and `target-entity` specifies example instances of the target entity type. We did collect 34 ERS topic versions based on the other tasks' 49 topics; after the selection process performed on the XER topics, 23 ERS topics are left as part of the final set. Unfortunately, only two participants submitted ERS runs (four in total). Given the very low number of runs, we did not pursue the assessments needed for the ERS task. A first discussion about ERS and its evaluation is described in [10].

3   Investigation on Sampling Strategies

In INEX-XER 2008, we decided to use sampling strategies for generating pools of documents for relevance assessments. The two main reasons for sampling are to reduce the judging effort and to include into the pools also documents from higher ranks.

The first aspect we have to analyse is how the comparative performances of systems is affected while we perform less relevance judgements. We used the 2007 INEX-XER collection simulating the situation of performing less relevance judgements. We compared three different sampling strategies, that is, a uniform random sampling from the top 50 documents retrieved by the IRSs, a sampling based on the relevance distribution among the different ranks, and a stratified sampling with strata manually defined by looking at the distribution of relevance from the previous year.

For the experimental comparison of the three different sampling approaches we used the 24 INEX-XER topics from the 2007 collection. As only data from 2007 could be used at the time of the study (to design the XER 2008 track setup), we used the leave-one-out approach for simulating the approach of learning from past data. That is, we considered all the topics but one as previous year data. In these topics the relevance assessments and, therefore, the relevance distribution over the ranks is known. The relevance distribution computed on all the topics but one is used for generating a random sample (based on such probabilities) of documents from the runs. The systems' ranking on the remaining topic (the one left out and therefore not used for learning) is then computed and compared with the original ranking. This process is iterated over all topics and the average correlation value[1] is taken.

3.1   Uniform Random Sampling

The first approach we decided to investigate is a Uniform Random Sampling of retrieved documents which would allow to compute metrics such as infAP [8]. In order to do so, we first randomly selected some ranks at which to take documents from the runs. Then, we considered only the assessments on those documents for ranking the systems, assuming that the other entities were not judged (and therefore not relevant). Finally, we measured the correlation with the original system ranking using the 24 XER topics from 2007. Figure 1 presents the result. We conclude that the desirable high correlation is feasible as long as sufficiently many assessments are made.

3.2   Relevance Based Random Sampling

In order to perform a sampling with a higher chance of selecting relevant documents into the pools, we used the distribution of relevance over ranks and learned from the 2007 data the probability of finding a relevant entity at each rank (up to 50 as the depth of 2007 pool) in the runs. We then sampled the 2008 runs

---

[1] We used Kendall's τ as measure for ranking correlation.

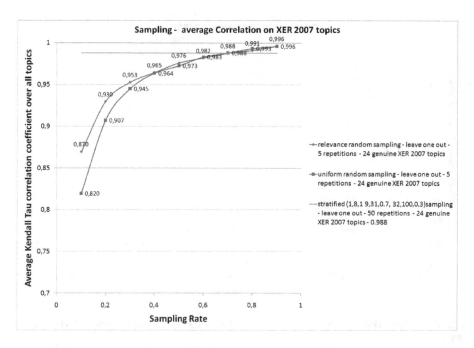

Fig. 1. Correlation results between the original system ranking and the ranking derived using the strata-based sampling strategy

using such probability distribution. The relevance distribution in 2007 for ranks up to 100 is displayed in Figure 2.

Figure 1 shows the correlation between the original system ranking and the ranking derived from either the relevance based or the uniform random sampling.

3.3   Stratified Sampling

A third option to performing a sampling in order to construct pools for relevance assessment is the stratified approach, which aims at including in the pools a big number of relevant results. The idea is to perform sampling within each stratum independently of the other. In this way it is possible to sample more documents in higher strata and less from strata which are down in the ranking. Using stratified sampling allows to compute $xinfAP$ [9] as evaluation metric, which is a better estimate of AP in the case of incomplete assessments. There is then the need to optimally selecting the strata and the sampling percentage for each strata, which is an open problem.

Considering the results shown in Figure 2, we decided to use the following strata for the pool construction of INEX-XER 2008:

- 1,8 100%
- 9,31 70%
- 32,100 30%

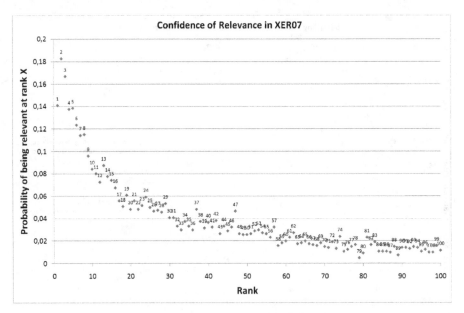

Fig. 2. Distribution of relevance over rank of top 100 retrieved results in INEX XER 2007

This means that we include in the pool 45 documents from each run. In order to compare this approach with the ones presented above, we computed the correlation using the strata-based sampling strategy. The result is presented in Figure 1.

Stratified sampling with the selected parameters performs, in terms of IRS ranking correlation, as well as uniform and relevance based sampling at 70%. The two 70% sampling approaches make each run contribute 35 documents to the pool while the stratified approach, by going down to rank 100 in the runs, make each run contribute 45 documents. Given that we used the 2007 collection for the experimental comparison we should notice that relevance assessments have been performed up to rank 50. Therefore, several documents ranked from 51 to 100 may not have been assessed; so they are considered not relevant in the experiments, even if they could be. If we want to fairly compare the judgement effort of the three sampling approaches we have to count the number of documents the stratified sampling approach make the runs contribute up to rank 50, which corresponds to 30 documents. In other words, stratified sampling gives a slightly lower judging effort than the uniform random sampling and the relevance based sampling for the same correlation in IRS rankings.

4   Results

At INEX XER 2008 six groups submitted 33 runs. The pools have been based on all submitted runs, using the stratified sampling strategy detailed in the previous Section. The resulting pools contained on average 400 entities per topic. The evaluation results for the ER task are presented in Table 1, those for the LC task in Table 2, both reporting xinfAP. In the LC task, the example entities provided in the topics are considered not relevant as the system is supposed not to retrieve them. The runs which name ends with "_fixed" have been corrected by the organizers removing the example entities present in the topics.

Most participants used language modelling techniques as underlining infrastructure to build their Entity Ranking engines. For both the ER and the LC task

Table 1. Evaluation results for ER runs at INEX XER 2008

Run	xinfAP
1_FMIT_ER_TC_nopred-cat-baseline-a1-b8:	0.341
1_cirquid_ER_TEC_idg.trec:	0.326
4_UAms_ER_TC_cats:	0.317
2_UAms_ER_TC_catlinksprop:	0.314
1_UAms_ER_TC_catlinks:	0.311
3_cirquid_ER_TEC.trec:	0.277
2_cirquid_ER_TC_idg.trec:	0.274
2_500_L3S08_ER_TDC:	0.265
1_L3S08_ER_TC_mandatoryRun:	0.256
3_UAms_ER_TC_overlap:	0.253
1_CSIR_ER_TC_mandatoryRun:	0.236
4_cirquid_ER_TC.trec:	0.235
4_UAms_ER_TC_cat-exp:	0.232
1_UAms_ER_TC_mixture:	0.222
3_UAms_ER_TC_base:	0.159
6_UAms_ER_T_baseline:	0.111

Table 2. Evaluation results for LC runs at INEX XER 2008

Run	xinfAP
1_FMIT_LC_TE_nopred-stat-cat-a1-b8:	0.402
1_FMIT_LC_TE_pred-2-class-stat-cat:	0.382
1_FMIT_LC_TE_nopred-stat-cat-a2-b6:	0.363
1_FMIT_LC_TE_pred-4-class-stat-cat:	0.353
5_UAms_LC_TE_LC1:	0.325
6_UAms_LC_TEC_LC2:	0.323
1_CSIR_fixed:	0.322
2_UAms_LC_TCE_dice:	0.319
5_cirquid_LC_TE_idg.trec.fixed:	0.305
1_L3S08_LC_TE_mantadoryRun:	0.288
2_L3S08_LC_TE:	0.286
5_cirquid_LC_TE_idg.trec:	0.274
6_cirquid_LC_TE.trec.fixed:	0.272
1_CSIR_LC_TE_mandatoryRun:	0.257
6_cirquid_LC_TE.trec:	0.249
5_UAms_LC_TE_baseline:	0.133

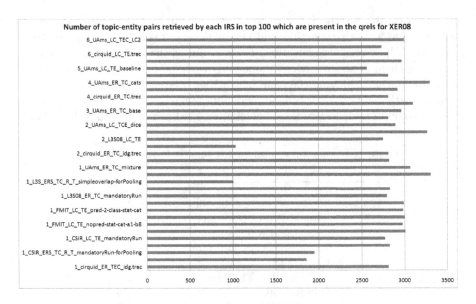

Fig. 3. Number of retrieved entities which are present in the pool for INEX XER 2008 submitted runs

the best performing approach uses topic difficulty prediction by means of a four-class classification step [7]. They use features based on the INEX topics definition and on the Wikipedia document collection obtaining 24% improvement over the second best LC approach. Experimental investigation showed that Wikipedia categories helped for easy topics and the link structure helped most for difficult topics. As also shown in last INEX-XER edition (best performing group at INEX-XER 2007), using score propagation techniques provided by PF/Tijah works in the context of ER [6]. The third best performing approach uses categories and links in Wikipedia [4]. They exploit distances between document categories and target categories as well as the link structure for propagating relevance information showing how category information leads to the biggest improvements.

For the LC tasks the same techniques performed well. Additionally, [4] also used relevance feedback techniques using example entities. In [3] they adapted language models created for expert search to the LC task incorporating category information in the language model also trying to understand category terms in the query text.

As for the use of the stratified sampling techniques we performed an analysis of possible bias due to the order in which IRS have been considered when constructing the pool. A potential drawback of the stratified sampling approach is that the order in which runs are considered for contributing to the pool could influence the strata information in the evaluation process, as we select the strata of the entity according to the run in which it was first encountered. Clearly, the order of the runs does not influence the number of entities contributed from each run in the pool, as Figure 3 shows that each run contributes an equal number

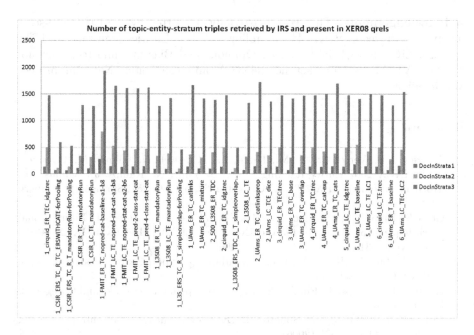

Fig. 4. Number of retrieved entities which are present in each stratum of the pool for INEX XER 2008 submitted runs

of entities to the pool set (except for the four ERS runs, that have the shorter bars). This conclusion is not valid anymore if we consider the run contribution to each stratum. As in the pooling process we considered an entity being part of the stratum where it was first encountered, the first runs considered had a bigger contribution as for the strata information. It is possible to see such bias in Figure 4 where run 1_FMIT_ER_TC_nopred-cat-baseline-a1-b8 has the most prominent presence in each stratum as it was the first run considered while creating the pool. Further research will have to show if this materializes into an effect on the system's ranking.

5   Conclusions

After the first edition of XER Track at INEX 2007 [1], the 2008 edition created additional evaluation material for IR systems that retrieve entities instead of documents. INEX XER 2008 created a set of 35 XER topics with relevance assessments for both the ER and LC tasks. Together with the 25 XER topics created in 2007, a collection of 60 topics is now available for evaluating Entity Retrieval systems. Future investigation will focus on evaluating the tasks on a more recent Wikipedia collection annotated with Information Extraction tools.

Acknowledgements. This work is partially supported by the EU Large-scale Integrating Projects OKKAM[2] - Enabling a Web of Entities (contract no. ICT-215032), LivingKnowledge[3] - Facts, Opinions and Bias in Time (contract no. 231126), VITALAS (contract no. 045389), and the Dutch National project MultimediaN.

References

1. de Vries, A.P., Vercoustre, A.-M., Thom, J.A., Craswell, N., Lalmas, M.: Overview of the INEX 2007 Entity Ranking Track. In: Fuhr, N., Kamps, J., Lalmas, M., Trotman, A. (eds.) INEX 2007. LNCS, vol. 4862, pp. 245–251. Springer, Heidelberg (2008)
2. Geva, S., Kamps, J., Trotman, A.: Advances in Focused Retrieval, 7th International Workshop of the Initiative for the Evaluation of XML Retrieval (INEX 2008). LNCS, vol. 5631. Springer, Heidelberg (2009)
3. Jiang, J., Lu, W., Rong, X., Gao, Y.: Adapting Expert Search Models to Rank Entities. In: Geva et al. [2]
4. Kaptein, R., Kamps, J.: Finding Entities in Wikipedia using Links and Categories. In: Geva et al. [2]
5. Craswell, N., de Vries, A.P., Soboroff, I.: Overview of the trec-2005 enterprise track. In: Proc. of TREC 2005 (2006)
6. Rode, H., Hiemstra, D., de Vries, A.P., Serdyukov, P.: Efficient XML and Entity Retrieval with PF/Tijah: CWI and University of Twente at INEX 2008. In: Geva et al. [2]
7. Vercoustre, A.-M., Pehcevski, J., Naumovski, V.: Topic Difficulty Prediction in Entity Ranking. In: Geva et al. [2]
8. Yilmaz, E., Aslam, J.A.: Estimating average precision with incomplete and imperfect judgments. In: Yu, P.S., Tsotras, V.J., Fox, E.A., Liu, B. (eds.) CIKM, pp. 102–111. ACM Press, New York (2006)
9. Yilmaz, E., Kanoulas, E., Aslam, J.A.: A simple and efficient sampling method for estimating AP and NDCG. In: Myaeng, S.-H., Oard, D.W., Sebastiani, F., Chua, T.-S., Leong, M.-K. (eds.) SIGIR, pp. 603–610. ACM, New York (2008)
10. Zhu, J., de Vries, A.P., Demartini, G., Iofciu, T.: Relation Retrieval for Entities and Experts. In: Future Challenges in Expertise Retrieval (fCHER 2008), SIGIR 2008 Workshop, Singapore (July 2008)

---

[2] http://fp7.okkam.org/
[3] http://livingknowledge-project.eu/

L3S at INEX 2008: Retrieving Entities Using Structured Information

Nick Craswell[1], Gianluca Demartini[2], Julien Gaugaz[2], and Tereza Iofciu[2]

[1] Microsoft Research Cambridge
7 JJ Thomson Ave
Cambridge, UK
nickcr@microsoft.com
[2] L3S Research Center
Leibniz Universität Hannover
Appelstrasse 9a D-30167 Hannover, Germany
{demartini,gaugaz,iofciu}@L3S.de

Abstract. Entity Ranking is a recently emerging search task in Information Retrieval. In Entity Ranking the goal is not finding documents matching the query words, but instead finding entities which match those requested in the query.

In this paper we focus on the Wikipedia corpus, interpreting it as a set of entities and propose algorithms for finding entities based on their structured representation for three different search tasks: entity ranking, list completion, and entity relation search. The main contribution is a methodology for indexing entities using a structured representation. Our approach focuses on creating an index of facts about entities for the different search tasks. More, we use the category structure information for improving the effectiveness of the List Completion task.

1 Introduction

Entity Ranking (ER) is an important step over the classical document search as it has been done so far. The goal is to find entities relevant to a query rather than just finding documents (or passages from documents) which contain relevant information. Ranking entities according to their relevance with respect to a given query is important in scenarios where the information load is bigger than what the user can handle. That is, with a correct ranking scheme the system can present the user with only entities of interest, and avoid the user having to analyze the entire set of retrieved documents.

As a step in this direction, we present, in this paper, our approaches to ranking entities in Wikipedia which are based on the usage of a structured representation of entities in order to enable search. Conceptually, we represent an entity as a set of attribute name / value pairs. For example an entity representing Albert Einstein can be represented as follows:

S. Geva, J. Kamps, and A. Trotman (Eds.): INEX 2008, LNCS 5631, pp. 253–263, 2009.
© Springer-Verlag Berlin Heidelberg 2009

first name Albert
last name Einstein
born 1879-03-14
died 1955-04-18
fields physics

We evaluate our approaches on the Wikipedia XML corpus provided within the INEX 2008 initiative. The main contribution of this paper is a set of methodologies for representing entities in a structured fashion for enabling entity retrieval. This is done using different representations for the different ER search tasks, namely entity ranking (XER), list completion (LC), and entity relation search (ERS).

The rest of the paper is organized as follows. In section 2 we describe the algorithms developed for the three different search tasks. In section 3 we present the experimental results on the INEX XER 2008 benchmark. In section 4 we compare our techniques with previous work and, finally, in section 5 we conclude the paper describing our future work.

2   Algorithms

In this section we describe the algorithms we designed and used for all the three tasks that have been run at XER 2008: Entity Ranking, List Completion, and Entity Relation Search. The official runs submitted for ER used state of the art NLP techniques for this task as proposed in [2]. After that, we developed and evaluated ER algorithms, which are described in Section 2.1, based on a structured index of entities. For the LC task we designed an algorithm that uses the structure of Wikipedia categories and, starting from example entities, exploits hard and soft categories for retrieving additional entities (see Section 2.2). For the ERS task we designed a search algorithm on top of a structured index based on sentence level entity co-occurrence (see Section 2.3).

2.1   Entity Ranking Task

For the Entity Ranking task we submitted two runs that have been built using state of the art techniques. Additionally, we developed techniques based on structured indexing representing entities as a set of attribute/value pairs. In the following we briefly describe the submitted runs which have been presented in previous work [2] and we then focus on the description of the structured indexing process for XER.

Entity Ranking using NLP techniques on Topic Title and Description.
The first XER run that we submitted to INEX 2008 (1_L3S08_ER_TC_mandatoryRun) uses only the title part of the topic and the category information. It uses a combination of NLP and IE techniques on the user query in order to improve and adapt it for the XER search task. The system extends the original keyword query adding also related words (i.e., all relations

from Wordnet, except antonyms) and synonyms, and it adds more weight on named entities and key-concepts (e.g., the type of result which is needed by the user) present in the original query by duplicating the respective terms. Additionally, the key-concepts which are present in the original query are used to search in the anchor-text of the outgoing links. Finally, the match of the category information in the query and the category information of the Wikipedia pages is merged by linear combination in the final ranking of the results.

The second XER run we submitted to INEX 2008 (2_500_L3S08_ER_TDC) is using also the description part of the topic. The entire title and description are used as a long query by the system in order to provide better results to the user. A TF.IDF vector is built out of the query and Wikipedia articles are ranked according to their cosine similarity to the query. Additionally, the title and the category information in the topic is used for searching the category information in the Wikipedia articles. This helps in ranking first the entities of the desired type. Finally, also the outgoing links of the Wikipedia documents are used for finding relevant entities. The topic title and the named entities present in the description are used together for searching in the anchor text of the outlinks of a page. The final score of the retrieved entities is computed as a linear combination of the cosine similarity scores deriving by the previous steps. A detailed description of the algorithms used for the submitted runs is presented in [2].

Entity Ranking using Structured Indexing. Even if we did not officially submitt them, we also performed entity ranking experiments using a structured index representation. We describe hereafter a first approach to generate a structured index from unstructured documents. By structured indexing we mean a set of attribute name / value pairs. It comprises of two steps: *entity reference resolution* and *attribute extraction*. We first identify the entity or entities referenced in the document. Those references are then used for *extracting attribute* name and values related to them from the text. Those two steps are described in the following.

Entity Reference Resolution. For building the structured representation, we first have to find references of the considered entity occurring in the Wikipedia page at hand. It is presently done in two different ways depending on whether the considered page is the page representing the entity, or a page linking to the page representing the entity (e.g., Wikipedia list-of pages).

- **On a page representing the considered entity:** we consider entity occurrences are all set of terms similar to the page title. The title might have two parts. The first part is called the *main title*, and is present in all titles. The second part of a title is called *sub-title* and it consists of the terms in the title occurring after a separating character like a '—' or in parentheses '()', '{}' or '[]'. For example, in the title "Napoleon (1995 film)", the main title is "Napoleon" and the sub-title is "1995 film". In this case we consider a set of tokens on the page to be similar to the title, and thus representing the considered entity, if it consists of either only the main title or the whole title (main plus sub).

- **On a page linking to the considered entity page:** we consider as entity
  occurrences the anchor text of the links.

Attribute Extraction. With help from a grammatical structure parser[1] and manually created rules on the grammatical structure, we extract attribute names and values representing the entity presented by a Wikipedia page. Those rules include:

- The *entity reference* is the subject of a sentence, the *attribute name* is an active verb plus prepositions, and the *attribute value* is the object of the verb.
- The *entity reference* is an object, the *attribute name* is a passive verb plus prepositions, and the *attribute value* is the subject of the passive verb.

We present below the rules used. The syntax is the one of the Tregex pattern matching tool[2]. The variable named "word" is replaced with the entity reference as described above. The variables beginning with "name" are concatenated to form the attribute name, and the variables beginning with "value" form the attribute value.

```
S < ((NP << word) $. (VP << (VBZ|VBD|VBG|VBN|VBP = name1 . (
    VP < (TO = name2 $. (VP = value))))))
S < ( (NP << (NP = name $. (PP < (IN $. (NP << word))))) $. (
    VP << (TO $. VP = value)))
S < ('' $. (NP << word $. (''$. (VP << (
    VBZ|VBD|VBG|VBN|VBP = name $. (
    NP = value1 ?$. (/,/ $. PP = value2)))))))
NP << word . (TO $. (VP < (VB = name $. NP = value)))
S < (NP << word $. (VP < (VBZ|VBD|VBG|VBN|VBP = name $. (
    PP = value1 $. PP = value2))))
word . (PP < (IN = name . NP = value))
S < (NP << word $. (VP < (VBZ|VBD|VBG|VBN|VBP = name1 $. (
    ADVP = name2 $. (VP < VBZ|VBD|VBG|VBN|VBP = name3 < (PP < (
    IN = name4 $. NP = value1 ?$. (/,/ $. PP = value2)))))))))
S < ((NP << word) ?$. VBZ|VBD|VBG|VBN|VBP = name1 ?$..
    ADVP = name2 $.. (VP << (VBZ|VBD|VBG|VBN|VBP = name2 $. (
    NP = value1 ?$. (/,/ $. PP = value2)))))
```

Other ad-hoc rules were also created based on a training set of Wikipedia pages, and the entity references in them. Those extracted attributes have the following particularities compared to attributes usually found in knowledge bases:

- Attribute names are less expressive and consist mostly of verbs plus modifiers or prepositions.
- Attribute values are mostly longer, and noisier in the sense that not all the terms of the attribute value are relevant. This is because the value consists mostly of a whole verbal phrase.

---

[1] http://nlp.stanford.edu/software/lex-parser.shtml
[2] http://nlp.stanford.edu/software/tregex.shtml

Table 1. Some representative example of extracted attributes

Entity	Attribute	Value
Albert Einstein	was	a Jewish German - born theoretical physicist of profound genius , who is widely regarded as the greatest scientist of the 20th century
Albert Einstein	described	the "predatory phase of human development"
Lausanne	is	a city in the French-speaking part of Switzerland, situated on the shores of Lake Geneva (French: Lac Lman), and facing vian-les-Bains (France)
Lausanne	located	some 60 km northeast of Geneva
Lausanne	follows	"La Nuit de Muses" (Museum's night) in the fall season
Lausanne	boasts	a dramatic panorama over the lake
Lausanne	is	the birthplace of: Umberto Agnelli, Anthony Bloom, Franois-Louis David, Bocion Johann, Ludwig Burckhardt, Benjamin Constant, Aloise Corbaz, Charles Dutoit, Egon von Furstenberg, Eugne Grasset, Bertrand Piccard, Charles Ferdinand, Ramuz Thophile, Steinlen Elizabeth, Thompson (Lady Butler), Bernard Tschumi, Flix Vallotton
Lausanne	has	some alternative culture
Lausanne	provide	a diverse and rich musical life
Lausanne	is	E 37 N 46 10 56 31 6
Lausanne	going to	become the first city in Switzerland to have a real metro system, with the m2 Line which will open in 2008

This allows to extract attribute for 368,788 entities (56% of the total of 659,385 entities). The distribution of the number of attributes per entity is close to a typical power-law with 29% of the entities having 80% of the attributes, and an average of 1.4 attribute per entity description; and a manual assessment of the attributes of 40 randomly picked entities showed a precision of 87%, i.e., in average 87% of the attributes of an entity are extracted correctly. Recall of the attribute extraction could not be assessed yet due to the considerable amount of time it requires—reference to an entity is not only found in the entity's page but in the rest of the collection as well. In Table 1 we show some meaningful examples of extracted attributes.

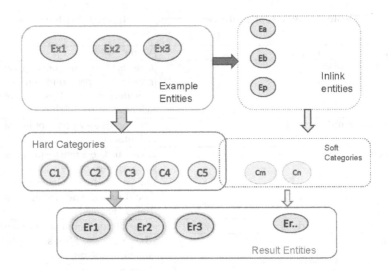

Fig. 1. List completion approach

In the future we will investigate how to leverage on already structured information such as info boxes in Wikipedia (see [1]) in order to enrich our index. We would then perform ER by retrieving relevant entities by performing a keyword (full-text) query on an inverted index built on top of the *attribute values* for all the identified entities. As future work we plan to extract a similar structure from the user query. By representing the query as a set of attribute-value pairs we can also take into account that two keywords belonging to a same attribute in the query should also appear in the same attribute in the structured index.

2.2   List Completion Task

In this task the information need is specified by the topic title and the type of the desired entities is given by example entities. From the INEX topics we can not use the category information anymore, this has to be learned from the examples entities with the help of the Wikipedia category structure.

Instinctively, one would say that if two entities are related by satisfying a query need, they should have at least one common category. But, as Wikipedia is a collaborative effort, entities usually belong to more than one category, depending on what the authors considered appropriate. The category information for entities is not always complete and sometimes is not entirely consistent and correct as well. Thus, from the topic example entities (usually between three or five) we need to discover which are the categories that will best satisfy the user need.

In our approach, for the example entities we extract two sets of categories from Wikipedia, as seen in Figure 1. The first set, which we call *main categories*, consists of the direct categories the example entities belong to. From these categories we filter out the ones that are too general, i.e., that have a high indegree (i.e., the number of entities that belong to that category) and we keep the ones with indegrees smaller than 1000. Each category from this set has a score computed based on the number of entities belonging to it. The score for a category is calculated as ten to the power of the number of occurring entities. Furthermore, the score is then divided by the category's specificity score, based on the indegree of the category, i.e. the number of Wikipedia entities belonging to that category. The assumption is that the higher the indegree of a category is, the less specific that category is. Therefore, we assume that the entities that belong to less specific categories do not have a strong relation. For example, for a category c_i with n example entities belonging to it, it's main score is computed based on the following formula:

$$mainCategoryScore(c_i) = 10^n/categorySpecificity(c_i) \qquad (1)$$

Where the category specificity is defined as $categorySpecificity(c) = min(5, \log(m))$, based on the Wikipedia link graph where category c has m entities. We built these heuristic by manual tuning of the parameters.

The second set of categories, which we called *soft categories*, consists of the top categories assigned to the linked entities. By linked entities we refer to entities the examples entities link to in their description. The score of these categories is given by the count of their occurrences in the linked entities. More, like the *main categories*, the *soft categories* with an indegree higher than 1000 are filtered out. This is done for avoiding the consideration of administrative Wikipedia categories, such as birth and death categories. These scores are then divided by the categories' specificity score. For example, for a category c_i with m linked entities belonging to it, it's soft score is computed based on the following formula:

$$softCategoryScore(c_i) = m/categorySpecificity(c_i) \qquad (2)$$

We then add the selected *soft categories* to the *main categories*, adding up the scores when there are duplicates, as shown in the formula given bellow. From the final set of categories we only keep top 30 categories, based on their final scores.

$$categoryScore(c_i) = mainCategoryScore(c_i) + softCategoryScore(c_i) \qquad (3)$$

For our first LC run (2_L3S08_LC_TE), we extract from Wikipedia all the entities that belong to the categories from the final set of categories built starting from the example entities. We order the retrieved entities based on their popularity and on the score of the categories (as defined above) they belong to. The score for an entity e_i belonging to n categories from the category set is obtained as seen in the following formula:

$$score(e_i) = \sum_{j=0}^{n} P(e_i) * categoryScore(c_j) \qquad (4)$$

where the entity popularity is defined as $P(e) = min(5, \log(L_e))$, based on the Wikipedia link graph where entity e has indegree L_e.

It is possible to see that our first run only uses the example entities ignoring other parts of the topic. For our second LC run (1_L3S08_LC_TE_mantadoryRun), we also used the topic title. In order to use together title and example entities, we search the collection (indexed using the vector space model with tf.idf weights and cosine similarity) using the title and we combine such ranking with the one obtained in our first run with the following formula:

$$rank(e_i) := (rank_{tf.idf}(e_i) * rank_{orig}(e_i))/2 \tag{5}$$

where $rank_{orig}(e_i)$ is the rank of entity e_i according to our first run which uses only the example entities, and $rank_{tf.idf}(e_i)$ is the rank produced by cosine similarity using tf.idf weighting scheme.

2.3   Entity Relation Search Task

Entity Relation Search is the step on top of Entity Ranking. After the system retrieved results for a XER query, the users can ask the system to provide entities which are related to the results with a given relation. For example, the user can first query for "american countries" and then ask for the capital of each of the results. For the Entity Relation Search task we have indexed relations between entities from Wikipedia in a structured fashion. We consider two entities to be related if they co-occur in a sentence sized window. For each two co-occurring entities we have indexed the sentences in which they appear together. In Table 2 we show a snippet from our entity relation index, with a few examples for the entities *Flint River* and *Albert Einstein*.

The Entity Relation Search runs that we submitted to INEX 2008 are based on our two submitted XER runs. The first run (1_L3S_ERS_TC_R_T_simpleoverlap) is based on the title-only Entity Ranking Run (i.e., 1_L3S08_ER_TC_mandatoryRun).

Table 2. Entity relation index

Entity	Predicate	Entity
flint river	the flint river with an area of 568 square miles is a tributary to the	tennessee river
flint river	much of the 342 sq	watershed
albert einstein	he proposed the and also made major contributions to the development of and the theory provided the foundation for the study of and gave scientists the tools for understanding many features of the universe that were discovered well after einstein death	cosmological models
albert einstein	he proposed the and also made major contributions to the development of and	theory of relativity

It is using the entities retrieved by the XER run as left entities LE. Then, for each left entity $l_i \in LE$ a search for relevant right entities is performed. Possible candidates are retrieved from the index: all triples containing as subject the left entities are considered. All the right entities candidates are then ranked according to a score which is computed as the overlap between the predicate in the index and the Entity Relation Search Title in the topic.

The second ERS run (2_L3S08_ER_TDC) is based based on our second XER run (i.e., 2_500_L3S08_ER_TDC) and it is computed as the first one.

3 Experiments

In this section we describe the experiments performed on the INEX XER 2008 benchmark in order to evaluate the proposed algorithms.

3.1 Experimental Setup

We first parse the document collection using standard Java libraries[3]. The next step is the creation of an inverted index out of the XML Wikipedia document collection. Starting from the parsed XML documents, we create a Lucene[4] index[5] with one Lucene document (i.e., a vector in the Vector Space) for each Wikipedia document. After this, we create an index with different fields (acting as separate inverted indexes, which can be combined for retrieval) for the title, text, and category of Wikipedia entities. Using such technology we then implemented the search algorithms described in Section 2.1. In the following we present official and additional evaluation results of the described approaches on the INEX XER 2008 benchmark.

3.2 Experimental Results

In Table 3 it is possible to see that by using NLP techniques also on the description part of the topic (which is text written in natural language) we improve search effectiveness. Evaluation results based on the structured index for the XER task shown low effectiveness (i.e., 0.01 xinfAP). This can be explained by

Table 3. Entity Ranking Results

Run	xinfAP
2_500_L3S08_ER_TDC	0.265
1_L3S08_ER_TC_mandatoryRun	0.256

the very preliminary extraction schemes that have been used. As we believe that

---

[3] We used the Java 6 `javax.xml.stream.*` classes.

[4] `http://lucene.apache.org/`

[5] The IR model used by Lucene is the term based Vector Space Model with standard cosine similarity.

the overall approach is promising, as future step, we will build our structured index based on better annotated Wikipedia collections such as [6,4].

As for the LC task, evaluation results[6] (see Table 4) show that when using also the ranking-by-title information we improve much over the simple use of category information from the example entities.

Table 4. List Completion Results

Run	xinfAP
2_L3S08_LC_TE	0.256
1_L3S08_LC_TE_mantadoryRun	0.314

Evaluation results for the ERS task have not yet been released.

4   Related Work

Previous approaches to rank entities in Wikipedia exploited the link structure between Wikipedia pages [5] or its category structure using graph based algorithms [7]. Other approaches used semantic and NLP techniques to improve effectiveness of ER systems [2,3].

With respect to previous approaches we based our algorithms on a structured representation of entities at indexing level. For both the XER and ERS tasks we used a structured index built using NLP techniques. For this reason, relevant to our work are projects aiming at extracting and annotating entities and structure in Wikipedia. For example, versions of Wikipedia annotated with state of the art NLP tools are available [6,4].

Another relevant work is [8] which also aims at retrieving entities in Wikipedia but without the assumption that an entity is represented by a Wikipedia page as done in INEX XER. They rather annotate and retrieve any passage of a Wikipedia article that could represent an entity. Our structured index allows such kind of retrieval as well.

5   Conclusions and Further Work

In this paper we proposed a first step towards a structured indexing approach for entity ranking. We proposed different indexing structures for the XER and ERS tasks. For the LC task we developed an algorithm based on the category structure of Wikipedia: by starting from the category information of the example entities we can identify the desired entity types. Experimental results show that the implemented approach for XER is not working well mainly because of

---

[6] Evaluation results are different than the official ones as we fixed a software bug after the runs submission.

the few annotations we performed on the corpus. We aim at improving search effectiveness of the XER task by using available collections annotated with state-of-the-art NLP tools.

Acknowledgements. This work is partially supported by the EU Large-scale Integrating Projects OKKAM[7] - Enabling a Web of Entities (contract no. ICT-215032) and LivingKnowledge[8] - Facts, Opinions and Bias in Time (contract no. ICT-231126).

References

1. Auer, S., Lehmann, J.: What Have Innsbruck and Leipzig in Common? Extracting Semantics from Wiki Content. In: Franconi, E., Kifer, M., May, W. (eds.) ESWC 2007. LNCS, vol. 4519, pp. 503–517. Springer, Heidelberg (2007)
2. Demartini, G., Firan, C.S., Iofciu, T., Krestel, R., Nejdl, W.: A Model for Ranking Entities and Its Application to Wikipedia. In: LA-WEB (2008)
3. Demartini, G., Firan, C.S., Iofciu, T., Nejdl, W.: Semantically Enhanced Entity Ranking. In: Bailey, J., Maier, D., Schewe, K.-D., Thalheim, B., Wang, X.S. (eds.) WISE 2008. LNCS, vol. 5175, pp. 176–188. Springer, Heidelberg (2008)
4. Ciaramita, M., Atserias, J., Zaragoza, H., Attardi, G.: Semantically Annotated Snapshot of the English Wikipedia. In: European Language Resources Association (ELRA) (ed.) Proceedings of the Sixth International Language Resources and Evaluation (LREC 2008), Marrakech, Morocco (May 2008)
5. Pehcevski, J., Vercoustre, A.-M., Thom, J.A.: Exploiting Locality of Wikipedia Links in Entity Ranking. In: Macdonald, C., Ounis, I., Plachouras, V., Ruthven, I., White, R.W. (eds.) ECIR 2008. LNCS, vol. 4956, pp. 258–269. Springer, Heidelberg (2008)
6. Schenkel, R., Suchanek, F.M., Kasneci, G.: YAWN: A Semantically Annotated Wikipedia XML Corpus. In: Kemper, A., Schöning, H., Rose, T., Jarke, M., Seidl, T., Quix, C., Brochhaus, C. (eds.) BTW. LNI, vol. 103, pp. 277–291. GI (2007)
7. Tsikrika, T., Serdyukov, P., Rode, H., Westerveld, T., Aly, R., Hiemstra, D., de Vries, A.P.: Structured Document Retrieval, Multimedia Retrieval, and Entity Ranking Using PF/Tijah. In: Fuhr, N., Kamps, J., Lalmas, M., Trotman, A. (eds.) INEX 2007. LNCS, vol. 4862, pp. 306–320. Springer, Heidelberg (2008)
8. Zaragoza, H., Rode, H., Mika, P., Atserias, J., Ciaramita, M., Attardi, G.: Ranking Very Many Typed Entities on Wikipedia. In: Silva, M.J., Laender, A.H.F., Baeza-Yates, R.A., McGuinness, D.L., Olstad, B., Olsen, Ø.H., Falcão, A.O. (eds.) CIKM, pp. 1015–1018. ACM, New York (2007)

---

[7] http://fp7.okkam.org/

[8] http://livingknowledge-project.eu/

Adapting Language Modeling Methods for Expert Search to Rank Wikipedia Entities

Jiepu Jiang, Wei Lu, Xianqian Rong, and Yangyan Gao

Center for Studies of Information Resources,
School of Information Management, Wuhan University, China
{jiepu.jiang,reedwhu,rongxianqian,gaoyangyan2008}@gmail.com

Abstract. In this paper, we propose two methods to adapt language modeling methods for expert search to the INEX entity ranking task. In our experiments, we notice that language modeling methods for expert search, if directly applied to the INEX entity ranking task, cannot effectively distinguish entity types. Thus, our proposed methods aim at resolving this problem. First, we propose a method to take into account the INEX category query field. Second, we use an interpolation of two language models to rank entities, which can solely work on the text query. Our experiments indicate that both methods can effectively adapt language modeling methods for expert search to the INEX entity ranking task.

Keywords: entity retrieval, entity ranking, language model, expert search.

1 Introduction

In this paper, we focus on how to adapt language modeling methods for expert search to the INEX entity ranking task (XER), which aims at finding a list of relevant entities according to a search query. A typical search query may involve several fields:

1. title: a text query field that describes the user's search needs;
2. category: a structural field specifying Wikipedia categories of relevant entities.

For example, a typical INEX XER search query can be:

<title>songs of Bob Dylan</title>
<categories>
<category id="40340">bob dylan songs</category>
</ categories >

The XER task shares a lot of similarities with the TREC expert search task, which can be considered as a special entity ranking task for persons only. Both tasks face the challenge of finding and utilizing descriptive information of entities in the documents. As a result, it is reasonable to adopt methods for expert search in the XER task.

Language modeling methods have been widely adopted in the expert search task. We have applied two widely used language modeling methods for expert search (i.e. model 1 and model 2 [1]) to the XER task. However, our experiments indicated that both methods cannot effectively distinguish entity types. As a result, we mainly focus on resolving this limitation.

S. Geva, J. Kamps, and A. Trotman (Eds.): INEX 2008, LNCS 5631, pp. 264–272, 2009.

First, we propose a method to take into account the INEX category query field, which can be applied to both model 1 and model 2. Second, we interpolate the entity model in model 1 with an entity category model, which solely works on the text query. In our experiments, it is indicated that both methods can effectively distinguish entity types. The first method was also adopted in our participation in INEX 2008. But our experiments indicate that the second method is much more effective.

Although the INEX entity-ranking track involves two tasks, i.e. the entity ranking task (XER) and the entity relation search task (ERS), we only discuss the XER task here due to the lack of evaluation for the ERS task. For our methods taken for the list completion task and ERS task, please refer to the pre-proceedings.

The remainder of this paper is organized as follows: section 2 reviews on language modeling methods for expert search and methods adopted in the INEX entity ranking task; in section 3, we describe our methods; section 4 evaluates the proposed methods; in section 5, we draw a conclusion.

2 Related Works

Language modeling methods are widely adopted for the expert search task. The most widely used language modeling framework for expert search was defined by Balog et al. [1] as model 1 and model 2. Further, refinements were made from various aspects. Petkova et al. [2] considered the dependency between candidates and terms. Balog et al. [3] elaborated candidate-document association. Serdyukov et al. [4] explored the relevance propagation. Balog et al. [5] used non-local information in the collection. For a complete review, please refer to [6].

Compared with expert search, less attention has been paid to the task of searching general entities of various types. In 2007, INEX provided the first collection for entity ranking, which is based on Wikipedia and involves a lot of useful features for entity ranking: entities are manually labeled with categories; the hierarchy of categories is given; entity occurrences are partly labeled in the documents.

Most of the methods adopted in INEX rely on the INEX category query field and Wikipedia category labels to distinguish entity types. Vercoustre et al. [7] used a set-based measure to calculate similarity between the INEX category query and the entity Wikipedia categories. Demartini et al. [8] expanded the category set using YAGO to improve the matching of entity types. Tsikrika et al. [9] adopted expert search model in [4] for entity ranking, and expanded category matching with child categories.

In section 3, we propose two methods to adapt language modeling methods for expert search to the INEX entity ranking task.

3 Models

In this section, we describe our methods. First, we propose a method to take into account the INEX category query field in both model 1 and model 2. Second, we interpolate the entity model estimated in model 1 with a category model, which can help model 1 better understand category query terms in the text query.

3.1 Language Modeling Methods for Expert Search

In section 3.1, we briefly describe two frequently used language modeling methods for expert search, i.e. model 1 and model 2 [1]. Both methods rank entities (experts) by $p(e|q)$, and use co-occurrence information of entities to estimate the probability. Assuming the same prior probability for each entity e, we can rank entities by $p(q|e)$.

For model 1, an entity model θ_e is inferred for each entity e. We can estimate $p(q|e)$ as Eq.(1):

$$p(q \mid e) = p(q \mid \theta_e) = \prod_{t \in q} p(t \mid \theta_e)^{tf(t,q)} \qquad (1)$$

In Eq.(1), $tf(t,q)$ is the frequency of t in the query q. Further, θ_e can be inferred using co-occurrence information of e in the collection.

For model 2, the estimation of $p(q|e)$ is divided into each sub event space of d:

$$p(q \mid e) = \sum_d p(q \mid d,e) \times p(d \mid e) \qquad (2)$$

Since there have been a lot of discussions on model 1 and model 2, we do not go further here. Please refer to [1] for details.

3.2 Considering the INEX Category Query Field

In section 3.2, we propose a method to consider the INEX category query field, which can be applied to both model 1 and model 2. We can represent the whole query as Q, which contains two parts: the text query q and the INEX category query q_{cat}. Then, we rank entities by $p(Q|e)$, which can be transformed as Eq. (3):

$$p(Q \mid e) = p(q, q_{cat} \mid e) = p(q \mid e) \times p(q_{cat} \mid e, q) \qquad (3)$$

Assuming q and q_{cat} are independent, $p(q_{cat}|e,q)$ can be simplified to $p(q_{cat}|e)$:

$$p(Q \mid e) = p(q, q_{cat} \mid e) = p(q \mid e) \times p(q_{cat} \mid e) \qquad (4)$$

In (4), $p(q|e)$ can be estimated using model 1 or model 2. As a result, the rest of the task is to estimate $p(q_{cat}|e)$.

In the INEX Wikipedia collection, entities are labeled with a list of categories. As a result, we can represent e's labeled categories as a category set, i.e. $CAT_e\{cat_i\}$. Also, we can represent the INEX category query field as a category set, i.e. $CAT_q\{cat_j\}$. Further, assuming that cat_j in CAT_q is generated independently, we estimate $p(q_{cat}|e)$ in Eq. (5):

$$p(q_{cat} \mid e) = p(CAT_q \mid CAT_e) = \prod_{cat_j \in CAT_q} p(cat_j \mid CAT_e) \qquad (5)$$

It should be noted that in (5) we adopt q_{cat} as a sequence of categories, although it is a set and may be more reasonable to be estimated in Eq. (6):

$$p(q_{cat} \mid e) = \left(\prod_{cat_j \in CAT_q} p(cat_j \mid CAT_e) \right) \times \left(\prod_{cat_j \notin CAT_q} \left\{ 1 - p(cat_j \mid CAT_e) \right\} \right) \quad (6)$$

Here, we adopt Eq.(5) for the following considerations: on the one hand, it is controversial to model categories that do not exist in CAT_q, since the category query field are not ensured to be accurate, and Wikipedia labels are also not completely accurate; on the other hand, a thorough estimation involving a large amount of unseen categories in (6) will consume a lot of computational resources.

In (5), $p(cat_j|CAT_e)$ is estimated using a maximum likelihood estimation with a Jelinek Mercer smoothing. Then, $p(cat_j|CAT_e)$ can be further considered using each cat_i in CAT_e:

$$p(cat_j \mid CAT_e) = (1 - \lambda_1) \times \sum_{cat_i \in CAT_e} \frac{p(cat_j \mid cat_i)}{\mid CAT_e \mid} + \lambda_1 \times p(cat_j) \quad (7)$$

In (7), $p(cat_j)$ is the probability of cat_j in the collection, which is estimated in (8). In Eq. (8), $ct(cat_j)$ is the number of entities in the collection that are labeled with cat_j.

$$p(cat_j) = \frac{ct(cat_j)}{\sum_{cat_i} ct(cat_i)} \quad (8)$$

For $p(cat_j|cat_i)$, we estimate it using some rule-based methods:

1. If $cat_j = cat_i$, or cat_j is cat_i's parent category, we set $p(cat_j|cat_i)$ to 1;
2. If cat_j is cat_i's child category, we set $p(cat_j|cat_i)$ to $1/|cat_i|$ ($|cat_i|$ is the number of child categories of cat_i);
3. For other circumstances, $p(cat_j|cat_i)$ is set to 0.

Using this method, we provide a solution to consider the category query field into current language modeling methods for expert search, which can help expert search models better distinguish entity categories. This method can be applied to both model 1 and model 2.

3.3 Understanding Category Terms in Search Query

In section 3.3, we use an interpolation of an entity model and a category model to understand category terms in the text query.

After manually checking all the search queries in INEX 07 and 08, we come to the following conclusion: text query for the INEX entity ranking task consist of two kinds of terms, i.e. *topic terms* and *category terms*.

We define topic terms as terms describing topical information of relevant entities, while category terms are used to specify categories of relevant entities. For example, for the query "songs of Bob Dylan", relevant entities are topically relevant with "Bob Dylan", and should be songs. So, "Bob" and "Dylan" are topic terms, while "songs" is a category term. Among the 95 queries in INEX 07 and 08, only 3 queries (Topic 50, 52 and 105) do not conform to our conclusion of entity ranking queries.

In contrast, queries for expert search only consist of topic terms. For example, the user will propose the query "wheel motor" to search for experts related to the topic "wheel motor". Category terms are omitted in expert search, since it is unnecessary to distinguish categories in the expert search task.

The difference between expert search queries and entity ranking queries is essential in explaining why language modeling methods for expert search are not effective in distinguishing entity categories. In language modeling methods for expert search, we infer entity (expert) models using co-occurrence information of entities. Although the entity models inferred are effective for expert search, considering that expert search queries only consist of topic terms, the entity models inferred may only indicate an approximation of probability distribution for topic terms. Thus, it is not surprising that expert search models are not very effective in understanding the category information need in the text query.

Thus, we infer two models for each entity e: T_e is the distribution model for topic terms, and C_e is the distribution model for category terms. Then, we can estimate $p(t|e)$ using an interpolation between T_e and C_e:

$$p(t \mid e) = \lambda_2 \times p(t \mid T_e) + (1 - \lambda_2) \times p(t \mid C_e) \tag{9}$$

In Eq.(9), λ_2 is a prior probability that a term will be generated from T_e. Though it is more reasonable to set different λ_2 for different entities, we adopt a constant value for λ_2 as a simplification. T_e can be inferred using model 1. In the INEX Wikipedia collection, we can infer C_e using labeled categories of the entities.

For each entity e, we represent its labeled categories in the Wikipedia as a category set $CAT_e \{cat_i\}$, in which cat_i is each labeled category in the set CAT_e. Then, we can use CAT_e to estimate C_e, which can be further considered using each cat_i in CAT_e:

$$p(t \mid CAT_e) = \sum_{cat_i \in CAT_e} p(t \mid cat_i, CAT_e) \times p(cat_i \mid CAT_e) \tag{10}$$

Assuming that the generation of t from cat_i is independent with CAT_e, $p(t|cat_i, CAT_e)$ can be simplified to $p(t|cat_i)$:

$$p(t \mid CAT_e) = \sum_{cat_i \in CAT_e} p(t \mid cat_i) \times p(cat_i \mid CAT_e) \tag{11}$$

For $p(t|cat_i)$, we simply estimate it using category name of cat_i by a maximum likelihood estimate in Eq.(12).

$$p(t \mid cat) \approx p_{mle}(t \mid cat) = \frac{tf(t)}{\sum_{t_i \in cat} tf(t_i)} \tag{12}$$

For $p(cat_i|CAT_e)$, we assign all categories with the equal weight and estimate it as $1/|CAT_e|$, where $|CAT_e|$ is the number of categories in the category set CAT_e. In the end, we can represent $p(t|C_e)$ as Eq.(13):

$$p(t \mid C_e) = \frac{\sum_{cat_i \in CAT_e} p_{mle}(t \mid cat_i)}{\mid CAT_e \mid} \tag{13}$$

Compared with the former method, this method can solely work on the text search query, which can resolve the limitation of using the INEX category query field. But a main limitation for this method is that C_e is estimated based on Wikipedia category labels. This limitation is left as a future work.

4   Evaluation

4.1   Experiment Settings

In our experiments, we adopt the INEX Wikipedia collection to evaluate our methods. Both INEX 2007 and 2008 queries are used. The INEX 2007 queries can be divided into two groups: one group consists of queries generated from the INEX ad hoc task (INEX 07 adhoc), and the other group consists of the genuine INEX 2007 XER query (INEX 07 xer).

The INEX Wikipedia collection is a subset of Wikipedia, which contains lots of semantic information. In this collection, entity occurrences are partly labeled in the documents. Thus, we do not further recognize named entities. Besides, each entity is also labeled with some categories. Category hierarchies are given. In our experiments, we have found 659,388 entities labeled with 75,601 Wikipedia categories (113,483 categories are provided in total, but some of them are not labeled with any entity).

In the pre-processing stage, we remove XML tags. The indexing process removes common stop words. Words are stemmed using Porter-Stemming algorithm.

Though official results in INEX 08 are evaluated using xinfAP, we use MAP as the main evaluation measure in order to be consistent with INEX 2007 (we do not have any method to evaluate xinfAP results for INEX 07 queries). The evaluation tool is trec_eval.

4.2   Expert Search Models

In section 4.2, we will evaluate the effectiveness of expert search models in the entity ranking task. In our experiments, we try to apply model 1 and model 2 to the INEX entity ranking task. Please refer to [1] for details about these models. In both models, we set the smoothing parameter λ to 0.5.

Table 1 shows evaluation results for model 1 and model 2 in the INEX 07 and 08 query sets, which are our baseline runs. It is indicated that both model 1 and model 2 are not very effective in the INEX entity ranking task. Besides, although previous researches indicated that model 2 is more effective than model 1 in the expert search task, model 1 apparently outperforms model 2 in all query sets of INEX.

Table 1. Evaluation results for model 1 and model 2 in the INEX entity ranking task

Query Set	Model 1		Model 2	
	MAP	**xinfAP**	**MAP**	**xinfAP**
INEX07	**0.2059**	--	0.1635	--
INEX07 adhoc	**0.2588**	--	0.1783	--
INEX07 xer	**0.1614**	--	0.1511	--
INEX08	**0.1189**	**0.1189**	0.0885	0.0885

4.3 Considering the INEX Category Query Field

In section 4.3, we evaluate the effectiveness of the method proposed in section 3.2, which considers the INEX category query fields into expert search language models. For a simplification, we set λ_1 in Eq.(7) to 0.5. For efficiency consideration, we only re-rank the top 500 entities returned by model 1 and model 2 when using the method proposed in section 3.2.

Table 2. MAP results for the method that considers the INEX category query field

Query Set	Mode 1	Model 1 + Method in 3.2		Mode 2	Model 2 + Method in 3.2	
INEX07	0.2059	**0.2522**	(+ 22.49%)	0.1635	0.2167	(+ 32.54%)
INEX07 adhoc	0.2588	**0.3374**	(+ 30.37%)	0.1783	0.2776	(+ 55.69%)
INEX07 xer	0.1614	**0.1806**	(+ 11.90%)	0.1511	0.1656	(+ 09.60%)
INEX08	0.1189	**0.2106**	(+ 77.12%)	0.0885	0.1627	(+ 83.84%)

Table 2 shows evaluation results for the method that considers the INEX category query field. It is indicated that, for both model 1 and model 2, this method can greatly enhance the effectiveness. In INEX 08, we adopted a combination of this method and model 1[1].

4.4 Considering Category Terms in Search Query

In section 4.4, we further consider category query terms into expert search model 1. For a simplification, the parameter λ_2 in (9) is set to 0.5. In our experiments, we try to investigate the following problem:

1. Can the adaptation method proposed in 3.3 help expert search model?
2. Compared with the method in 3.2, can the method using only text query be more effective?

Table 3 shows evaluation results for the method proposed in section 3.3 (the INEX category query field is not used). In Table 3, it is indicated that the method proposed in section 3.3 can also greatly enhance the effectiveness. Besides, compared with the

---

[1] Due to a coding error, our officially submitted run 1_CSIR_ER_TC_mandatoryRun had used a measure of $p(cat_j|CAT_e)$ different from the method proposed in 3.2. But we mean to use the method in 3.2. Results in Table.2 strictly conform to the method in section 3.2.

Table 3. MAP results for the method proposed in section 3.3

Query Set	Model 1	Model 1 + Method in 3.3		Model 1 + Method in 3.2	
INEX07	0.2059	**0.2952**	(+ 43.37%)	0.2522	(− 14.57%)
INEX07 adhoc	0.2588	**0.3585**	(+ 38.53%)	0.3374	(− 05.89%)
INEX07 xer	0.1614	**0.2420**	(+ 49.94%)	0.1806	(− 25.37%)
INEX08	0.1189	**0.2942**	(+147.43%)	0.2106	(− 28.42%)

method that considers the INEX category query field (in section 3.2), the interpolation of two models is evidently more effective in all query sets.

It may indicate some problems of using the INEX category query field in the entity ranking task. First, for a large collection containing a huge amount of entity categories (such as the INEX Wikipedia collection), it is difficult and impractical for the user to specify precisely all possible categories of relevant entities. Thus, when the user fails to select out some possible categories for relevant entities, some relevant entities will be excluded. Second, since the categories of relevant entities are specified in the text query, it is also unnecessary to learn it using the structural category query field.

Further, we combine both methods into model 1. Table 4 shows evaluation results of considering both methods into model 1, which means to estimate $p(q|e)$ in Eq.(3) using Eq.(8). However, in the experiments, it is indicated that the combination of two methods is not ensured to receive better effectiveness than using the method proposed in 3.3 only. This problem is left as a future work for us to discover.

In table 5, we gives out xinfAP for each method in INEX 08 query set.

Table 4. MAP results of considering both methods into model 1

Query Set	Model 1	Model 1 + Method in 3.2	Model 1 + Method in 3.3	Model 1 + Method in 3.2 & 3.3
INEX07	0.2059	0.2522	**0.2952**	0.2838
INEX07 adhoc	0.2588	0.3374	0.3585	**0.3755**
INEX07 xer	0.1614	0.1806	**0.2420**	0.2067
INEX08	0.1189	0.2106	0.2942	**0.3042**

Table 5. xinfAP results

Query Set	Model 1 + Method in 3.2	Model 2 + Method in 3.2	Model 1 + Method in 3.3	Model 1 + Method in 3.2 3.3
INEX08	0.2106	0.1627	0.2942	**0.3042**

5   Conclusion

In this paper, we describe two methods to adapt language modeling methods for the expert search task to the INEX entity ranking task. First, we propose a method to take into account the INEX category query field, which can be applied to both model 1 and model 2. Second, we use an interpolation between the entity model and the category model to understand category terms in the text query.

In our experiments, it is indicated that both methods can effectively adapt language modeling methods for expert search to the INEX entity ranking task. Compared with the method that considers the INEX category query field, the method using category terms (section 3.3) is more effective. However, a combination of both methods is not ensured to further enhance the effectiveness.

References

1. Balog, K., Azzopardi, L., de Rijke, M.: Formal Models for Expert Finding in Enterprise Corpora. In: Proceeding of the 29th annual international ACM SIGIR conference on Research and development in information retrieval (SIGIR 2006), Seattle, Washington, USA, pp. 43–50 (2006)
2. Petkova, D., Croft, W.B.: Proximity-Based Document Representation for Named Entity Retrieval. In: Proceedings of the 16th ACM conference on information and knowledge management (CIKM 2007), Lisbon, Portugal, pp. 731–740 (2007)
3. Balog, K., de Rijke, M.: Associating People and Documents. In: Macdonald, C., Ounis, I., Plachouras, V., Ruthven, I., White, R.W. (eds.) ECIR 2008. LNCS, vol. 4956, pp. 296–308. Springer, Heidelberg (2008)
4. Serdyukov, P., Rode, H., Hiemstra, D.: Modeling multi-step relevance propagation for expert finding. In: Proceedings of 17th ACM conference on Information and knowledge management (CIKM 2008), Napa Valley, California, USA, pp. 1133–1142 (2008)
5. Balog, K., de Rijke, M.: Non-Local Evidence for Expert Finding. In: Proceedings of the 17th ACM conference on information and knowledge management (CIKM 2008), Napa Valley, California, USA, pp. 731–740 (2008)
6. Vercoustre, A., Thom, J.A., Pehcevski, J.: Entity Ranking in Wikipedia. In: Proceedings of the 2008 ACM symposium on Applied computing (SAC 2008), Fortaleza, Ceara, Brazil (2008)
7. Demartini, G., Firan, C.S., Iofciu, T.: L3S Research at INEX 2007: Query Expansion for Entity Ranking Using a Highly Accurate Ontology. In: Fuhr, N., Kamps, J., Lalmas, M., Trotman, A. (eds.) INEX 2007. LNCS, vol. 4862, pp. 252–263. Springer, Heidelberg (2008)
8. Tsikrika, T., Serdyukov, P., Rode, H., Westerveld, T., Aly, R., Hiemstra, D., de Vries, A.P.: Structured Document Retrieval, Multimedia Retrieval, and Entity Retrieval Using PF/Tijah. In: Fuhr, N., Kamps, J., Lalmas, M., Trotman, A. (eds.) INEX 2007. LNCS, vol. 4862, pp. 306–320. Springer, Heidelberg (2008)

Finding Entities in Wikipedia
Using Links and Categories

Rianne Kaptein[1] and Jaap Kamps[1,2]

[1] Archives and Information Studies, Faculty of Humanities, University of Amsterdam
[2] ISLA, Faculty of Science, University of Amsterdam

Abstract. In this paper we describe our participation in the INEX Entity Ranking track. We explored the relations between Wikipedia pages, categories and
links. Our approach is to exploit both category and link information. Category
information is used by calculating distances between document categories and
target categories. Link information is used for relevance propagation and in the
form of a document link prior. Both sources of information have value, but using
category information leads to the biggest improvements.

1 Introduction

In the entity ranking track, our aim is to explore the relations and dependencies between
Wikipedia pages, categories and links. For the entity ranking task we have looked at
some approaches that proved to be successful in previous entity ranking and ad hoc
tracks. In these tracks it has been shown that link information can be useful. Kamps and
Koolen [2] use link evidence as document priors, where a weighted combination of the
number of incoming links from the entire collection and the number of incoming links
from the retrieved results for one topic is used. Tsikrika et al. [4] use random walks
to model multi-step relevance propagation from entities to their linked entities. For the
entity ranking track specifically also the category assignments of entities can be used.
Vercoustre et al. [5] use the Wikipedia categories by defining similarity functions between the categories of retrieved entities and the target categories. The similarity scores
are estimated using lexical similarity of category names. We combined and extended
the aforementioned approaches.

2 Model

In this section we describe how we use category information for entity ranking and list
completion, how we exploit link information and finally how we combine these sources
of information.

Category information. Although for each topic one or a few target categories are provided, relevant entities are not necessarily associated with these provided target categories. Relevant entities can also be associated with descendants of the target category

S. Geva, J. Kamps, and A. Trotman (Eds.): INEX 2008, LNCS 5631, pp. 273–279, 2009.

or other similar categories. Therefore, simply filtering on the target categories is not sufficient. Also, since Wikipedia pages are usually assigned to multiple categories, not all categories of an answer entity will be similar to the target category. We calculate for each target category the distances to the categories assigned to the answer entity. To calculate the distance between two categories, we tried three options. The first option (binary distance) is a very simple method: the distance is 0 if two categories are the same, and 1 otherwise. The second option (contents distance) calculates distances according to the contents of each category, and the third option (title distance) calculates a distance according to the category titles. For the title and contents distance, we need to calculate the probability of a term occurring in a category. To avoid a division by zero, we smooth the probabilities of a term occurring in a category with the background collection:

$$P(t_1, ..., t_n|C) = \sum_{i=1}^{n} \lambda P(t_i|C) + (1 - \lambda)P(t_i|D)$$

where C, the category, consists either of the category title to calculate title distance, or of the concatenated text of all pages belonging to that category to calculate contents distance. D is the entire wikipedia document collection, which is used to estimate background probabilities. We estimate $P(t|C)$ with a parsimonious model [1] that uses an iterative EM algorithm as follows:

E-step: $$e_t = tf_{t,C} \cdot \frac{\alpha P(t|C)}{\alpha P(t|C) + (1 - \alpha)P(t|D)}$$

M-step: $$P(t|C) = \frac{e_t}{\sum_t e_t}, \text{i.e. normalize the model}$$

The initial probability $P(t|C)$ is estimated using maximum likelihood estimation. We use KL-divergence to calculate distances, and calculate a category score that is high when the distance is small as follows:

$$S_{cat}(C_d|C_t) = -D_{KL}(C_d|C_t) = -\sum_{t \in D} \left(P(t|C_t) * \log\left(\frac{P(t|C_t)}{P(t|C_d)} \right) \right)$$

where d is a document, i.e. an answer entity, C_t is a target category and C_d a category assigned to a document. The score for an answer entity in relation to a target category $S(d|C_t)$ is the highest score, or shortest distance from any of the document categories to the target category.

In contrast to Vercoustre et al. [5], where a ratio of common categories between the categories associated with an answer entity and the provided target categories is calculated, we take for each target category only the shortest distance from any answer entity category to a target category. So if one of the categories of the document is exactly the target category, the distance and also the category score for that target category is 0, no matter what other categories are assigned to the document. Finally, the score for an answer entity in relation to a query topic $S(d|QT)$ is the sum of the scores of all target categories:

$$S_{cat}(d|QT) = \sum_{C_t \in QT} \underset{C_d \in d}{\mathrm{argmax}}\, S(C_d|C_t)$$

Besides the entity ranking task, the second task in the entity ranking track is list completion. Instead of the target category, for each topic a few relevant examples entities are given. We treat all categories assigned to the example entities as target categories. Our approach for using the category information is the same as before. But to get the final score of an article in relation to a topic, we use two variants. The first one is:

$$S_{Sum}(d|QT) = \sum_{ex \in QT} \sum_{C_{ex} \in ex} \underset{C_d \in d}{\arg\max}\, S_{cat}(C_d|C_{ex})$$

In the second variant $S_{Max}(d|QT)$, instead of summing the score of each example category of the entity examples to one of the categories of the document. Furthermore, we apply explicit relevance feedback based on the text of the example entities to expand the query.

Link information. We implement two options to use the link information: relevance propagation and document link degree prior. For the document link degree prior we use the same approach as in [2]. The prior for a document d is:

$$P_{Link}(d) = 1 + \frac{Indegree_{Local}(d)}{1 + Indegree_{Global}(d)}$$

The local indegree is equal to the number of incoming links from within the top ranked documents retrieved for one topic. The global indegree is equal to the number of incoming links from the entire collection.

The second use of link information is through relevance propagation from initially retrieved entities, as was done last year in the entity ranking track by Tsikrika et al. [4].

$$P_0(d) = P(q|d)$$
$$P_i(d) = P(q|d)P_{i-1}(d) + \sum_{d' \to d}(1 - P(q|d'))P(d|d')P_{i-1}(d')$$

Probabilities $P(d|d')$ are uniformly distributed among all outgoing links from the document. Documents are ranked using a weighted sum of probabilities at different steps:

$$P(d) = \mu_0 P_0(d) + (1 - \mu_0) \sum_{i=1}^{K} \mu_i P_i(d)$$

For K we take a value of 3, which was found to be the optimal value last year. We try different values of μ_0 and distribute $\mu_1 ... \mu_K$ uniformly, i.e. $\mu_1 ... \mu_K = 1/3$.

Combining information. Finally, we have to combine our different sources of information. We start with our baseline model which is a standard language model. We have two possibilities to combine information. We can make a linear combination of the probabilities and category score. All scores and probabilities are calculated in the log space, and then a weighted addition is made. Alternatively, we can use a two step model. Relevance propagation takes as input initial probabilities. Instead of the baseline probability, we can use the scores of the run that combines the baseline score with the category information. Similarly, for the link degree prior we can use the top results of the baseline combined with the category information instead of the baseline ranking.

3 Experiments

In this section we describe our experimental results on the training and the test data.

3.1 Training Results

For our training data we use the 25 genuine entity ranking test topics that were developed for the 2007 entity ranking track. For our baseline run and to get initial probabilities we use the language modeling approach with Jelinek-Mercer smoothing, Porter stemming and pseudo relevance feedback as implemented in Indri [3] to estimate $P(d|q)$. We tried different values for the smoothing λ. We found $\lambda = 0.1$ gives the best results, with a MAP of 0.1840 and a P10 of 0.1920. For the document link degree prior we have to set two parameters: the number of top documents to use, and the weight of the document prior. For the number of top documents to use, we try 50, 100, 500 and 1,000 documents. For the weight of the prior we try all values from 0 to 1 with steps of 0.1. Only weights that give the best MAP and P10 are shown in Table 1.[1]

Table 1. Document link degree prior results

# docs	Weight	MAP	P10
Baseline		0.1840	0.1920
50	0.6	0.1898⁻	**0.2040**⁻
50	0.5	0.1876⁻	0.2000⁻
100	0.7	0.1747⁻	0.2000⁻
100	0.3	0.1909⁻	0.1920⁻
500	0.5	**0.1982**°	0.2000⁻
500	0.3	0.1915⁻	**0.2040**°
1,000	0.5	0.1965⁻	0.1960⁻
1,000	0.4	0.1965°	0.2000⁻

Table 2. Category distances results

Dist.	Weight	MAP	P10
Binary	0.1	0.2145⁻	0.1880⁻
Cont.	0.1	0.2481°	0.2320°
Title	0.1	0.2509°	0.2360°
Cont.	0.05	**0.2618**°	**0.2480**°
Title	0.05		

The results of using category information are summarized in Table 2. The weight of the baseline score is 1.0 minus the weight of the category information. For all three distances, a weight of 0.1 gives the best results. In addition to these combinations, we also made a run that combines the original score, the contents distance and the title distance. When a single distance is used, the title distance gives the best results. The combination of contents and title distance gives the best results overall.

In our next experiment we combine all information we have, the baseline score, the category and the link information. Firstly, we combine all scores by making a linear combination of the scores and probabilities (shown in Table 3). Secondly, we combine the different sources of information by using the two step model (see Table 4). Link information is mostly useful to improve early precision, depending on the desired results we can tune the parameters to get optimal P10, or optimal MAP. Relevance propagation performs better than the document link degree prior in both combinations.

---

[1] Significance of increase over the baseline according to the t-test, one-tailed, at significance levels 0.05 (°), 0.01 (°), and 0.001 (•).

Table 3. Results linear combination

Link Info	Weight	MAP	P10
Prior	0.3	0.2682°	0.2640°
Prop.	0.1	**0.2777°**	**0.2720°**

Table 4. Results two step model

Link info	Weight	MAP	P10
Prior	0.5	0.2526°	0.2600°
Prop.	0.2	0.2588°	**0.2960** •
Prop.	0.1	**0.2767°**	0.2720°

For the list completion task, we use the examples for relevance feedback. To evaluate the list completion results, example entities are removed from our ranking. Applying explicit and pseudo relevance feedback leads to the results given in Table 5. Additional

Table 5. Feedback results

RF	PRF	MAP	P10
No	No	0.1409	0.1240
Yes	No	**0.1611**	0.1600
Yes	Yes	0.1341	**0.1960**

Table 6. List Completion results

Dist.	Weight	$S(A|QT)$	C_t	MAP	P10
Baseline LC				0.1611	0.1600
Cont.	0.1	Sum	No	0.2385°	0.2520°
Cont.	0.9	Sum	Yes	0.2467•	0.2560°
Cont.	0.2	Max	No	0.1845⁻	0.2360⁻
Title	0.1	Sum	No	0.2524°	0.2640°
Title	0.9	Sum	Yes	**0.2641•**	**0.2760°**
Title	0.5	Max	No	0.1618⁻	0.2080⁻
Cont.	0.05	Sum	No	0.2528•	0.2640°
Title	0.05				

pseudo relevance feedback after the explicit feedback, only improves early precision, and harms MAP. We take the run using only relevance feedback as our baseline for the list completion task.

When we look at the previous entity ranking task, the largest part of the improvement comes from using category information. So here we only experiment with using the category information, and not the link information. We have again the different category representations, content and category titles. Another variable here is how we combine the scores, either add up all the category scores $S_{Sum}(A|QT)$ or taking only the maximum score $S_{Max}(A|QT)$. Not part of the official task, we also make some runs that use not only the categories of the example entities, but also the target category(ies) provided with the query topic. In Table 6 we summarize some of the best results. The combination of contents and title distance, does not lead to an improvement over using only the title distance. The maximum score does not perform as well as the summed scores. We use all categories assigned to the entity examples as target categories, but some of these categories will not be relevant to the query topic introducing noise in the target categories. When the scores are summed, this noise is leveled out, but when only the maximum score is used it can be harmful. Comparing the list completion and the entity ranking task, the list completion task has a slightly lower baseline score, but the results of both tasks when category information is used, are very similar.

3.2   Test Results

The test data consists of 35 new entity ranking topics.We use the parameters that gave the best results on the training data, i.e. baseline with pseudo-relevance feedback and $\lambda = 0.1$, weights of contents and title category information is 0.1, or 0.05 and 0.05 in the combination. For the link prior we use the top 100 results, and the two-step model is used to combine the information. In Table 7 our results on the test topics are shown. Using the category information leads to an improvement of 100% over the baseline, the score is doubled! Even when we rerank the top 500 results retrieved by the baseline using only the category information, the result are significantly better than the baseline, with a MAP of 0.2405. Since the category information is so important, it is likely that relevant pages can be found outside the top 500. Indeed, when we rerank the top 2500, but still evaluating the top 500, our results improve up to a MAP of 0.3519. Furthermore, we found that on the test data doubling the weights of the category information leads to slightly better results. Similar to the training results, relevance propagation performs better than the link prior, and leads to small additional improvements over the runs using category information.

Table 7. Results on the 2008 test topics

# Results	Category info				Link info		MAP	P10
			Baseline				0.1586	0.2257
500	Title	0.1			No		0.3059•	0.4171•
	Title	0.2			No		0.3164•	0.4400•
			Cont.	0.1	No		0.3031•	0.4086•
			Cont.	0.2	No		0.3088•	0.4200•
	Title	0.05	Cont.	0.05	No		0.3167•	0.4343•
	Title	0.1	Cont.	0.1	No		0.3189•	0.4400•
	Title	0.05	Cont.	0.05	Prior	0.5	0.3196•	0.4371•
	Title	0.05	Cont.	0.05	Prop.	0.1	0.3324•	0.4543•
2500	Title	0.1			No		0.3368•	0.4343•
	Title	0.2			No		0.3504•	0.4514•
	Title	0.2			Prop.	0.1	0.3519•	0.4629•

For the list completion task we submitted two runs. These runs use only the category information, in the form of category titles and summing scores over categories. Curiously, the run using only the examples scores slightly better than the run that uses also the specified target categories, with MAP of 0.325 and 0.323 respectively.

4   Conclusion

We have presented our entity ranking approach where we use category and link information. Category information is the factor that proves to be most useful and we can do more than simply filtering on the target categories. Category information can both be extracted from the category titles and from the contents of the category. Link information can also be used to improve results, especially early precision, but these improvements are smaller. In future research, we will look in more detail at the list completion task to derive more focused target categories from the example entities.

Acknowledgments. This research is funded by the Netherlands Organization for Scientific Research (NWO, grant # 612.066.513).

References

[1] Hiemstra, D., Robertson, S., Zaragoza, H.: Parsimonious language models for information retrieval. In: Proceedings SIGIR 2004, pp. 178–185. ACM Press, New York (2004)

[2] Kamps, J., Koolen, M.: The importance of link evidence in Wikipedia. In: Macdonald, C., Ounis, I., Plachouras, V., Ruthven, I., White, R.W. (eds.) ECIR 2008. LNCS, vol. 4956, pp. 270–282. Springer, Heidelberg (2008)

[3] Strohman, T., Metzler, D., Turtle, H., Croft, W.B.: Indri: a language-model based search engine for complex queries. In: Proceedings of the International Conference on Intelligent Analysis (2005)

[4] Tsikrika, T., Serdyukov, P., Rode, H., Westerveld, T., Aly, R., Hiemstra, D., de Vries, A.P.: Structured document retrieval, multimedia retrieval, and entity ranking using PF/Tijah. In: Focused Access to XML Documents, pp. 306–320 (2007)

[5] Vercoustre, A.M., Pehcevski, J., Thom, J.A.: Using wikipedia categories and links in entity ranking. In: Focused Access to XML Documents, pp. 321–335 (2007)

Topic Difficulty Prediction in Entity Ranking

Anne-Marie Vercoustre[1], Jovan Pehcevski[2], and Vladimir Naumovski[2]

[1] INRIA, Rocquencourt, France
anne-marie.vercoustre@inria.fr
[2] Faculty of Management and Information Technologies, Skopje, Macedonia
{jovan.pehcevski,vladimir.naumovski}@mit.edu.mk

Abstract. Entity ranking has recently emerged as a research field that aims at retrieving entities as answers to a query. Unlike entity extraction where the goal is to tag the names of the entities in documents, entity ranking is primarily focused on returning a ranked list of relevant entity names for the query. Many approaches to entity ranking have been proposed, and most of them were evaluated on the INEX Wikipedia test collection. In this paper, we show that the knowledge of predicted *classes* of topic difficulty can be used to further improve the entity ranking performance. To predict the topic difficulty, we generate a classifier that uses features extracted from an INEX topic definition to classify the topic into an experimentally pre-determined class. This knowledge is then utilised to dynamically set the optimal values for the retrieval parameters of our entity ranking system. Our experiments suggest that topic difficulty prediction is a promising approach that could be exploited to improve the effectiveness of entity ranking.

1 Introduction

The INitiative for Evaluation of XML retrieval (INEX) started the XML Entity Ranking (XER) track in 2007 [4] with the goal of creating a test collection for entity ranking using the Wikipedia XML document collection [6]. The XER track was run again in 2008, introducing new tasks, topics and pooling strategies aiming at improving the XER test collection [5]. The objective of the two INEX XER tracks was to return names of entities rather than full documents that correspond to those entities as answers to an INEX topic. Different approaches to entity ranking have been proposed and evaluated on the two INEX Wikipedia XER test collections, which resulted in many advances to this research field. However, little attention has been put on the impact of the different types (or classes) of topics on the entity ranking performance.

Predicting query difficulty in information retrieval (IR) has been the subject of a SIGIR workshop in 2005 that focused both on prediction methods for query difficulty and on the potential applications of those predicted methods [2]. The applications included re-ranking answers to a query, selective relevance feedback, and query rewriting. On the other hand, the distinction between easy and difficult queries in IR evaluation is relatively recent but offers important insights and new perspectives [8,23]. The difficult queries are defined as the ones on which

S. Geva, J. Kamps, and A. Trotman (Eds.): INEX 2008, LNCS 5631, pp. 280–291, 2009.

the evaluated systems are the less successful. The initial motivation for query difficulty prediction was related to the need for evaluation measures that could be applied for robust evaluation of systems across different collections [19,20]. Recently, Mizzaro [13] also advocated that the evaluation measures should reward systems that return good results on difficult queries more than on easy ones, and penalise systems that perform poorly on easy queries more than on difficult ones.

In this paper, we build on the above arguments and develop a method for query (topic) difficulty prediction in the research field of XML entity ranking. Our approach is based on the generation of a topic classifier that can classify INEX XER topics from a number of features extracted from the topics themselves (also called static or a-priori features) and possibly from a number of other features calculated at run time (also called dynamic or a-posteriori features). The main goal is to apply the topic difficulty prediction to improve the effectiveness of our entity ranking system that was evaluated as one of the best performing XER systems at INEX 2007 and 2008 [4,5].

2   Topic Difficulty Classification

In this section, we classify the INEX XER topics by their difficulty. After a brief description of the two INEX XER tracks, we present our methodology to identifying the different classes of topics.

2.1   XML Entity Ranking at INEX

The two INEX XER test collections comprise 46 topics (2007) and 35 topics (2008) with corresponding relevance assessments available, most of which were proposed and assessed by the track participants [4,5]. Figure 1 shows an example of an INEX 2007 XER topic definition. In this example, the `title` field contains the plain content only query, the `description` provides a natural language description of the information need, and the `narrative` provides a detailed explanation of what makes an entity answer relevant. In addition to these fields, the `categories` field provides the target category of the expected entity answers (used in the entity ranking task), while the `entities` field provides a few examples of the expected entity answers (used in the list completion task).

Eight participating groups submitted in total 35 XER runs in 2007, while six participants submitted another 33 XER runs in 2008. The INEX XER track does not offer a large test collection, however this is currently the only test collection available for the purposes of our topic difficulty classification. In this paper we focus on the list completion task, corresponding to a training data set comprising a total number of 46 topics and 17 runs, and a testing data set comprising a total number of 35 topics and 16 runs.

2.2   Identifying Classes of Topics

The classification of topics into groups is based on how well the runs submitted by participating systems answered to the topics. For each topic, we calculate the

```
<inex_topic>
<title>circus mammals</title>
<description>
I want a list of mammals which have ever been tamed to perform in circuses.
</description>
<narrative>
Each answer should contain an article about a mammal which can be a part
of any circus show.
</narrative>
<categories>
   <category id="138">mammals</category>
</categories>
<entities>
   <entity id="379035">Asian Elephant</entity>
   <entity id="4402">Brown Bear</entity>
</entities>
</inex_topic>
```

Fig. 1. INEX 2007 XER topic definition

topic difficulty using the Average Average Precision (AAP) measure [14]. AAP is the average AP of all the submitted runs for a given topic: the higher the AAP, the easier the topic.

We define two methods for grouping topics into classes depending on the number of groups we want to build, either two or four classes to experiment with two different types of classes. For grouping the topics into two classes (Easy and Difficult), we use the mean AAP measure as a splitting condition: if AAP for a given topic is superior to the mean AAP (calculated across all topics) then the topic is classified as Easy otherwise it is classified as Difficult. For grouping the topics into four classes (Easy, Moderately_Easy, Moderately_Difficult, and Difficult), we use the mean AAP and the standard deviation around the mean as a splitting condition:

> if AAP >= (mean AAP + stDev) **then** Easy topic
> if AAP >= (mean AAP) **and** AAP < (mean AAP + stDev) **then** Moderately_Easy topic
> if AAP >= (mean AAP - stDev) **and** AAP < (mean AAP) **then** Moderately_Difficult topic
> if AAP < (mean AAP - stDev) **then** Difficult topic

The above two or four classes of INEX XER topics are then used as a basis for evaluating our automatic feature-based topic classification algorithm, as described in Section 4.

3   Our Entity Ranking System

Our approach to identifying and ranking entities combines: (1) the full-text similarity of the entity page with the query; (2) the similarity of the page's categories with the categories of the entity examples; and (3) the link contexts found in the top ranked pages returned by a search engine for the query.

We developed an XER system that involves the following modules.

1. The *topic module* takes an INEX topic as input and generates the corresponding full-text query and the list of entity examples.
2. The *search module* sends the query to a search engine[1] and returns a list of scored Wikipedia pages. The assumption is that a good entity page is a page that answers the query.
3. The *link extraction module* extracts the links from a selected number of highly ranked pages, together with the information about the paths of the links (XML paths). The assumption is that a good entity page is a page that is referred to by a page answering the query; this is an adaptation of the HITS [9] algorithms to the problem of entity ranking.
4. The *linkrank module* calculates a weight for a page based (among other things) on the number of links to this page. The assumption is that a good entity page is a page that is referred to from contexts with many occurrences of the entity examples. A coarse context would be the full page that contains the entity examples. Smaller and better contexts may be elements such as paragraphs, lists, or tables [16].
5. The *category similarity module* calculates a weight for a page based on the similarity of the page categories with the categories attached to the entity examples. The assumption is that a good entity page is a page associated with a category close to the categories of the entity examples [18].
6. The *full-text module* calculates a weight for a page based on its initial search engine score.

The global score $S(t)$ for a target entity page is calculated as a linear combination of three normalised scores, the linkrank score $S_L(t)$, the category similarity score $S_C(t)$, and the full-text score $S_Z(t)$:

$$S(t) = \alpha S_L(t) + \beta S_C(t) + (1 - \alpha - \beta)S_Z(t) \tag{1}$$

where α and β are two parameters whose values can be tuned differently depending on the entity retrieval task.

Details of the global score and the three separate scores can be found in previous publications [16,18]. In this paper we are interested in automatically adapting the values of α and β parameters to the topic, depending on the topic class. By predicting the optimal values for α and β parameters that correspond to each class of topic difficulty, we aim at improving the performance score of our system over the current best performance score that uses pre-defined static values for the α and β parameters.

---

[1] We used Zettair, an open source search engine: http://www.seg.rmit.edu.au/zettair/

4   Topic Difficulty Prediction

In this section we present our methodology for predicting topic difficulty. Our approach is based on generating a classifier to classify topics in two or four classes (as described in Section 2). The classifier is built using features extracted from the INEX topic definition. We use the open source data mining software Weka [21] developed by the University of Waikato. Weka is a collection of machine learning algorithms for data mining tasks that, given a training set of topics, can generate a classifier from the topics and their associated features.

4.1   Topic Features

From the specific structure of the INEX 2007 XER topics we developed 32 different a-priori features. We call these *a-priori* (or static) features because each of them can be calculated by using only the topic definition before any processing is made by our system. The features include the number of words (excluding stop-words) found in the topic title, the topic description and the topic narrative, respectively; the number of verbs and sentences found in the description or narrative part, as well as the number of words in the union or intersection of these parts of the topic. Additional features are built using the ratio between previous features, for example the ratio between the number of words in the title and the description, or the ratio between the number of sentences in the description and the narrative. The idea is that if the narrative needs more explanation than the title or the description, it may be because good answers could be difficult to identify among many irrelevant pages. Counting verbs required some natural language processing. We used the NLTK toolkit software [12] which especially helped with the different features concerning verb forms.

Other a-priori features are related to the target categories and the entity examples listed in the topic. The target categories can be either very broad or very specific and they represent indications about the desired type of answer entities, not hard constraints. There could be a correlation between the number of target categories and the topic performance that we wanted to identify. Other features involve not just the topic description but also the INEX Wikipedia test collection, for example, the number of Wikipedia pages attached to the target categories. We also count the number of different Wikipedia categories attached to the entity examples in the topic. Finally we create features that represent the union or intersection of target categories and categories attached to the entity examples.

We also defined a few *a-posteriori* (dynamic) features that could be calculated at run time, i.e. when sending the topic (query) to our system. These features include the number of links from the highly ranked pages returned by the search engine, the number of contexts identified in those pages and the number of common categories attached to the entity examples and the answer entities.

Table 1. Nine topic features that correlated well with the topic difficulty prediction. In the table, w stands for words, narr for narrative, desc for description, t_cat for target categories, and e_cat for categories attached to entity examples. For example, #sent_narr is the number of sentences in the narrative part of the INEX topic.

Number	Description	Features
1	Topic definition features	#sent_narr
2 − 6	Ratio of topic definition features	#w_title/#w_narr, #w_intersec(title,narr)/#w_union(title,narr) #w_intersec(desc,narr)/#w_union(desc,narr), #w_intersec(title,desc)/#w_union(title,desc,narr), #w_intersec(desc,narr)/#w_union(title,desc,narr)
7 − 8	Topic definition and Wikipedia features	#pages_per_t_cat, #intersec(e_cat)
9	Ratio of topic definition and Wikipedia features	#intersec(e_cat)/#union(e_cat)

4.2   Topic Classifier

The next step was to identify among the 32 features those that best correlated with the topic difficulty, i.e. the features that would be usable by a classifier to predict between different classes of topics. We first generated many classifiers, each one associated with a random subset of the 46 INEX 2007 XER topics. The classifiers were generated using the Weka j48 classifier based on the well known Quinlan's C4.5 statistical classifier [17], with each training topic subset classified using the topic classification explained in Section 2. We then manually analysed all the decision trees generated by Weka to identify the features that were actually used by the generated classifiers. As a result, we could extract a small subset of nine features that correlated well with topic difficulty.

Table 1 shows the nine features used to generate the training topic subsets. We discovered that the dynamic (a-posteriori) features had no influence on the generated classifiers, and so we only used the a-priori features.

4.3   Training and Testing Topic Sets

For each training subset of INEX 2007 XER topics that we used previously for generating a classifier, we used the testing set comprising all the 35 INEX 2008 XER topics for evaluating the performance of this classifier. We tried many different mostly random combinations of training topic subsets, but because of their relatively small sizes on average the accuracy of the correctly classified instances was around 71%.

To improve the accuracy, we used a well known approach that combines several decision trees, each generated from slightly different topic sets. This is known as Random Forests [1] and was used in query prediction by Yom-Tov et al. [22]. Before implementing the combined classifier, we carefully built the training topic set for each individual predictor so that the included topics were representative of different features.

We manually divided the training set of 46 INEX 2007 XER topics into four subsets of around 35 topics each. We had to do it manually in order to get as

Table 2. Accuracy achieved by the six two-class classifiers on the 35 INEX 2008 topics

	Class	
	2	
Classifier	Correct	Incorrect
1	24/35 (68%)	11/35 (32%)
2	25/35 (71%)	10/35 (29%)
3	25/35 (71%)	10/35 (29%)
4	25/35 (71%)	10/35 (29%)
5	24/35 (68%)	11/35 (32%)
combined	26/35 (**74%**)	9/35 (**26%**)

much different and representative topics as possible, especially because of the small topic subset sizes. So those four subsets and the one with all 46 topics made five different training sets from which we built five separate classifiers. For these and the final combined classifier we also had to build a testing topic set that does not include any of the training topics. The INEX 2008 XER topic set was used for this purpose.

4.4   Validation of Topic Difficulty Prediction

The final topic difficulty prediction classifier was built using a simple voting system which is the reason why we needed an odd number of classifiers. For building a two-class classifier the voting algorithm is trivial: for a topic we get a prediction from the five classifiers and count the number of predictions as Easy and the number of predictions as Difficult; the majority gives the prediction for the final classifier. For example, if the predictions from the five classifiers are [diff, easy, easy, diff, diff], the combined prediction is Difficult.

The combined two-class classifier resulted in a precision of 74% on our testing set which is better than what we could achieve with a single classifier. Table 2 shows the accuracy achieved by each of the six classifiers.

We also considered the possibility of building a four-class classifier (Easy, Moderately_Easy, Moderately_Difficult, and Difficult). A similar voting algorithm is used by simply choosing diff : easy = 0:5 and diff : easy = 5:0 to be predicted as Easy and Difficult topics, respectively, with the diff : easy = (1:4 | 2:3) and diff : easy = (4:1 | 3:2) combinations resulting in Moderately_Easy and Moderately_Difficult topics, respectively. The combined four-class classifier resulted in an accuracy of 31% on our testing set which was much less than that achieved by the two-class classifier.

5   Applying Topic Difficulty Prediction in Entity Ranking

Our objective with topic difficulty prediction was to improve the performance of our XER system by dynamically tuning the values for system parameters

Table 3. Estimated values for optimal α/β system parameters, as measured by MAP using the 46 topics of the INEX 2007 XER training collection

	Class					
	2		4			
	Easy	Diff	Easy	modEasy	modDiff	Diff
Measure	α β	α β	α β	α β	α β	α β
MAP	0.2 0.6	0.1 0.8	0.0 0.7	0.2 0.6	0.1 0.8	0.5 0.0

according to the predicted topic class. Specifically, for each of the 35 INEX 2008 topics in the testing set, we adapt the values for the α and β system parameters in accordance with the estimated optimal values observed on the training set.

5.1 Choosing Optimal System Parameters by Topic Difficulty

We first estimated the optimal values for the system parameters by using the 46 INEX 2007 XER training topics and by also taking into account their corresponding topic difficulty classes. We generated all the possible 66 runs by respectively varying the values of α and β from 0 to 1 by increment of 0.1. For each topic, we then measured the average precision (AP) for each run and ordered the runs by decreasing value of AP. This way we could identify the values of the two (α, β) parameters that performed *best* for each individual topic. To estimate which (α, β) pair would be *optimal* for a given topic difficulty class, we used the topics that belong to a particular class (such as Easy), and calculated the mean AP (MAP) for each run that appeared at least once among the ten highly ranked runs for a given topic.[2] We then ordered the runs by decreasing scores and identified the highest ranked run as the optimal (α, β) pair for each topic difficulty class. We did this both for the two and the four classes of topic difficulty.

Table 3 shows the estimated optimal values for (α, β) as measured by MAP, when using two or four classes of topic difficulty. Interestingly, with the four-class prediction the optimal parameter values for the Easy topics are $(\alpha = 0.0, \beta = 0.7)$, i.e. the link score is ignored. For the Difficult topics the opposite effect is observed with the category score ignored and a high weight spread evenly on the link score α and the Zettair score $(1 - \alpha - \beta)$.

5.2 Evaluation of the Predicted Topic Performance

We now use our combined topic difficulty prediction algorithm (described in Section 4) to tune and evaluate the performance of our XER system on the 35 INEX 2008 XER testing topics. According to the estimated prediction, we aim at dynamically setting the α and β parameters to their optimal values shown in Table 3. We did two sets of experiments, respectively with two and four classes of topic difficulty. We use MAP as our choice of evaluation measure.

---

[2] The value ten was determined experimentally on the XER training topic set.

Table 4. Evaluation of the predicted topic performance, as measured by MAP using the 35 topics of the INEX 2008 XER testing collection. The † symbol shows statistical significance over the Baseline run ($p < 0.05$).

	Class						
	2		**4**			
	Easy	Diff	Easy	modEasy	modDiff	Diff
Run	0.2 0.6	0.1 0.8	0.0 0.7	0.2 0.7	0.1 0.8	0.6 0.0
Baseline	N/A	0.36280	N/A			
Predicted	**0.38085**		0.30769			
Optimal	0.38705		0.38431			
Perfect	N/A	0.45746†	N/A			

To evaluate the benefits of using topic difficulty prediction, we compare the performances of four different runs:

1. *Baseline* run that does not use topic prediction with parameter vales set to (α=0.2, β=0.6). This was the best performing entity ranking run at INEX 2007 (for the list completion task) when using the MAP measure.
2. *Predicted* run with parameter values set according to the estimated topic difficulty prediction on the training collection. The difficulty of a particular INEX 2008 XER testing topic was first predicted by our topic prediction algorithm, and the system parameter values were then set to the estimated optimal values for that topic difficulty class (as shown in Table 3).
3. *Optimal* run with parameter values set to the estimated optimal values on the training collection. Given the previously determined difficulty of a particular INEX 2008 XER testing topic (by applying the AAP measure on all the INEX 2008 submitted runs), the system parameter values were set to the estimated optimal values for that topic difficulty class. This is the best run we could aim at by using our topic difficulty prediction algorithm.
4. *Perfect* run with parameter values set to the best values (out of all the 66 value combinations) that can be achieved for each topic on the INEX 2008 XER testing collection. This is the run that produces the absolute best performance with our current XER system.

The results are presented in Table 4. The table shows that a two-class prediction of topic difficulty is performing better than the baseline (our last year best run), although the difference in performance is not statistically significant. These two runs were among the top four best performing runs at INEX 2008, all of which were submitted by our participating group. On the other hand, the four-class prediction of topic difficulty resulted in decreased performance, which is mainly due to the fact that the topic prediction algorithm is specifically designed for two-class rather than for four-class prediction. Although the results are promising, we recognise that the small size of the training and testing topic sets do not allow for very conclusive evaluation.

6   Related Work

The approaches to query prediction can generally be grouped into two types: *static prediction* approaches, based on intrinsic characteristics of the query and possibly the document collection [8]; and *dynamic prediction* approaches, which use characteristics of the top ranked answers to the query [23].

Hao Lang et al. [11] evaluate query performance based on the covering topic score that measures how well the topic of the query is covered by documents retrieved by the system (dynamic prediction). Cronen-Townsend et al. [3] propose to predict query performance by computing the relative entropy (clarity score) between a query language model and the corresponding collection language model (static prediction).

Mothe and Tanguy [15] predict query difficulty based on linguistic features, using TreeTagger for part-of-speech tagging and other natural language processing tools. Topic features include morphological features (number of words, average of proper nouns, average number of numeral values), syntactical features (average conjunctions and prepositions, average syntactic depth and link span) or semantic features (average polysemy value). They found that the only positively correlated feature is the number of proper nouns, although the average syntactic link span and the average polysemy value also have some correlation with topic difficulty. We use some morphological or syntactic features in our topic prediction algorithm, but we also take advantage of the structure of the topic (title, description, narrative).

Kwok [10] uses Support Vector Machine (SVM) regression to predict the weakest and strongest queries in the TREC 2004 topic set (static prediction). Their choice of features include inverse document frequency of query terms and average term frequency in the collection. They found that features based on term frequencies could predict correctly even with short queries.

Yom-Tov et al. [22] predict query difficulty by measuring the contribution of closely related query terms, using features such as the overlap between the k-top answers to a sub-query (a query based on one query term) and to the full query. They experimented with two different query predictors: an histogram-based predictor and a modified decision tree. The difficulty predictor was used for selective query expansion or reduction.

Grivolla et al. [7] propose several classifiers to predict easy and difficult queries. They use decision tree and SVM types of classifiers, and select useful features among a set of candidates; the classifiers are trained on the TREC 8 test collection. The features are computed from the query itself (static features), the retrieval results and the knowledge about the retrieval process (dynamic features). They tested many different classifiers but did not combine them. Our approach is very similar to theirs, although we use different topic-specific features, a combined classifier and different test collection (INEX instead of TREC).

7   Conclusion and Future Work

We have presented our experiments in predicting topic difficulty and its application to the system we have developed for XML entity ranking. We demonstrated that it is possible to predict accurately a two-class level of topic difficulty with a classifier generated from a selected number of static features extracted from the INEX topic definition and the Wikipedia document collection. We also experimented with dynamic features from the intermediary results of the query processing but those features did not correlate well with the topic prediction. The more interesting result is related to the analysis of four classes of topic difficulty and their impact on the optimal parameter values for our XER system: for the Easy topics, the use of Wikipedia categories is very important while for the Difficult topics the link structure plays a very important role.

The application of topic prediction in tuning our system has shown encouraging improvement over our last year best result but we need a larger test collection to confirm the significance of our findings. The major limitation of our topic prediction approach is that it relies on the INEX topic definition that is much richer than standard Web queries. In the future we plan to develop a dynamic query prediction approach based (among other things) on the query similarity scores of the relevant entities retrieved by our XER system.

Acknowledgements

Most of this work was completed while Vladimir Naumovski was doing his internship at INRIA in 2008.

References

1. Breiman, L.: Random forests. Machine Learning 45(1), 5–32 (2001)
2. Carmel, D., Yom-Tov, E., Soboroff, I.: Predicting query difficulty - methods and applications. SIGIR Forum 39(2), 25–28 (2005)
3. Cronen-Townsend, S., Zhou, Y., Croft, W.B.: Predicting query performance. In: Proceedings of the 25th ACM SIGIR conference on Research and development in information retrieval (SIGIR 2002), Tampere, Finland, pp. 299–306 (2002)
4. de Vries, A.P., Vercoustre, A.-M., Thom, J.A., Craswell, N., Lalmas, M.: Overview of the INEX 2007 entity ranking track. In: Fuhr, N., Kamps, J., Lalmas, M., Trotman, A. (eds.) INEX 2007. LNCS, vol. 4862, pp. 245–251. Springer, Heidelberg (2008)
5. Demartini, G., de Vries, A.P., Iofciu, T., Zhu, J.: Overview of the INEX 2008 entity ranking track. In: Geva, S., Kamps, J., Trotman, A. (eds.) INEX 2008. LNCS, vol. 5631. Springer, Heidelberg (2009)
6. Denoyer, L., Gallinari, P.: The Wikipedia XML corpus. SIGIR Forum 40(1), 64–69 (2006)
7. Grivolla, J., Jourlin, P., de Mori, R.: Automatic classification of queries by expected retrieval performance. In: Proceedings of the SIGIR workshop on predicting query difficulty, Salvador, Brazil (2005)

8. He, B., Ounis, I.: Query performance prediction. Information Systems 31(7), 585–594 (2006)
9. Kleinberg, J.M.: Authoritative sources in hyperlinked environment. Journal of the ACM 46(5), 604–632 (1999)
10. Kwok, K.: An attempt to identify weakest and strongest queries. In: Proceedings of the SIGIR workshop on predicting query difficulty, Salvador, Brazil (2005)
11. Lang, H., Wang, B., Jones, G., Li, J.-T., Ding, F., Liu, Y.-X.: Query performance prediction for information retrieval based on covering topic score. Journal of Computer Science and technology 23(4), 590–601 (2008)
12. Loper, E., Bird, S.: NLTK: The natural language toolkit. In: Proceedings of the ACL 2002 Workshop on Effective tools and methodologies for teaching natural language processing and computational linguistics, Philadelphia, Pennsylvania, pp. 63–70 (2002)
13. Mizzaro, S.: The good, the bad, the difficult, and the easy: Something wrong with information retrieval evaluation? In: Macdonald, C., Ounis, I., Plachouras, V., Ruthven, I., White, R.W. (eds.) ECIR 2008. LNCS, vol. 4956, pp. 642–646. Springer, Heidelberg (2008)
14. Mizzaro, S., Robertson, S.: HITS hits TREC: Exploring IR evaluation results with network analysis. In: Proceedings of the 30th ACM SIGIR conference on Research and development in information retrieval (SIGIR 2007), Amsterdam, The Netherlands, pp. 479–486 (2007)
15. Mothe, J., Tanguy, L.: Linguistic features to predict query difficulty. In: Proceedings of the SIGIR workshop on predicting query difficulty, Salvador, Brazil (2005)
16. Pehcevski, J., Vercoustre, A.-M., Thom, J.A.: Exploiting locality of Wikipedia links in entity ranking. In: Macdonald, C., Ounis, I., Plachouras, V., Ruthven, I., White, R.W. (eds.) ECIR 2008. LNCS, vol. 4956, pp. 258–269. Springer, Heidelberg (2008)
17. Quinlan, J.R.: C4.5: Programs for Machine Learning. Morgan Kaufmann, San Francisco (1993)
18. Thom, J.A., Pehcevski, J., Vercoustre, A.-M.: Use of Wikipedia categories in entity ranking. In: Proceedings of 12th Australasian Document Computing Symposium (ADCS 2007), Melbourne, Australia, pp. 56–63 (2007)
19. Voorhees, E.M.: The TREC robust retrieval track. In: Proceedings of the Thirteenth Text Retrieval Conference (TREC 2004) (2004)
20. Webber, W., Moffat, A., Zobel, J.: Score standardization for inter-collection comparison of retrieval systems. In: Proceedings of the 31st ACM SIGIR conference on Research and development in information retrieval (SIGIR 2008), Singapore, pp. 51–58 (2008)
21. Witten, I.H., Frank, E.: Data Mining: Practical Machine Learning Tools and Techniques (2/E). Morgan Kaufmann, San Francisco (2005)
22. Yom-Tov, E., Fine, S., Carmel, D., Darlow, A., Amitay, E.: Juru at TREC 2004: Experiments with prediction of query difficulty. In: Proceedings of the Thirteenth Text Retrieval Conference (TREC 2004) (2004)
23. Zhou, Y., Croft, W.B.: Query performance prediction in web search environments. In: Proceedings of the 30th ACM SIGIR conference on Research and development in information retrieval (SIGIR 2007), Amsterdam, The Netherlands, pp. 543–550 (2007)

A Generative Language Modeling Approach for Ranking Entities

Wouter Weerkamp, Krisztian Balog, and Edgar Meij

University of Amsterdam, ISLA, Science Park 107, 1098XG Amsterdam, The Netherlands
{w.weerkamp,k.balog,edgar.meij}@uva.nl

Abstract. We describe our participation in the INEX 2008 Entity Ranking track. We develop a generative language modeling approach for the entity ranking and list completion tasks. Our framework comprises the following components: (i) entity and (ii) query language models, (iii) entity prior, (iv) the probability of an entity for a given category, and (v) the probability of an entity given another entity. We explore various ways of estimating these components, and report on our results. We find that improving the estimation of these components has very positive effects on performance, yet, there is room for further improvements.

1 Introduction

The Enitity Ranking track of this year's INEX features three tasks: *entity ranking, list completion*, and *entity relation search* [3]. In our participation, we focus on the first and second tasks, leaving entity relation search for coming years. Both tasks (entity ranking and list completion) are aimed at retrieving entities from a semi-structured document collection. The document collection at hand is Wikipedia [4], and an entity is a Wikipedia article by definition.

The entity ranking task aims at retrieving entities given a certain topic and Wikipedia category: the goal is to identify the entities that are relevant given the topic and fit within the given category. The list completion task is slightly different and aims at adding entities of the same type to a small sample set of entities. Again, we also have the topic and category available, but as additional information we get one or more entities as examples.

In our participation we use a generative language modeling approach to model both tasks. This approach has been successfully applied in many information retrieval tasks [1, 5, 7, 8, 10]. Language models are attractive because of their foundations in statistical theory, the great deal of complementary work on language modeling in speech recognition and natural language processing, and the fact that very simple language modeling retrieval methods have performed quite well empirically.

A large portion of this paper is directed to the modeling of entity ranking and the estimation of the various components this framework offers us. We submitted a total of six runs, again with a focus on the entity ranking task (four runs).

The remainder of this paper introduces the modeling of the entity ranking task in Section 2, and of the list completion task in Section 3. Next, we discuss the estimation of the various components of both models in Section 4 and the experimental setup

S. Geva, J. Kamps, and A. Trotman (Eds.): INEX 2008, LNCS 5631, pp. 292–299, 2009.

in Section 5. The submitted runs (Section 6) and their results and discussion of these results (Section 7) follow, and we conclude in Section 8.

2 Modeling Entity Ranking

Entities are ranked by their probability of being relevant given a query q and a set of categories C, that is $P(e|q, C)$. We assume that q and C are conditionally independent and, moreover, that each of the categories $c \in C$ are mutually independent. Formally:

$$P(e|q, C) = P(e|q) \cdot P(e|C) \tag{1}$$
$$= P(e|q) \cdot \prod_{c \in C} P(e|c).$$

To estimate the probability of an entity given the query $P(e|q)$, we apply Bayes' rule and drop the denominator $P(q)$ which is constant for all entities and, thus, does not influence the ranking:

$$P(e|q) \propto P(q|e) \cdot P(e). \tag{2}$$

Here, $P(q|e)$ expresses the probability that q is generated by entity e, and $P(e)$ is the *a priori* probability of an entity being relevant (independent of the query). For the sake of simplicity, $P(e)$ is assumed to be uniform, and is not included in the equations from now onwards.

We infer an entity model for each entity e, such that the probability of a term given the entity model is $P(t|\theta_e)$. This model is then used to predict how likely the entity would produce query q. Each query term is assumed to be sampled identically and independently. Thus, the query likelihood is obtained by taking the product of the individual term probabilities across all terms in the query:

$$P(q|\theta_e) = \prod_{t \in q} P(t|\theta_e)^{n(t,q)}, \tag{3}$$

where $n(t, q)$ denotes the number of times t is present in q. Putting together our choices so far (Eqs. 1, 2, and 3) we obtain the following:

$$P(e|q, C) = \prod_{t \in q} P(t|\theta_e)^{n(t,q)} \cdot \prod_{c \in C} P(e|c). \tag{4}$$

For computational reasons, we move to the log domain, and use the following formula for ranking entities:

$$\log P(e|q, C) \propto \left(\sum_{t \in q} P(t|\theta_q) \cdot \log P(t|\theta_e) \right) + \sum_{c \in C} \log P(e|c). \tag{5}$$

Note that $n(t, q)$ has been replaced with $P(t|\theta_q)$, where θ_q is referred to as the *query model*. This allows us more flexible weighting of query terms. Three important components remain to be defined: the entity model θ_e, the query model θ_q, and the probability of an entity given a category $P(e|c)$. These will be introduced in the following sections.

3   Modeling List Completion

The list completion task is modeled similarly to the entity ranking task, with the addition that the probability of the entity is also conditioned on a set of example entities, E. We assume that example entities are conditionally independent from the query and the categories, as well as mutually independent from each other. Formally:

$$P(e|q, C, E) = P(e|q, C) \cdot \prod_{e' \in E} P(e|e'). \qquad (6)$$

Again, we perform this computation in the log domain:

$$\log P(e|q, C, E) = \log P(e|q, C) + \sum_{e' \in E} \log P(e|e'). \qquad (7)$$

The estimation of $P(e|q, C)$ has already been discussed in Section 2. A new component to be defined is the probability of an entity e given entity e': $P(e|e')$. In other words, this probability expresses the similarity of two entities.

4   Estimating the Components

In this section we detail how various components of the models introduced in the previous sections are estimated. Specifically, we discuss the implementation of the entity model θ_e, the query model θ_q, the probability of an entity given a set of categories $P(e|C)$, and finally, the probability of an entity given another entity $P(e|e')$.

4.1   Entity Model

The entity is represented as a multinomial probability distribution over terms. To estimate $P(t|\theta_e)$ we smooth the empirical entity model with the background collection to prevent zero probabilities:

$$P(t|\theta_e) = (1 - \lambda_e) \cdot P(t|e) + \lambda_e \cdot P(t) \qquad (8)$$

Since entities correspond to Wikipedia articles, this way of modeling an entity is identical to constructing a smoothed document model for each Wikipedia page. The choice of the smoothing parameter λ_e is discussed in Section 5.

4.2   Query Model

Our baseline query model $P(t|\theta_q)$ is set to $n(t, q) \cdot |q|^{-1}$, where $n(t, q)$ is the number of occurrences of term t in query q, and $|q|$ is the query length. Essentially, the probability mass is distributed uniformly across query terms. Since this representation of the query is quite sparse, we would like to add more terms to the original query. By mixing new terms and original query terms, we end up with the following equation:

$$P(t|\theta_q) = (1 - \lambda_q) \cdot P(t|\hat{q}) + \lambda_q \cdot P(t|q), \qquad (9)$$

where $P(t|\hat{q})$ is the probability of the term given the expanded query.

Balog et al. [2] introduce various methods for constructing expanded query models by sampling terms from a set of example documents (complementing the textual query). Based on the information provided with the topic statement, we have three straightforward ways of applying these methods to our current scenario, by (i) treating all entities belonging to the target categories as examples (both for entity ranking and list completion), (ii) employing a blind-relevance feedback approach, in which we perform a baseline run and look at the categories which are assigned to the 10 highest-ranked entities (category feedback for entity ranking), or (iii) using the example entities (only list completion). Specifically, we use the best performing method, maximum likelihood (ML), from [2] to estimate the expanded query model $P(t|\hat{q})$.

4.3 Entity-Categories Probability

The probability of an entity e given a set of target categories C, $P(e|C)$, is computed as follows. Let cat(e) denote the set of categories e is assigned to. The overlap ratio between cat(e) and the set of target categories C is used as an estimate of $P(e|C)$:

$$P(e|C) = \frac{|\text{cat}(e) \cap C|}{|C|}, \tag{10}$$

where $|C|$ is the size of the set of target categories. We experiment further with this way of estimating $P(e|C)$, by introducing a parameter δ to control the weight of the overlap between the two sets C and cat(e) and dropping the term in the denominator:

$$P(e|C) \propto \delta \cdot |\text{cat}(e) \cap C|. \tag{11}$$

Based on initial experiments we set $\delta = 6$.

Further, we hypothesize that the target categories for each topic, as used in Eqs. 10 and 11, are not exhaustive. Therefore, in order to amend this set of categories, we apply a simple expansion strategy. We leverage the hierarchical structure of Wikipedia categories, by expanding the set of target categories with their subcategories, up to certain depth. Based on preliminary experiments, we set the maximum depth to three.

4.4 Entity-Entity Probability

Our model for the list completion task involves the estimation of the similarity between two entities. This is expressed as $P(e|e')$, the probability of an entity e given another entity e' (see Eq. 6). We estimate this probability based on set overlap between the categories assigned to each of the entities. To this end, we employ a standard set-based similarity measure, Dice's coefficient, calculated as follows:

$$P(e|e') = \frac{2 \cdot |\text{cat}(e) \cap \text{cat}(e')|}{|\text{cat}(e)| + |\text{cat}(e')|}, \tag{12}$$

where cat(e) and cat(e') are the set of categories assigned to entities e and e', respectively.

5 Experimental Setup

5.1 Document Representation

Besides representing the entity (Wikipedia page) by its entire textual content (referred to as *full representation*), we opted for a second representation. Assuming that most valuable information on a Wikipedia page is presented at the beginning of the article, we select only the first paragraph of each article. This paragraph is the new representation of the entity (refered to as *paragraph representation*). For the sake of comparability of runs, we use the full representation in almost all cases.

5.2 Document Preprocessing

Document preprocessing consisted of removing stopwords only. Besides the "standard" English stopwords, we added several Wikipedia-specific stopwords to the stopword list (e.g. *disambiguation*, *category*, and *stub*).

5.3 Smoothing Parameter

For the smoothing parameter λ_e in Eq. 8, we set λ_e equal to $\frac{|e|}{\beta+|e|}$, where $|e|$ is the length of the entity (i.e., the number of terms in the entity's representation). Essentially, the amount of smoothing is proportional to the length of the entity (and is like Bayes smoothing with a Dirichlet prior [6]). If there is very few content available for the entity (i.e., the corresponding article is very short) then the model of the entity is more uncertain, leading to a greater reliance on the background probabilities. We set β to be the average entity length, i.e. $\beta = 409$ for the full representation and $\beta = 42$ for the paragraph representation.

5.4 Query Modeling Parameter

For the construction of the new query model (Eq. 9), we need to set λ_q and decide on the number of terms in the new query. For the entity ranking task, in which we select our expansion terms from the category feedback approach, we set $\lambda_q = 0.5$ and select the 20 terms with the highest probability. For the list completion task we set $\lambda_q = 0.2$ and again select the top 20 terms to be included in the new query.

6 Submitted Runs

This section lists our submitted runs (six in total) and the configuration used for each (note that all runs use the full representation, unless stated otherwise). For the entity ranking task the following four runs were submitted:

6_UAms_ER_T_baseline: Our baseline run using Eq. 3.

3_UAms_ER_TC_overlap: Overlap run using Eq. 5; we estimate $P(e|C)$ as in Eq. 11, and use an expanded category set C up to depth three.

4_UAms_ER_TC_cat-exp: Expanded overlap run; similar to run *3_UAms_ER_T_overlap*, except that we model the query according to Eq. 9, where expansion terms are selected using the category feedback method. We select the top 2 categories and use the entities within these categories as examples.

1_UAms_ER_TC_mixture: Mixture run; we construct two runs using Eq. 5, one on the full representation and one on the paragraph representation. Each run estimates $P(e|C)$ as in Eq. 11, and uses an expanded category set C up to depth three (paragraph representation) or two (full representation). Both runs are combined using a linear rank combination with a weight of 0.1 for the paragraph representation and 0.9 for the full representation.

The remaining two runs were submitted for the list completion task:

5_UAms_LC_TE_baseline: Our baseline run using Eq. 7; we model the query according to Eq. 9 and select expansion terms from the provided example entities.

2_UAms_LC_TCE_dice: Our overlap run; similar to run *5_UAms_LC_T_baseline*. We estimate $P(e|e')$ using Eq. 12 and $P(e|C)$ as in Eq. 11, and use an expanded category set C up to depth two.

7 Results and Discussion

We report on the results of our submissions and the best performing run among all submitted runs for each task. The metric used for measuring performance is *xinfAP* [9]. Significance is tested against our best performing run, using a two-tailed paired t-test and $\alpha = .05$.

Table 1. Results for the entity ranking task. Significant differences against our best performing run are marked with *.

Run	xinfAP
1_FMIT_ER_TC_nopred-cat-baseline-a1-b8	0.341
3_UAms_ER_TC_overlap	0.253
4_UAms_ER_TC_cat-exp	0.232
1_UAms_ER_TC_mixture	0.222 *
6_UAms_ER_T_baseline	0.111

Table 1 presents the results for the entity ranking task, in decreasing order of performance. Our main finding are as follows. Taking category information into account (*6_UAms_ER_T_baseline* vs. the other runs) improves performance. Despite the apparent increase over the baseline, in terms of xinfAP scores, these differences are not significant. We leave the investigation of this to further work. Adding additional features (besides using category information) seems less beneficial; query expansion using the category feedback method (method (ii) in Section 4.2) has a slight negative effect on performance (*4_UAms_ER_TC_cat-exp* vs. *3_UAms_ER_TC_overlap*). Mixing the two document representations, *full* and *paragraph*, hurts performance (*1_UAms_ER_TC_mixture* vs. *3_UAms_ER_TC_overlap*), in this case significantly so.

Table 2. Results for the list completion task. Significant differences against our best performing run are marked with *.

Run	xinfAP
1_FMIT_LC_TE_nopred-stat-cat-a1-b8	0.402
2_UAms_LC_TCE_dice	0.319
5_UAms_LC_TE_baseline	0.133 *

Results for the list completion task are shown in Table 2. We observe that adding estimates of $P(e|e')$ and $P(e|C)$ to the baseline yields in substantial and significant improvements over our (naive) baseline. Further analysis is needed to see which of the components makes up for most of this improvement.

8   Conclusion

We described the approach, submitted runs, and results of our participation in this year's INEX Entity Ranking track. Our focus lied on the entity ranking and list completion tasks, and our chief aim was to develop a general language modeling framework to model these in a uniform and theoretically sound way. Given the models we introduced, we are left with plenty of choices on how to estimate the various components these models offer. For most of these components we applied simple options which mainly make use of the category information available in Wikipedia. Results show that using category and entity information in different ways leads to increases in performance over simple baselines, yet there is room for further improvements. Based on these initial findings, in further work we aim at investigating additional, more elaborate ways of estimating the various components, and exploring their impact on performance. Since we made quite strong independence assumptions regarding our input variables, in another line of work we are planning on examining these dependencies.

References

[1] Balog, K., Azzopardi, L., de Rijke, M.: Formal models for expert finding in enterprise corpora. In: SIGIR 2006, pp. 43–50 (2006)
[2] Balog, K., Weerkamp, W., de Rijke, M.: A few examples go a long way: constructing query models from elaborate query. In: SIGIR 2008, pp. 371–378 (2008)
[3] Demartini, G., de Vries, A.P., Iofciu, T., Zhu, J.: Overview of the INEX 2008 entity ranking track. In: Geva, S., Kamps, J., Trotman, A. (eds.) INEX 2008. LNCS, vol. 5631. Springer, Heidelberg (2009)
[4] Denoyer, L., Gallinari, P.: The Wikipedia XML corpus. SIGIR Forum 40, 64–69 (2006)
[5] Hiemstra, D.: Using Language Models for Information Retrieval. PhD thesis, University of Twente (2001)
[6] Mackay, D.J.C., Peto, L.: A hierarchical dirichlet language model. Natural Language Engineering 1(3), 1–19 (1994)

[7] Miller, D., Leek, T., Schwartz, R.: A hidden Markov model information retrieval system. In: SIGIR 1999, pp. 214–221 (1999)

[8] Ponte, J.M., Croft, W.B.: A language modeling approach to information retrieval. In: SIGIR 1998, pp. 275–281 (1998)

[9] Yilmaz, E., Kanoulas, E., Aslam, J.A.: A simple and efficient sampling method for estimating ap and ndcg. In: SIGIR 2008, pp. 603–610 (2008)

[10] Zhai, C.: Statistical language models for information retrieval a critical review. Foundations and Trends in Information Retrieval 2, 137–213 (2008)

Overview of the INEX 2008 Interactive Track

Nils Pharo[1], Ragnar Nordlie[1], and Khairun Nisa Fachry[2]

[1] Faculty of Journalism, Library and Information Science, Oslo University College, Norway
nils.pharo@jbi.hio.no, ragnar.nordlie@jbi.hio.no
[2] Archives and Information Studies, University of Amsterdam, The Netherlands
k.n.fachry@uva.nl

Abstract. This paper presents the organization of the INEX 2008 *interactive track*. In this year's iTrack we aimed at exploring the value of element retrieval for two different task types, fact-finding and research tasks. Two research groups collected data from 29 test persons, each performing two tasks. We describe the methods used for data collection and the tasks performed by the participants. A general result indicates that test persons were more satisfied when completing research task compared to fact-finding task. In our experiment, test persons regarded the research task easier, were more satisfied with the search results and found more relevant information for the research tasks.

1 Introduction

The INEX interactive track (iTrack) is a cooperative research effort run as part of the INEX Initiative for the Evaluation of XML retrieval [1]. The overall goal of INEX is to experiment with the potential of using XML to retrieve relevant parts of documents through the provision of a test collection of XML-marked Wikipedia articles. The main body of work within the INEX community has been the development and testing of retrieval algorithms. Interactive information retrieval (IIR) [2] aims at investigating the relationship between end users of information retrieval systems and the systems. This aim is approached partly through the development and testing of interactive features in the IR systems and partly through research on user behavior in IR systems. In the INEX iTrack the focus has been on how end users react to and exploit the potential of IR systems that facilitate the access to *parts* of documents in addition to the full documents.

The INEX interactive track (iTrack) was run for the first time in 2004 [3], repeated in 2005 [4] and again in 2006/2007 [5] (due to technical problems the tasks scheduled for 2006 were actually run in early 2007). Although there has been variations in task content and focus, some fundamental premises has been in force throughout:

- a common subject recruiting procedure
- a common set of user tasks and data collection instruments such as questionnaires
- a common logging procedure for user/system interaction
- an understanding that collected data should be made available to all participants for analysis

S. Geva, J. Kamps, and A. Trotman (Eds.): INEX 2008, LNCS 5631, pp. 300–313, 2009.
© Springer-Verlag Berlin Heidelberg 2009

This has ensured that through a manageable effort, participant institutions have had access to a rich and comparable set of data on user background and user behavior, of sufficient size and level of detail to allow both qualitative and quantitative analysis. This has already been the source of a number of papers and conference presentations ([6], [7], [8], [9], [10], [11], [12]).

In 2008, we wanted to preserve as much of the "common effort" quality of the previous years as possible. We invited the participants to participate in a minimum experimental effort using the system and data provided and described below. Within the framework of the track, participants could then design their own investigations under certain constraints, such as:

- The collection of documents was the same as the one used for the INEX ad hoc retrieval task [13], i.e., in 2008 a collection of xml-coded Wikipedia articles.
- The IR system developed for the 2006 track was made available for the participants to use, either alone or in comparison with participants' own system(s).
- Each participating site was responsible for recruiting a minimum of 8 (but preferably more) test persons to participate in the study as searchers.
- The participants were required to make their data available to all participating groups, and describe their collection process and experimental procedure in a way which would make it possible for others to interpret and use the data.

2 Tasks

For the 2008 iTrack the experiment was designed with two categories of tasks, from each of which the searchers were instructed to select one out of three alternative search topics constructed by the track organizers. The original intention was to also give the searchers the opportunity to perform one self-generated task, but it was unfortunately not possible to implement this in our IR system. The two categories of tasks were, respectively, fact-findings tasks (category 1) and research tasks (category 2). The tasks were intended to represent information needs believed to be typical for Wikipedia users. In order to ensure a certain amount of user-system interaction, we also wanted the tasks to be so complex that searchers needed to access more than one individual article to solve them. In order to diminish system learning effect, the order of tasks performed by searchers was rotated by category.

The Fact-Finding Tasks

sto1.   As a frequent traveler and visitor of many airports around the world you are keen on finding out which is the largest airport. You also want to know the criteria used for defining large airports.

sto2.   The "Seven summits" are the highest mountains on each of the seven continents. Climbing all of them is regarded as a mountaineering challenge. You would like to know which of these summits were first climbed successfully.

sto3.   In the recent Olympics there was a controversy over the age of some of the female gymnasts. You want to know the minimum age for Olympic competitors in gymnastics.

The Research Tasks

sto4.   You are writing a term paper about political processes in the United States and Europe, and want to focus on the differences in the presidential elections of France and the United States. Find material that describes the procedure of selecting the candidates for presidential elections in the two countries.

sto5.   Every year there are several ranking lists over the best universities in the world. These lists are seldom similar. You are writing an article discussing and comparing the different ranking systems and need information about the different lists and what criteria and factors they use in their ranking.

sto6.   You have followed the news coverage of the conflict between Russia and Georgia over South Ossetia. You are interested in the historic background for the conflict and would like to find as much information about it as possible. In particular you are interested in material comparing this conflict with the parallel border conflict between Georgia and Abkhazia.

3   Participating Groups

Originally 7 groups expressed their interest in participating in the i-Track experiments. Unfortunately, in the end only two groups were able to perform experiments; University of Amsterdam and Oslo University College. Fifty-six sessions, 14 in Amsterdam and 42 in Oslo, performed by 29 test persons were recorded successfully (i.e. without system failure and with completed questionnaires).

4   Research Design

4.1   Search System

The experiments were conducted on a java-based retrieval system built within the Daffodil framework [14], which resides on a server at and is maintained by the University of Duisburg. The search system interface is quite similar to the one used in the 2005 and 2006 i-Tracks.

The system returns elements of varying granularity (full Wikipedia articles, sections or sub-sections of articles) based on the hierarchical xml-coded document structure. Figure 1 shows the result list interface of the program. In the top left corner is the query box, below it we see the result list. Relevant elements are grouped by document in the result list and up to three high ranking elements are shown per document. To help searchers select query terms, the system has a related term feature which presents the searcher with a set of potential query terms, generated through analysis of term frequency in the top-ranked elements. These appear in a box showing terms related to the current query. Using mouse-over, searchers can view the context from which the related terms were generated.

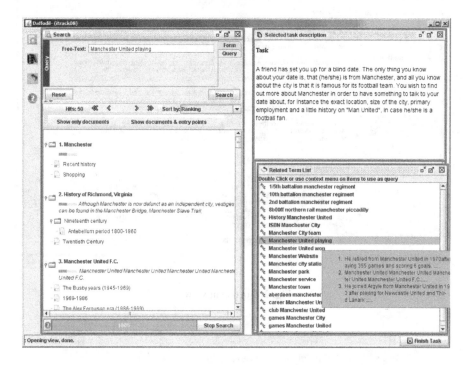

Fig. 1. Daffodil result list view

When a searcher clicks on the result list to examine a document, the system enters document view mode, where the entire full text of the document is shown, with background highlighting for high ranking elements (Figure 2). In addition to this, the document view screen shows a Table of Contents generated from the XML formatting of the documents. From the ToC, the searcher can choose individual sections and subsections for closer examination. In the ToC, the system's relevance estimation is also indicated through color-coding of relevant elements. In addition, the ToC shows elements that the searcher has viewed (indicated by an eye - 👁) and/or relevance assessed (coded as shown in Figure 3).

4.2 Document Corpus

The document corpus used was the same as the one used in the 2006 i-track and in the other 2008 INEX tracks. It consists of more than 650,000 encyclopedia articles extracted from Wikipedia [15]. The articles are structurally formatted in XML.

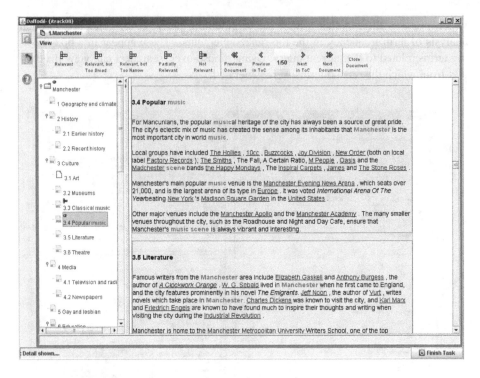

Fig. 2. Document view

Fig. 3. Relevance scores

4.3 Online Questionnaires

During the course of the experiment, searchers were issued brief online questionnaires to support the analysis of the log data. Before the search tasks were introduced, the searchers were given a pre-experiment questionnaire, with demographic questions such as searchers' age, education and experience in information searching, particularly in searching and using Wikipedia. Each search task was preceded with a pre-task questionnaire, which concerned searchers' perceptions of the difficulty of the search task, their familiarity with the topic etc. After each task, the searcher was asked to fill out a post-task questionnaire. The intention of the post-task questionnaire was to learn about the searchers' use of and their opinion on various features of the search system, in relation to the task they had just completed. The experiment was closed with a post-experiment questionnaire, which elicited the searchers' general opinion of the search system. The responses to the questionnaires were logged in a database.

4.4 Relevance Assessments

The system was designed to have searchers assess the relevance of each item they looked at. These could be either full articles or article elements. The relevance scale (see fig. 3) was similar to the one used in the 2006 interactive track, based on work by Pehcevski [16]. It aims to balance the need for information on the perceived granularity of retrieved elements and their degree of relevance, and is intended to be simple and easy to visualize [5]. The system did not oblige searchers to perform relevance judgments, but in the instructions for the experiment they were told to "select an assessment for each viewed piece of information with regards to how you consider it to be of help in solving the task." Searchers were not given any more specific instructions on how to perform the relevance judgments; they were, for instance, not required to view each retrieved element as independent from other components viewed. Experiences from user studies (e.g. [17]) clearly show that users learn from what they see during a search session. To impose a requirement on searchers to discard this knowledge were thought to create an artificial situation and restrain the searchers from interacting with the retrieved elements in a natural way.

Five different relevance scores were defined. The scores express two aspects or dimensions in relation to solving the task:

1. How much **relevant information** does the part of the document contain? It may be *highly relevant, partially relevant* or *not relevant.*
2. How much **context is needed** to understand the element? It may be *just right, more* or *less.*

This is combined into the five scores:

Relevant, but too broad, contains relevant information, but also a substantial amount of other information.

Relevant, contains highly relevant information, and is just the right in size to be understandable.

Relevant, but too narrow, contains relevant information, but needs more context to be understood.

Partially relevant, has enough context to be understandable, but contains only partially relevant information.

Not relevant, does not contain any relevant information that is useful for solving the task.

4.5 Logging

All search sessions were logged and saved to a database. The logs registered and time stamped the events in the session and the actions performed by the searcher, as well as the responses from the system.

5 Experimental Procedure

Each experiment was performed following the standard procedure outlined below. Steps 7 to 10 were repeated for each of the two tasks performed by the searcher. The tasks were automatically assigned according to a Latin square design to secure a balanced distribution of the order of the research and fact-finding tasks.

1. Experimenter briefed the searcher, and explained format of study. The searcher read and signed the Consent Form.
2. The experimenter logged the searchers into the experimental system. Tutorial of the system was given with a training task provided by the system. The experimenter handed out and explained the system features document.
3. Any questions answered by the experimenter.
4. The control system administered the pre-experiment questionnaire.
5. Topic descriptions for the first task category administered, and a topic selected
6. Pre-task questionnaire administered.
7. Task began by clicking the link to the search system. Maximum duration for a search was 15 minutes, at which point the system issued a "timeout" warning. Task ended by clicking the "Finish task" button.
8. Post-task questionnaire administered.
9. Steps 5-8 repeated for the second task.
10. Post-experiment questionnaire administered.

6 Data Analysis

In this section, we summarize our preliminary analysis of the questionnaire data and the transaction log files. More detailed analyses will be the subject of further research from the participating institutions.

Table 1. Distribution of tasks and sessions

Task Type	Task	Sessions
Fact-finding	Sto1	13
	Sto2	8
	Sto3	5
Research	Sto4	9
	Sto5	9
	Sto6	12
Total		56

Table 1 shows the distribution of tasks and sessions, due to a technical error one searcher performed two research tasks and one searcher performed only one task (also a research task) thus it is not a completely even distribution of task types (26 fact-finding tasks and 30 research tasks).

6.1 Questionnaire Data

Questionnaire results reported in this report are based on the data of test persons who completed the questionnaire.

Pre-Experiment Questionnaire

A total number of 27 test persons completed the questionnaire (9=Male, 18=Female). Test persons had a mean age of 30.33 years and with the exception of six test persons, all were students. Test persons' mean experience with searching for information using the Web was 8.22 years. When asked about how often they search, our test persons' mean search experience using digital libraries was 3.60, using search engines was 4.81, and using Wikipedia was 3.81 (where 1=never, 2 = once or twice a year, 3 = once or twice a month, 4 = once or twice a week and 5 = once or more times a day).

As we were using Wikipedia, we administered test persons' experiences with Wikipedia in detail. First, we asked about the test persons' search purposes with Wikipedia. Out of 27 test persons, 25 of them mentioned that they used Wikipedia for fact-finding purposes, none of them used Wikipedia for decision making, 10 test persons used Wikipedia for research and 9 test persons used Wikipedia for entertainment. When asked if they generally found what they were looking for when using Wikipedia, they responded positively (their mean experience was 3.96), and when asked if they trust the information in Wikipedia, subjects mean experience was 3.41 (where 1=strongly disagree, 2=disagree, 3=not sure, 4=agree and 5=strongly agree). Lastly, our pre-experiment questionnaire result indicated that only 1 out of 27 test persons mentioned that he or she occasionally has edited articles in Wikipedia and none of our users ever have created new articles in Wikipedia.

Pre-Task Questionnaire

Table 2. Pre-task questionnaire, with answers on a 5-point scale (1-5)

Q2.1:	How familiar are you with the topic of the search task?
Q2.2:	How interesting do you find the topic of the search task?
Q2.3:	How easy do you think it will be to find information for this task

Table 3. Pre-task responses on searching experience: mean scores and standard deviations (in brackets)

Type	Q2.1	Q2.2	Q2.3
All tasks	1.96 (0.78)	3.43 (0.73)	3.22 (0.68)
Fact Finding	1.81 (0.84)	3.26 (0.68)	3.59 (0.53)
Research	2.11 (0.74)	3.59 (0.69)	2.85 (0.72)

Each task was preceded with a pre-task questionnaire, collecting information regarding test persons' familiarity, level of interest and easiness of the search topic. Table 2 shows the items asked in the pre-task questionnaire. The answer categories used a 5-point scale (1=not at all, 3=somewhat and 5=extremely). Test persons' responses are presented in table 3.

As shown in table 3, the research task was rated slightly higher compared to fact-finding task in terms of test person's familiarity with the topic (Q2.1) and level of interest (Q2.2) of the search task. Only in terms of perceived easiness to find information for the task (Q2.3), the fact-finding task was rated higher.

Post-Task Questionnaire

Table 4 shows the items asked in the post-task questionnaire. The answer categories used a 5-point scale (1=not at all, 3=somewhat and 5=extremely). Test persons' responses are summarized in Table 5. If we look at the responses over all tasks, the average response varies from 2.83 to 4.46 signaling that the test persons rated the tasks positively.

We also looked at the responses for each task type. As shown in table 5, for all questions asked with the exception of Q3.1 and Q3.8, the research task was rated higher than the fact-finding task. Here, we see that test persons understood both tasks very well (Q3.1). Fact-finding received higher responses on average, which makes sense given the nature of the simulated tasks and thereby confirms that the chosen simulated tasks represent the particular task types. The research task was regarded easier (Q3.2) and more similar to the searching task that our test persons typically perform (Q3.3), compare to the fact-finding task. This may be a result of our selection of test persons who all had an academic education. Moreover, test persons were more satisfied with the search results provided by the system (Q3.6) for the research task. A possible explanation is that the research tasks are more open-ended than the fact-finding tasks where test persons need to find specific and precise answers. Hence, additional material provided by the system may be more useful in the research task context. This explanation is supported by the response when asked about the relevancy of the found information (Q3.7). Test persons believed that they found more relevant results for the research tasks. This finding is also coherent with the relevance assessment results where searchers found more articles and more elements to be relevant when completing research tasks compare to when they performed fact-finding tasks (see Section 6.2).

Table 4. Post-task questionnaire, with answers on a 5-point scale (1-5)

Q3.1:	How understandable was the task?
Q3.2:	How easy was the task?
Q3.3:	To what extent did you find the task similar to other searching tasks that you typically perform?
Q3.4	Was it easy to perform the search for this task?
Q3.6:	Are you satisfied with your search results?
Q3.7:	How relevant was the information you found?
Q3.8:	Did you have enough time to do an effective search?
Q3.9:	How certain are you that you completed the task?
Q3.10	How well did the system support you in this task?*

Table 5. Post-task responses on searching experience: mean scores and standard deviations (in brackets)

Type	Q3.1	Q3.2	Q3.3	Q3.4	Q3.6	Q3.7	Q3.8	Q3.9	Q3.10
All tasks	4.46 (0.64)	3.13 (1.27)	3.46 (1.04)	3.31 (1.06)	3.02 (1.51)	3.50 (1.28)	3.04 (1.45)	2.83 (1.46)	3.02 (1.22)
Fact Finding	4.63 (0.56)	3.00 (1.47)	3.30 (1.10)	3.19 (1.11)	2.56 (1.63)	3.07 (1.38)	3.07 (1.57)	2.63 (1.64)	2.70 (1.05)
Research	4.30 (0.67)	3.26 (1.06)	3.63 (0.97)	3.44 (1.01)	3.48 (1.25)	3.93 (1.04)	3.00 (1.36)	3.04 (1.26)	3.33 (0.91)

Next, we look at the time test persons spent on each task. On the question of whether there was enough time for an effective search (Q3.8), responses for the fact-finding tasks were higher than for the research tasks. This is also consistent with the log result where test persons spent less time completing fact-finding tasks compared to research tasks (see Section 6.2). This means that test persons had enough time for the fact-finding task, but they stopped searching before the maximum allocated time ran out. This could be because the system did not support them well enough in finding relevant results (Q3.10) or they expected the system to do better in retrieving relevant results (Q3.7) for fact-finding tasks. This is consistent with the assessment of task completion (Q3.9) where, on average, test persons were less certain that they completed the fact-finding task compared to the research task. Also note that the standard deviations for fact-finding tasks for almost all questions are larger than for the research tasks. A possible explanation is again that several test persons were not satisfied with the results they found when completing the fact-finding task.

6.2 Log Statistics

In total 118 assessments were made of full articles, Table 6 shows the distribution of assessment on the different relevance levels.

Table 6. Article relevance assessments

Fully relevant	Relevant, but too broad	Relevant, but too narrow	Partially relevant	Not relevant
45 (38 %)	14 (12 %)	12 (10 %)	17 (14 %)	30 (25 %)

In Table 7, we see relevance distribution of articles for each topic, the results show that the sessions generated by task sto6 (on the South Ossetia conflict), which is the most popular research task, has returned more than half of the articles found to be

Table 7. Distribution of article relevance assessments per task

Topic	Fully relevant	Relevant, but too broad	Relevant, but too narrow	Partially relevant	Not relevant
sto1	0	2	6	5	11
	.0%	14.3%	14.3%	29.4%	34.4%
sto2	2	0	0	2	2
	4.4%	.0%	.0%	11.8%	6.3%
sto3	2	0	0	4	11
	4.4%	.0%	.0%	23.5%	34.4%
sto4	7	1	2	1	3
	15.6%	7.1%	15.4%	15.4%	9.4%
sto5	9	1	3	4	3
	20.0%	7.1%	23.1%	23.5%	9.4%
sto6	25	10	2	1	2
	55.6%	71.4%	15.4%	5.9%	6.3%
Total	45	14	13	17	32
	100.0%	100.0%	100.0%	100.0%	100.0%

fully relevant. Even more interesting to see is that sessions dealing with the most popular fact-finding task (sto1 – large airports) has not returned any fully relevant articles.

Table 8 shows the distribution of relevance assessments on element level, i.e. assessments of sections and subsections. Interestingly we also see that task sto6 also on the element level has returned the highest number of fully relevant scores and that sto1 only has returned 3 fully relevant elements.

Table 8. Distribution of element relevance assessments per task

Topic	Fully relevant	Relevant, but too broad	Relevant, but too narrow	Partially relevant	Not relevant
sto1	3	2	1	3	5
	2.7%	25.0%	3.3%	7.5%	10.6%
sto2	6	1	7	1	1
	5.3%	12.5%	23.3%	2.5%	2.1%
sto3	0	0	0	0	1
	.0%	.0%	.0%	.0%	2.1%
sto4	5	2	4	7	3
	4.4%	25.0%	13.3%	17.5%	6.4%
sto5	44	1	5	18	33
	38.9%	12.5%	16.7%	45.0%	70.2%
sto6	55	2	13	11	4
	48.7%	25.0%	43.3%	43.3%	8.5%
Total	113	8	30	40	47
	100.0%	100.0%	100.0%	100.0%	100.0%

We have performed further analysis to investigate if there are any significant differences between the two task types. A T-test shows a significant difference between fully relevant assessment on both article (p=0.000) and element level (P=0.011) when comparing fact-findings tasks and research tasks, but then one needs to be aware of the heavy influence of relevance assessments for tasks sto1 and sto6. For fact-finding tasks searchers found 0.15 fully relevant articles per session and 0.35 fully relevant elements, compared to 1.37 fully relevant articles and 3.47 elements per session for research tasks. Also fact-finding sessions resulted in significantly more non relevant articles (1.197 compared to 0.583 for research tasks). This supports the findings from the questionnaire analysis that searchers were more familiar with the research tasks and found them easier to solve, and also that they believed they found more relevant information for the research tasks.

Table 9. Queries per task

	Task type	N	Mean
Number of queries	Fact	26	5.88
	Research	30	4.83

Table 10. Time per task

	Task type	N	Mean
Time in seconds	Fact	26	653.15
	Research	30	767.10

We have also compared the task types with respect to number of queries (Table 9) performed and time invested (Table 10). As can be seen the searchers performed more queries in fact-finding sessions but, but spent more time to solve research tasks. In other words research task sessions are characterized by searchers being more thorough in their interaction with the individual article/element. A T-test did not report significant difference between the two task categories in these matters, but the mean time per task was very close to being significant (p=0.064).

7 Conclusions

We have reported the experimental design of the 2008 Inex interactive track and the analysis of data related to the difference between searchers performing fact-finding and research tasks. Although the number of participating institutions was low, we have been able to collect a set of data that shows interesting results related to the two task categories.

In general, searchers were more satisfied when completing the research task compared to fact-finding task. We found that test persons regarded the research task easier, were more satisfied with the search result and found more relevant information for the research task. This is plausibly related to the task type, where test persons regard more information as relevant or useful when searching for a more open-ended research task. Fact-finding tasks require a more specific and precise answer, which may diminish the additional value of exploring a wide range of search results.

This finding is consistent with the relevance assessment results where searchers found more relevant articles and elements when completing the research task compared to the fact-finding task. Also fact-finding sessions resulted in significantly more non-relevant articles than research sessions. Test persons reported that they were less certain that they had completed the fact-finding task compared to the research task.

A general result seems to be that the system was better at supporting research tasks than fact-finding tasks. This is particularly interesting since the participants claimed to use Wikipedia more for fact-finding than for research tasks.

Acknowledgments

We would like to thank Ingo Frommholz, Norbert Fuhr, Claus-Peter Klas and Saadia Malik from the University of Duisburg-Essen for their administration of the Daffodil system. Khairun Nisa Fachry was supported by the Netherlands Organization for Scientific Research (NWO) under grant # 639.072.601.

References

[1] Malik, S., Trotman, A., Lalmas, M., Fuhr, N.: Overview of INEX 2006. In: Fuhr, N., Lalmas, M., Trotman, A. (eds.) INEX 2006. LNCS, vol. 4518, pp. 1–11. Springer, Heidelberg (2007)

[2] Ruthven, I.: Interactive Information Retrieval. Annual Review of Information Science and Technology 42, 43–91 (2008)

[3] Tombros, A., Larsen, B., Malik, S.: The Interactive Track at INEX 2004. In: Fuhr, N., Lalmas, M., Malik, S., Szlávik, Z. (eds.) INEX 2004. LNCS, vol. 3493, pp. 410–423. Springer, Heidelberg (2005)

[4] Larsen, B., Malik, S., Tombros, A.: The interactive track at INEX 2005. In: Fuhr, N., Lalmas, M., Malik, S., Kazai, G. (eds.) INEX 2005. LNCS, vol. INEX 2005, pp. 398–410. Springer, Heidelberg (2006)

[5] Larsen, B., Malik, S., Tombros, A.: The Interactive track at INEX 2006. In: Fuhr, N., Lalmas, M., Trotman, A. (eds.) INEX 2006. LNCS, vol. 4518, pp. 387–399. Springer, Heidelberg (2007)

[6] Pharo, N., Nordlie, R.: Context Matters: An Analysis of Assessments of XML Documents. In: Crestani, F., Ruthven, I. (eds.) CoLIS 2005. LNCS, vol. 3507, pp. 238–248. Springer, Heidelberg (2005)

[7] Hammer-Aebi, B., Christensen, K.W., Lund, H., Larsen, B.: Users, structured documents and overlap: interactive searching of elements and the influence of context on search behaviour. In: Ruthven, I., et al. (eds.) Information Interaction in Context: International Symposium on Information Interaction in Context: IIiX 2006. Proceedings, Copenhagen, Denmark, October 18-20, pp. 80–94. Royal School of Library and Information Science, Copenhagen (2006)

[8] Malik, S., Klas, C.-P., Fuhr, N., Larsen, B., Tombros, A.: Designing a user interface for interactive retrieval of structured documents: lessons learned from the INEX interactive track? In: Gonzalo, J., Thanos, C., Verdejo, M.F., Carrasco, R.C., et al. (eds.) ECDL 2006. LNCS, vol. 4172, pp. 291–302. Springer, Heidelberg (2006)

[9] Kim, H., Son, H.: Users Interaction with the Hierarchically Structured Presentation in XML Document Retrieval. In: Fuhr, N., Lalmas, M., Malik, S., Kazai, G. (eds.) INEX 2005. LNCS, vol. 3977, pp. 422–431. Springer, Heidelberg (2006)

[10] Kazai, G., Trotman, A.: Users' perspectives on the Usefulness of Structure for XML Information Retrieval. In: Dominich, S., Kiss, F. (eds.) Proceedings of the 1st International Conference on the Theory of Information Retrieval, pp. 247–260. Foundation for Information Society, Budapest (2007)

[11] Larsen, B., Malik, S., Tombros, A.: A Comparison of Interactive and Ad-Hoc Relevance Assessments. In: Fuhr, N., Kamps, J., Lalmas, M., Trotman, A. (eds.) INEX 2007. LNCS, vol. 4862, pp. 348–358. Springer, Heidelberg (2008)

[12] Pharo, N.: The effect of granularity and order in XML element retrieval. Information Processing and Management 44(5), 1732–1740 (2008)

[13] Kamps, J., Geva, S., Trotman, A., Woodley, A., Koolen, M.: Overview of the INEX 2008 Ad Hoc Track. In: Proceedings from the 7th International Workshop of the Initiative for the Evaluation of XML Retrieval, INEX 2009. Springer, Berlin (2009)

[14] Fuhr, N., Klas, C.P., Schaefer, A., Mutschke, P.: Daffodil: An integrated desktop for supporting high-level search activities in federated digital libraries. In: Agosti, M., Thanos, C. (eds.) ECDL 2002. LNCS, vol. 2458, pp. 597–612. Springer, Heidelberg (2002)

[15] Denoyer, L., Gallinari, P.: The Wikipedia XML corpus. SIGIR Forum 40(1), 64–69 (2006)

[16] Pehcevski, J.: Relevance in XML retrieval: the user perspective. In: Trotman, A., Geva, S. (eds.) Proceedings of the SIGIR 2006 Workshop on XML Element Retrieval Methodology, Seattle, Washington, USA, August 10, pp. 35–42. Department of Computer Science, University of Otago, Dunedin, New Zealand (2006)

[17] Pharo, N.: The SST Method Schema: a tool for analyzing work task-based Web information search processes. Doctoral Thesis. University of Tampere (2002)

Overview of the INEX 2008 Link the Wiki Track

Wei Che (Darren) Huang[1], Shlomo Geva[1], and Andrew Trotman[2]

[1] Faculty of Science and Technology, Queensland University of Technology, Brisbane,
Australia
[2] Department of Computer Science, University of Otago, Dunedin, New Zealand
w2.huang@student.qut.edu.au,
s.geva@qut.edu.au,
andrew@cs.otago.ac.nz

Abstract. The Link the Wiki track at INEX 2008 offered two tasks, file-to-file link discovery and anchor-to-BEP link discovery. In the former 6600 topics were used and in the latter 50 were used. Manual assessment of the anchor-to-BEP runs was performed using a tool developed for the purpose. Runs were evaluated using standard precision & recall measures such as MAP and precision / recall graphs. 10 groups participated and the approaches they took are discussed. Final evaluation results for all runs are presented.

Keywords: Wikipedia, Link Discovery, File-to-File, Anchor-to-BEP, Assessment, Evaluation.

1 Introduction

Trotman & Geva [1] introduced the Link the Wiki task in 2006. It ran at INEX for the first time in 2007 [2]. This contribution discusses the track as it was run in 2008. The track provides an independent evaluation forum for approaches to link discovery in the Wikipedia. In 2007 the track examined file-to-file linking in the Wikipedia, but in 2008 this was extended to include anchor to best entry point (anchor-to-BEP) link discovery. The goal is to investigate the linking methods and to develop a forum for evaluation and application. A test set including document collection, qrels, metrics, and tools for evaluating submissions [3] was constructed and is now provided for future experimenters. The document collection was the INEX Wikipedia collection, the topics (known as orphans) were documents from within the collection.

Ten groups from eight different organizations participated in the track. 25 runs were submitted to the file-to-file task and 31 runs to the anchor-to-BEP task. All runs were evaluated against a ground truth of those links already presented in the collection. Anchor-to-BEP runs were additionally evaluated against a ground truth determined through manual assessment. These manual assessments allow for file-to-BEP, anchor-to-file, anchor-to-BEP and also file-to-file assessment; something that was essential because many submitted runs were file-to-file runs despite the task being defined as anchor-to-BEP; that is, the anchor texts were the document title and the best entry point was the beginning of the target document.

S. Geva, J. Kamps, and A. Trotman (Eds.): INEX 2008, LNCS 5631, pp. 314–325, 2009.

Anchor-to-BEP link discovery differs from traditional link discovery by pointing from anchors directly to relevant material within the target document, rather than pointing to simply the document [4][5]. The purpose of *focused link discovery* is to identify anchors together with the corresponding *best entry points* such that the link is relevant *with respect to the anchor's specific context*. This simplifies the way people discover the relevant information without browsing through the entire document. Automated linking to a best entry point is especially useful for restricted screen devices such as mobile. Few clicks to go through relevant information can be easily achieved.

2 Document Collection

The document collection was the INEX Wikipedia collection of 659,388 articles. For the file-to-file task 6600 documents were randomly selected from the collection. For anchor-to-BEP assessment each participating group was asked to nominate 5 candidate documents, 10 groups participated which resulted in 50 documents for manual assessment. These documents were separated from the collection by removing all outgoing links from the documents into the collection as well as all incoming links from the collection into the documents. The separated documents are known as orphans.

The orphaning process itself was performed by the track participants. The exact method was left to the participant however the requirement was that the process should be equivalent to: orphaning one document; identifying the links to and from that document; then returning the (original) document to the collection. In this way each orphan was linked against the remainder of the collection as it would have been if that orphan was presented for insertion into the collection.

3 Task Specification and Submission Format

The task was specified as twofold: the identification of links from the orphan into the document collection; and the identification of links from the collection into the orphan.

In the anchor-to-BEP scenario the best 50 anchors within the orphan could be identified, and for each no more than 5 BEP destinations could be specified. Alongside these the best 250 incoming links from anchor texts in the remaining collection to BEPs within the orphan could be specified. For file-to-file evaluation the task was to identify the best 250 outgoing and best 250 incoming links.

The specification of a file-to-file link is a special case of the specification of an anchor-to-BEP link. A file-to-file link is from the start of the source to the start of the target. This reduction of the complex task to the less complex task provided a low-cost entry into the track for those who had not participated before.

For submission purposes the orphans were identified by the triplet (topic-id, file name, title). Although each is unique for each orphan (and are thus any one could have been used) all three were used for clarity's sake. Both the INEX ad hoc XPath syntax and the INEX File-Offset-Length (FOL) formats were used for submissions.

All file offsets and lengths were specified as character offsets with respect to the text content of the files; counting from zero; and ignoring all mark-up. An anchor might be specified, for example, as (*23816.xml, 1234, 8*) but a BEP has no length so it would be specified (*23816.xml, 672*). Examples of the anchor-to-BEPs formats are given in Figure 1.

In order to facilitate the position identification of anchor and BEP, various resources were made available to participants including: a text-only version of the collection so that file-offset-lengths could be computed by counting characters from the start of the file; *XML2FOL*, a program that produces a list of all the offsets and lengths of all elements in an XML file; *XML2FOLpassage*, a program to convert any INEX XPath specification into the FOL format.

```
<link>                              <link>
  <anchor>                            <anchor>
    <offset>234</offset>                <start>/article[1]/p[5]/text()[3].12</start>
    <length>24</length>                 <end>/article[1]/p[5]/text()[3].32</end>
  </anchor>                            </anchor>
  <linkto>                            <linkto>
    <file>123.xml</file>                <file>43768.xml</file>
    <bep>334</bep>                      <bep>/article[1]/p[3]/text()[4].40</bep>
  </linkto>                            </linkto>
  ... <multiple links>                ... <multiple links>
</link>                             </link>
```

Fig. 1. Sample Anchor-to-BEP Submission Format

4   Preparation of Qrels

For the file-to-file evaluation of the 6600 orphans the ground truth was constructed without manual assessment. The links from the pre-orphan to the remaining collection were extracted and used as the outgoing ground truth. The links from the collection into the pre-orphans was used as the incoming ground truth. For anchor-to-BEP assessment this is not possible because BEPs are rarely specified in the Wikipedia.

There are known problems with using the Wikipedia itself as the ground-truth: some Wikipedia links are topically-obsolete or have been incorrectly assigned; linking is not exhaustive; articles are unlikely to link to very recently added content; and some links are inserted by bots. As a consequence, evaluation results may be biased various ways. On the one hand, results may appear optimistic because some links are trivial to discover (such as year-links). On the other hand, results may appear pessimistic because useful links not already in the Wikipedia are considered non-relevant. However, evaluation based on the Wikipedia ground-truth does measure performance relative to what is present, and so it is reasonable to believe it is useful.

Although the Wikipedia does contain anchor-to-BEP links, in practice they are rarely used. In order to evaluate anchor-to-BEP link discovery an evaluation result-set was created through manual assessment. A special case of pooling was used in the track – all links for a given orphan were pooled, then for each anchor, all BEPs were pooled. The pool was assessed to completion.

5 Assessment

An assessment tool (see Figure 2) was used in 2008 to facilitate the assessment of link discovery in both the anchor-to-BEP and file-to-file scenarios. As assessment is laborious and time consuming, special care was taken to minimize the amount of mouse motion and clicking – a single click could, for example, be used to specify a relevant link.

Overlapping links were also addressed by the tool. Links for each anchor were grouped for easy and clear presentation (for example, the anchor *Modern Information Retrieval* may appear as the anchors: *Information Retrieval* or *Modern Information Retrieval*), but the tool also captured each sub-anchor explicitly so that the assessor could differentiate and judge with respect to each sub-anchor. File-to-file links were presented as linking from the title of the document to the beginning of the target document.

No concerns were raised about the tool. The number of links identified per topic varied from 405 to 1722. An average of about 5 hours was spent assessment an orphan. Most links were file-to-file. Less than 10% of the links identified for each orphan were judged relevant.

6 Evaluation

6.1 The Evaluation Tool

An evaluation tool (*ltwEval*, see Figure 3) was developed for the track. Performance measures including Mean Average Precision (MAP), precision at the number of relevant documents (R-Prec), and precision at given retrieval cut-offs (P@5, P@10, P@20, P@30, P@50 and P@250) were computed. The tool draws Interpolated Precision / Recall plots allowing graphical comparative analysis of multiple runs. *LtwEval* is GUI driven and was written in Java for platform independence.

The tool gives the number of outgoing and incoming links in the qrels as well as in each run (duplicate links being eliminated). Performance measures can be calculated using all topics in the qrels or just the topics in the run. From the evaluation result table (that displays all metrics), the color and used in graphing can be specified.

The Wikipedia ground-truth qrels (for both the 6600 file-to-file topics and the 50 anchor-to-BEP topics) can only be used to evaluate the submission runs in *file-to-file* mode while the manual assessment results can be used to perform the evaluation in several different modes. Besides evaluating file-to-file links, the anchor-to-BEP submission runs are also evaluated at *file-to-BEP, anchor-to-file* and *anchor-to-BEP* modes. The file-to-BEP evaluation considers the entry point, weighting the link score by BEP proximity in a similar manner to that used in the ad hoc track: the score drops linearly to 0 over a distance of 1000 characters; an exact match is given a score of 1 while 0 is given to the BEP beyond 1000 characters. If more than one BEP is specified in the target document, the closest is used. The evaluation in anchor-to-file and anchor-to-BEP mode considers only the first 50 anchors, and only the first BEP of each anchor.

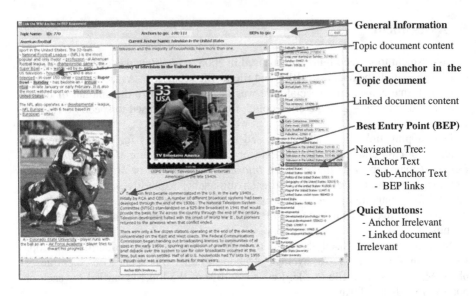

Fig. 2. The Assessment Tool

6.2 Metrics

In the INEX use case of link discovery it is important to rank the discovered links for presentation to the page author. A typical scenario might involve a user who wishes to inspect and then accept or reject recommended links. This use case was modeled in the manual assessment evaluation where assessors did exactly this. In a realistic link discovery setting the user is unlikely to trudge through hundreds of recommended anchors, so the best anchors should be presented first. The link discovery system must also balance extensive linking against link quality.

Traditional measures such as MAP, R-Prec, P@n and Interpolated Precision-Recall plots address the problem of file-to-file link discovery well, but do not address the performance of anchor-to-BEP methods at all well (because anchor and BEP near misses are not considered); it is necessary to adapt metrics to the problem.

For evaluation purposes runs must be of a finite length (and quite short for manual assessment purposes). Often there are more known relevant links in the qrels than the assessment imposed submission length – in short, there are sometimes more than 50 relevant anchors in an orphan despite the submission requirements capping the number of anchors that can be identified at 50.

To address this problem MAP was altered so that it now corresponds to the maximum point of recall in a run or the actual number of relevant links, whichever is smaller. That is, as the run length was limited to 50, the calculation of MAP was based on a maximum recall of 50 relevant links. Because of this, a run consisting of 50 relevant links scores a MAP of 1.0 and the RP curve depicts a line-at-1.

An anchor may be defined by a user in several slightly different ways. For instance, *The Theory of Relativity*, *Theory of Relativity*, and *Relativity* may be conceptually identical anchors. Furthermore, if the anchor text occurred several times in a document only one instance is likely to be anchored (according to the Wikipedia guidelines) and so the location of an anchor may vary without becoming semantically incorrect (we leave for further work the question of which occurrence of an anchor is *best*). During assessment anchors were explicitly assessed as either relevant or irrelevant. Only relevant anchors contributed to the score of a submission – through the score assigned to the relevant links, if exist. In a quick pass over the orphan the assessor could reject all anchors that were trivially irrelevant – even without looking at the linked documents. Year links, for instance, could be rejected outright without the need for inspection of the target.

Similarly to anchors, a BEP cannot be defined with absolute accuracy. Some reasonable proximity to a designated BEP in the assessments must be allowed. So a BEP might be considered relevant if, when viewed on a screen, it is no more than some distance (N characters) away from a point chosen by an assessor. The track defined the BEP score of a link as:

$$\text{bep score} = \text{file score} \times [1 - (|\text{bep\_position}_{Run} - \text{bep\_position}_{qrel}| \: / \: N)]. \tag{1}$$

So in summary, an anchor-to-BEP link was assessed as relevant on the basis of approximately matching both the anchor and the BEP of a relevant link in the assessments. Anchors were either accepted or rejected. Having computed all individual anchor-to-BEP link scores for accepted anchors, the document score can be derived using the Average Precision in the usual manner. The MAP can then be computed over the entire set of topics.

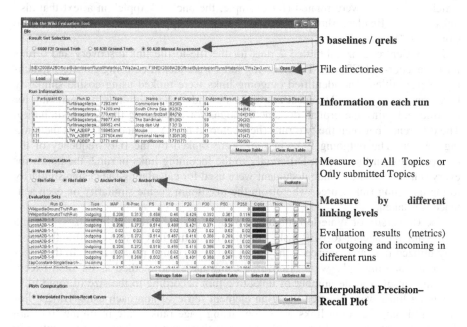

Fig. 3. The Evaluation Tool

7 Approaches to Link Discovery

This section describes some of approaches taken by track participants. In all, there were 10 participating groups (including 2 independent groups from each of: the University of Amsterdam and from Queensland University of Technology).

The University of Amsterdam (de Rijke) submitted 3 runs for the file-to-file link discovery. For the outgoing links, they selected anchors with LLR (Link Likelihood Ratio) > 1 and used the anchor text as a query to retrieve target pages (searching in the title field). For the incoming links, the topic title was used as a query to retrieve the top 250 source pages within the language modeling framework. In anchor-to-BEP link discovery, outgoing links were discovered by selecting anchors with LLR > 1 and then retrieving the target page whose title matched (exact or partial) the anchor text. The target pages were ranked according to the likelihood of the target title in the topic page (p ($Title \mid D$)). Incoming links were retrieved by using the topic title to find exact matches in the collection. In their third submission, the topic title was used as a query to retrieve 250 candidate target pages (ranked by cosine similarity) and the pages whose rsv was greater than 0.15 were selected as the outgoing links. Incoming links used the same strategy to select the source pages with similarity greater than 0.026.

Lycos Europe GmbH submitted 2 runs for the file-to-file link discovery and 5 runs for the anchor-to-BEP task. The approach used by Lycos is derived from Itakura & Clarke's approach in 2007 [14]. The difference is that Lycos dynamically selected the best-matching target for a given anchor text based on content similarity. For example, in a text about computers, the anchor "Apple" is more likely to refer to the page "Apple Computers" than to the page "Apple Records". Moreover, the system also analyzed the links between the potential targets for all anchor texts so that they could see which set of links were related (for example, the anchor "Apple" in a text that also links to "The Beatles" should most likely link to "Apple Records" and not "Apple Computers").

Know-Centre Graz submitted 2 runs for the file-to-file link discovery and 6 runs for the anchor-to-BEP task. The outgoing links were identified using gazetteer matching of page titles. The identified outgoing links were ranked using cosine similarity based on noun vectors. The incoming links were identified similarly by searching for the title and using the orphan documents nouns for calculating the cosine similarity. The difference between the two runs (here referred to as run1 and run2) was the ranking scheme. The outgoing links in run2 were ranked by the IDF frequency of the occurring text in the corpus. Differently to the incoming links in run2, the nouns for every sentence in the orphan document were used for calculating the cosine similarity to the incoming link source, wherefrom the maximum cosine similarity on each sentence was taken.

The University of Waterloo submitted 3 runs for the file-to-file link discovery and 3 runs for the anchor-to-BEP task. For the file-to-file link discovery, thir first run, they utilized the same approach they used last year (which placed first). In their second run, outgoing links were discovered using the same method as thr first, except with the cut-off for the number of links to return according to the size of topic files. Incoming links were selected using an element retrieval approach using BM25. For their third run, Outgoing links were done using page rank while incoming links are done using topic oriented page rank assuming that what was found for the outgoing links was correct. The algorithms used for the anchor-to-BEP task was the same, except for finer granularity in specifying sources and destinations.

The Queensland University of Technology submitted 5 runs for the file-to-file task and 6 runs for the anchor-to-BEP task. Several runs used the GPX search engine using the same approach they used in 2007. Several runs used the Terrier search engine out of the box to find document to document links was also tried. Finally, several runs used frequent phrase mining to identify suitable anchors and links.

The University of Otago submitted 3 runs for the file-to-file link discovery and 3 runs for the anchor-to-BEP task. These runs were based on the Itakura & Clarke approach from 2007, but with particular attention paid to parsing issues.

8 Results and Conclusion

The tables 1 and 2 present the final assessment results using Mean Average Precision (MAP). The figures 4-11 present the Interpolated Precision/Recall graphs of each run.

This is the second year of the Link-the-Wiki track at INEX, and the year the anchor-to-BEP link discovery task was introduced. Since the anchor-to-BEP link discovery can be applied in different scenarios to enhance the efficiency of the interaction it is important to build a standard procedure to measure the performance and tools to facilitate the evaluation and assessment. This attempt has opened a door for participants to share their suggestions and opinions for the track, which will improve the capability of the track to facilitate further the link discovery research. Several qrels sets for evaluating runs at different granularity levels were produced and used to measure the performance of various approaches. The GUI-based tools balance the time-consuming assessment and evaluation.

Because this was the first year for the anchor-to-BEP link discovery task it was expected (and seen) that some runs would contain invalid positions for anchors and BEPs (some contained file-to-file links). Because of this, the relative comparison of runs may be biased towards correctly formatted runs (at the expense of better but incorrectly formatted runs).

Table 1. MAP of 6600 File-to-File topics link discovery evaluated by Wikipedia Ground Truth

Runs	Out	In	Runs	Out	In
AmsterdamDeRijke_ltw01	0.2924	0.4800	Amsterdam_a2a_1	0.1071	0.3392
AmsterdamDeRijke_ltw02	0.3475	0.5249	Amsterdam_a2a_2	0.1088	0.2879
AmsterdamDeRijke_ltw03	0.1041	0.3345	Amsterdam_a2a_3	0.1017	0.3575
Otago_capConstSingleSearchWei	0.3045	0.4314	LycosF2F-1-1	0.2360	0.3266
Otago_capConstTitleOnly	0.3045	0.4869	LycosF2F-1-5	0.2379	0.3266
Otago_nonCap-FirstPara	0.7343	0.2228	Waterloo_f2f#1	0.3345	0.5540
KnowCenterGraz_globTFIDFSen	0.1407	0.5369	Waterloo_f2f#2	0.2920	0.5350
KnowCenterGraz_WordLvldisam	0.1129	0.5299	Waterloo_f2f#3	0.2053	0.5563
QUT_F-F_1	0.1026	0.0925	CMIC_F2F_01	-	0.4579
QUT_F-F_2	0.1026	0.2915	CMIC_F2F_02	-	0.5116
QUT_LTW_F2F_01	-	0.4322	CSIR_LTW_F2F_1	-	0.1645
QUT_GPXF2Ftitle	0.0566	-	CSIR_LTW_F2F_2	0.0082	0.2940
QUT_GPXF2FnameInOut	0.1440	0.5713			

Table 2. MAP of 50 Anchor-to-BEP topics evaluated by manual and Wikipedia ground-truths

Submission Runs	F2F	F2B	A2F	A2B	Out	In
WikipediaGroundTruthRun	0.2765	0.2079	0.3945	0.3888	1	1
LycosA2B-1-5	0.2463	0.2078	0.4973	0.4918	0.1193	0.1753
LycosA2B-1-1	0.2431	0.2050	0.4930	0.4876	0.1172	0.1753
LycosA2B-5-1	0.2427	0.2050	0.4931	0.4876	0.1169	0.1753
LycosA2B-1-0	0.2387	0.2008	0.4708	0.4656	0.1148	0.1753
Otago_capConst-SingleSearch	0.1745	0.1365	0.3952	0.3910	0.3810	0.2389
Otago_capConst-TitleOnly	0.1745	0.1365	0.3952	0.3910	0.3810	0.2408
Otago_nCapConst-WholeDoc	0.1724	0.1352	0.3896	0.3853	0.3769	0.0745
KnowCenterGrazdisamDocNoneSen	0.1546	0.1077	0.1764	0.1453	0.2370	0.1435
KnowCenterGrazdisamDocNoneTopic	0.1546	0.0603	0.2131	0.1968	0.2370	0.1429
KnowCenterGrazdisamTopicNonSen	0.1522	0.1058	0.2076	0.1662	0.2091	0.1695
KnowCenterGrazdisamTopicNonTopic	0.1522	0.0620	0.2643	0.2384	0.2091	0.1676
KnowCenterGrazglobalIDFSentence	0.1371	0.1222	0.2309	0.1895	0.2200	0.1725
KnowCenterGrazglobalIDFTopic	0.1371	0.0688	0.2873	0.2619	0.2200	0.1725
Waterloo_a2a#1	0.1282	0.1004	0.4111	0.4071	0.2191	0.2165
LycosA2B-0-1	0.1200	0.1051	0.3291	0.3249	0.0432	0.1753
QUT_LTWA2BnameRerank	0.1196	0.0946	0.3042	0.3012	0.1816	0.4615
Amsterdam_a2bep_5	0.1127	0.0847	0.2079	0.2058	0.1426	0.2349
QUT_GPXA2Bname	0.1110	0.0882	0.2912	0.2882	0.1522	0.4236
Waterloo_a2a#2	0.1071	0.0823	0.3355	0.3325	0.1854	0.1804
Waterloo_a2a#3	0.0882	0.0656	0.3874	0.3835	0.1710	0.2044
CMIC_LTW_01	0.0763	0.0576	0.1760	0.1740	0.1004	-
CSIR_LTW_A2BEP_2	0.0760	0.0478	0.1307	0.1237	0.0647	0.1577
Amsterdam_a2bep_1	0.0746	0.0556	0.1271	0.1261	0.0973	0.2349
Amsterdam_a2bep_3	0.0685	0.0518	0.0983	0.0975	0.0911	0.1566
Amsterdam_a2bep_2	0.0671	0.0491	0.1127	0.1115	0.0872	0.2349
QUT_Anchor-BEP_1	0.0524	0.0424	0.1149	0.1141	0.0729	0.0710
QUT_P9_GPXA2Btitle)	0.0487	0.0388	0.1725	0.1712	0.0533	0.4511

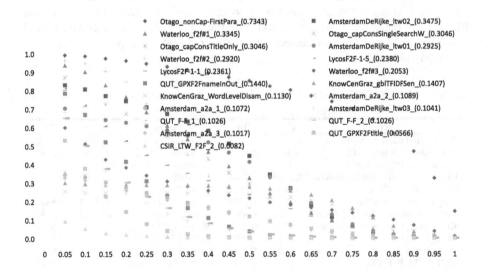

Fig. 4. 6600 File-to-File Topics Outgoing link discovery evaluated by Wikipedia Ground Truth

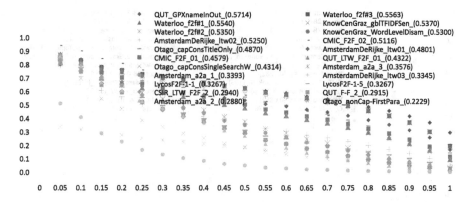

Fig. 5. 6600 File-to-File Topics Incoming link discovery evaluated by Wikipedia Ground Truth

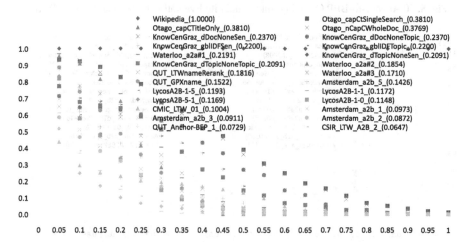

Fig. 6. 50 Anchor-to-BEP Outgoing link discovery evaluated by Wikipedia Ground Truth

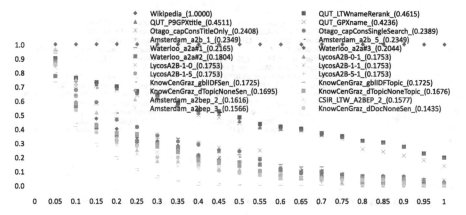

Fig. 7. 50 Anchor-to-BEP Incoming link discovery evaluated by Wikipedia Ground Truth

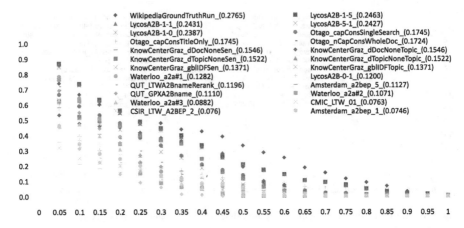

Fig. 8. 50 Anchor-to-BEP Outgoing links: File2File Evaluation by Manual Ground Truth

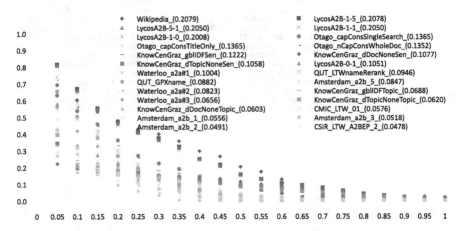

Fig. 9. 50 Anchor-to-BEP Outgoing links: File2BEP Evaluation by Manual Ground Truth

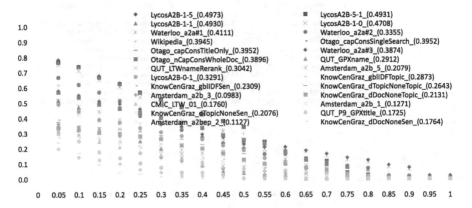

Fig. 10. 50 Anchor-to-BEP Outgoing links: Anchor2File Evaluation by Manual Ground Truth

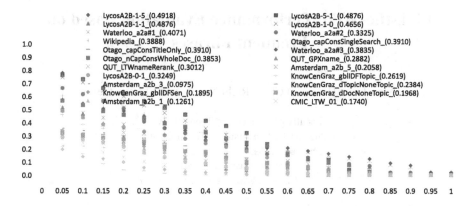

Fig. 11. 50 Anchor-to-BEP Outgoing links: Anchor2BEP Evaluation by Manual Ground Truth

References

1. Trotman, A., Geva, S.: Passage Retrieval and other XML-Retrieval Tasks. In: The SIGIR 2006 Workshop on XML Element Retrieval Methodology, pp. 48–50 (2006)
2. Huang, W.C., Xu, Y., Trotman, A., Geva, S.: Overview of INEX 2007 Link the Wiki Track. In: Fuhr, N., Kamps, J., Lalmas, M., Trotman, A. (eds.) INEX 2007. LNCS, vol. 4862, pp. 373–387. Springer, Heidelberg (2008)
3. Huang, W.C., Xu, Y., Trotman, A., Geva, S.: Experiments and Evaluation of Link Discovery in the Wikipedia. In: The SIGIR 2008 Focused Retrieval Workshop, Singapore (2008)
4. Voss, J.: Measuring Wikipedia. In: The 10th International Conference of the International Society for Scientometrics and Informetrics (ISSI 2005) (2005)
5. Adafre, S.F., de Rijke, M.: Discovering missing links in Wikipedia. In: The SIGIR, Workshop on Link Discovery: Issues, Approaches and Applications (2005)

Link-the-Wiki: Performance Evaluation Based on Frequent Phrases

Mao-Lung (Edward) Chen, Richi Nayak, and Shlomo Geva

Faculty of Science and Technology
Queensland University of Technology
chen@student.qut.edu.au,
r.nayak@qut.edu.au,
s.geva@qut.edu.au

Abstract. In this paper, we discuss our participation to the INEX 2008 Link-the-Wiki track. We utilized a sliding window based algorithm to extract the frequent terms and phrases. Using the extracted phrases and term as descriptive vectors, the anchors and relevant links (both incoming and outgoing) are recognized efficiently.

1 Introduction

With the information boom on the Internet, there are many encyclopaedia-like websites for gathering and sharing knowledge. One of the leading website is Wikipedia, which is a collaborative repository written and contributed by Internet users. With rich articles and features in Wikipedia, the INEX (Initiative for the Evaluation of XML Retrieval) organisers have collected and presented the documents and articles into an XML dataset, named as INEX Wikipedia corpus. The corpus is large in size, about 650,000 documents, and useful for various ranges of information retrieval and data mining research. One of the research tracks organized by INEX is Link-the-Wiki, which was introduced on 2006 [1]. The objective of this track is to automatically discover the hyperlinks among Wikipedia web pages. The Link-the-Wiki track offers many interesting challenges. One of them is related to the size and nature of the Wikipedia data corpus. This corpus has more than 659,000 XML documents and is about 5GB in size. The challenge includes performance on large dataset, handling high dimensional, complex and noise-full data source.

This research utilises frequent phrases for link discovery. The assumption is that a word or a phrase is linked with other documents (Web page) only if it is important in its own document. In this research, the importance is measured by the frequency of the word or the phrase in the document. Non-frequent words or phrases can be ignored for linking purposes. This is also a way to deal with such a large dataset. We first attempt to reduce the size, complexity and dimensionality of the dataset by extracting the frequent terms and phrases from the corpus. We then discover the hyperlinks between Web pages according to the extracted frequent phrases and terms. Empirical analysis shows that the anchors and relevant links are recognized efficiently using the extracted phrases and term as descriptive document vectors.

S. Geva, J. Kamps, and A. Trotman (Eds.): INEX 2008, LNCS 5631, pp. 326–336, 2009.

This paper details the proposed approach. Section 2 provides an overview of the proposed multi-stage approach. Section 3 describes the data pre-processing steps including stop-words removal and stemming. Section 4 explains the Frequent Phrase Extraction algorithm. The link discovery process including both incoming and outgoing links is discussed in section 5. Section 6 gives the detail of empirical analysis. The conclusion section summaries the research and offers some future extensions and applications of this research.

2 Overview of the Proposed Approach

Figure 1 illustrates the proposed approach undertaken in this research. It includes four main stages including data preparation, frequent phrase recognition, link discovery and validation.

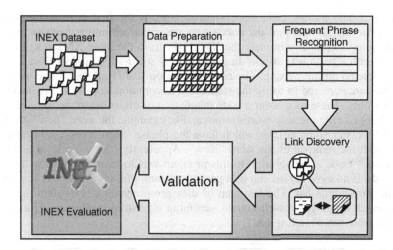

Fig. 1. The proposed approach

Data Preparation: In the first stage of this approach, data cleaning, transformation and preparation are performed. The Wikipedia documents in the INEX corpus require a series of data cleaning process to get them ready for data mining. All the 659,000 documents of the Wikipedia corpus are stored into a relational database. By gathering all the documents into data tables in a database, all the Wikipedia articles will be well aligned. Any relational database, such as Microsoft SQL Server, Oracle and MySQL, etc is an appropriate selection to reside the data. Once the articles have been arranged into database, each document is processed with data parsing, word stemming and stop-words removal.

Frequent Phrase Recognition: Initial data preparation steps including stop-word removal and stemming are able to eliminate a certain amount of noise and reduce the size of the corpus. However, the database is still very large in size and has redundant information. In order to further reduce the size and complexity, the second stage in this research employs an algorithm to recognise and extract the frequent phrases from each document. Each document is represented as a vector of descriptive phrases or terms. It is hoped that after this step, database size and article complexity would be remarkably decreased.

Link Discovery and Validation: With each document represented as frequent terms and phrases, the link discovery step becomes straightforward. Each orphan document is processed to recognize the appropriate anchors according to the existing frequent phrases. Anchors of the orphan document are linked with the other documents (or frequent phrases) in that they are present. The links of the recognized anchors can be ranked and sorted according to the frequency of extracted frequent phrases. As to the validation step, certain percentages of recognized links are examined manually for evaluating their accuracy.

3 Data Pre-processing

Similar to a data mining task, data pre-processing is the first step of the proposed approach. In this research, all the documents are organized into a database. The first pre-processing step is to eliminate the XML tags from the input XML document and transfer it into a plain text article. In addition, any word which is less than 2 characters was deleted during parsing. The next step is stop-word removal. There were some difficulties encountered in using the standard and common stop-word lists to identify the stop-words. These lists cover a wide range; as a result, some of the meanings were lost or changed after the stop-words removal. For example, the word "new" is covered in these lists. For some articles which have the phrase "New York", it became only "York" with the removal of the word "New". Apparently, "New York" is totally different from "York". The solution to this problem was to manually review the list of stop-words. If a keyword that can be a part of a meaningful phrase should be excluded from the stop-words list. The last step of data pre-processing is to stem the words. This research employed a well-known stemming algorithm by Porter [2] to remove the suffix from words in English.

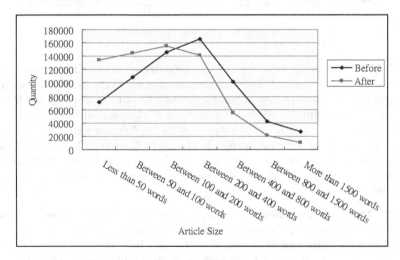

Fig. 2. Article size before and after pre-processing

Figure 2 shows that the article size was effectively reduced by pre-processing. After pre-processing, the distribution of article size was drift to left and distributed more evenly than before pre-processing. The average size of articles was reduced almost 40% after pre-processing. It was average 389 words in a document before pro-processing. It has condensed to average 234 words per document after preprocessing.

4 Frequent Phrase Extraction

A sliding-window based algorithm is used to extract the frequent phrases from each single document. This algorithm recognizes and extracts the frequent terms and phrases from each document independently in the corpus. Let D be the set of documents in the Wikipedia corpus. At this point we assume that each document is independent from each other.

$$D = \{ d_1, d_2, d_3, \ldots\ldots, d_n \} \tag{1}$$

Considering each document independently, a set of frequent phrases from each document is extracted. Frequent phrases are extracted from a document at the level of sentences and paragraphs. Figure 3 shows the process of frequent phrase extraction in which a document is modeled as a set of sentences. The algorithm applies a window on a number of words. Several window sizes are used during the experiments. Using the moving window, 1-term, 2-terms to n-terms frequent phrases are extracted (several n-sizes are used in experiments). Output of this algorithm is a set of 1-term to n-terms frequent phrases which belongs to that particular document.

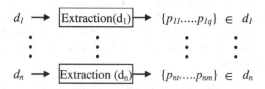

Fig. 3. The inputs and outputs of frequent phrase extraction

5 Link Discovery

The link discovery task is to recognize the anchors in a set of orphan documents and to recognize the appropriate incoming and outgoing links to these orphan documents via these anchors. An incoming link is to identify a potential anchor term or phrase in the corpus documents and refer the anchor to this particular orphan document. In contrast, an outgoing link is to find out the potential anchor text within this orphan document and point the anchor to an appropriate document. We utilized the extracted frequent phrases to recognize the anchors and identify both incoming and outgoing links as shown in figure 4.

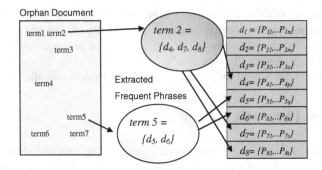

Fig. 4. The link discovery design with extracted frequent phrases

Each orphan document is scanned for a common term or phrase identified in the corpus. Each term (or combinations of the terms in the window) of an orphan document is matched with the extracted frequent phrases in the corpus. If a term (or phrase) is matched with a frequent phrase, the link can be created by tracking back to the original document. The sections below explain the detail process of identifying outgoing and incoming links.

5.1 Outgoing Links

The first task in identifying outgoing links of an orphan document is to recognize the anchors in the document which are phrases. A phrase is composed of multiple single terms and is a set of element terms. Consider the following example. Assume that the orphan document has an anchor that is *"Australian Open Tennis Championship"*. This 4-terms phrase, P_1, is a set of 4 elements, including t_1= *"Australian"*, t_2= *"Open"*, t_3= *"Tennis"* and t_4 is *"Championship"*. The first challenge of outgoing link discovery is how to identify "Australian Open Tennis Championship" as an anchor phrase present in the orphan document.

$$P_1 = \{t_1, t_2, t_3, t_4\} = \{\text{``Australian''}, \text{``Open''}, \text{``Tennis''}, \text{``Championship''}\} \qquad (2)$$

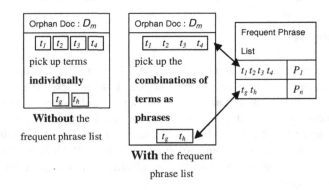

Fig. 5. Comparing with / without the assistance of frequent phrase list

Let us first consider the outgoing link discovery without the assistance of the Frequent Phrase List. The link discovery algorithm considers t_1 ("*Australian*") individually and search the link for t_1. As shown in figure 5 (left part), the program can only pick up *t1* ("*Australian*"), *t2* ("*Open*"), *t3* ("*Tennis*") and *t4* ("*Championship*") individually. Without the Frequent Phrase List, all these terms are independent from each other and there exist no relationship among them.

In contrast, with the assistance of extracted frequent phrases, the procedure recognizes the anchors $\{t_1, t_2, t_3, t_4\}$ as a single phrase. As shown in figure 5 (right part), $\{t_1, t_2, t_3, t_4\}$ ("*Australian Open Tennis Championship*") in the orphan document D_m is recognized according to P_1 in Frequent Phrase List. In this example, t_g and t_h would also be identified as a phrase P_n. In other words, the relationships among the individual terms are identified and stored in the Frequent Phrase List. This procedure achieves a simulation of natural language and recognizes phrase anchors.

After the anchors have been recognized, the next task is locating the documents which contain information about this anchor. For example, the articles containing information about previous winners of "Australian Open Tennis Championship" would be a good candidate. By exploiting the Frequent Phrase List, this link discovery procedure executes a series of queries against the documents which contain the query phrase. Figure 6 shows the link discovery procedure that first obtains the query phrase (anchor) P_1, and filters the list of documents. In this example, there are 3 documents returned by this query, including D_u, D_v, D_y. For example, if P_1 is the "*Australian Open Tennis Championship*", the D_u may be an article regarding the "*The history of Australian Open Tennis Championship*".

In summary, the phrase anchors from the orphan document are first recognized according to a frequent phrase extraction algorithm. These anchors are then used to identify the documents that have them to source the outgoing links.

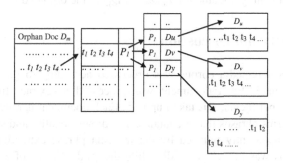

Fig. 6. Finding the outgoing links for anchor P₁

5.2 Incoming Links

The incoming link discovery uses the same concept as outgoing link discovery, but the direction is reversed. The first task is identifying anchors in the orphan document. The frequent n-terms phrases are extracted from the orphan document. As shown in figure 7, the frequent phrases, P_1, P_2, P_3 and P_4 are extracted and viewed as descriptive vectors of this particular document D_m. The descriptive vectors can indicate the topics of this orphan document. For instance, the possible frequent phrases from D_m are "*Australian Open Tennis Championship*", "*Melbourne Park*", "*hard court*", "*Grand Slam*". The combination of these frequent phrases represents and describes this orphan document to some extent.

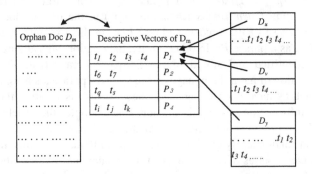

Fig. 7. Finding the incoming links for document D_m

The next step of incoming link discovery is to scan through the Wikipedia corpus and find out the articles which have information about the "Australian Open Tennis Championship". These articles become the incoming links to this orphan document. In figure 7 for example, P_1 ("*Australian Open Tennis Championship*") is a descriptive vectors of document D_m. The last step of creating an incoming link is to store the information of incoming document ID $\{D_u, D_v, D_y\}$ to the orphan document D_m.

6   Experiments and Discussion

The INEX 2008 Wikipedia corpus was processed according to the procedures explained in section 3. Each document was processed to extract frequent n-terms phrases. In the experiments, n is taken up to 5. Each document is now represented by the frequent n-terms phrases that it contains. The dimensionality and size of the original corpus was apparently reduced by the frequent phrase extraction. The original Wikipedia corpus was more than 5GB, while the total file size of extracted phrases was only 1.2GB.

Table 1 gives some instances of recognised frequent (stemmed) phrases with their frequency in an orphan document. The pre-processed document with a title "violin" had a total of 7258 words. On the right hand part of Table1, it shows there are 492 frequent 1-term, 326 frequent 2-term and 79 recognized frequent 5-terms. This document is now represented as a vector of 1528 terms/phrases.

Table 1. Some instances of the extracted frequent phrases

Doc ID	Freq	Phrase	n-terms
Doc 1	7	europ	1
Doc 1	9	violin	1
Doc 1	4	music instrument	2
Doc 1	6	standard pitch	2
Doc 1	12	finger posit	2
Doc 1	4	violin mak techniqu	3
Doc 1	5	type harmon artifici natur	4
Doc 1	6	vibrato common techniqu pitch	4
Doc 1	3	plai violin tune twist peg.	5

n-terms	Quantity
1	492
2	326
3	381
4	250
5	79

Table 2 shows that when the article size (word count) was increased, the average number of extracted phrases in that particular document was also raised as well. This shows that the extracted phrases were sufficient enough to describe the original document.

Table 2. The average frequency of every phrase

Document Size (Word Count)	Average Frequency
Less than 200	3
Between 200 and 400 words	8
Between 400 and 700 words	16
Between 700 and 1000 words	29
Between 1,000 and 3,000 words	62
More than 3,000 words	214

By investigating the extracted Frequent Phrase List, some interesting observations were made. For example, as shown figure 8, the comparison between total phrases and unique phrases revealed the features of the natural language, English. There are a total of 270,826 unique words (1-term phrases) in the corpus. It can be said that a dictionary with about 270,000 words would explain almost everything in this world.

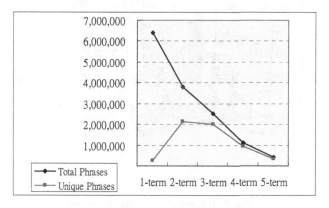

Fig. 8. The comparison between total phrases and unique phrases

However, there are many more 2-terms and 3-terms unique phrases in the corpus as compared to 1-term, 4-term and 5-term unique phrases. It reveals that the 2-terms and 3-terms phrases are more descriptive and representative to explain a concept and more accurate to describe meanings in the English language.

The link discovery procedures were applied on orphan documents to recognize the potential anchor text and possible links within. This procedure is based on the integrated n-terms which is the combination of 1-term to 5-terms. Frequent phrases including 1-term to 5-terms are extracted from an orphan document and all of them are considered as anchors. Any overlapping among the n-terms is removed. For example, if a 2-term phrase is a sub-string of a 3-term phrase, this 2-term phrase will be removed from n-term collection. Results in Table 3 show that this approach, representing the documents with frequent phrases only and then using these frequent phrases for recognising links, is able to allocate sufficient quantity of links in those orphan documents.

Table 3. The quantity of links discovered

Discovered Links	Minima		Maxima	
	Incoming	Outgoing	Incoming	Outgoing
Small docs (less than 500 words)	16	21	90	106
Medium docs (500 ~ 2000 words)	36	54	278	295
Large docs (more than 2000 words)	127	176	523	610

As shown in Figure 9, the average of links discovered from small, medium and large documents are 111, 302 and 631, respectively. The next task will be to rank these links so the high quality links can only be reported. Moreover, the INEX Link-the-Wiki evaluation can accept up to 250 incoming and 50 outgoing links for each

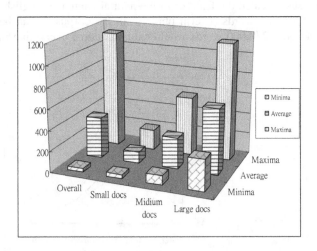

Fig. 9. The quantity of links discovered

orphan document. The threshold for filtering the potential links is based on the ranking of frequency of that particular phrase. For example, frequency of "finger posit" in Table 1 is 12; while the "standard pitch" has frequency of 6 in the same Table 1. In this scenario, the links of "finger posit" will be considered of higher importance than the links of "standard pitch".

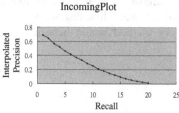

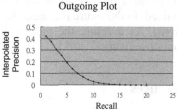

Fig. 10. The plot of Incoming Links   **Fig. 11.** The plot of Outgoing Links

The quality of links can be evaluated using Precision and Recall. Figure 10 and Figure 11 show the quality of identified incoming links and outgoing links respectively with the proposed approach. For the incoming links, this approach gives a fair result with the precision up to almost 0.7. However, this approach did not perform very well for identifying outgoing links. The precision of outgoing links only reached a bit higher than 0.4.

There is one interesting issue raised by comparing the different results of incoming and outgoing links. Both the incoming and outgoing links used the same tokenization method. The Frequent Phrase Extraction (mentioned in section 4) is the fundamental of both incoming and outgoing links. However, the recognition process of incoming links is different from the process of recognizing outgoing links. Due to the difference in nature of incoming and outgoing links, different procedures are implemented to find these links. As a result, it is likely to improve the outgoing links and reach a good result similar as incoming links.

7 Conclusions and Future Work

This paper presents our approach of discovering the incoming and outgoing links based on frequent phrases. Through data pre-processing, Frequent Phrase Recognition, Link Discovery and Validation, this proposed approach was able to discover the links automatically with certain accuracy. The precision of incoming links was found to be 0.7; while the precision of outgoing links did not perform as well as incoming links, only reached 0.4.

After conducting a series of experiments, results were found to support the hypothesis and assumptions made in the research. For example, the complexity and dimensions were effectively reduced by extracting the frequent phrases. The descriptive information collected from frequent phrases was sufficient to a certain level to undertake the Link-The-Wiki link discovery tasks.

There are some possible future extensions of this research. From the perspective of text mining, the recognition of frequent phrases is a difficult issue. In this research, we did not consider the named entity recognition as a part of pre-processing. It is hoped that the use of known entities such as nouns may improve the quality of anchors and consequently the links.

On the other hand, hyperlink is a particular feature of hypertext and web pages. The hyperlinks discovered in the research were almost as meaningful as manually maintained. In the future research, the precision of automatic link discovery would be improved. As a result, the generic link discovery method would benefit the huge amount of websites.

References

1. Trotman, A., Geva, S.: Passage Retrieval and other XML-Retrieval Tasks. In: Proceedings of SIGIR 2006 Workshop on XML Element Retrieval Methodology, Seattle, Washington, USA, pp. 48–50 (2006)
2. Porter, M.F.: An algorithm for suffix stripping. Automated Library and Information Systems 14, 130–137 (1980)
3. Kostoff, R.N., Tshiteya, R., Pfeil, K.M., Humenik, J.A.: Electrochemical power text mining using bibliometrics and database tomography. Journal of Power Sources 110, 163–176 (2002)
4. Myat, N.N., Hla, K.H.S.: A Combined Approach of Formal Concept Analysis And Text Mining For Concept Based Document Clustering. In: IEEE/WIC/ACM International Conference on Web Intelligence 2005, p. 4 (2005)
5. Girju, R., Badulescu, A., Moldovan, D.: Learning semantic constraints for the automatic discovery of part-whole relations. In: 2003 Conference of the North American Chapter of the Association for Computational Linguistics on Human Language Technology, vol. 1. Association for Computational Linguistics, Edmonton (2003)
6. Hideo, J., Mark, S.: Retrieving descriptive phrases from large amounts of free text. In: 9th international conference on Information and knowledge management. ACM, McLean (2000)
7. Parisut, J., Worapoj, K.: Dimensionality reduction of features for text categorization. In: 3rd conference on IASTED International Conference: Advances in Computer Science and Technology. ACTA Press, Phuket (2007)
8. Beil, F., Ester, M., Xu, X.: Frequent term-based text clustering. In: 8th ACM SIGKDD international conference on Knowledge discovery and data mining. ACM, New York (2002)
9. Yanjun, L., Soon, M.C.: Text document clustering based on frequent word sequences. In: 14th ACM international conference on Information and knowledge management, pp. 293–294. ACM, New York (2005)
10. Shen, D., Chen, Z., Yang, Q.Z.: H., Zhang, B., Lu, Y., Ma, W.: Web-page classification through summarization. In: SIGIR 2004: Proceeding of the 27th ACM Int. Conference on Research and development in information retrieval, Sheffield, pp. 242–249 (2004)
11. Lei, Z., Debbie, Z., Simeon, J.S., John, D.: Weighted kernel model for text categorization. In: 5th Australasian conference on Data mining and analytics, vol. 61. Australian Computer Society, Inc., Sydney (2006)

CMIC@INEX 2008: Link-the-Wiki Track

Kareem Darwish

Cairo Microsoft Innovation Center,
Bldg B115, Smart Village
Km. 28 Cairo/Alexandria Desert Rd.
Abu Rawash, Egypt
kareemd@microsoft.com

Abstract. This paper describes the runs that I submitted to the INEX 2008 Link-the-Wiki track. I participated in the incoming File-to-File and the outgoing Anchor-to-BEP tasks. For the File-to-File task I used a generic IR engine and constructed queries based on the title, keywords, and keyphrases of the Wikipedia article. My runs performed well for this task achieving the highest precision for low recall levels. Further post-hoc experiments showed that constructing queries using titles only produced even better results than the official submissions. For the Anchor-to-BEP task, I used a keyphrase extraction engine developed in-house and I filtered the keyphrases using existing Wikipedia titles. Unfortunately, my runs performed poorly compared to those of other groups. I suspect that this was the result of using many phrases that were not central to articles as anchors.

Keywords: Document Linking, keyphrase extraction.

1 Introduction

This paper presents the experiments I conducted at the Cairo Microsoft Innovation Center (CMIC) for the INEX Link-the-Wiki track. I participated in the outgoing Anchor-to-BEP (A2B) and the incoming File-to-File (F2F) tasks only. For the A2B task, the task was reduced to an Anchor-to-File task by setting all the best entry points to 0. The focus for the A2B task was on the identification of possible anchors via performing keyphrase extraction on the text of the orphan pages. The keyphrase extraction algorithm that I used attempted to find all possible phrases, but neglected to determine if the keyphrases are central to the page. Such a determination of centrality is crucial for identifying good anchors. For the F2F task, I used generic information retrieval techniques without any special processing for Wikipedia articles. I performed further runs for the F2F task, and the results suggest that using the titles of articles only produced the best results.

This paper is organized as follows: Section 2 presents my keyphrase extraction technique and survey existing techniques; section 3 presents my methodology for the A2B and F2F tasks and reports on the results; and Section 4 concludes the paper.

S. Geva, J. Kamps, and A. Trotman (Eds.): INEX 2008, LNCS 5631, pp. 337–342, 2009.

2 Keyphrase Extraction

Identifying a word sequence consisting of one or more words that represents a *valuable concept* in text is an important NLP problem. Such valuable concepts, which are henceforth referred to as keyphrases, are often called keywords (if they are single words), collocations and multi-word expressions, and are assumed to obey "semantic non-compositionality, syntactic non-modifiability, and non-substitutability of components by semantically similar words" [2, 4]. Conventionally, keyphrases represent the central concepts in an article, and hence, a sequence of words can be a keyphrase in one article and not in another. Another application is identifying salient words or phrases that can serve as hypertext to link from one article to another. Depending on the desired level of linking, a sequence of words may not have to be central to the article, which was my target of the work presented in this paper. Perhaps the fundamental difference between the two aforementioned applications is that the first is concerned with the top *n* valuable word sequences and the other is concerned with "all" such word sequences.

Subsections 2.1 and 2.2 describe related work on keyphrase extraction and my keyphrase extraction algorithm respectively.

2.1 Related Work

Much effort has gone in defining what keyphrases and there variants are [2, 4]. There are many approaches to keyphrase extraction including approaches that use phrase occurrence counts and part of speech patterns and word collocations, in which words that co-occur with a mean distance that has low variance [4]. Other approach are based on supervised learning in which a classifier is trained on features such as phrase location in a text segment, a phrase term frequency and document frequency [6]. Another approach is based on constructing a directed graph where the nodes represent tokens from a reference corpus and weighted links between nodes indicating the count of subsequent occurrences in text. After constructing the graph, graph walks over the highest weighted links are used to extract keyphrases [3]. The list of approaches listed above is by no means exhaustive, but provides a flavor of the most popular approaches.

2.2 Keyphrase Extraction System

I developed a keyphrase extraction technique that uses supervised machine learning in which a support vector machine (SVM) classifier is trained on the following features:

1. The probability of sequence occurrence. Keyphrases are expected to have a high probability of occurrence. The probability is computed using a language model that is trained on a reference corpus. For this work, I trained the language model on the Link-the-Wiki Wikipedia corpus.
2. The unigram occurrence probability of head and tail words. Head and tail words are typically expected to be valuable words, which would indicate that they have a low occurrence probability. The probability is computed using a language model that is trained on Link-the-Wiki corpus.

3. The sequence probability of words between head and tail words. These words are assumed to connect between the head and tail words and hence should have high probability of occurrence. For example, for the keyphrase "Department of Energy" the connect sequence is just "of" and is expected to be a common sequence. Again the probability is computed using a language model that is trained on the Link-the-Wiki corpus.

4. The probability of Part-of-Speech (POS) sequence being a keyphrase. The probability is computed using a language model that is trained on a list of POS tagged keyphrases. The POS tagging was done using an in-house POS tagger.

5. The percentage of digits.

6. The percentage of words with upper case letters.

7. The percentage of words that are a part of noun phrase chunk. The chunking was done using an in-house chunker.

8. The number of words in a sequence.

My keyphrase extractor can be tuned to be recall or precision oriented. For the submitted runs, I tuned the system to be more precision oriented, because a user would generally be willing to tolerate missing hyperlinks but would generally not tolerate incorrectly assigned hyperlinks. My system achieves 40% recall when tuned to be approximately 99% precise, as measured on a reference corpus. An important feature that was omitted is a feature that measures the importance of the sequence in the article. Such a feature can be the term frequency of the term, some combination of the term frequency and inverse document frequency, or some other feature such as the binomial log likelihood ratio [1].

3 Approach to Link-the-Wiki and Results

For the F2F task, I used the Indri search toolkit for indexing and searching the Wikipedia articles. I used Indri with stop-word removal and no stemming or blind relevance feedback. Indri combines inference network model with language modeling [5]. I submitted two runs, namely CMIC_F2F_01 and CMIC_F2F_02. I used three items to construct queries, namely the titles of Wikipedia articles (with the phrase operator if a title was longer than 1 word), the keyphrases extracted from the articles, and the top 20 terms from each article as ranked by term frequency only. For CMIC_F2F_01 run, I constructed the queries from the titles and the keyphrases. As for the CMIC_F2F_02 run, I constructed the queries using titles, keyphrases, and top 20 terms. The resultant mean average precision for the CMIC_F2F_01 and CMIC_F2F_02 runs was 0.46 and 0.51 respectively. It is also noteworthy that CMIC_F2F_02 achieved the highest precision among all the submitted runs for low recall levels (recall < 0.25), which suggests that my approach is more precision oriented and more suitable for generating a good small list of suggestions.

I ran additional post-hoc experiments to identify the effects of the titles and the top n-terms on retrieval effectiveness. For the ad-hoc runs, the queries were constructed using: the titles only with and without the phrase operator (Raw and Phrase respectively) or in combination of both; the Top n terms from the document where n was either 10, 20, or 30; and a combination of the Top 10 terms and the titles. Table 1 and Figure 1 report on the results of the submitted runs and the post-hoc runs. The results show that the titles of the articles were the most effective for the F2F task. In fact, constructing the queries using the titles with the phrase operator in combination with the titles without the phrase operator yielded results that were better than the official submitted runs. Also, adding the Top 10 terms to the titles improved retrieval effectiveness when using the titles with the phrase operator only, while adding more terms, namely Top 20 terms, degraded retrieval effectiveness. This is not surprising given that searching using the Top 10 terms only performed better than the Top 20, which in turn performed better than the Top 30 terms. Thus, I opted to perform no post-hoc experiments with more than Top 10 terms.

When using the titles with and without the phrase operator, adding Top 10 terms resulted in no improvement in retrieval effectiveness. The lack of improvement due to adding top n terms to the queries constructed from titles is a bit counter intuitive because intuitively adding the top n terms to the queries should have a query expansion effect. Further, this result contradicts experiments that I have run on other collections.

As the results suggest, constructing the queries for the F2F task using the titles of the articles produces the best results and expanding the resultant queries using other terms from the articles is unlikely to produce better retrieval effectiveness.

For the A2B task, I submitted one run, namely CMIC_LTW_01. For the run, I extracted the keyphrases in the orphan article and I filtered the keyphrases using the titles of the articles in Wikipedia articles that I was allowed to link to. The filtering

Table 1. Results of the ad-hoc and post-hoc runs

Run	Description	MAP
Phrase Titles	Titles w/ phrase operator	0.46
Raw Titles	Titles w/o phrase operator	0.41
Raw + Phrase Titles	Titles w & w/o phrase operator	0.54
Top 10	Top 10	0.32
Top 20	Top 20	0.27
Top 30	Top 30	0.25
Title + Top 10	Title w/phrase operator + Top 10	0.50
	Title w/ & w/o phrase operator	0.54
CMIC_F2F_01	Title w/phrase operator + Top 20	0.46
CMIC_F2F_012	Title w/phrase operator + Top 20 + Keyphrases	0.51

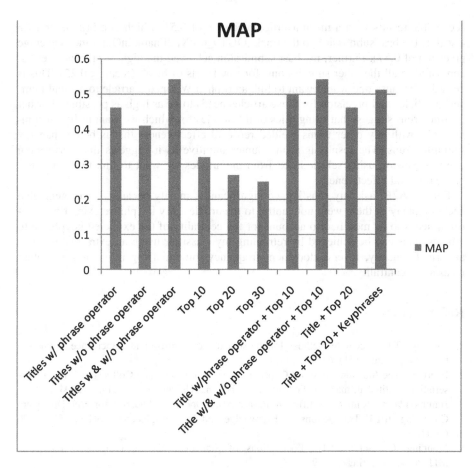

Fig. 1. Results of the ad-hoc and post-hoc runs

involved allowing a keyphrase to match any title that was either an exact match or one that subsumes the keyphrase completely. For example, in article 100011 entitled Otago, the keyphrase "Firth of Clyde" was linked to the article 144233 entitled "Firth of Clyde", and the keyphrase "Free Church of Scotland" was allowed to link to articles 554606 and 909535 entitled "Free Church of Scotland" and "Free Presbyterian Church of Scotland" respectively. Unfortunately, my results were dismal with a mean average precision of 0.05. I suspect that my runs performed poorly because many of the keyphrases that were chosen to be anchors were not central to the articles and were hence deemed irrelevant by assessors.

4   Conclusion

This paper presented the CMIC runs to the INEX Link-the-Wiki track. I did well in the F2F task, but dismally in the A2B task. For the F2F, using generic information retrieval techniques in combination with keyphrase and key word extraction produced acceptable results with a mean average precision of 0.51, which is a little over 10% less than the best submission to the track (QUT9_GPXF2FnameInOut – mean average precision of 0.57). Further, my best submission achieved the highest precision levels, compared to all the other submissions, for low levels of recall (recall $\leq$ 0.25). This is desirable because one would want to link an orphan Wikipedia article to a small number of articles and precision for those articles needs to be as high as possible. Further post-hoc runs suggest that using titles of the articles for which we want to find incoming links with any other terms yielded retrieval effectiveness that is better than my submitted runs. The results is a bit counter intuitive as it suggests that performing query expansion using selected text from the articles we want to link to did not improve retrieval effectiveness.

For the A2B task, my runs lagged significantly, mostly because I over generated anchors, many of them were not central to the articles. My keyphrase extraction algorithm needs to be modified to account for the centrality of the extracted keyphrase to articles. This can be achieved by retraining my classifier using an extra feature such as term frequency, inverse document frequency, binomial log ratio, or some other measure of centrality.

References

1. Dunning, T.E.: Accurate methods for the statistics of surprise and coincidence. Computational Linguistics 19(1), 61–74 (1993)
2. Evert, S.: The Statistics of Word Cooccurrences: Word Pairs and Collocations. Ph.D. Dissertation, Institut für maschinelle Sprachverarbeitung, University of Stuttgart (2004)
3. Hammouda, K., Kamel, M.: Efficient Phrase-based Document Indexing for Web Document Clustering. IEEE Transactions on Knowledge and Data Engineering 16(10), 1279–1296 (2004)
4. Manning, C., Schutze, H.: Foundations of Statistical Natural Language Processing. MIT Press, Cambridge (1999)
5. Metzler, D., Croft, W.B.: Combining the Language Model and Inference Network Approaches to Retrieval. Information Processing and Management Special Issue on Bayesian Networks and Information Retrieval 40(5), 735–750 (2004)
6. Turney, P.D.: Coherent keyphrase extraction via web mining. In: Proceedings of IJCAI, Acapulco, Mexico, pp. 434–439 (2002)

Stealing Anchors to Link the Wiki

Philipp Dopichaj, Andre Skusa, and Andreas Heß

Lycos Europe GmbH
Carl-Bertelsmann-Str. 29
P. O. Box 315
33311 Gütersloh
Germany
dopichaj@acm.org, andre.skusa@googlemail.com, mail@andreas-hess.info

Abstract. This paper describes the Link-the-Wiki submission of Lycos Europe. We try to learn suitable anchor texts by looking at the anchor texts the Wikipedia authors used. Disambiguation is done by using textual similarity and also by checking whether a set of link targets "makes sense" together.

1 Introduction

In this paper, we describe the Link-the-Wiki submission of Lycos Europe. Details about INEX and the Link-the-Wiki track are given elsewhere in these proceedings, so we do not repeat them here. In this paper, we use *new text* to refer to the text which should be linked (conceptually, this is a text entered by a user of the platform without any links; the aim of the system is to support the user to find suitable links). We use *anchor text* or *anchor* to refer to the link label, that is, the clickable part of the text that links to a *target page*.

Our approach to the Link-the-Wiki task is based on that described by Itakura and Clarke [2]: All existing anchor texts from the training collection are indexed along with their link targets, and the new text is scanned for these anchor texts to find links. In accordance with the Link-the-Wiki task specification, the training collection does not include any of the topics or references to them.

The main difference is that we try to select the best-matching target dynamically whereas Itakura and Clarke use a static mapping from anchor text to target – the target is always the page most frequently referenced by the anchor. For example, in a text about computers, the anchor *Apple* is more likely to refer to the page *Apple Computers* than to the page *Apple Records*. We use heuristics based on text similarity and link structure to determine which of the potential targets is the most likely real target.

Finding outgoing links is done in the following steps:

1. The potential anchor texts are identified. The chosen anchor texts do not overlap, and each anchor text has one or more potential targets associated with it.

S. Geva, J. Kamps, and A. Trotman (Eds.): INEX 2008, LNCS 5631, pp. 343–353, 2009.

2. For each potential anchor text, a ranking of the potential targets *in the context of the new text* is performed. Furthermore, general statistical information obtained at indexing time – like the absolute frequency of the anchor text/target in the training collection – is used.

Our main focus is finding outgoing links, as opposed to finding incoming links from existing documents to the newly-added content. Outgoing links are determined in two main steps that will be described in the following sections:

1. Finding the parts of the texts that should serve as links to other documents (*anchor texts*).
2. Finding the correct target pages for the anchor texts in case of ambiguities.

The second point means that even if a given anchor text is known to refer to *some* other document, it is not necessarily known to *which* article it refers.

2   Preparations for Finding Outgoing Links

This section describes how potential anchor texts are found in the new text and also what index structures are needed to support this.

2.1   Finding Potential Anchor Texts

The first step toward identifying links in a new document is to find potential anchors; this is done by searching for occurrences of the training anchors in the new text. We give preference to longer anchor texts: For example, in the example text from figure 1a, we have the sequence *Mac OS X v10.2*. Potential anchors include *Mac*, *Mac OS*, and *Mac OS X v10.2*; here, the last one is the longest anchor text, so it is selected. In case of overlapping anchor texts, the anchor occurring earlier is selected.

Using word boundaries as implemented in the Java regex package for anchor detection does not work for two reasons:

– Due to the idiosyncrasies of the INEX Wikipedia collection, spurious spaces are inserted or removed around markup, even in the middle of words, so word boundaries cannot be trusted.
– Anchors may only partly cover a given word; this is bad style, but there are instances where *child* as part of *children* is linked to the corresponding article. For other languages like German, compound words can be formed without spaces, so this might happen more frequently.

Although the second case is rare – especially in the English version of Wikipedia used for INEX –, the first reason is sufficient to justify the decision not to analyze word boundaries.

The result of this stage is a collection of non-overlapping anchor texts that might be turned into links. Based on the training data set, we know for each anchor set the possible target pages as well as the absolute frequency of references to a certain target page under the given name. We now have to develop a ranking of the targets for every potential anchor.

Apple bundled a similar program, Sherlock 3 , with Mac OS X v10.2 .

(a) Input text.

[[Apple]] bundled a similar program, [[Sherlock 3]] , with [[Mac OS X v10.2]] .

(b) Selected anchors.

Apple:	Apple Computer, APPLE, Apple Records, Apple (album), Apple II family, Malus, Apple Store (retail), Apple (super mario), Yabluko, Apple I
Sherlock 3:	Sherlock 3
Mac OS X v10.2:	Mac OS X v10.2

(c) Possible targets of the selected anchors

[[Apple Computers|Apple]] bundled a similar program, [[Sherlock 3]] , with [[Mac OS X v10.2]] .

(d) Final linked text, with the *Apple* anchor directed to *Apple computers*.

Fig. 1. Processing of an input text from the Wikipedia article *Karelia Watson*

2.2   Reducing the Size of the Anchor Index

Our approach requires statistics about the existing links in the training collection. We examine every link in the collection and store the anchor text along with the target page's ID. Then, we count the number of occurrences for each anchor text/target page pair to see how often a given anchor text is used to refer to the given page.

This information is sufficient input for our approach, but to both keep the index size small and remove spurious entries, we remove all anchor text/target page pairs with one of the following properties:

- The length of the anchor text is less than 5 or greater than 60 characters. Very long anchors include anchors like *Best Writing, Story and Screenplay Based on Factual Material or Material Not Previously Published or Produced*; they mostly refer to very specific page titles that are unlikely to occur in normal text. Short anchors are removed because they are usually ambiguous and they can lead to false positives.
- The anchor text refers to ten or more different pages. This implies that the anchor text is very general like, for example, *her father*.
- The anchor text occurs less than five times in the collection.

The numbers used were chosen in a rather ad-hoc fashion; further research is required to determine whether these numbers are good (or even whether the filtering is needed at all). We will test this once the results and evaluation tools are available.

3   Link Target Disambiguation

In many cases, anchor texts refer to only one possible target, like *Sherlock 3* in the example in figure 1c. However, the anchor text *Apple* from the same example shows that there is not always a one-to-one mapping of anchor texts to target pages, so the link detector has to make a choice. Furthermore, it may be necessary to remove spurious anchors.

One obvious problem is that anchor texts are frequently only sensible in the context in which they occur; for example, the anchor text "her father" refers to different persons depending on who "her" refers to. Since low-level information about the document frequency of terms is not available in our setup, we could not use Itakura and Clarke's formula for selecting anchors to index, so we implemented the simple heuristics from section 2.2.

The remainder of this section is based on the following values that influence the choice of which targets to use for a given anchor:

1. The rank of this target for this anchor, based on the total number of references;
2. the rank of the target page when doing a full-text search for the new article's title; and
3. the rank of the target page when doing a full-text search for the new article's full text (optional).

We chose to use a linear combination of these factors to obtain the final rank of a target.

3.1   Analysis of Anchor/Link Frequency

In absence of any other information, the link finder can still look at the *prior probability* of a given anchor text referring to a given target. This information can be obtained by analyzing the frequencies of the different target pages for a certain anchor text. For example, in the INEX collection, the anchor *Apple* refers to *Apple Computer* 399 times, to *APPLE* 83 times and to *Apple Records* 65 times, so in absence of any further information, *Apple Computer* is most likely the correct target.

3.2   Analysis of the Target Text

Simply using the frequency of targets in the training collection, however, does not take into account the context provided by the new document: for example, the text of the document should already give a strong indication whether the article is about computers or music. Thus, a straightforward approach is to calculate the *textual similarity* of the new text and the possible targets; if the new text and a target have a high similarity, it is likely that they are about the same general topic (like computers or music).

In our implementation, we implement this by doing a single full-text search for the complete new text respectively its title on an index that comprises the

Table 1. Target distribution of the anchor *Apple* (case sensitive)

Rank	Count	Target page
1	399	Apple Computer
2	83	APPLE
3	65	Apple Records
4	7	Apple (album)
5	2	Apple II family
6	2	Malus
7	1	Apple Store (retail)
7	1	Apple (super mario)
7	1	Yabluko
7	1	Apple I

full texts of all articles in the test collection. This results in a single ranked list of articles that are somehow related to the new text; for every anchor text that is found in the new text, the highest-ranked article from this list is chosen.

3.3 Analysis of the Link Structure

According to our observation, it is likely that the documents that are linked from the same source document are connected. This is because these pages typically share a main topic, so if two topics are mentioned (or pages are referenced) on the same source page, these topics are more likely to be connected than two randomly chosen topics. We can exploit this to find the correct link target among a set of candidates; for every such set, we determine how many links to the target pages for the *other* anchor texts exist. The more links exist, the more likely the target is to be the correct link target for this anchor.

Figure 2 demonstrates that the pages linked from a single page tend to be heavily connected. We can see that *APPLE* is not connected to the pages that are actually referenced from the source page at all and that *Apple Records* only has one link, whereas *Apple Computer* has many links in this cluster of pages.

The link analysis will not work properly if there is a very low number of targets (or, more generally, if the potential targets are mostly unconnected). In this case, the link finder should select potential targets even if they are isolated. The exact mechanism and threshold for this are the subject of future research.

3.4 Combination of These Approaches

Of course, it is possible to combine the evidence to obtain better quality. Since each of the approaches can be used to find a ranked list of possible targets for a given anchor text, we chose to use a weighted combination of the different ranks as the basis for the final decision. Given the example rankings from table 2, and the weights $w_1 = 1$ (anchor/link frequency), $w_2 = 5$ (text similarity), and $w_3 = 2$ (link analysis) results in a final value of 12 for *Apple Computer*, of 13 for *Apple Records*, and of 24 for *Apple I*. Thus, in this case, the link target *Apple*

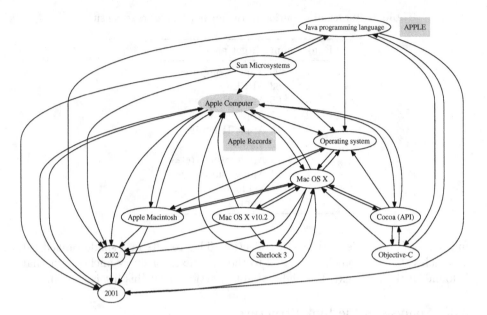

Fig. 2. The link network for pages linked from the page *Karelia Watson*. Our focus is on the shaded items, which are potential targets for the anchor text *apple*. The *Apple Records* and *APPLE* pages (in rectangles) are not linked from this page, but shown here for demonstration.

Table 2. Example for combining the different aspects of target rankings

Target	Anchor/link freq.	Text sim.	Link analysis
Apple Computer	1	2	1
Apple Records	2	1	3
Apple I	3	3	2

Computers has the lowest combined rank and is selected as the final target. (Note that a higher weight value decreases the influence of the corresponding factor.)

Since we did not finish the implementation of link-based target disambiguation in time, we only submitted runs using anchor/link frequency and text similarity. For text similarity, we search the full text of all articles for occurrences of the title of the new page to be linked. From a quality point of view, it would probably be better to search for the complete body text of the new article – otherwise we implicitly assume that the concept is already mentioned in the existing articles, although it does not have an article of its own. Unfortunately, the cost for doing this was prohibitive on our setup, so we had to settle for searching for the titles only. We used the different combinations of text similarity–anchor/link frequency weight, from equal weights for both (run LycosA2B-1-1), a weight of 5 for one and 1 for the other (runs LycosA2B-1-5 and LycosA2B-5-1). Furthermore, we submitted runs using only one of the two factors (runs LycosA2B-0-1 and LycosA2B-1-0).

Note that we do not actually calculate a best entry point in the target file – we always use the start of the document instead.

3.5   Limitations

One base limitation of our work is that we assume that the collection already contains a large number of related articles. As Huang et al. [1] note, this assumption does not hold for a batch upload of related articles where links between the articles are at least as important as links to or from the collection. Another potential problem is that the anchor texts that have been used by the authors might not be meaningful (for example, "click here").

We believe, however, that the approach can work well in the right circumstances. We plan to use it on a community platform about German history, with the anchors from the German Wikipedia as a training set. The results from preliminary tests are quite promising.

4   Finding Incoming Links

Our current implementation for finding incoming links is simplistic: we simply search for the new document's title in the full-text index to determine a ranked list of candidate sources. (Note that no phrase search is performed, so in effect the results may contain pages where the terms from the title occur out of order.) Next, the title of the new document is searched for in each candidate's text, and the first occurrence is added to the list of links, ordered by the rank of the text search. Finally, all pages in which the title is *not* found verbatim – this may happen if the title comprises several words – are added to the end of the list. For example, if the title is *Apple Computer*, pages containing only "Apple" are ranked behine pages containing the full title.

5   Results and Discussion

At the time of writing, the evaluation tools have not been made public yet, so our evaluation only includes the official results; this means that we cannot discuss the effect of the network analysis.

5.1   Anchor to File

Although the original task was to find the best entry points in the link targets, many participants (including Lycos) always used the start of the document as the best entry point. Because of this, the anchor-to-best-entry-point results were also evaluated as anchor-to-file results, ignoring the best entry points if available.

Figure 3 shows that it pays to use a combination of text similarity and anchor/link frequency; the runs using both features better the runs using only one of the features. Interestingly, the exact weights used do not affect the results significantly, an equal weight for both factors performs as well as a 1:5 weight ratio in favor of either factor. On the other hand, omitting text similarity leads to a much higher loss in precision than omitting anchor/link frequency weights.

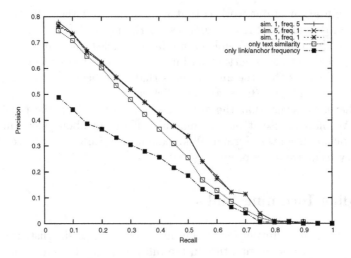

Fig. 3. Comparison of the Lycos runs for Anchor2BEP. "Best run" takes the maximum precision of all other runs for every recall (thus this is not an actual run). Clearly it pays to use both text similarity and anchor/link frequency.

As figure 4 shows, it betters both the best submitted runs by other organizations and even the Wikipedia ground truth for most of the precision-recall curve. (The Wikipedia ground truth does not get perfect results because apparently the assessors disagreed with the article authors about what constitutes a good link.) Minor deficits can be seen in the high-recall regions (starting around 0.6), where our method trails the maximum of the other submissions by a significant margin.

The results for the global measures mean average precision (MAP) and R-precision (see table 3) are inconclusive: whereas the MAP of our method is significantly higher than that of all other methods (arounf 0.49 compared to at most 0.42 for the others), including the Wikipedia ground truth, our R-precision (0.40448) is lower than that of the best run, Amsterdam_a2bep_5 (0.42146). The reason for this is unclear; these measures have generally been shown to be highly correlated in information retrieval.

Table 3. Official results for anchor-to-file evaluation. The highest numbers in each column are highlighted; for all given recall points as well as MAP, our run LycosA2B-1-5 has the best result. For R-precision, this run is surpassed by run Amsterdam_a2bep_5 and the Wikipedia ground truth.

Run ID	MAP	R-Prec	P5	P10	P20	P30	P50	P250
LycosA2B-1-5	*0.4973*	*0.40498*	*0.64400*	*0.61600*	*0.52100*	*0.44133*	*0.37840*	*0.07568*
LycosA2B-5-1	0.4931	0.40448	0.63600	0.61000	0.51900	0.44067	0.37800	0.07560
LycosA2B-1-1	0.4930	0.40448	0.63600	0.61000	0.51900	0.44133	0.37800	0.07560
LycosA2B-1-0	0.4708	0.37015	0.63200	0.59600	0.49900	0.41667	0.34280	0.06856
Waterloo_a2a#1	0.4111	0.33201	0.55600	0.50600	0.43100	0.36467	0.30560	0.06112
Otago_capConstant-SingleSearch-A2B	0.3952	0.35800	0.44400	0.45200	0.41400	0.37933	0.33240	0.06648
Otago_capConstant-TitleOnly-A2B	0.3952	0.35800	0.44400	0.45200	0.41400	0.37933	0.33240	0.06648
WikipediaGroundTruthRun	0.3945	0.40634	0.47600	0.46600	0.43500	0.39667	0.36520	0.07304
Otago_nCapConstant-WholeDocument-A2B	0.3896	0.35234	0.45600	0.46400	0.40200	0.36933	0.32040	0.06408
Waterloo_a2a#3	0.3874	0.34910	0.42800	0.44600	0.43200	0.39600	0.30560	0.06112
Waterloo_a2a#2	0.3355	0.39324	0.55600	0.50600	0.42600	0.35533	0.26680	0.05336
LycosA2B-0-1	0.3291	0.31201	0.35600	0.35800	0.32200	0.31933	0.31280	0.06256
QUT_LTWA2BnameRerank	0.3042	0.24854	0.39200	0.37200	0.32200	0.27733	0.22200	0.04440
QUT_GPXA2Bname	0.2912	0.22597	0.40000	0.37600	0.31100	0.26200	0.20480	0.04096
KnowCenterGraz_globalIDF_topic	0.2873	0.31495	0.25600	0.24600	0.26000	0.29533	0.35080	0.07016
KnowCenterGraz_disamTopic_IL_None_topic	0.2643	0.27409	0.24000	0.24400	0.25400	0.26533	0.30760	0.06152
KnowCenterGraz_globalIDF_sentence	0.2309	0.26539	0.22800	0.20800	0.20100	0.23400	0.28760	0.05752
KnowCenterGraz_disamDoc_IL_None_topic	0.2131	0.25125	0.14400	0.17000	0.20600	0.22267	0.28640	0.05728
Amsterdam_a2bep_5	0.2079	*0.42146*	0.37600	0.40800	0.35600	0.31400	0.22240	0.04448
KnowCenterGraz_disamTopic_IL_None_sentence	0.2076	0.22194	0.18000	0.19400	0.19700	0.21467	0.24480	0.04896
KnowCenterGraz_disamDoc_IL_None_sentence	0.1764	0.21270	0.11200	0.14000	0.16900	0.18800	0.24120	0.04824
CMIC_LTW_01	0.1760	0.18689	0.13200	0.18600	0.18700	0.17200	0.18280	0.03656
QUT_P9_GPXA2Btitle	0.1725	0.13739	0.21600	0.20600	0.18000	0.16467	0.12640	0.02528
CSIR_LTW_A2BEP_2	0.1307	0.10081	0.19600	0.17400	0.14000	0.11267	0.08480	0.01696
Amsterdam_a2bep_1	0.1271	0.25973	0.14000	0.14200	0.18600	0.18800	0.18520	0.03704
QUT_Anchor-BEP_1	0.1149	0.11075	0.11200	0.10400	0.11300	0.11333	0.11080	0.02216
Amsterdam_a2bep_2	0.1127	0.23963	0.12400	0.14800	0.19000	0.18400	0.16080	0.03216
Amsterdam_a2bep_3	0.0983	0.34507	0.13200	0.18800	0.23100	0.22000	0.16280	0.03256

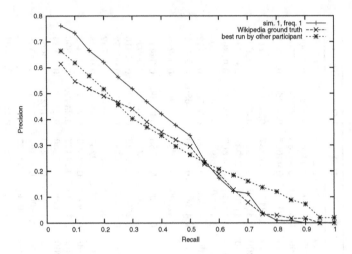

Fig. 4. Lycos versus Wikipedia ground truth and best of other submissions for Anchor2BEP

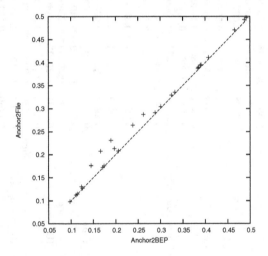

Fig. 5. Mean average precision for anchor-to-best-entry-point and anchor-to-file results. There is a strong correlation, only a few runs stand out.

5.2   Anchor to Best Entry Point

Surprisingly, the results for the anchor-to-best-entry-point evaluation do not differ much from the results for the anchor-to-file results (see figure 5). The main reason is presumably that most participants actually submitted anchor-to-file results to this task; furthermore, in many cases, the best entry point in a link target will in fact be at the very start of the document. Since the results are virtually identical, we will not discuss them here.

5.3   File to File

Although our main focus was on the anchor-to-file runs, we also submitted runs to the file-to-file task. Instead of sorting the targets for every found anchor, all targets for all anchors were put in a single list, ordered by the calculated scores. The submission was formed by taking the first unique entries from this list. Unsurprisingly, our runs did not perform well in this task, both for the outgoing and the incoming links.

6   Conclusions and Future Work

We have confirmed that the basic approach of Itakura and Clarke [2] works very well as a baseline for new methods. We have shown that this method can be improved significantly by incorporating textual similarity to disambiguate anchor texts that could refer to several articles (the original method only used frequency statistics). Unfortunately, the run-time penalty for this can be rather high, since a similarity search on all articles is required for each file (but not for every anchor!). State-of-the-art search engines are quite fast, so this should not be a major problem in all but the most time-critical settings.

Our submission should be regarded as a first attempt at the problem; in particular, we have not yet evaluated using network analysis for disambiguation. In preliminary experiments, we have found this to be quite successful, but we still need to perform more elaborate experiments on the INEX corpus. In future work, we plan to address this by taking more factors into account.

Acknowledgements

The research presented in this paper was partially funded by the German Federal Ministry of Economy and Technology (BMWi) under grant number 01MQ07008. The authors are solely responsible for the contents of this work. We thank our colleagues at Lycos Europe as well as the anonymous reviewers who gave valuable feedback.

References

1. Huang, D.W.C., Xu, Y., Trotman, A., Geva, S.: Overview of INEX 2007 link the wiki track. In: Fuhr, N., Kamps, J., Lalmas, M., Trotman, A. (eds.) INEX 2007. LNCS, vol. 4862, pp. 373–387. Springer, Heidelberg (2008)
2. Itakura, K.Y., Clarke, C.L.A.: The University of Waterloo at INEX2007: Adhoc and Link-the-Wiki tracks. In: Fuhr, N., Kamps, J., Lalmas, M., Trotman, A. (eds.) INEX 2007. LNCS, vol. 4862, pp. 417–425. Springer, Heidelberg (2008)

Context Based Wikipedia Linking

Michael Granitzer[1], Christin Seifert[2], and Mario Zechner[2]

[1] Knowledge Management Institute
Graz University of Technology
Inffeldgasse 21a, 8010 Graz
mgranitzer@tugraz.at
http://kmi.tugraz.at/
[2] Know-Center Graz
Inffeldgasse 21a, 8010 Graz
{mzechner,cseifert}@know-center.at
http://www.know-center.at/

Abstract. Automatically linking Wikipedia pages can be done either content based by exploiting word similarities or structure based by exploiting characteristics of the link graph. Our approach focuses on a content based strategy by detecting Wikipedia titles as link candidates and selecting the most relevant ones as links. The relevance calculation is based on the context, i.e. the surrounding text of a link candidate. Our goal was to evaluate the influence of the link-context on selecting relevant links and determining a links best-entry-point. Results show, that a whole Wikipedia page provides the best context for resolving link and that straight forward inverse document frequency based scoring of anchor texts achieves around 4% less Mean Average Precision on the provided data set.

Keywords: INEX, Link-the-Wiki, Context Exploitation.

1   Introduction

This paper outlines the efforts taken by the Know-Center Graz in the Link-the-Wiki Track of INEX 2008. The track focuses on automatically linking an orphan page to already existing Wikipedia pages (outgoing links; out-links) and from already existing Wikipedia pages to the orphan page (incoming links; in-links). In contrast to last years focus on identifying source and target pages of a link, this years track also includes the identification of anchor position and best-entry-points (BEP). Anchor positions mark the character position of a link in the source page; best-entry-points in the target page.

In last years Link-the-Wiki Track [6], matching page titles for identifying link candidates have been quite successful [4]. It was shown that without considering contextual information around the link, reasonable results could be achieved; a fact supported also from outside the INEX community [10]. Besides page titles, link structure provides valuable information. In [7] an algorithm using anchor

S. Geva, J. Kamps, and A. Trotman (Eds.): INEX 2008, LNCS 5631, pp. 354–365, 2009.

texts and link structures achieved a very high accuracy. However, non of these approaches took the context of a link, i.e. its surrounding text, into account, while [4] argued on the potential of such approaches.

In our approach we evaluate the potential of different context types to calculate the relevance of a possible link candidate. Link candidate identification itself utilizes word sequence matching based on a finite state machine gazetteers. Thereby, entries of the gazetteer contain not only the title of a Wikipedia page, but also anchor texts, similar to work reported in [7]. Link candidates via gazetteer matching are ranked subsequently using different context types, i.e. different ranges of words surrounding the anchor. This context based relevance should allow a more precise selection of correct hyperlinks hopefully removing high frequent, irrelevant links like for example "The" or "Are".

Our major contribution is an in depth analysis of different context types compared to straight forward, context free scoring mechanisms. Besides the official runs we also present a detailed parameter study using the "trec_val" evaluation tool and t-tests for estimating the influence of different parameter sets and syntactic matching properties like case sensitivity.

Experiments are evaluated on the Wikipedia XML Corpus consisting of 659,413 Wikipedia pages and split into two test sets by the track organizers. The *file-to-file test set*, has around 6.600 test documents with existing Wikipedia links as ground truth. The *anchor-to-bep test set* consists of around 50 manually annotated topics. The candidate page for automatic linking, a wiki page having all links removed from, is called an *orphan page*. In the following we refer to the corpus without the test set as the *Wikipedia corpus*. The two runs are distinguished as *file-to-file run*, having 6.600 test documents and *anchor-to-bep run* having 50 topics.

In the following, section 2 outlines the corpus preparation and preprocessing and defines the anchor context types. This anchor context is used in section 3 to explain our link selection and scoring mechanism. Experiments, official results as well as internal parameter studies are shown in section 4, followed by the conclusion in section 5.

2   Preprocessing and Context Types

The Wikipedia corpus is indexed using the open source search engine Lucene [5] applying standard stop word removal and stemming. For each Wikipedia page the title and all anchors of links pointing to this page are extracted and stored as gazetteer list. For matching, this list transform into a finite state machine (FSM) consisting of three states. A start state serving as entry point, intermediate states retaining the structure of the FSM, and final states containing the URL of the Wikipedia page. Transitions between states of the FSM consist of words occurring in a gazetteer entry. Beginning at the start state, transitions are followed recursively if the transitions word occurs in the word sequence to match. If upon matching a final state is reached, an annotation pointing to the particular Wikipedia page is added. In this way gazetteer matching allows us to annotate word sequences with hyperlinks for a large number of possible link targets at a reasonable speed.

Orphan pages are preprocessed using the OpenNLP toolkit [1]. Preprocessing includes tokenization, sentence detection and part-of-speech tagging. Afterwards, the document is segmented into non-overlapping parts, defining the context for the following relevance calculations. In our experiments we distinguish between the following context types:

- *Document:* Most straightforward, the whole document is taken as context.
- *Section:* Sections are provided via the XML-Schema and correspond to the Wikipedia sections of articles.
- *Paragraph:* Similar to sections, paragraphs are also provided via the XML Schema.
- *Topics:* Topics are automatically annotated based on sentence clustering. Blocks of similar sentences are found and annotated as topic using the well known C99 segmentation algorithm[2].
- *Sentences:* Sentences are obtained from the sentence detector of the OpenNLP pipeline and serve as smallest possible context.

The context based linking strategies introduced in the next section exploit those context types in order to determine the relevance of a link.

3   Linking Strategies

For a given orphan page d_o, our system determines a set of n possible in-links $I = \{< l_1, s_1 > \ldots < l_n, s_n >\}$ and a set of m possible out-links $O = \{< l_1, s_1 > \ldots < l_m, s_m >\}$. Each out-link/in-link is assigned a score s_i determining the confidence of the system in generating such a link. One link is - as defined in the LTW result set specification - a quadruple $l_h = < s_h, t_h, sp_h, b_h >$ where for link l_h, s_h denotes the source page, t_h the target page, sp_h the span (i.e. character based start and end position) of the link in the source document and b_h the best-entry-point in the target document.

In the following we present how the different properties have been determined, differentiating between out-link and in-link generation. While both follow the same conceptual approach, their implementation varies for reducing time complexity in the in-link generation step.

3.1   Out-Link Generation

Out-link generation starts with preprocessing the orphan document d_o as outlined in section 2. Matching the content of the document with the FSM- gazetteer returns a set of possible out-link candidates O, whereby for each link $l_i \in O$ we know its source s_i, its target t_i and its span sp_i. For each link we determine the *anchor context*, that is the context the link span is contained in. All nouns of the anchor context are extracted and fed into the retrieval backend as Boolean

OR query. To speed up this potentially large OR query we restrict the result set to pages pointed to by all links in the anchor context simply by adding all link target identifiers (i.e. the file name of the page) as AND query part. Thus, for all links contained in the span of the current anchor context we are receiving a score s. In particular the query is formulated as

$$(ID = t_1 \ OR \dots OR \ ID = t_n) \ AND \ (w_1 \ OR \ w_2 \ \dots \ OR \ w_k)$$

with $\{w_1 \dots w_k\}$ as the nouns of the anchor context and t_k as unique identifier for the kth link target and $"ID = "$ specifying the search on the metadata field containing the unique identifiers of a Wikipedia page. Formally, the score (named *anchor context score* in the following) returned is obtained from standard Lucene ranking as

$$s_i = coord_{w,i} * norm(w) * \sum_{t \in w} \frac{\sqrt{tf_{t,i}} * idf_t^2}{norm(i)} \tag{1}$$

where

- $tf_{t,i}$ is the frequency of term t in document i
- $idf_t = 1 + \log \frac{\#D}{\#D_t + 1}$ is the inverse document frequency with $\#D$ as the number of documents in the corpus and $\#D_t$ the number of documents containing term t
- $norm(w)$ is the norm of the query calculated as $\sqrt{\sum_k idf_k^2}$
- $norm(i)$ is the length norm of document i, namely the number of terms contained in document i
- and $coord_{w,i}$ is a overlapping factor increasing the score the higher the number of overlapping terms between query and documents are.

The Lucene scoring equation has been proven as reliable heuristic for full text searching. It can be seen as an heuristic version of a cosine similarity between anchor context and target document with emphasize towards the number of overlapping words. This assumption is quite naturally for resolving the context of a link. For example "tree" in computer science will occur more frequently with terms describing data structures than the "tree" in nature. Thus, depending on the position of a link in the document and its surrounding text we receive different scores hopefully disambiguating the tree data structure from the forest tree.

Besides context based scoring method an evaluation scheme solely based on the inverse document frequency of an anchor text is used for comparison reasons. The rational behind is that high frequent anchor texts like "The" or "Are" occur in nearly every document and therefore provide no additional information independent whether they are a true links or not. In particular the score, named *anchor IDF* in the following, is calculated as

$$s_i = \log \frac{\#D}{\#D_a + 1}$$

where $\#D$ is the number of wiki pages in the corpus and $\#D_a$ the number of wiki pages containing the anchor text of the link.

For the file-to-file task links pointing to the same target t but having different spans sp are merged. We distinguish three different merging strategies, namely the highest score of the link, the average score of the link or simply by counting the number of links to a target t.

3.2   In-Link Generation

In-link generation is in principle similar to out-link generation with the difference that in a first step we have to determine the source document d_j of a particular link. Again we utilize title matching for doing so, but in contrast to out-link generation the title is used as search string instead of gazetteer matching. Similarly to out-link generation we are determining different contexts in the orphan document to assign a score to a link. Given the nouns of this context as sequence $< w_1, \ldots, w_k >$ we are sending the following query to the backend:

$$\text{``title''} \; AND \; (w_1 \; OR \; w_2 \ldots OR \; w_k)$$

where "*title*" indicates a phrase query for the title of the orphan page. Again the score is calculated as outlined in equation 1.

From the result set we obtain a ranked list of possible link source candidates. If the context is different than the whole document, merging strategies are required to merge the ranked lists of the different contexts. As for out-link generation, we calculate the relevance either as the highest relevance of a link, the average relevance of a link or simply by counting the occurrences of a link. Taking the n best source candidates is either the input for determining the best-entry-points or gives us already the result for the file-to-file linking task.

3.3   Best-Entry-Point Detection

Both in-link and out-link generation provides a list of best matching links including target page, source page and the span of a link. In the final step, best-entry-points are determined again based on the link context. Our hypothesis is that the best-entry-point in the link target has to be similar to the anchor context. Furthermore, if the title of the source page is contained in the link target, those parts of the target document are preferred entry points. Since we obtain a score for each entry point, results are ranked and the best five entry points are taken as result.

In particular, similarity is calculated using a simple vector space model with local TFIDF weighting. Given the link target t, the textual content of the target is preprocessed and decomposed into segments $t_{r,1} \ldots t_{r,k}$. Segments are either sentences or topics and correspond to the context defined in section 2. After filtering out all non-noun words, each segment is converted into a term vector. The weight of a term is calculated according to the TFIDF scheme, but based on the extracted segments, as:

$$w_{r,l} = tf_{r,l} * \log(\frac{(\#R + 1)}{\#R_l + 1})$$ (2)

where $w_{r,l}$ is the weight of term l in segment r, $tf_{r,l}$ is the number of times a term l occurs in segment r divided by all terms in segment s, $\#R$ is the number of segments in the target document and $\#R_l$ is the number of segments containing term l.

Similarly to the target segments, the anchor context in the source document - denoted as a - is also converted into a term vector by filtering all non-nouns and applying equation 2.

The ranking of best-entry-points is obtained by calculating the cosine similarity between anchor context $\vec{a}$ and all target segments $\vec{t}_{r,1} \dots \vec{t}_{r,k}$ and rank them accordingly. Segments containing the title of the anchor page are favored by increasing the similarity as follows:

$$s(\vec{a}, \vec{t}_{r,i}) = \begin{cases} title \in t_{r,i} : & (1 + \frac{\vec{a} \cdot \vec{t}_{r,i}}{\|\vec{a}\| * \|\vec{t}_{r,i}\|})/2 \\ title \notin t_{r,i} : & \frac{\vec{a} \cdot \vec{t}_{r,i}}{\|\vec{a}\| * \|\vec{t}_{r,i}\|} \end{cases}$$ (3)

Best entry points are returned as starting point of the text segment since we assume that a reader does not want to start reading in the middle of a sentence or paragraph.

4   Implementation and Evaluation Details

As outlined above, Lucene [5] has been used as search backend and OpenNLP [1] for preprocessing. All algorithms are developed in Java, including the gazetteer component. Since our approach, at least for out-link detection, heavily relies on gazetteer matching the question is whether a gazetteer with low runtime and low memory resource consumption is feasible. In our FSM approach the gazetteer with titles and anchors consisted of around 1.7 million entries and used up around 800 MB main memory. Additionally, gazetteer entries may be distributed using distributed computing techniques like Map & Reduce [3] and thus scaling up is possible in our approach.

Runtime behavior also satisfies interactive requirements. On a dual core laptop with 4GB of main memory file-to-file runs took around 64 minutes using the more complex anchor context scoring - that is around 1.7 documents per second. After finding the link candidates, best-entry-point matching does not increase runtime complexity. Thus, the overall process can be seen as computational tractable and scalable.

The runs can be differentiated in file-to-file in-link/out-link generation, anchor detection and best-entry-point detection. File-to-file runs are evaluated on the 6.600 topics defined by the organizers. Anchor detection and best-entry-point detection are conducted on the 50 topics defined by the participants. After the development of our algorithms we did an in depth parameter analysis by taking the available ground truth of the 6.660 topics test set and evaluated file-to-file and anchor-to-file runs on it. This allowed an in depth evaluation of all runs but the manually assessed 50 topic based anchor-to-bep runs.

4.1 Parameter Analysis

Basically our experiments are focused on analysing the following parameters:

- Case sensitive (CS) matching distinguished between considering the case in gazetteer matching or not.
- Longest Common Sequence Matching (LCS) removed overlapping gazetteer annotations by taking those annotations with the longest common sequence of tokens.
- *Title only* matching only considers page titles in the gazetteer while otherwise anchors of links are also included in the gazetteer list.
- The *context* level determined the type of context to use for the anchor context scoring scheme. If no context was provided the anchor IDF scoring scheme was used.
- For the file-to-file runs 3 different merging strategies - maximum, average and count- for aggregating anchors on the file level have been considered

Permutation of the different parameters yielded 120 test runs for the file-to-file task and 40 test runs for the anchor-to-file task. Since in-link creation is conceptually similar, we restricted the parameter analysis task to out-link detection only. In order to cope with the large number of runs, statistical significance testing was used to determine the influence of the different parameters.

For determining the most influential parameters, we started determining statistically significant differences between runs using a one-sided paired t-test [9]. Statistically significant differences allow us to calculate a parameter value's "success rate", defined as how often a run with the particular parameter value is significantly better than all other runs. More formally, given $B(r_i)$ resp. $W(r_i)$ as the number of runs where run i is significantly better resp. more worse and given $R_{p_a=v}$ as the set of runs where parameter p_a has value v, the success rate $S_{p_a=v}$ of value v for parameter p_a is calculated as

$$S_{p_a=v} = \frac{\sum_{r_i \in R_{p_a=v}} B(r_i)}{\sum_{r_i \in R_{p_a=v}} B(r_i) + \sum_{r_i \in R_{p_a=v}} W(r_i)} \tag{4}$$

By ranking parameter values according to their success rate the most influential parameter value, i.e. those parameter values most often participating in successful run, can be estimated. In other words, by selecting a parameter value with a high success rate it is very likely that this run will perform good, independent of the other parameter values.

By analysing file-to-file runs it turned out that context based evaluation strategies had an overall success rate of around 67% outperforming all other parameters. Also, only using page titles yields to a higher success rate of around 62% than using gazetteers based on anchor texts. Merging links using the maximum score also turned out to outperform the average score and count based merging strategy. Case sensitive vs. case insensitive matching as well as longest common sequence matching did not have a huge impact on the performance of a run. Analyzing the context parameter for file-to-file runs more closely showed that taking the whole document gives a success rate of 96%. Thus, nearly every time the whole document is used as anchor context the run outperforms all other runs. Topic detection also turned out to have a high success rate (82%), outperforming sentences, paragraph and sections as topic. However, for the later two it must be noted that a large number of queries did not have sections or paragraphs assigned, thereby biasing the results. Similar results are achieved by the anchor-to-file task. However, different to the file-to-file runs anchor idf based scoring turned out to be as good as context based scoring.

Table 1. Results for out-link Generation for the file-to-file run with 6600 orphan test pages and the anchor-to-file task with 50 orphan pages. Runs with no context used the anchor IDF scoring method. NA depicts measures not available due to missing ground truths.

Task	Title Only	LCS	CS	Context	MAP_{intern}	$MAP_{official}$	MAP_{reeval}
file-to-file (6.600)	true	false	false	document	0.548	0.1129	0.516
file-to-file (6.600)	true	true	false	none	0.5038	0.1407	0.475
file-to-file (6.600)	true	false	true	document	0.471	NA	NA
file-to-file (6.600)	true	true	true	none	0.4508	NA	NA
file-to-file (6.600)	false	true	true	none	0.4392	NA	NA
file-to-file (6.600)	true	false	true	topic	0.4258	NA	NA
file-to-file (6.600)	false	true	false	none	0.4215	NA	NA
file-to-file (6.600)	false	false	true	document	0.3827	NA	NA
anchor-to-file (50)	true	false	false	document	NA	0.2131	0.2350
anchor-to-file (50)	true	false	false	topic	NA	0.2643	0.2908
anchor-to-file (50)	true	false	false	sentence	NA	0.2309	0.309
anchor-to-file (50)	true	true	false	none	NA	0.2873	0.3130

4.2   Official Results

We submitted 2 runs for the file-to-file task for comparing the best anchor context method with the context free anchor IDF approach. For anchor-to-bep we submitted a combination of 3 different out-link generation and 4 different in-link generation approaches again distinguishing between context based and context free approaches. For the remaining parameters we took the best choices obtained by the parameter analysis.

MAP of the official and the best internal runs for *out-link generation* are depicted in table 1. Due to an error in the submission format, our official runs scored much more worse than our internal benchmarks. We corrected the submission error on the submitted files and re-evaluated the results. Those re-evaluated mean average precisions are depicted as MAP_{reeval} in the tables. Figure 1 shows the precision recall curve for the out-going links comparing official with the inofficial results for the submitted runs and comparing the official runs with the

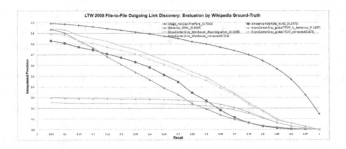

Fig. 1. Precision-Recall curve of our out-link file-to-file runs comparing the official runs with the corrected runs and with the official results of the best LTW 08 runs

Table 2. Results for in-link generation file-to-file

Parameters		file-to-file			anchor-to-bep
Title as OR Query	context	MAP_{intern}	$MAP_{official}$	$MAP_{reevaluated}$	$MAP_{official}$
false	document	0.6355	0.5300	0.625	0.2384
false	sentence	0.5938	NA	NA	0.1895
false	topic	NA	NA	NA	0.2619
false	no context	0.5938	0.5369	0.606	0.1968
true	document	0.5066	NA	NA	NA
true	sentence	0.4088	NA	NA	NA
true	no context	0.4088	NA	NA	NA

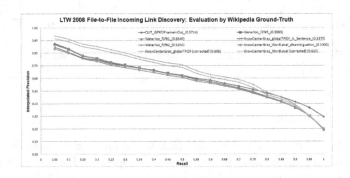

Fig. 2. Precision-Recall curve of our in-link file-to-file runs comparing the official runs with the corrected runs and with the official results of the best LTW 08 runs

best runs in the Link-The-Wiki track. By correcting the submission format our runs performed quite well and would be ranked 2nd. It can be observed that considering the context of a link improves mean average precision by around 4%. While the increase is significant, we would have expected a larger increase through the more complex anchor context scoring mechanisms. Also, if the anchor context is other than the whole document, the differences becomes smaller and is nearly negligible.

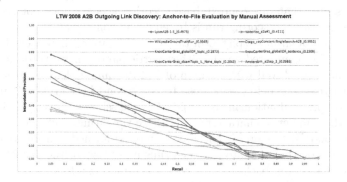

Fig. 3. Precision-Recall curve of our anchor-to-file runs compared to the best runs of other participants

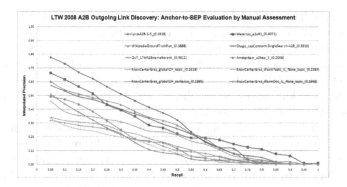

Fig. 4. Precision-Recall curve of our anchor-to-bep runs compared to the best runs of other participants

Similar performance figures can be observed for *in-link detection*, as shown in table 2. The difference between the best context scoring method - again a document based context - and the context free scoring method is smaller, with around 2% on the re-evaluated runs. Overall file-to-file in-link generation did quite well compared to the best runs of the track (see figure 2). The original runs have been ranked third, while the re-evaluated runs achieved the highest map. Also, precision-recall curves provide high precision values over large parts of the recall.

Anchor-to-file results of the manually assessed 50 topic task are provided in table 1 and precision-recall curves compared to the best other runs is shown in figure 3 . Overall, the performance of anchor detection was lower than what could be expected from file-to-file matching. Also, re-evaluation did not provide a huge increase due to the smaller size of links per topic. For the manually assessed anchors it seems that our context based scoring scheme does not score well, especially since our top scoring run is based on no context at all.

Anchor-to-bep evaluation, depicted in table 2 and figure 4, shows very low mean average precision compared to the other participants. In contrast, file-to-bep evaluation performed well with an map of 12.219 compared to the best map of 20.79 from Lycos and being very close to the second best group of runs from Otago. Since the evaluation measure penalizes missing the exact position linearly with the number of characters, only those runs using sentences as context achieved a good BEP score. Topic based runs performed considerable worse. Overall, results on the manually assessed runs point toward the hypothesis that vector space based approaches using words surrounding a link are not discriminative enough for achieving reasonable accuracy values.

5   Conclusion

In this paper we have outlined context based methods for automatically detecting links between Wikipedia pages. Experiments showed that considering the context of an link increases precision by around 4%. However, the choice of the type of context is critical. The whole document seems to be best suited as anchor context, followed by automatically detected topics. Predefined document structures like sections and paragraph are bad context choices, decreasing accuracy below the straightforward IDF approach. Constructing gazetteers from page titles only seem to be more appropriate than using anchor texts, from which follows that using context based scoring schemes hardly resolves noisy links introduced by anchor texts. Results obtained by the experiment point toward the hypothesis that vector space based approaches using words surrounding a link are not powerful enough, especially for anchor and BEP detection. Hence, sequence based approaches, language models or link based methods (c.f. [7]) may be required for achieving reasonable accuracies.

In the future we plan to focus more on machine learning based approaches. As shown in recent work [8], machine learning can achieve rather high user judged accuracy while retaining parameter robustness. Another fruitful future challenge is the automatic labeling of link types. For example the page "Berlin" linking to "Germany" marks a part-of relationship while a link between "Berlin" and "Capital" marks a is-a relationship. Automatically identifying such relationship types may have both, a huge practical as well as a huge theoretical impact in the context of semantic wikis.

Acknowledgement. The Know-Center is funded within the Austrian COMET Program - Competence Centers for Excellent Technologies - under the auspices of the Austrian Federal Ministry of Transport, Innovation and Technology, the Austrian Federal Ministry of Economy, Family and Youth and by the State of Styria. COMET is managed by the Austrian Research Promotion Agency FFG.

References

1. Baldridge, T.M.J., Bierner, G.: Opennlp: The maximum entropy framework (2001), http://maxent.sourceforge.net/about.html (last visited June 2008)
2. Choi, F.Y.Y.: Advances in domain independent linear text segmentation. In: Proceedings of the first conference on North American chapter of the Association for Computational Linguistics, pp. 26–33. Morgan Kaufmann Publishers Inc., San Francisco (2000)
3. Dean, J., Ghemawat, S.: Mapreduce: simplified data processing on large clusters. Commun. ACM 51(1), 107–113 (2008)
4. Geva, S.: Gpx: Ad-hoc queries and automated link discovery in the wikipedia, pp. 404–416 (2008)
5. Hatcher, E., Gospodnetic, O.: Lucene in Action (In Action series). Manning Publications (December 2004)
6. Huang, D.W.C., Xu, Y., Trotman, A., Geva, S.: Overview of inex 2007 link the wiki track. In: Fuhr, N., Kamps, J., Lalmas, M., Trotman, A. (eds.) INEX 2007. LNCS, vol. 4862, pp. 373–387. Springer, Heidelberg (2008)
7. Itakura, K.Y., Clarke, C.L.: University of waterloo at inex2007: Adhoc and link-the-wiki tracks, pp. 417–425 (2008)
8. Milne, D., Witten, I.H.: Learning to link with wikipedia. In: CIKM 2008: Proceeding of the 17th ACM conference on Information and knowledge mining, pp. 509–518. ACM, New York (2008)
9. Smucker, M.D., Allan, J., Carterette, B.: A comparison of statistical significance tests for information retrieval evaluation. In: CIKM 2007: Proceedings of the sixteenth ACM conference on Conference on information and knowledge management, pp. 623–632. ACM, New York (2007)
10. Wu, F., Weld, D.S.: Autonomously semantifying wikipedia. In: CIKM 2007: Proceedings of the sixteenth ACM conference on Conference on information and knowledge management, pp. 41–50. ACM, New York (2007)

Link Detection with Wikipedia

Jiyin He

University of Amsterdam, Science Park 107, 1098 XG Amsterdam, The Netherlands
j.he@uva.nl

Abstract. This paper describes our participation in the INEX 2008 Link the Wiki track. We focused on the file-to-file task and submitted three runs, which were designed to compare the impact of different features on link generation. For outgoing links, we introduce the anchor likelihood ratio as an indicator for anchor detection, and explore two types of evidence for target identification, namely, the title field evidence and the topic article content evidence. We find that the anchor likelihood ratio is a useful indicator for anchor detection, and that in addition to the title field evidence, re-ranking with the topic article content evidence is effective for improving target identification. For incoming links, we use exact match and retrieval method with language modeling approach, and find that the exact match approach works best. On top of that, our experiment shows that the semantic relatedness between Wikipedia articles also has certain ability to indicate links.

1 Introduction

In this paper, we describe our participation in the INEX 2008 Link-The-Wiki (LTW) track. The goal of the LTW track is to automatically identify hyperlinks between documents. The 2006 Wikipedia collection is used as the development and test data, which contains the ground truth of linked documents. This year's LTW track consists of two sub-tasks, namely, the *file-to-file* (f2f) task and the *anchor-text to Best Entry Point* (BEP) task. We focused on the f2f task and submitted three runs. The task is formulated as follows: *a set of 6600 Wikipedia articles are randomly picked from the collection as topic pages; the participants are supposed to discover at most 250 incoming and 250 outgoing links between the topic pages and the rest of the collection.*

In our participation in the Link-the-Wiki track, the goal is to explore different features that indicate links betweenWikipedia articles, as well as to develop a generative approach to automatic link generation. The runs we submitted were designed to compare the impact of different features on link generation, which we will specify in detail in the following sections.

The rest of the paper is organized as follows. Section 2 describes the approaches we use to generate the outgoing links and Section 3 describes the approaches for incoming links. Section 4 presents the experimental settings and the runs we submitted. In section 5 the results of the official runs are discussed and analyzed. Section 6 concludes the paper.

S. Geva, J. Kamps, and A. Trotman (Eds.): INEX 2008, LNCS 5631, pp. 366–373, 2009.

2 Outgoing Links

Our approach to identifying outgoing links can be seen as a two-step procedure: anchor text detection (i.e., where to start a link) and target identification (i.e., which article should be linked). Although in the f2f task, only the target article is required, i.e., the exact anchor text or the position of an anchor text is not required, it is useful to identify the anchor texts, since the identified anchor texts have strong indication on which articles should be selected as target. In addition to the two-step procedure, we also experiment with the semantic relatedness between Wikipedia articles as an indicator of existing links, which does not involve the identification of anchor text. We discuss this kind of feature in 2.4.

2.1 Anchor Detection - Anchor Likelihood Ratio

For anchor text detection, we introduce the anchor likelihood ratio measure (ALR). Since the Wikipedia articles are well-structured and interconnected, the existing links provide an indication of the patterns about linked articles. Mihalcea and Csomai [1] proposed the link probability measure which measures the likelihood of a word sequence being an anchor text by calculating the ratio between the number of times a word sequence is used as an anchor text and the number of times this word sequence occurs in the collection. The likelihood estimation is a reasonable measure and is proved to be useful. However, the value of the likelihood is a continuous number between 0 and 1, which needs a threshold to determine wether the word sequence is an anchor text and this threshold is heuristically set. By modifying this measure, we try to model it without a "magic" threshold.

For a given word sequence w, we assume that it can be sampled from two different underlying models: the anchor model θ_A and the background collection model θ_C. To select the model for generating the word sequence, we take the likelihood ratio between the two models, which is formulated as:

$$ALR_w(\theta_A||\theta_C) = \frac{P(w|\theta_A)}{P(w|\theta_C)} \tag{1}$$

Given Eq. 1, it is obvious that the larger the value of the likelihood ratio, the more likely that the word sequence w is generated by the anchor model. Particularly, when $ALR_w > 1$, it expresses that the anchor model is preferred over the collection model, and therefore we can obtain a *non-magic* threshold at $ALR_w = 1$.

To estimate the probability of $P(w|\theta_A)$ and $P(w|\theta_C)$, we simply use the maximum likelihood, i.e., $P(w|\theta_A) = \frac{|w \in A|}{|A|}$, $P(w|\theta_C) = \frac{|w \in C|}{|C|}$, where A is the anchor collection, C is the collection of all ngrams and $|.|$ denotes the number of elements in a collection.

In practice, for a given Wikipedia article, we extract all possible n-grams and rank them according to the ALR score.

2.2 Target Identification - Title Field Evidence

For target identification, we use a language modeling approach. In Wikipedia, the title of an article is the main concept of the article and is usually the same as or similar to the anchor texts that are linked with it. We model this relationship by assuming that a given anchor text can be generated by the language models that generate the titles. Thus the problem boils down to estimating the probability that the given anchor text a is generated from a given title model θ_t by applying the Bayes' Theorem.

$$P(\theta_t|a) = \frac{P(a|\theta_t)P(\theta_t)}{P(a)}, \tag{2}$$

where the $P(\theta_t)$ is assumed to be uniformly distributed and $P(a)$ is the same for all anchors, which can be dropped since it does not affect the ranking. For estimating the probability $P(a|\theta_t)$, we take the joint probability of the terms in an anchor text given the title model, and assume that those terms are sampled identically and independently.

$$P(a|\theta_t) = \prod_{w_a \in a} P(w_a|\theta_t) \tag{3}$$

To estimate $P(w_a|\theta_t)$, we simply use the maximum likelihood (ML) of the anchor term w_a generated by the title model. To avoid zero probabilities, we smooth $P(w_a|t)$ with the background collection of all Wikipedia titles C_T using the Jelinek-Mercer method to obtain $P(w_a|\theta_t)$:

$$P(w_a|\theta_t) = (1 - \lambda) \cdot P(w_a|t) + \lambda \cdot P(w_a|C_T), \tag{4}$$

where

$$P(w_a|\theta_t) = \frac{n(w_a, t)}{\sum_{w'_a} n(w'_a, t)} \tag{5}$$

In this equation, $n(w_a, t)$ and $n(w'_a, t)$ are the number of times terms w_a and w'_a occur in a candidate target article's title t.

2.3 Target Identification - Topic Article Content Evidence

Since most Wikipedia titles are short, it is very likely to end up with equal probabilities for different target candidates, especially in cases where disambiguation articles are involved. In order to solve this disambiguation problem, we try to incorporate an additional evidence source, the topic article content evidence. The underlying assumption is that if a candidate target article is the real target article for a given topic article, the content of the topic article should semantically relate to the terms in the title field of the candidate target article. Based on this assumption, we model the problem as to estimate the probability that the language model of the topic article θ_d generates the title of the candidate target article t. Similar as before, we apply Bayes' Theorem to estimate this probability.

$$P(\theta_d|t) = \frac{P(t|\theta_d)P(\theta_d)}{P(t)}, \tag{6}$$

where $P(t)$ is ignored for ranking, and $P(\theta_d)$ is assumed to be uniformly distributed. The same ML estimation is applied to calculate $P(t|\theta_d)$:

$$P(t|\theta_d) = \prod_{w_t \in t} P(w_t|\theta_d) \tag{7}$$

Again, JM smoothing is applied to avoid zero probabilities.

2.4   Semantic Relatedness between Articles

An earlier work in link generation with Wikipedia [2] involves clustering highly similar articles around a given topic article and suggest links between the topic article and those similar articles in the cluster. The assumption behind this approach is that, linked articles should be semantically similar. One of our submitted runs is designed to test this assumption, of which the details are to be discussed in Section 4.

Before actually generating the official run, we did a sanity check on this assumption. We construct two sets of Wikipedia articles. The first set is a set of 500 articles which are randomly sampled from the Wikipedia collection. In order to have articles with valid content, we only sample from the un-redirected articles (non-article pages such as image pages and category pages are not included). Then we calculate the pairwise cosine similarity scores between the 500 samples. The second set is obtained by collecting the target articles that are linked with the 500 samples and we only calculate the similarity scores between the linked pairs. Figure 1(a) and 1(b) shows the histogram of the similarity scores between random articles and that between linked articles. It is clear from the plot that in the first set, the similarity scores between two random articles are very low, mostly close to 0 and below 0.1. In the second situation, the histogram is more widely spread and has a large portion of article pairs whose similarity scores are higher than 0.1. The difference exhibits in the distribution of the similarity scores among random Wikipedia articles and that of the linked articles suggests that: *linked articles are more likely to be semantically similar.* We submit a run to test the actually effectiveness using the semantic relatedness between articles for link detection.

3   Incoming Links

Our approach for identifying incoming links is quite straightforward. We experiment with two types of methods: The first method is based on exact matching of titles. We get the articles that contain the exact matches of the title of the topic article and select random 250 articles as linked articles. In this case, we simply ignore the context of the matched terms and assume that the existence of a link is context-independent. The second method is a rank-based approach. In this

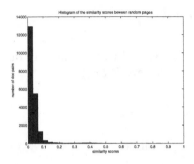

(a) Similarity scores between random articles

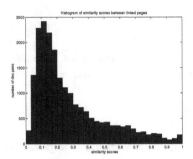

(b) Similarity scores between linked articles

Fig. 1. Histogram of similarity scores

case, we perform a retrieval run using the title of the topic article as query. The top 250 articles are selected as the linked articles. Further thresholding on these 250 articles is also explored, and we describe this later in detail in our submitted runs section.

4   Experimental Settings and Runs

4.1   Experiment Settings

In this section, we discuss the experimental settings and the design of our submitted runs. We use the content only Wikipedia articles as input. All the XML-tags are removed and the XML-structures of the documents are ignored in our experiments. For preprocessing, we use Porter stemmer for both topic articles and target articles without stop words removal. All the topic articles are virtually deleted from the collection. For all retrieval experiments for incoming links, we use the Lemur toolkit[1]. We use the KL model with JM smoothing. Smoothing parameter λ (e.g., in Eq. 4) is heuristically set to 0.1 for all experiments (i.e. the background statistics have little impact).

4.2   Submitted Runs

We submitted three runs to the LTW track, which are designed to compare the impact of different features on generating the links.

ltw01

Outgoing links: For a given topic article, we select n-grams whose ALR is larger than 1 as anchor texts ; use the anchor text as query to retrieve target articles from the collection of all Wikipedia titles; the top-ranked article is selected as the target article.

---

[1] http://www.lemurproject.org

Incoming links: We use the title of the topic article as query and retrieve top 250 articles from the collection as the source articles that are linked to the given topic.

Run description: We use this run as our baseline run.

ltw02

Outgoing links: For a given topic article, we select n-grams with ALR larger than 1 as anchor texts; retrieve the target article whose title matches (exact or partial) the anchor text; re-rank the target articles based on the topic article content evidence (i.e., Eq. 6), and select the top-ranked article as the target article.

Incoming link: use the topic title to find exact matches in the collection, select random 250 articles as the incoming source article.

Run description: We compare this run with ltw01. For outgoing links, we use this run to test if re-ranking with topic-article content evidence would help disambiguate the target articles. For incoming links, this run compares the exact match method and the retrieval method.

ltw03

Outgoing links: We use the topic title as query to retrieve 250 candidate target articles, rank them by cosine similarity to the topic article, and select the articles whose similarity score is larger than 0.15 as target.

Incoming links: We use the same strategy as that of the outgoing links, but select the articles whose similarity score is larger than 0.026 as linked articles.

Here, 0.15 is the average cosine similarity between linked articles that are sampled from the collection; 0.026 is the threshold for "exceptionally" high similarity between two random articles[2]. Different thresholds indicate that we would like outgoing links to emphasize on precision, and incoming links to emphasize on recall.

Run description: This run checks an assumption: the linked articles should be semantically similar. Here we do not try to identify anchor text.

5   Results and Discussion

In this section, we report on the results of our submitted runs and our observations from the results. Table 1 lists the evaluation results of the submitted runs. The metrics used for measuring the performance are MAP, R-Prec, Precision at 5(P@5) and Precision at 10(P@10). Besides, figure 2 shows the precision-recall plots for both incoming and outgoing links.

---

[2] We rank the pairwise similarity scores among the random set of articles in descending order, and take the score at top 5% as the threshold.

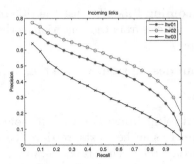
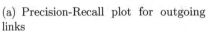
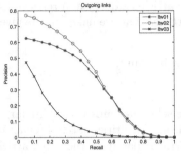

(a) Precision-Recall plot for outgoing links

(b) Precision-Recall plot for incoming links

Fig. 2. Precision-Recall plots for the submitted runs

We see that for both outgoing links and incoming links, run ltw02 has better performance than ltw01, in terms of all metrics. From this observation, we have following conclusions. For outgoing links, the re-ranking with topic article content evidence does help to improve the performance, especially in early precision, which is shown in 2(a). Since we apply the same method for anchor detection in both ltw01 and ltw02, the improvement in the final results suggests that the topic article content evidence has an impact on target identification, in terms of disambiguation of the target articles. For incoming links, the fact that ltw02 is better than ltw01 suggests that simple title match works better than retrieval-based method.

On the other hand, Run ltw03 performs consistently worse than the other two runs in both outgoing and incoming links. However, early precision for incoming links does not completely fail, suggesting that the semantic relatedness between articles could be a reasonable feature for indicating a link. Further experiments on combining the features, i.e., the ALR and the semantic similarity between articles are needed.

Table 1. Evaluation of the submitted runs; highest scores are marked as boldface

Outgoing links				
Runs	**MAP**	**R-Prec**	**P@5**	**P@10**
ltw01	0.2925	0.4288	0.4063	0.46578
ltw02	**0.3475**	**0.4461**	**0.6369**	**0.6130**
ltw03	0.1041	0.1590	0.3748	0.2890
Incoming links				
Runs	**MAP**	**R-Prec**	**P@5**	**P@10**
ltw01	0.4801	0.4933	0.6161	0.5733
ltw02	**0.5250**	**0.5478**	**0.6648**	**0.6264**
ltw03	0.3345	0.3604	0.5244	0.4540

6   Conclusion

We described our participation in INEX 2008 link-the-wiki track. We submitted three runs in the file-to-file task. For outgoing links we based our runs on a two-step procedure, i.e., anchor detection and target identification. For anchor detection, we use the anchor likelihood ratio (ALR), and for target identification, we apply the language modeling approaches to model different types of evidence. For incoming links, we apply two very simple method, i.e., exact match and retrieval with the title of the topic page. On top of that, we also submit a run that only use the semantic similarity between pages as an indication of a link. The result shows that it is a reasonable feature but it alone is not powerful enough for link detection. For future work, we plan to experiment with combining the features and develop a unified framework for combination. Particularly, we would like to experiment with machine learning techniques, which make less assumptions on the model form so that the features can be incorporated flexiblely.

References

1. Mihalcea, R., Csomai, A.: Wikify!: linking documents to encyclopedic knowledge. In: CIKM 2007: Proceedings of the sixteenth ACM conference on information and knowledge management, pp. 233–242. ACM Press, New York (2007)
2. Fissaha Adafre, S., de Rijke, M.: Discovering missing links in Wikipedia. In: Proceedings of LinkKDD 2005 Workshop (2005)

Wikisearching and Wikilinking

Dylan Jenkinson, Kai-Cheung Leung, and Andrew Trotman

Department of Computer Science
University of Otago, Dunedin, New Zealand
{dylan,kcleung,andrew}@cs.otago.ac.nz

Abstract. The University of Otago submitted three element runs and three passage runs to the Relevance-in-Context task of the ad hoc track. The best Otago run was a whole-document run placing 7th. The best Otago passage run placed 13th while the best Otago element run placed 31st. There were a total of 40 runs submitted to the task. The ad hoc result reinforced our prior belief that passages are better answers than elements and that the most important aspect of the focused retrieval is the identification of relevant documents. Six runs were submitted to the Link-the-Wiki track. The best Otago run placed 1st (of 21) in file to file automatic assessment and 6th (of 28) with manual assessment. The Itakura & Clarke algorithm was used for outgoing links, with special attention paid to parsing and case sensitivity. For incoming links representative terms were selected from the document and used to find similar documents.

1 Introduction

Otago participated in the Relevance-in-Context task of the ad hoc track submitting six runs, three passage and three element runs. The passage runs compared the Otago 2007 algorithm to a previous algorithm examined by Otago, the Kullback-Leibler model, and to whole document retrieval. The result suggests that whole document is better than passage retrieval and that there is little difference between the other two algorithms.

Otago also participated in the Link-the-Wiki track, preferring a variant of the Itakura & Clarke algorithm for outgoing links, and searching for the orphan title for documents that should link to the orphan. The results suggest that the Itakura & Clarke algorithm is a solution to the linking problem when measured against the ground truth of the Wikipedia itself.

2 Wikisearching

2.1 The Otago 2007 Passage Algorithm

The approach taken by Otago at INEX 2007 [1] was two step. First, relevant documents were identified using BM25. Second, all the occurrences of all the search terms with a document were identified (stemming with Porter's algorithm) and a fixed sized

S. Geva, J. Kamps, and A. Trotman (Eds.): INEX 2008, LNCS 5631, pp. 374–388, 2009.
© Springer-Verlag Berlin Heidelberg 2009

window of 300 words placed on the centroid. The centroid was defined as the mean of the term locations within the document, or alternatively the mean of those within one standard deviation of the true mean.

2.2 The Kullback-Leibler Passage Algorithm

In earlier experiments at Otago, Huang et al. [2] examined techniques for identifying relevant passages within a relevant document and converting those into elements by taking the smallest element that fully enclosed the passage. Of the passage selection methods examined, the Kullback-Leibler model was the most effective:

$$KL(W|Q) = \sum_{t \in Q} P(t|W) \log \left(\frac{p(t|W)}{p(t|D)} \right)$$

where W is a window within a document, D, and t is a search term of query, Q, and

$$p(t|W) = \frac{tf_W + 0.5}{|W| + 1}$$

and

$$p(t|D) = \frac{tf_D + 0.5}{|D| + 1}$$

where tf_D is the number of occurrences of t in D and $|D|$ is the length of document D (and likewise for tf_W with respect to the window, W).

Several strategies for choosing the window were examined. The sliding non-overlapping window of size 400 words was shown to be effective on the INEX IEEE document collection (measured with MAep and iMAep).

Itakura and Clarke [3] suggest that methods of identifying elements from passages are not as effective as methods of identifying elements directly. This is, in part, because the conversion from a passage to an element usually involves increasing the size of the passage and the extra text is often non-relevant text. That is, the conversion from a passage to an element is unlikely to affect recall but is likely to decrease precision. If this is the case then the prior result of Huang et al. is understated. This motivates our comparison to Kullback-Leibler to Otago 2007 in INEX 2008 where all results are measured as if passages.

2.3 The Beigbeder Element Algorithm

Beigbeder [4] proposes a method of scoring elements based on fuzzy proximity. If a document contains one occurrence of one search term, then the fuzzy proximity (*fp*) to term occurrence t, for location p is

$$fp = max \left(\frac{k - (p - t)}{k}, 0 \right)$$

where k is a controlling parameter.

If the document contains more than one term occurrence of the same term then the fuzzy proximity is defined as the fuzzy proximity to the closest term occurrence (that is, *max(fp)* with respect to that term). If the document contains multiple search terms then the fuzzy proximity is defined as the minimum fuzzy proximity to all search terms.

The fuzzy score of an element in a document is computed as the sum of fuzzy proximity scores for each term in the element, normalized by the length of the element. However, as the documents are hierarchically structured, if a search term occurs in the title of a section then the fuzzy proximity of a term in the element to the search term in the title is defined as 1.

2.4 Small Improvements

Beigbeder's algorithm treats all terms as equal whereas it is usual for scoring algorithms to weight terms differently. The algorithm is thus extended to include some aspect of the strength of a search term (IDF). The IDF weighted fuzzy proximity, *fp'* is given by

$$fp' = IDF * max\left(\frac{k - (p - t)}{k}, 0\right)$$

the variant of IDF chosen is

$$IDF = \frac{N - n + 0.5}{n + 0.5}$$

where N is the number of documents in the collection and n is the number of documents in which the term occurs. We set k=200.

Problematically, if a search term is missing from the document then the fuzzy proximity to that term is always zero and so no part of the document is considered relevant (due to the *min()* function). Using the sum of fuzzy proximity weights in place of the minimum overcomes this problem.

The Beigbeder algorithms is of general interest as it is a method of identifying relevant elements as a function of term proximity, and can be extended to identify relevant passages. A comparison of the original Beigbeder algorithm and the Otago variant; as well as to the Otago passage runs will help answer the question of whether passages or elements are the best result to the Relevance-in-Context task.

2.5 Documents

At INEX 2007 an RMIT University ad hoc submission demonstrated that a full-document run could be more effective at focused retrieval than a focused run [5]. This is because the F measure of recall and precision pre-selects choosing whole documents as 100% recall within a document can be easily realized. Whole document runs were, therefore, submitted for comparison to the focused retrieval runs.

2.6 Otago Ad Hoc 2008 Runs and Results

Three runs were submitted to the Relevance-in-Context passage task. In all cases documents were identified using BM25 (k_1=1.2, k_3=7.0, b=0.75) and then one passage was identified for each document in the top 1500 documents. The rank order of the final results was BM25. Stemming was not used.

WHOLEDOC_PASSAGE: The whole document was returned as the passage.

DYLAN_200: A fixed sized window of 200 words was placed on the centroid of the search terms within the document. The standard deviation method was used to compute the centroid.

SW_KL_200: The Kullback-Leibler method with a sliding window of 200 words was used to identify a relevant passage.

Three runs were submitted to the Relevance-in-Context element task, BM25 was used to identify the top 1500 documents, one element was identified, and the results re-ranked based on the Beigbeder score. For these experiments k=200.

WHOLEDOC: The whole document was returned as an element (this run is identical to WHOLEDOC_PASSAGE and was submitted as a sanity check).

BEIGBEDER_ORIG: Elements were scored using Beigbeder's algorithm.

BEIGBEDER_IDF: Elements were scored using the IDF weighed version of Beigbeder's algorithm. Due to a bug in our code we actually implemented the product of the sum of the IDF and fp scores in place of the sum of the product.

2.7 Wikisearching Results

The results are presented in Table 1 where it can be seen that WHOLEDOC and WHOLEDOC_PASSAGE do, indeed, score the same thus passing the sanity check.

Table 1. Ad hoc Relevance-in-Contest task results

Run	Type	MAgP
WHOLEDOC_PASSAGE	Passage	0.192
WHOLEDOC	Element	0.192
SW_KL_200	Passage	0.183
DYLAN_200	Passage	0.182
BEIGBEDER_IDF	Element	0.149
BEIGBEDER_ORIG	Element	0.107

The passage algorithms are superior to the element algorithms with the Kullback-Leibler approach bettering the Otago 2007 approach by a very small amount. The IDF enhancement to Beigbeder's algorithm increases the precision substantially, but not sufficiently to better the passage runs.

3 Wikilinking

The Link-the-Wiki task, first included in INEX in 2007, requires participants to automatically identify hypertext links between documents in the Wikipedia. The user model is that of a user who creates a new Wikipedia entry and would like to link that entry to pre-existing entries in the Wikipedia (and *vice versa*).

The production of a new article can be simulated by taking an existing Wikipedia document and removing all trace of it from the collection. Link identification software can then be applied to the collection and the orphaned document. A comparison of the automatically generated links to the original collection gives some measure of the quality of the link detection system – that is, the original links are considered to be the gold-standard by which systems are compared.

Exactly this approach was taken in the INEX 2007 Link-the-Wiki track, and was used again for document-to-document linking in 2008. In 2008, 6600 documents (about 1% of the document collection) were randomly selected and orphaned for document-to-document link detection.

New in 2008 is the anchor-to-BEP linking task, in which the task is to identify the best orphan anchor from which to link from and the best-entry-point (BEP) in the target document from which to link to. Unlike document-to-document linking, anchor-to-BEP linking requires manual assessment because the Wikipedia documents are typically not *a priori* marked-up in this way. For 2008, 50 anchor-to-BEP documents were suggested by task participants and were orphaned for the experiment. A limit of 50 anchors per document was imposed (for practical reasons) and at most each anchor could link to 5 locations in the Wikipedia.

Two separate problems exist with identifying links, the identification of outgoing links (from the orphan to the collection) and the identification of incoming links (from the collection to the document).

3.1 Outgoing Links

Although the Otago runs in 2007 were adequate, those of Itakura & Clarke [6] were substantially better – effort was, therefore, spent investigating methods of improving their technique. It should be noted that the Itakura & Clarke algorithm relies on a pre-existing heavily interlinked document collection (such as the Wikipedia). In the case where no prior links exist in the collection the techniques of Geva [7] which were also successful in INEX 2007 can be used.

3.1.1 The Itakura and Clarke Algorithm

The Itakura & Clarke algorithm relies entirely on pre-existing links between documents within the document collection. Of the link types available in the collection, only the <collectionlink> type is utilized because the other link types do not link between two documents in the collection (for example, a <wikipedialink> links from a document in the collection to a document in the Wikipedia that is not in the INEX collection).

Initially a list of all the links within the document collection is created. This is generated by parsing each document in the collection and extracting the anchor text of the link and the target document id.

Next from the output of the previous stage, a list of target documents is created for each unique anchor text in the collection. For a given anchor text in the collection, the most frequent target is most likely to be the correct target.

For each anchor text / target pair a strength value (γ) is constructed

$$\gamma = \frac{np}{af}$$

where *np* is the number of documents that link from the anchor to the target and *af* is the number of documents in which the anchor text occurs.

An orphaned document is then parsed and the first location of each anchor in the pre-generated list is located. For overlapping anchors (for example, "Lennon" and "John Lennon") the longest possible anchor is chosen as a longer anchor is more likely to be correct than a short anchor. A limit of 250 anchors per document was enforced by the Link-the-Wiki track definition.

3.1.2 Small Improvements

After implementing the Itakura & Clarke algorithm verbatim a small number of improvements were identified.

The algorithm defines the anchor text as all text occurring between the tags, converted to lowercase, and including punctuation. Anchor texts often contain punctuation at the end thus creating a distinction between "John Lennon" and "John Lennon.". We stripped punctuation from the anchors thus conflating these two cases.

Anchor texts beginning at the start of a sentence are capitalized for grammatical reasons so the algorithm converts the text into lower case. Unfortunately this results in the loss of the distinction between "unfinished music" and "Unfinished Music" (the two part experimental work by John Lennon and Yoko Ono). Geva [7] identifies the importance of case in link detection so the case conversion step was dropped.

Finally, over-weighting γ for capitalized terms in the orphan will help identify proper noun conflicts (such as Unfinished Music). A capitalization constant, Π, is added to γ where terms in the orphan were found capitalized.

Figure 1 compares the improvements to the original algorithm using the INEX 2007 Link-the-Wiki topics. The line labeled "Waterloo" is the Itakura & Clarke run as submitted. Removing punctuation (Alphanumeric) from the anchor list improves the algorithm, removing case folding (Case Sensitive) leads to further improvements. Weighting (Weighed) includes punctuation removal, case sensitivity, and weighted γ, and was the best experimental run on the 2007 orphans.

Figure 2 shows the effect of Π on precision, a value of 0.3 is best for early precision, but a value of 0.1 holds the precision longer resulting in the highest mean average precision.

3.1.3 Best Entry Points

Several studies have shown the best entry point for Wikipedia documents is the start of the document. [1, 8]. No further investigation was performed on BEPs.

3.1.4 Multiple Targets

The Link-the-Wiki task specification for 2008 allowed at most 5 targets for each anchor point. The Itakura & Clarke algorithm was, consequently, extended so that the γ value was computed for not just the most common target, but also for all targets of an anchor text. The γ values represent the probability of the target document being the correct target; consequently choosing the top five documents (by γ) for each anchor text satisfies the track requirements.

3.2 Incoming Links

The best Otago run at INEX 2007 achieved an excellent early precision (P@5) score of 0.751. The experiments described in this section were conducted in an effort to improve the overall performance (MAP) and were conducted on the 2007 Link-the-Wiki oprhans.

3.2.1 The Otago 2007 Algorithm

The algorithm for detecting incoming links relies on a simple theme extraction technique used to identify the semantic content of the document.

For each unique term (excluding stop words) in the orphaned document the Otago 2007 algorithm [1] computes the actual frequency of that term, af

$$af = \frac{tf}{dl}$$

where tf is the number of occurrences of the term in the orphan and dl is the length of the orphan (in terms); to the expected frequency, ef

$$ef = \frac{cf}{df * ml}$$

where cf is the number of occurrences of the term in the collection, df is the number of documents containing the term and ml is the mean length of a document. Ranking the terms in the orphan by ratio of af to ef (st),

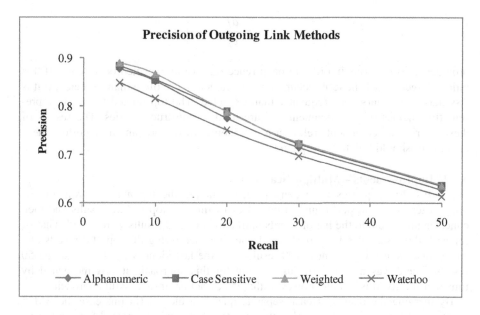

Fig. 1. Small improvements on the Itakura & Clarke algoritm (Waterloo) are seen when punctuation is removed (Alphanumeric), when case folding is removed (Case Sensitive) and when uppercase anchors are preferred over lowercase anchors (Weighted)

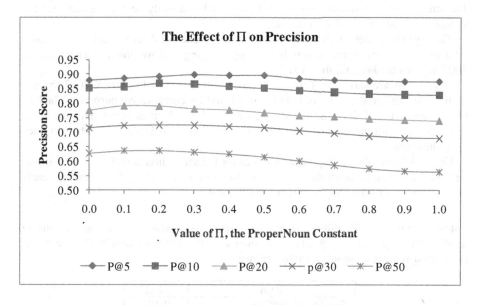

Fig. 2. Effect of varying Π on the precision. Small value of Π (0.3) is best for early precision but a very small score (0.1) holds the precision higher longer (best for MAP).

$$st = \frac{af}{ef}$$

provides a list of terms in order of occurrence relative to expected occurrence. If this ratio is larger than 1 the term occurs in the document more often than expected, if it is less than 1 it occurs less frequently than expected. The top ranked terms are representative themes of the document and are used to construct queries. The results of these queries are documents relevant to the themes of the orphan and therefore these documents should link to the orphan.

3.2.2 Improvements – Multiple Searches

For INEX 2007, queries were constructed by taking the top n terms from the st-ordered term-list and performing a query, extracting the top $n * 50$ results and then concatenating them to the list of results until a total of 250 results were found. That is, for $n=2$, three searches were performed, the first identifying the top 100 results and the second identifying the next 100 results, and the last identifying the remaining 50 results. There was no theoretic justification for this approach; it was motivated by time constraints. It is of note, however, that it was not an unsuccessful approach.

By merging the results of each separate query on the rsv (in this case the BM25 score) good targets that match other than the top theme will be placed high in the results list. This approach might also place documents that are good matches for non-key themes high in the results list because of a high rsv with respect to a non-key term.

To alleviate this problem the BM25 score for each search term can be weighted. The strength of a term with respect to the orphan has already been computed (st) and so that is a reasonable value to choose.

The best Otago run at INEX 2007 used two searches of 4 terms each, producing a total of 250 results in the results list. Using merging and weighted merging on the 2007 orphans the best number was 2.

The results are shown in Table 2. The best runs submitted to INEX 2007 (by any participant) achieved a score of 0.484 and is listed for comparative purposes. The best Otago run at INEX 2007 achieved a score of 0.339 which is better than the score achieved by result merging (0.319) but not as good as the 0.350 achieved by weighted result merging.

Figure 3 shows the early precision scores for the same three techniques. Of particular interest is that although the MAP score for weighted merging is highest, the early precision scores of the Otago 2007 run are highest.

Table 2. MAP scores for different approaches to multiple searches. The weighted merging of queries containing 2 terms each achieved a better score than the best Otago 2007 run, however not as good as the best run submitted by any institute.

Run	MAP
Top INEX 2007 run	0.484
Weighted merge	0.350
Otago 2007	0.339
Merged	0.319

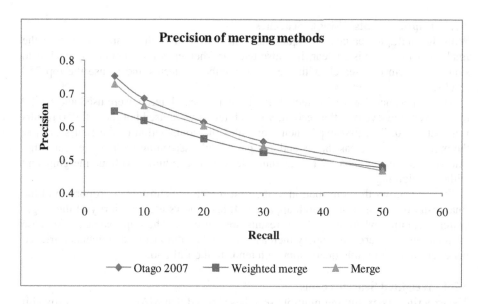

Fig. 3. Early precision scores for the three merging techniques. Although the MAP of weighted merge is highest, the early precision of Otago 2007 is highest.

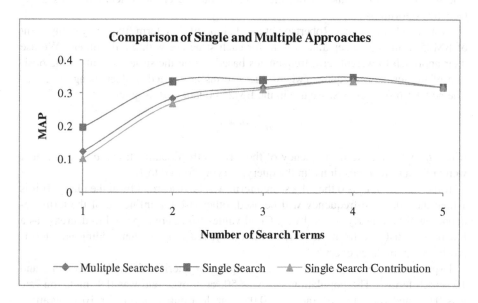

Fig. 4. A comparison of the multiple search technique to the single search technique suggests that the single search technique is best

3.2.3 Improvements – Single Searches

With the multiple search technique the contribution of each separate search to the final precision score is unclear. It is also unclear whether or not a better approach is to simply perform one search with the given number of terms and to use the top 250 results.

Two experiments were conducted: in the first, n search terms were used and $n * 50$ results were retrieved; in the second, n search terms were used but the full 250 results were retrieved. The first experiment computes the contribution of the first search to the multi-search whereas the second compares multi-searching with single-searching. The results were compared to the multiple search technique without merging and without weighting.

Figure 4 shows the contribution of the first search is a substantial proportion of the final result of the multiple search approach. It also shows the superiority of the single search technique when the full 250 results are retrieved. The improvements decrease as the number of terms per query increases to 5 as the number of documents retrieved per query in the multiple query approach tends to the full 250.

3.2.4 Weighted Search Terms

The experiments examining multiple searches showed that MAP could be improved if the search terms were weighted by st. Improvements are therefore expected in the single search approach if the individual search terms in a single query are weighted. The weights could be taken from the st score, but we chose to learn weights using Genetic Algorithms [9].

Trotman [10] and later Robertson et al. [11] modify the term frequency component of BM25 to include a separate weight for each structure within a document. We use their approach to weight term frequencies based not on the structure, but on the position of the term in the query (where query terms are sorted in decreasing st score). The new term frequency score use in the BM25 equation, tf, is given by

$$tf = tft * c_q$$

where tft is the true term frequency of the term in the document; and c_q is a constant weight for a term at position q in the query, varying from 0 to 1.

If the weight of c_q is 0 then the search term will be discarded from the query. If it is 1 then the true term frequency will be used, otherwise the influence of the term frequency will be linearly scaled by c_q. Good values for c_q are expected to decrease as a function of distance from the start of the query, reaching 0 when adding new terms creates an ambiguous query.

Experiments were conducted to learn weights for queries of lengths between 2 and 10 search terms[1]. The population size was 50, crossover rate was 0.9, mutation rate was 0.05, and reproduction rate was 0.05. The learning was run for 10 generation. Elitism was used. Many iterations of the learning were conducted and the best weights of the best run were recorded.

---

[1] In the case of a single search term the weight has a scaling effect which does not affect the relative rank order of the results; and so has no effect on MAP.

For the best MAP score achieved for queries ranging from 2 to 8 search terms, Table 3 shows the weights that were learned. It can be seen that the first two terms are responsible for the majority of the performance.

Figure 5 shows that weighting search terms results in an increase in precision for all tested cases (with the exception of a single search term). It should be noted that the experiments over-fit the weights to the orphan documents; unfortunately there is an insufficient number of orphans (in the 2007 set) to conduct a traditional learn / validate / evaluate experiment.

Table 3. Best learned weights for different queriy lengths

Search Terms	Weights (from first to last term)
2	0.96, 0.95
3	0.99, 0.96, 0.04
4	0.97, 0.73, 0.05, 0.06
5	0.95, 0.83, 0.14, 0.1, 0.01
6	0.89, 0.97, 0.44, 0.41, 0, 0.06
7	0.8, 0.95, 0.75, 0.29, 0, 0.07, 0.25
8	1, 0.88, 0.14, 0.05, 0, 0.22, 0.08, 0.19

Table 4. MAP scores of the runs using terms from different parts of the document

Run	MAP
Title	0.410
Overview	0.143
Document	0.080
Otago 2007	0.339
Weighted merge	0.350

3.2.5 Other Sources of Search Terms

The experiments thus far suggest that the best approach is to perform a single search using a small number (two or three) highly representative search terms to identify document that should point to the orphan. The approach to identifying terms involved identifying document themes by simple text processing techniques. Wikipedia documents, however, are structured and include a title as well as a brief overview of the content of the document. These document structures might be used as a method of identifying good representative document-thematic terms, or the whole document (as seen by others [12]) might be used.

The title of the Wikipedia document is held between <name> tags. These were processed to remove duplicate search terms and stop words, and then used as queries.

The overview of the Wikipedia document occurs as an untitled section before the first titled section. It was extracted by using all text before the first <title> tag of the document, stop words and duplicate terms removed and used as the query.

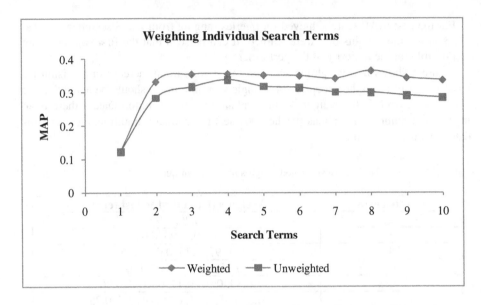

Fig. 5. Effect of weighting individual search terms in the query

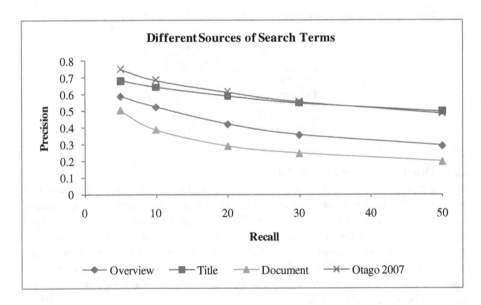

Fig. 6. Different sources of search terms. The title is a more effective source of terms than the overview which is better than the whole document. For early precision the best source was the approach used by Otago at INEX 2007.

The full-text of the Wikipedia document can easily by extracted by removing all XML tags from the document, removing stop and duplicate words, and used as the query.

Figure 6 shows the effect on early recall of the different techniques. Selecting terms from the whole document is better than using the title which is better than the overview which in turn is better than the whole document. However, the result is somewhat different when the MAP scores are compared; Table 4 presents the MAP scores and it can be seen that using the title is better overall than the other approaches, even bettering the weighted merge approach from above.

3.3 Otago Link-the-Wiki 2008 Runs

3.3.1 File-to-File Linking

Three runs were submitted, each used BM25 (k_1=0.421, k_3=242.61, b=0.498)

- capConstant-SingleSearchWeighted: outgoing links identified using the Otago version of Itakura & Clarke with Π = 0.1. Incoming links identified using the weighted merge method with 4 terms and weights of 0.97, 0.73. 0.05 & 0.06.
- capConstant-TitleOnly: outgoing links identified using Otago Itakura & Clarke with Π = 0.1. Incoming links we identified using the orphan title.
- nonCap-FirstPara: outgoing links were identified using Otago Itakura & Clarke without Π. Incoming links were identified using the outline of the orphan.

3.3.2 Anchor-to-BEP Linking

- capConstant-SingleSearch-A2B: same as capConstant-SingleSearchWeighted.
- capConstant-TitleOnly-A2B: same as capConstant-TitleOnly.
- nCapConstant-WholeDocument-A2B: same as nonCap-FirstPara, but using the whole document for the query.

3.4 Wikilinking Results

Noncap-FirstPara performed exceptionally well (1st place of 21) in file-to-file automatic assessment, scoring an outgoing MAP of 0.7343. The next best run was from Amsterdam with a MAP of 0.3475. This result raised questions about the validity of the implementation. A cold-room re-implementation resulted in MAP scores very similar to those reported. Investigation as to why the other Otago runs did not perform as well in file-to-file ranking suggests that case-sensitivity can have a catastrophic effect; if the index is built case-sensitive and the orphan is parsed case-insensitive then a substantial number of links can be missed and the performance degrades substantially. For incoming links, run capConstant-TitleOnly placed 9th (of 4) with a MAP of 0.4870. Little conclusion can be drawn from an ineffective run.

In anchor-to-BEP outgoing linking, run capConstant-SingleSearch-A2B placed 6th (of 28, MAP=0.3910) and capConstant-TitleOnly-A2B placed 7th with the same MAP (to 4 decimal places). Of particular note, the highly effective algorithm for file-to-file linking was not so for anchor-to-BEP. This appears to be because many links in the Wikipedia documents are assessed as not-relevant by the assessor. Preliminary experiments suggest that identifying year-links and not including them in a manual run will improve performance.

4 Conclusions

Experiments were conducted to gain insights into effective method of searching for the Relevance-in-Context task. In passage retrieval the Otago 2007 algorithm was compared to the Kullback-Leibler model, and virtually no difference was seen in the performance in the 2008 topics. This suggests the simpler Otago algorithm may be an effective alternative algorithm, especially when efficiency is an issue. In element retrieval the Beigbeder algorithm was compared to an IDF weighted variant and substantial improvements were seen on the 2008 topics – suggesting there is further room for improvement on Beigbeder's work.

In the Link-the-Wiki task the Itakura & Clarke algorithm was used for outgoing links. It was extended by removing punctuation from the anchors, and adding case sensitivity weighting. Results show that a careful implementation of this algorithm can produce near-perfect results in file-to-file linking, but not for manual assessment. In further work we will investigate methods to improve the algorithm's performance when assessed by a manual assessor. Our algorithms were not effective for outgoing links (placing 9th of 24); we will be investigating incoming links in further work.

Acknowledgements

Funded in part by a University of Otago Research Grant.

References

1. Jenkinson, D., Trotman, A.: Wikipedia Ad Hoc Passage Retrieval and Wikipedia Document Linking. In: Fuhr, N., Kamps, J., Lalmas, M., Trotman, A. (eds.) INEX 2007. LNCS, vol. 4862, pp. 426–439. Springer, Heidelberg (2008)
2. Huang, W., Trotman, A., O'Keefe, R.A.: Element Retrieval Using a Passage Retrieval Approach. Australian Journal of Intelligent Information Processing Systems 9, 80–83 (2006)
3. Itakura, K., Clarke, C.: From Passages into Elements in XML Retrieval. In: Trotman, A., Geva, S., Kamps, J. (eds.) SIGIR 2007 Workshop on Focused Retrieval, pp. 17–22 (2007)
4. Beigbeder, M.: ENSM-SE at INEX 2007: Scoring with proximity. In: Preproceedings of INEX 2007, pp. 53–55 (2007)
5. Fuhr, N., Kamps, J., Lalmas, M., Malik, S., Trotman, A.: Overview of the INEX 2007 Ad Hoc Track. In: Fuhr, N., Kamps, J., Lalmas, M., Trotman, A. (eds.) INEX 2007. LNCS, vol. 4862, pp. 1–23. Springer, Heidelberg (2008)
6. Itakura, K.Y., Clarke, C.L.: University of Waterloo at INEX2007: Adhoc and Link-the-Wiki Tracks. In: Fuhr, N., Kamps, J., Lalmas, M., Trotman, A. (eds.) INEX 2007. LNCS, vol. 4862, pp. 417–425. Springer, Heidelberg (2008)
7. Geva, S.: GPX: Ad-Hoc Queries and Automated Link Discovery in the Wikipedia. In: Fuhr, N., Kamps, J., Lalmas, M., Trotman, A. (eds.) INEX 2007. LNCS, vol. 4862, pp. 404–416. Springer, Heidelberg (2008)
8. Kamps, J., Koolen, M., Lalmas, M.: Where to Start Reading a Textual XML Document? In: 30th SIGIR (2007)
9. Holland, J.H.: Adaptatio. In: Natural and Artificial Systems. Univ. Michigan Press (1975)
10. Trotman, A.: Choosing Document Structure Weights. IP&M 41, 243–264 (2005)
11. Robertson, S., Zaragoza, H., Taylor, M.: Simple BM25 extension to multiple weighted fields. In: 13th CIKM, pp. 42–49 (2004)
12. Fachry, K.N., Kamps, J., Koolen, M., Zhang, J.: Using and Detecting Links in Wikipedia. In: Fuhr, N., Kamps, J., Lalmas, M., Trotman, A. (eds.) INEX 2007. LNCS, vol. 4862, pp. 388–403. Springer, Heidelberg (2008)

CSIR at INEX 2008 Link-the-Wiki Track

Wei Lu, Dan Liu, and Zhenzhen Fu

Center for Studies of Information Resources,
School of Information Management, Wuhan University, P.R. China
{reedwhu,DanLiu.whu,zhenzhenfu}@gmail.com

Abstract. In this paper, we describe methods taken by CSIR in the INEX 2008 Link-the-Wiki track. For the incoming link detection, we use $p(d|t)$, the probability to generate a document, when given the topic file, to judge which documents are proper link sources for the given topic. For the file-to-file task of outgoing link detection, we take a two-step approach: first, we identify a group of candidate target documents by literally matching the topic file title and document content; then, candidate documents are ranked by the number of incoming links. For the anchor-to-BEP task, we use $p(d|a,t)$, the probability to generate a document, when given the topic file and an anchor name, to select anchors and link targets for a given topic.

1 Introduction

Link discovery between documents aims at discovering the potential links between given documents by analyzing the texts, which will be very helpful for the access of relevant document when acquiring information.

INEX introduced link discovery into the Wikipedia corpus as the Link-the-Wiki track in 2007. The task is to recommend incoming and outgoing links for a given Wikipedia document automatically.

There are two separate tasks in INEX 2008, i.e. file-to-file task and anchor-to-BEP task, both of which have to find incoming and outgoing links. For the file-to-file task, links are at file level. For the anchor-to-BEP task, anchor texts in the source document and the destination offsets of links in the target document have to be specified.

This is the first year for the Center for Studies of Information Resources to participate in the INEX Link-the-Wiki track. We have submitted 4 runs, 2 for the file-to-file task and 2 for the anchor-to-BEP task. For each task, the two submitted runs differ in the approaches of incoming link detection.

For the incoming link detection, we rank candidate documents by $p(d|t)$, the probability to generate a document, when given the topic file. The same incoming link detection method is used in the two sub tasks.

For the outgoing link detection, in the file-to-file task, we firstly identify all the candidate target documents by matching the content of the topic file with all the document titles of the Wikipedia collection, and then rank the candidate documents by the number of incoming links. Finally, we output the top 250 anchors as results. In the anchor-to-BEP task, we rank the candidate documents by $p(d|a,t)$, the probability

S. Geva, J. Kamps, and A. Trotman (Eds.): INEX 2008, LNCS 5631, pp. 389–394, 2009.

to generate a document, when given the topic file and an anchor name. The anchors are selected by a window-based technique. The top 50 anchors with the highest $p(d|a,t)$ score are outputted as final results.

The rest of this paper is structured as follows: in section 2, we briefly review related works; in section 3, we explain our methods; section 4 evaluates our methods; in section 5, we draw a conclusion.

2   Related Work

In INEX 2007's Link-the-Wiki track, 4 participants were involved in [1]. Jenkinson et al. [2] identified over-represented terms in topic files, and used BM25 to identify potential relevant documents in order to create link between topic file and relevant documents. Itakura et al. [3] firstly created a list of outgoing links specified by an anchor and the destination file for each topic and then selected the most frequent target file for each anchor. Fachry et al. [4] firstly retrieved the top 100 similar documents, and then detected incoming links by line-matching for every line of each of the 100 documents with all lines of the topic file, outgoing links by iterating over all lines of the topic file and matching the lines top-down with the 100 documents. Geva et al. [5] identified incoming links by searching elements that were about the topic name element, and identified outgoing links by a systematic search for anchor text in topic file that matches existing document names.

Besides, Huang et al. improved their approaches used in the outgoing link found task, they reordered candidate anchors by some assumed principles. For example, they take it as a principle that numbers and single terms have less probability as an anchor [6]. Zhang et al. focused on outgoing links and investigated link density, and especially repeated occurrences of links with the same anchor text and destination. They used link density/anchor distance and repeated candidate links to assist link discovery [7].

3   Approaches

3.1   File-to-File Task

The file-to-file task needs to find out 250 outgoings and 50 incomings for each topic. For this task, we submitted 2 runs: LTW_F2F_1 and LTW_F2F_2. Our submitted two runs use different approaches in finding incomings, and adopt the same method in finding outgoings.

3.1.1   Outgoing

Our method of finding outgoings in the file-to-file task is based on the following assumption: outgoings are generated in a document which contains terms of the topic title. So, we suppose that if a title of a document A is in the current topic file B, there should be an outgoing link from B to A. Additionally, not only the title of the document but also its different variants is used to match. We consider names of anchors which link to a document A as variants of the title of A, and extract these names in the experiment's pre-processing stage.

After finding a group of candidate outgoings, we rank them by their probability as an outgoing. Those who have more incoming links are considered to have more possibility as an outgoing and are treated as outgoing links for a given topic. In the preprocess stage, the number of incoming links of each document has been counted. The possibility of a document d as an outgoing is defined as Eq. (1):

$$Possibility\ as\ an\ outgoing = \frac{Number\ of\ incomings\ of\ d}{Max\ incomings\ of\ all\ docs} \qquad (1)$$

3.1.2 Incoming

We assume that if two documents are about similar themes, they can link to each other. So, we use $p(d|t)$, the probability of a topic file generating a document, to judge whether a document is an incoming link.

$$p(d|t) = \begin{cases} p(content|title) & \text{Language model} \\ p(content|main\ content\ of\ topic\ file) & \text{VSM} \end{cases}$$

We use two different approaches to estimate $p(d|t)$. For the first run we use language model, and for the second run we use the default vector space model of Lucene.

We define the main content of the topic file as sequence of important words in the file. In our runs, we simply generate the main content of the topic by the following steps: replace some symbols by blank, such as "\","&amp" and so on, delete words less than 5 letters for efficiency consideration, and the remaining terms constitute the main content.

3.2 Anchors to BEP Task

The anchor-to-BEP task needs to find out 50 anchors and 5 target files per anchor, totally 250 outgoings, and 50 incomings.

We submitted 2 runs: LTW_A2BEP_1 and LTW_A2BEP_2. The two runs differ from each other in that their incomings are generated by different approaches.

3.2.1 Outgoing

We assume that an anchor can be linked to a document if the anchor and the topic file are both relevant to the document. Then, we can rank the documents by $p(d|a,t)$, the probability of a specified anchor and topic file to generate the document.

Assuming t and a are independent, $p(d|a,t)$ can be transformed into Eq. (2):

$$p(d \mid a,t) = \frac{p(da,dt))}{p(a,t)} = \frac{p(da)p(dt)}{p(a)p(t)} = p(d \mid a)p(d \mid t) \qquad (2)$$

Firstly, we choose those candidate anchors. After replace some symbols by blank, such as "\", "&amp" and so on, and delete words less than 3 letters for efficiency consideration, we use a window-based method to determine the candidate anchors. The size of the windows differs from 1 to 7. For example, for a sentence in topic file 18845.xml:

A stay mouse is a part of the standing rigging...

stay mouse, stay mouse part, stay mouse part the, stay mouse part the standing, stay mouse part the standing rigging, mouse, mouse part are all valid candidate anchors.

Secondly, for every candidate anchor, we calculate $p(d|a)$. This process is as follows: (1) Determine whether there are documents which are named as the candidate anchor. If the candidate anchor is exactly the same as the topic or one of the incoming anchor names of the document, we assign $p(d|a)$ to 1. (2) Use the candidate anchor as query and search for the query in the "incoming anchor names" field of the index. In this step, we output a maximum of 50 documents for the next stage. (3) Use the candidate anchor as query and search for the query in the "content" field. Finally, we output a maximum of 50 documents for the next stage.

Thirdly, we calculate $p(d|t)$ and $p(d|a,t)$. In this process, we use the main content of the topic file as query and search for the query in the "content and incoming anchor names" field. Then, we output the top 1000 documents to estimate $p(d|a,t)$.

Fourthly, we rank all the anchors by the score of the candidate anchors, choose the top 50 anchors. The score of a candidate anchor is calculated by the highest target files' score of the anchor in our runs. For anchors with the same offset, which means that they are overlapped, we choose either the one with highest score or longest length if they have the same score.

At last, the BEP of the anchor in the target file is assigned by the index of the anchor in the target file or 0 if the index is -1.

3.2.2 Incoming

The approach used in finding incoming is the same as what we use in finding incoming for the file-to-file task. We set all the BEP of the topic file as 0. For the offset and length of the link anchor, we simply calculate the index of the topic title in the target link file and its length.

4 Result and Discussion

Table 1 gives out the evaluation results of our submitted runs. For the file-to-file task, results are evaluated based on the original links that exist in the Wikipedia articles. For the anchor-to-BEP task, automatic assessment of file-to-file level is performed for

Table 1. The evaluation result of link-the-Wiki track

Run ID	Link Type	Eval Type	MAP	R-Prec	P@5
CSIR_LTW_F2F_1	In	Automatic	0.16456	0.18356	0.2580.
CSIR_LTW_F2F_2	**In**	**Automatic**	**0.29403**	**0.33810**	**0.60348**
CSIR_LTW_F2F_2	Out	Automatic	0.00820	0.02116	0.02967
CSIR_LTW_A2BEP_1	In	Automatic	0.12454	--	--
CSIR_LTW_A2BEP_2	**In**	**Automatic**	**0.15773**	**0.19124**	**0.63600**
		Automatic	0.06468	0.14368	0.32800
		Manual at F2F	**0.07595**	**0.15473**	**0.31200**
CSIR_LTW_A2BEP_2	Out	Manual at F2BEP	0.04780	0.10460	0.25240
		Manual at A2F	**0.13070**	**0.10081**	**0.19600**
		Manual at A2BEP	0.12370	0.09610	0.18316

both incoming and outgoing finding, while manual assessment of file-to-file and file-to-BEP level and manual assessment of anchor-to-file and anchor-to-BEP level are performed for outgoing finding. The manual assessments are based on those topics assessed as relevant by a human assessor.

Our observations from the results are as follows:

For finding incomings, using the main content of the topic file as query is better than just using the title of the topic file either for file-to-file or anchor-to-BEP task.

For finding outgoings of the file-to-file task, we have ignored the fact that those document which have lots of incomings are mostly those about countries, places, years or lists. Moreover, learning from the statistical result of table 1, we can find that they take account for only a small percentage of the Wikipedia corpus. Hence, our assumption that those who have lots of incomings have a greater possibility as outgoing links for a given topic is in some sense unreasonable.

For finding outgoings of the anchor-to-BEP task, file-to-file level and anchor-to-file assessment are better than file-to-BEP and anchor-to-BEP level assessment respectively, which means our BEP assigning approach needs to be improved.

5 Conclusion and Future Work

In this year's Link-the-Wiki track, we submitted 2 runs for each of the sub-task. For the incoming link detection, we expressed it as $p(d|t)$, the probability to generate a document, when given the topic file. For the outgoing links of the file-to-file task, we take a two step approach—first identifying a cluster of candidate target documents by doing string matching of topic file title and document content, and then ordering these documents by the number of incoming links they already have, and for the outgoing links of the anchor-to-BEP task, we expressed it as $p(d|a,t)$, the probability to generate a document, when given the topic file and an anchor name, to select anchors and link targets for a given topic.

From the official evaluation result, we find that there are still some problems in our approaches, and the way we processing the topic file needs further consideration. As Link-the-Wiki can be deemed as a specific application of link discovery in the Wikipedia corpus, we would try to adopt more automatic link generation methods in next year's Link-the-Wiki track.

References

1. Huang, D.W.C., Xu, Y., Trotman, A.: Overview of INEX 2007 Link the Wiki Track. In: Fuhr, N., Kamps, J., Lalmas, M., Trotman, A. (eds.) INEX 2007. LNCS, vol. 4862, pp. 373–387. Springer, Heidelberg (2008)
2. Jenkinson, D., Trotman, A.: Wikipedia Ad Hoc Passage Retrieval and Wikipedia Document Linking. In: Fuhr, N., Kamps, J., Lalmas, M., Trotman, A. (eds.) INEX 2007. LNCS, vol. 4862, pp. 426–439. Springer, Heidelberg (2008)
3. Itakura, K.Y., Clarke, C.L.A.: University of Waterloo at INEX2007:Ad Hoc and Link-the-Wiki Tracks. In: Fuhr, N., Kamps, J., Lalmas, M., Trotman, A. (eds.) INEX 2007. LNCS, vol. 4862, pp. 417–425. Springer, Heidelberg (2008)

4. Fachry, K.N., Kamps, J., Koolen, M., Zhang, J.: The University of Amsterdam at INEX 2007. In: Focused Access to XML Documents, 6th International Workshop of the Initiative for the Evaluation of XML Retrieval, INEX 2007, Dagstuhl Castle, Germany, December 17-19, pp. 388–402 (2007)
5. Geva, S.: GPX@INEX 2007:Ad-Hoc Queries and Automated Link Discovery in the Wikipedia. In: Fuhr, N., Kamps, J., Lalmas, M., Trotman, A. (eds.) INEX 2007. LNCS, vol. 4862, pp. 404–416. Springer, Heidelberg (2008)
6. Huang, D.W.C., Trotman, A., Geva, S.: Experiments and Evaluation of link discovery in the wikipedia. In: Proceedings of SIGIR workshop on Focused Retrieval (2008)
7. Zhang, J., Kamps, J.: Link Detection in XML Documents: What about repeated links? In: Proceedings of SIGIR workshop on Focused Retrieval (2008)

A Content-Based Link Detection Approach Using the Vector Space Model

Junte Zhang[1] and Jaap Kamps[1,2]

[1] Archives and Information Studies, Faculty of Humanities, University of Amsterdam
[2] ISLA, Faculty of Science, University of Amsterdam

Abstract. Link detection can be seen as a special application of Focused Retrieval. This paper presents a content-based link detection approach using the Vector Space Model. We present our results, and conclude by discussing the merits and deficiencies of our approach.

1 Introduction

This paper reports on our participation in the Link The Wiki (LTW) track of INEX. LTW is aimed at detecting or discovering missing links between a set of Wikipedia topics, and the remainder of the collection, hence effectively establishing cross-links between those documents using IR techniques. Existing links were removed from the topics, making these documents 'orphans' that could be linked to potential 'fosters'. This means that hypertext has be constructed automatically. Many hypertext systems have been based on the *Dexter Hypertext Reference Model* [2], and subsequently our system as outlined in this paper is also compliant with this model.

LTW consisted of two tasks. The first task was a continuation of the track of last year with the detection of links between whole files. The second task used 50 selected orphan topics, and went further than link detection on the document level, as links had to be established between spans of characters within one document and spans of characters with another document. The latter is here a *Best Entry Point* (BEP), i.e. the best point where the user can start reading in a document, which makes link detection particularly a special application of Focused Retrieval. A maximum of 5 BEPs per anchor value was allowed. What both tasks had in common was that it consisted of 2 sub-tasks; the detection of links from an 'orphan' (outgoing) and to an 'orphan' (incoming).

Detected links are treated as uni-directional hyperlink arcs. The issue of link density and link repetition as mentioned in [3] has not been addressed, henceforth we restricted our experimentation to detecting unique cross-links between documents. In Sections 2 and 3, we present our approaches. The results are presented and discussed in Section 4, and we conclude with our findings in Section 5.

2 Detection of Document-to-Document Links

We employ a content-based (and thus collection-independent) approach with IR techniques as previously outlined in [1]. This means we do not rely on learning,

S. Geva, J. Kamps, and A. Trotman (Eds.): INEX 2008, LNCS 5631, pp. 395–400, 2009.

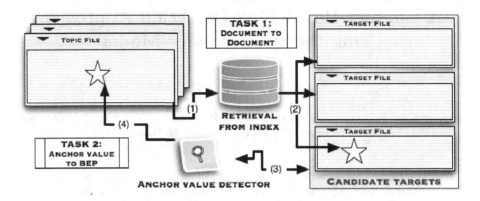

Fig. 1. System overview of a content-based link detection approach

heavy heuristics or existing link structures in the Wikipedia, and *only* use the orphaned topics as evidence. An overview of our system is depicted in Fig. 1.

We adopt a *breadth m–depth n* technique for automatic text structuring for identifying the anchor values and links, i.e. a fixed m number of documents accepted in response to a query (step 1) and a fixed n number of iterative searches (step 2). The similarity on the document level and text segment level (substrings of a line) is used as evidence. We used the whole document (i.e. full-text content) as a query (not only title), because in prior experiments we found that this performed best. The standard Vector Space Model (VSM) implementation of Lucene was used for retrieval, i.e., for a collection D, document d, and query q:

$$sim(q, d) = \sum_{t \in q} \frac{tf_{t,q} \cdot idf_t}{norm_q} \cdot \frac{tf_{t,d} \cdot idf_t}{norm_d} \cdot coord_{q,d} \cdot weight_t, \qquad (1)$$

where $tf_{t,X} = \sqrt{freq(t,X)}$; $idf_t = 1 + \log \frac{|D|}{freq(t,D)}$; $norm_q = \sqrt{\sum_{t \in q} tf_{t,q} \cdot idf_t^2}$; $norm_d = \sqrt{|d|}$; and $coord_{q,d} = \frac{|q \cap d|}{|q|}$.

Before the actual link detection starts, some pre-processing is done by extracting for each topic the title enclosed within the `<name>` tag and storing that in a hash-table for substring matching. We do not apply case-folding, but we do remove any existing disambiguation information put between brackets behind the title. Only titles of > 3 characters length are considered.

We do not assume that links are reciprocal or bi-directional, so we have different approaches for detecting outgoing and incoming links. A threshold of 250 was set for both types of links, and repeated links were not allowed. Links also appear locally within a document to improve the navigation there, but this was outside the scope of the LTW track. So there is (a) an *outgoing link* for an 'orphan' topic when the title of a 'foster' document occurs in the orphan topic. and (b) there is an *incoming link* for an orphan when the title of the orphan occurs in a foster document. We describe the following 2 runs:

a2a_1. The whole orphan document is used as a query. The pool of plausible 'foster' (candidate) documents is the top 300 returned by this query.

a2a_3. The whole orphan document is used as a query. The pool of plausible candidate links is the top 500 of the ranked list.

3   Detection of Anchor-to-BEP Links

The Anchor-to-BEP task is based on a hypertext mechanism called *anchoring* [2]. The actual *anchor value* had to be specified using the File-Offset-Length (FOL) notation, which at the same time serves as the *anchor identifier* [2]. At the same time, the BEP of the outgoing link had to be provided. For all of these runs, we assume that the BEP is always the start of the document (i.e. offset $= 0$). Multiple links per anchor were only computed for the run **a2bep_5**.

a2bep_1. The whole orphan document is used as query, and the top 300 results is used to find potential cross-links.

a2bep_3. The whole orphan document is used as query. The top 50 ranking documents is harvested. Each of these documents is used again as a query to retrieve its top 6 results; resulting in 300 foster documents.

a2bep_5. This run is similar to the first Anchor-to-BEP run, but we expanded this run by allowing more than 1 BEP for each anchor. We use the depth-first strategy, and the broader-narrower conceptualization of terms by re-grouping the extracted list of titles based on a common substring. For example, the anchor value *"Gothic"* could refer to the document "Gothic", but also to documents with the titles *"Gothic alphabet"*, *"Gothic architecture"*, *"Gothic art"*, *"Gothic Chess"*, and so on.

4   Experimental Results and Discussion

Our results are evaluated against the set of existing links (in the un-orphaned version of the topics) as ground truth, both for the sample of 6,600 topics in the first task, as well as the 50 topics in the second task. The results of our runs are depicted in Table 1 for links on the document-level and in Table 2 for the Anchor-to-BEP links. Additionally, the outgoing Anchor-to-BEP links were assessed manually (see Table 3), so there are no results for the incoming links.

Table 1. Document-to-Document runs with Wikipedia as ground truth

Run	Links	MAP	R-Prec	P@10	Rank
a2a_1	In	0.33927	0.35638	0.57082	15/24
a2a_3		0.35758	0.37508	0.58585	14/24
a2a_1	Out	0.10716	0.17695	0.19061	15/21
a2a_3		0.10174	0.16301	0.17073	19/21

Table 2. Anchor-to-BEP runs with Wikipedia as ground truth

Run	Links	MAP	R-Prec	P@10	Rank
a2bep_1		0.23495	0.25408	0.80400	7/27
a2bep_3	In	0.15662	0.16527	0.77400	23/27
a2bep_5		0.23495	0.25408	0.80400	8/27
a2bep_1		0.09727	0.20337	0.27400	20/30
a2bep_3	Out	0.09106	0.18296	0.32800	23/28
a2bep_5		0.14262	0.24614	0.47000	14/30

Table 3. Anchor-to-BEP runs based on manual assessments

Run	Links	MAP	R-Prec	P@10	Rank
Wikipedia		0.20790	0.31258	0.45996	1/28
a2bep_1		0.05557	0.12511	0.14195	23/28
a2bep_3	Out	0.05181	0.13368	0.18699	24/28
a2bep_5		0.08472	0.16822	0.31773	16/28

Table 4. Number of Document-to-Document links

Links	Measure	Qrel	a2a_1	a2a_3
In	Mean	35.66	17.04	19.66
	Median	21	9	9
Out	Mean	36.31	109.09	123.76
	Median	28	95	110

Table 5. Number of Anchor-to-BEP links

Links	Measure	Qrel (auto)	a2bep_1	a2bep_3	a2bep_5
In	Mean	278.32	62.96	23.2	62.96
	Median	134	29.5	17	29.5
Out	Mean	79.18	36.82	25.72	26.02
	Median	62	41	24	26.5

Generally, our approach performed better for detecting incoming links than outgoing ones. We achieved the highest early precision for incoming links detection. Table 3 suggests that the existing links in the Wikipedia do not suffice or is a spurious ground truth given the user assessments, where MAP = 0.27653 for

Document-to-Document links, and MAP = 0.20790 for Anchor-to-BEP links. For example, when we compare the scores of automatic (Table 2) vs manual evaluation (Table 3) of outgoing links, we see that the actual set of detected links is only a small subset of what users really want.

These results, especially the sub-optimal results for the outgoing links and the general results on the document-level, warrant some reflection on several limitations of our approach. We did exact string matching with the titles of the candidate foster topics and did not apply case-folding or any kind of normalization. This means we could have incorrectly discarded a significant number of relevant foster documents (false negatives). Moreover, we could missed a significant number of linkable candidates in step 1 due to the limitations of the VSM. Conversely, this means effectively under-generating the incoming and outgoing links, however, for task 1 we over-linked the outgoing links in the topics (see Tables 4 and 5). Interestingly, we found that we can significantly improve the accuracy of the detection of our outgoing links by generating multiple BEPs for an anchor, which partly deals with the issue of underlinking.

5 Conclusions

In summary, we continued with our experimentation with the Vector Space Model and simple string processing techniques for detecting missing links in the Wikipedia. The link detection occurred in 2 steps: first, a relevant pool of foster (candidate) documents is collected; second, substring matching with the list of collected titles to establish an actual link. We used entire orphaned documents (full-text) as query, with the idea to use all textual content as maximal evidence to find 'linkable' documents.

Clearly, we showed the limitations of this full-text approach based on the VSM, especially on the document level. A content-based full-text approach is not competitive against anchor-based approaches, however, a content-based approach adheres most strictly to an obvious assumption of link detection, namely that documents do not already have existing links as evidence and these cannot be used to 're-establish' links, which is not necessarily equal to 'detection'.

A competitive content-based link detection approach that discovers high quality links is needed, for example for detecting links in legacy or cultural heritage data. The impact of link detection on those datasets and domains will be large (for users and systems), since there are no such links yet (which would enable new navigation and search possibilities), and the alternative is expensive manual linking. To improve our approach, we are considering to experiment more with the granularity in a document to find focusedly link candidates (besides title and whole document), such as on the sentence level.

Acknowledgments. This research is supported by the Netherlands Organization for Scientific Research (NWO) under project number 639.072.601.

References

[1] Fachry, K.N., Kamps, J., Koolen, M., Zhang, J.: Using and detecting links in wikipedia. In: Fuhr, N., Kamps, J., Lalmas, M., Trotman, A. (eds.) INEX 2007. LNCS, vol. 4862, pp. 388–403. Springer, Heidelberg (2008)
[2] Halasz, F., Schwartz, M.: The Dexter hypertext reference model. Commun. ACM 37(2), 30–39 (1994)
[3] Zhang, J., Kamps, J.: Link detection in XML documents: What about repeated links? In: SIGIR 2008 Workshop on Focused Retrieval, pp. 59–66 (2008)

Overview of the INEX 2008 XML Mining Track
Categorization and Clustering of XML Documents in a Graph of Documents

Ludovic Denoyer and Patrick Gallinari

LIP6 - University of Paris 6

Abstract. We describe here the XML Mining Track at INEX 2008. This track was launched for exploring two main ideas: first identifying key problems for mining semi-structured documents and new challenges of this emerging field and second studying and assessing the potential of machine learning techniques for dealing with generic Machine Learning (ML) tasks in the structured domain i.e. classification and clustering of semi structured documents. This year, the track focuses on the supervised classification and the unsupervised clustering of XML documents using link information. We consider a corpus of about 100,000 Wikipedia pages with the associated hyperlinks. The participants have developed models using the content information, the internal structure information of the XML documents and also the link information between documents.

1   Introduction

The XML Document Mining track[1] was launched for exploring two main ideas: first identifying key problems for mining semi-structured documents and new challenges of this emerging field and second studying and assessing the potential of machine learning techniques for dealing with generic Machine Learning (ML) tasks in the structured domain i.e. classification and clustering of semi structured documents.

This track has run for four editions during INEX 2005, 2006, 2007 and 2008 and the fifth phase is currently being launched. The three first editions have been summarized in [1] and [2] and we focus here on the 2008 edition.

Among the many open problems for handling structured data, the track focuses on two generic ML tasks applied to Information Retrieval: while the preceding editions of the track concerned supervised *classification/categorization* and unsupervised *clustering* of independent document, this track is about the classification and the clustering of XML documents organized in a graph of documents. The goal of the track was therefore to explore algorithmic, theoretical and practical issues regarding the classification and clustering of interdependent XML documents. In the following, we first describe the task and the corpus used in this edition, we then describe the different models submitted this year and comment the results. We then conclude by explaining the future of the XML Mining track for INEX 2009.

---

[1] http://xmlmining.lip6.fr

S. Geva, J. Kamps, and A. Trotman (Eds.): INEX 2008, LNCS 5631, pp. 401–411, 2009.

2   Categorization/Clustering of a Graph of XML Documents Organized

Dealing with XML document collections is a particularly challenging task for ML and IR. XML documents are defined by their logical structure and their content (hence the name semi-structured data). Moreover, in a large majority of cases (Web collections for example), XML documents collections are also structured by links between documents (hyperlinks for example). These links can be of different types and correspond to different information: for example, one collection can provide hierarchical links, hyperlinks, citations, ... Most models developed in the field of XML categorization/clustering simultaneously use the content information and the internal structure of XML documents (see [1] and [2] for a list of models) but they rarely use the external structure of the documents i.e the links between documents.

In the 2008 track, we focus on the problem of classification/clustering of XML documents organized in graph. More precisely, this track was composed of:

- a *single label classification* task where the goal was to find the single category of each document. This task consider a tansductive context where, during the training phase, the whole graph of documents is known but the labels of only a part of them are given to the participants (Figure 1).
- a *single label clustering* task where the goal was to associate each document to a single cluster, knowing both the documents and the links between documents (Figure 2).

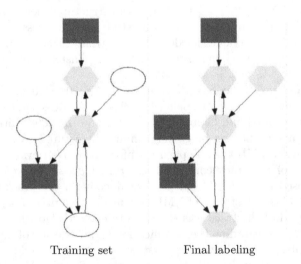

Training set   Final labeling

Fig. 1. The supervised classification task. Shapes correspond to categories, circle nodes are unlabeled nodes.

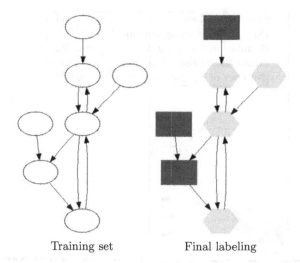

Training set   Final labeling

Fig. 2. The unsupervised clustering task. Shapes correspond to categories, circle nodes are unlabeled nodes. The categories are discovered automatically.

3 Corpus

3.1 Graph of XML Documents

The corpus provided is a subset of the *Wikipedia XML Corpus* [3]. We have extracted a set of 114,336 documents and the links between documents. These links corresponds to the links provided by the authors of the Wikipedia articles or automatically generated by Wikipedia. Note that we have only kept the links that concern the 114,333 documents of the corpus and we have removed the links that point to other articles.

XML Documents. The documents of this corpus are the original XML documents of the Wikipedia XML Corpus. Table 3.1 gives some statistics about the documents. Note that:

- The number of documents is important and classification/clustering models with a low complexity have to be used.
- The documents are large and the number of distinct words is greater than 160,000 (with a classical Porter Stemmer preprocessing) which does not correspond to the classical Text Categorization case where the vocabulary is usually smaller.

Graph Topology. The provided corpus is composed 636,187 directed links that correspond to hyperlinks between the documents of the corpus. Each document is pointed by 5.5 links on average and provide 5.5 links to other documents. Figure 3 (a) gives the distribution of the number of in-links and out-links and shows that a large majority of documents are concerned by less than 10 links.

Number of documents	114,336
Number of training documents	11,437
Number of test documents	102,929
Mean size of each document	408 words
Number of distinct words	166,619

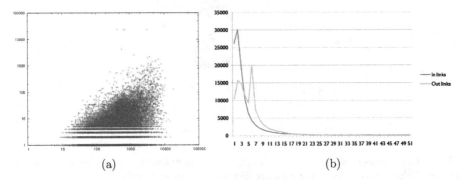

(a)   (b)

Fig. 3. (a) In-links, Out-links distribution: Each point corresponds to a document. The X-axis is the number of in-links and the Y-axis is the number of out-links **(b) The distribution of the number of documents w.r.t in-links and out-links:** The Y-axis corresponds to the number of documents while the X-axis corresponds to the number of in-links/out-links. For example, about 8000 documents have 5 in-links.

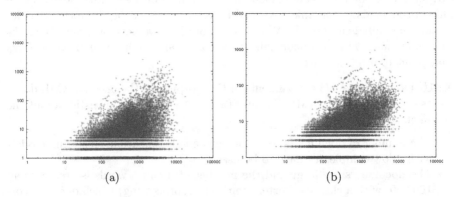

(a)   (b)

Fig. 4. (a) In-links distribution: Each point corresponds to a document. The X-axis is the number of in-links and the Y-axis is the size of each document **(b) Out-links distribution:** Each point corresponds to a document. The X-axis is the number of out-links and the Y-axis is the size of each document.

Figure 3 (b) shows the correlation between the number of in-links and out-links by document. Figure 4 shows the correlation between the size (the number of words) of the documents and the number of in-links and out-links. These figures clearly show that the number of links (in-links and out-links) directly depends on the sizes of the documents. This means that larger documents are more cited than smaller ones. This characteristic is specific to Wikipedia and does not fit well with Web graph for examples.

The global corpus topology is dense: the corpus is composed of one central component where a large majority of documents are linked to and some very small "islands" of documents that are not linked to this component. This is due to the sampling method we have used in order to extract this corpus from the whole Wikipedia XML Corpus. The underlying idea was to only keep documents with links because the goal of the task was to use the links structure as a relevant information for classification and clustering. Note that this topology does not correspond to the whole Wikipedia topology and a future work could be to try to sample a subset of Wikipedia keeping the same topology.

3.2   Labels

In order to provide a single label classification/clustering benchmark, we have labeled the documents with a subset of the original Wikipedia categories. These categories have not been chosen randomly in the whole set of categories[2] and we have kept a subset of categories that allow reasonable performances for the supervised classification task using a Naive Bayes classifier. Table 1 describes the set of the 15 categories kept here and provide some statistics about the distribution of these categories among the corpus.

Table 1. The set of 15 categories and the number of documents for each category

Category	Number of documents
reference	14905
social institutions	8199
sociology	1165
sports	9435
fiction	6262
united states	29980
categories by nationality	6166
europe	991
tourism	2880
politics by region	7749
urban geography	7121
americas	6088
art genres	2544
demographics	3948
human behavior	6933

---

[2] The Wikipedia XML Corpus provide more than 20,000 possible categories , one document can belong to many categories.

Labels and Graph Structure. In figure 5 , we measure the correlation between the categories and the graph structure. This figure shows that for a large majority of categories, a document that belongs to category c is connected to documents that belongs to the same category. The graph provided in the track is smooth in the sense that the label information does not mainly change among the graph structure. The graph structure seems very relevant for classification/clustering in the sense that, if one is able to find the category of a document, it can propagate this information among the out-links of this document to find the labels of the neighbors. Note that for some categories, the graph is also relevant, but not smooth: 83.52 % of the out-links of "United states" documents point to "Demographics" documents. We hope that the proposed models will capture this correlation between categories and links.

80.66	6.15	1.06	2.47	0.36	4.88	0.93	0.91	2.7	1.25	1.83	4.44	1.53	1.56	2.05
0.11	20.63	0.13	0.02	0.12	0.38	0.07	0.07	0.21	0.28	0.97	0.07	0.38	0.4	0.47
0.44	7.11	58.61	2.59	0.15	7.89	0.84	0.27	0.62	0.42	1.09	4.96	0.33	1.23	1.69
3.75	3.21	15.46	80.42	0.65	10.43	4.22	83.52	2.89	3.07	2.38	7.77	3.45	2.23	2.93
0.43	3.44	0.64	0.4	77.03	1.32	0.44	0.12	2.92	2.62	4.38	1.06	4.91	5.51	1.99
2.53	5.75	6.77	0.58	0.65	43.35	1.36	1.13	1.63	2.12	3.56	2.15	2.07	1.82	7.69
0.41	3.95	2.41	2.65	0.3	3.36	48.24	1.32	18.88	8.25	1.39	2.15	4.75	2.37	2.42
1.66	3.67	1.34	1.74	0.3	3.37	0.99	9.44	0.74	1.66	1.01	1.77	1.38	0.86	2.33
3.75	6.91	1.99	2.11	4.8	4.4	19.49	0.73	47.71	5.4	6.65	5.45	7.67	4.82	5
2.47	13.68	2.43	2.96	6.79	8.15	18.9	1.27	11.31	64.2	11.79	5.99	17.47	9.22	12.25
0.29	3.76	0.45	0.31	1.09	1.13	0.41	0.07	1.12	1.2	51.12	0.76	1.47	1.75	2.47
1.62	5.01	3.82	1.99	0.48	1.86	0.62	0.39	1.97	1.48	1.5	58.18	1.79	1.43	1.81
0.07	1.85	0.13	0.17	0.72	0.31	0.38	0.08	1.08	1.41	1.02	0.62	40.77	1.61	0.64
0.8	8.73	2.32	0.8	5.72	3.75	2.23	0.24	4.34	4.32	5.97	2.4	9.29	63.67	2.36
1.01	6.15	2.43	0.78	0.85	5.42	0.89	0.44	1.88	2.33	5.35	2.23	2.73	1.52	53.89

Fig. 5. The number in cell line i and column j corresponds to the percentage of links of out-links starting from a document of category i that points to a document of category j. the diagonal of the matrix corresponds to the percentages of smooth links that link two documents of the same category.

3.3   Train/Test Splits

For the categorization task, we have provide the labels of 10 % of the documents as a training set. These labels have been choosen randomly among the documents of the corpus. Figure 6 shows a piece of corpus.

3.4   Evaluation Measures

The track was composed of one supervised categorization task and one unsupervised clustering task. The organizers have made a blind evaluation on the testing corpus.

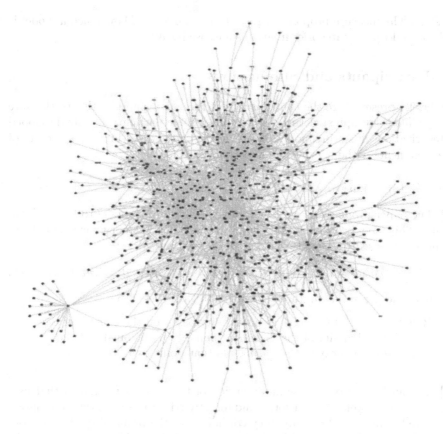

Fig. 6. Piece of the corpus. Only 10 % of the nodes are labeled.

Categorization: For categorization, we have asked the participants to submit one category for each of the documents of the testing set. We have then evaluated the correlation between the categories found by the participants and the real categories of the documents. For each category, we have computed a *recall* that corresponds to the percentage of documents of the category that have been correctly classified.

Clustering: For the clustering task, the participants have submitted a cluster index for each of the documents of the testing set. We have then evaluated if the obtained clustering corresponds to the real categories of the documents. For each submitted cluster, we have computed a *purity* measure that is a recall of the cluster considering that the cluster belongs to the category of the majority of its documents. We have also used a *micro average purity* and a *macro average purity* in order to summarize the performances of the different models over all the documents and all the clusters. Note that the evaluation of clustering is very difficult and it is still an open problem particularly with semi-structured document where clusters can correspond to structural clusters or to thematic clusters. The measure proposed here just gives an idea of how much a model is able to find the 15 categories in an unsupervised way.

4   Participants and Submissions

We briefly present here the contributions of the participants to this track. Note that, while the graph structure was provided, some participants only developed categorization/clustering models based on the content and internal structure of the documents without using the hyperlinks information.

4.1   Categorization Models

[4] The paper proposes an extension of the Naive Bayes model to linked documents. This model considers that the probability of a category knowing a document is composed by a mixture of three different categories:

$$P(cat|d) = \alpha P_0(cat|d) + \beta P_1(cat|d) + (1 - \alpha - \beta)P_2(cat|d) \qquad (1)$$

where α and β are the mixture coefficients $0 \leq \alpha + \beta \leq 1$ and

- P_0 is the classical naive bayes probability based on the content of d
- P_1 is the probability of the document pointed by the out-links of d
- P_2 is the probability of the documents that point to d

[5] The authors also propose an extension of the naive bayes model that uses the relationship between categories and directly tries to capture the information presented in figure 5. Basically, they consider that the probability of a category is weighted by the probability of this category knowing the categories of the neighbors of the document to classify.

[6] In this article, the authors use a pure graph-based semi-supervised method based on the propagation of the labels of the labeled nodes. This method directly learns a decision function that correctly classifies training examples and that is smooth among the whole graph-structure - considering both labeled and unlabeled examples. The learning algorithm is a label expansion algorithm that simultaneously learn a content classifier and propagate the labels of labeled nodes to unlabeled ones.

[7] The method submitted is a pure content-based method that does not use any content information. This method is a vector-space model where the size of the index is reduced by a features selection algorithm. While this method does not use the graph structure, it performs surprisingly well on the categorization task.

[8] Here, the authors propose to use a classical SVM with a TF-IDF representation of documents where the term frequencies have been replaced by link frequencies (LF-IDF). The SVM used for classification is a multi class SVM.

4.2   Clustering Models

[8] The paper proposes to use K-trees for document clustering. K-trees are Tree Structured Vector Quantizers which correspond to trees of clusters where the clusters are based on a vectorial representation of documents. K-trees are interesting because they scale very well with a large number of examples.

[9] The authors propose to make clustering based on the extraction of frequent subtrees using both the content and the internal structure information. The idea is to represent each document as a vector in both the space of a subset of frequent subtrees extracted from the whole corpus and the space of words. This representation is then used with classical clustering methods.

[10] The model proposed is based on the computation of a latent semantic kernel that is used as a similarity to perform clustering. This kernel only used content information but it is computed on a set of clusters extracted using the structural information.

[11] The article uses an extension of Self Organizing Map called GraphSOM and a new learning algorithm from document clustering. The new learning methods is an improvement of the classical one in term of stability and complexity and allows us to deal with large scale collections.

5   Official Results

We present here the results obtained by the different participants of the track. The results have been computed by the organizers on the test labels.

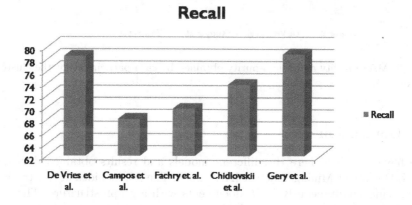

Fig. 7. Best recall obtained by each participant for the classification task

5.1   Categorization

Figure 7 shows the best results for each submitted models[3]. We show that the two best models (more than 78% recall) are obtained using classical vectorial classifiers (SVMs) with an appropriated document representation that mainly only uses the content information (and link frequencies for De Vries et al.). The three other models that better use the graph structure perform between 73.8 % and 68.1 % in term of recall. Note that the model by Romero et al. and the model by Fachry et al. are based on Naive Bayes extensions and perform better than a simple Naive Bayes model (but worse than a SVM method). At last, the propagation model by Chidlovskii et al. give reasonnably good results on this task.

5.2   Clustering

Figure 8 shows the results of the best submitted models for 15 clusters in term of macro and micro purity. One can see that almost all these models perform around 50% micro-purity which is a nice performance for a clustering model. Note that these 50% can directly be compared to the 78% recall obtained by the supervised methods showing that supervision improves unsupervised learning by 28%.

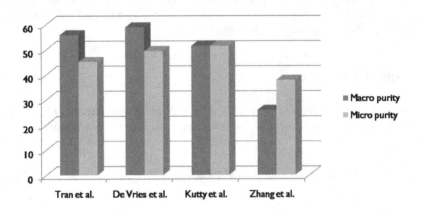

Fig. 8. Macro-purity and Micro-purity obtained by each participant for the clustering task

6   Conclusion

We have presented here the different models and results obtained during the XML Document Mining Track at INEX 2008. The original idea of this track was to provide simultaneously XML documents with a graph structure. The graph labeling task is a promising task that corresponds to many real applications

---

[3] Additional results can be found in the participants publications and on the website: http://xmlmining.lip6.fr

(classification on the Web, classification on Social networks, ...) and the XML Mining track is a first step to develop new models for text categorization/ clustering in a graph structure.

Acknowledgments

We would like to thank all the participants for their efforts and hard work.

References

1. Denoyer, L., Gallinari, P.: Report on the xml mining track at inex 2005 and inex 2006: categorization and clustering of xml documents 41(1), 79–90 (2007)
2. Denoyer, L., Gallinari, P.: Report on the xml mining track at inex 2007 categorization and clustering of xml documents 42(1), 22–28 (2008)
3. Denoyer, L., Gallinari, P.: The Wikipedia XML Corpus (2006)
4. Fachry, K.N., Kamps, J., Kaptein, R., Koolen, M., Zhang, J.: The University of Amsterdam at INEX 2008: Ad Hoc, Book, Entity Ranking, Interactive, Link the Wiki, and XML Mining Tracks. In: Workshop of the INitiative for the Evaluation of XML Retrieval (2008)
5. de Campos, L.M., Fernandez-Luna, J.M., Huete, J.F., Romero, A.E.: Probabilistic Methods for Link-based Classification at INEX 2008. In: Workshop of the INitiative for the Evaluation of XML Retrieval (2008)
6. Chidlovskii, B.: Semi-supervised Categorization of Wikipedia collection by Label Expansion. In: Workshop of the INitiative for the Evaluation of XML Retrieval (2008)
7. Mathias Gery, C.L., Moulin, C.: UJM at INEX 2008 XML mining track. In: Workshop of the INitiative for the Evaluation of XML Retrieval (2008)
8. Vries, C.M.D., Geva, S.: Document Clustering with K-tree. In: Workshop of the INitiative for the Evaluation of XML Retrieval (2008)
9. Kutty, S., Tran, T., Nayak, R., Li, Y.: Combining the structure and content of XML documents for Clustering using frequent subtress. In: Workshop of the INitiative for the Evaluation of XML Retrieval (2008)
10. Tran, T., Kutty, S., Nayak, R.: Utilizing the Structure and Data Information for XML Document Clustering. In: Workshop of the INitiative for the Evaluation of XML Retrieval (2008)
11. Zhang, S., Hagenbuchner, M., Tsoi, A., Sperduti, A.: Self Organizing Maps for the clustering of large sets of labeled graphs. In: Workshop of the INitiative for the Evaluation of XML Retrieval (2008)

Semi-supervised Categorization of Wikipedia Collection by Label Expansion

Boris Chidlovskii

Xerox Research Centre Europe
6, chemin de Maupertuis, F–38240 Meylan, France

Abstract. We address the problem of categorizing a large set of linked documents with important content and structure aspects, for example, from Wikipedia collection proposed at the INEX XML Mining track. We cope with the case where there is a small number of labeled pages and a very large number of unlabeled ones. Due to the sparsity of the link based structure of Wikipedia, we apply the spectral and graph-based techniques developed in the semi-supervised machine learning. We use the content and structure views of Wikipedia collection to build a transductive categorizer for the unlabeled pages. We report evaluation results obtained with the label propagation function which ensures a good scalability on sparse graphs.

1 Introduction

The objective of the INEX 2008 XML Mining challenge is to develop machine learning methods for structured data mining and to evaluate these methods for XML document mining tasks. The challenge proposes several datasets coming from different XML collections and covering a variety of classification and clustering tasks.

In this work, we address the problem of categorizing a very large set of linked XML documents with important content and structural aspects, for example, from Wikipedia online encyclopedia. We cope with the case where there is a small number of labeled pages and a much larger number of unlabeled ones. For example, when categorizing Web pages, some pages have been labeled manually and a huge amount of unlabeled pages is easily retrieved by crawling the Web. The semi-supervised approach to learning is motivated by the high cost of labeling data and the low cost for collecting unlabeled data. Withing XML Mining challenge 2008, the Wikipedia categorization challenge has been indeed set in the semi-supervised mode, where only 10% of page labels are available at the training step.

Wikipedia (http://www.wikipedia.org) is a free multilingual encyclopedia project supported by the non-profit Wikipedia foundation. In April 2008, Wikipedia accounted for 10 million articles which have been written collaboratively by volunteers around the world, and almost all of its articles can be edited by anyone who can access the Wikipedia website. Launched in 2001, it is currently the largest and most popular general reference work on the Internet. Automated analysis, mining and categorization of Wikipedia pages can serve to improve its internal structure as well as to enable its integration as an external resource in different applications.

S. Geva, J. Kamps, and A. Trotman (Eds.): INEX 2008, LNCS 5631, pp. 412–419, 2009.
© Springer-Verlag Berlin Heidelberg 2009

Any Wikipedia page is created, revised and maintained according to certain policies and guidelines [1]. Its edition follows certain rules for organizing the content and structuring it in the form of sections, abstract, table of content, citations, links to relevant pages, etc.. In the following, we distinguish between four different aspects (or views) of a Wikipedia page:

Content - the set of words occurred in the page.
Structure - the set of HTML/XML tags, attributes and their values in the page. These elements control the presentation of the page content to the viewer. In the extended version, we may consider some combinations of elements of the page structure, like the root-to-leaf paths or their fragments.
Links - the set of hyperlinks in the page.
Metadata - all the information present in the page Infobox, including the template, its attributes and values. Unlike the content and structure, not all pages include infoboxes [3].

We use these alternative views to generate a transductive categorizer for the Wikipedia collection. One categorizer representing the content view is based on the text of page. Another categorizer represents the structural view, it is based on the structure and Infobox characteristics of the page.

Due to the transductive setting of the XML Mining challenge, we test the graph-based semi-supervised methods which construct the similarity graph $W = \{w_{ij}\}$ and apply a function propagating labels from labeled nodes to unlabeled ones. We first build the content categorizer, with weights w_{ij} being the textual similarity between two pages. We then build the structure categorizer, where weights w_{ij} are obtained from the structure and Infobox similarity between the pages. Finally, we linearly combine the two categorizers to get the optimal performance.

2   Graph-Based Semi-supervised Learning

In the semi-supervised setting, we dispose labeled and unlabeled elements. In the graph-based approach [4,5] to linked documents, one node in the graph represents one page. We assume a weighted graph G having n nodes indexed from 1 to n. We associate with graph G a symmetric weight matrix W where all weights are non-negative ($w_{ij} > 0$), and weight w_{ij} represents the similarity between nodes i and j in G. If $w_{ij} = 0$, there is no edge between nodes i and j.

We assume that the first l training nodes have labels, $y_1, y_2, \ldots, y_l$, where y_i are from the category label set C, and the remaining $u = n - l$ nodes are unlabeled. The goal is to predict the labels $y_{l+1}, \ldots, y_n$ by exploiting the structure of graph G. According to the *smoothness* assumption, a label of an unlabeled node is likely to be similar to the labels of its neighboring nodes. A more strongly connected neighbor node will more significantly affect the node.

Assume the category set C includes c different labels. We define the binary label vector Y_i for node i, where $Y_i = \{y_{ij}|y_{ij} = 1 \text{ if } j = y_i, 0 \text{ otherwise}\}$. We equally introduce the category prediction vector $\hat{Y}_i$ for node i. All such vectors for n nodes define a $n \times c$-dimensional score matrix $\hat{Y} = (\hat{Y}_1, \ldots, \hat{Y}_n)$. At the learning step, we determine $\hat{Y}$ using all the available information. At the prediction step, the category labels are predicted by thresholding the score vectors $\hat{Y}_{l+1}, \ldots, \hat{Y}_n$.

The graph-based methods assume the following:

1. the score $\hat{Y}_i$ should be close to the given label vectors Y_i in training nodes, and
2. the score $\hat{Y}_i$ should not be too different from the scores of neighbor nodes.

There exist a number of graph-based methods [5]; we test some of them and report on one called the *label expansion* [4]. According to this approach, at each step, node i in graph G receives a contribution from its neighbors j weighted by the normalized weight w_{ij}, and an additional small contribution given by its initial value. This process can be expressed iteratively using the graph Laplacian matrix $L = D - W$, where $D = diag(d_i)$, $d_i = \sum_j w_{ij}$. The normalized Laplacian $\mathcal{L} = D^{-1/2}LD^{-1/2} = I - D^{-1/2}WD^{-1/2}$ can be used instead of L to get a similar result. The process is detailed in Algorithm 1 below.

Algorithm 1. Label expansion

Require: Symmetric matrix W, $w_{ij} \geq 0$ (and $w_{ii} := 0$)
Require: Labels y_i for $x_i, i = 1, \ldots, l$
Ensure: Labels for $x_{l+1}, \ldots, x_n$
1: Compute the diagonal degree matrix D by $d_{ii} := \sum_j w_{ij}$
2: Compute the normalized graph Laplacian $\mathcal{L} := I - D^{-1/2}WD^{-1/2}$
3: Initialize $\hat{Y}^{(0)} := (Y_1, \ldots, Y_l, 0, 0, \ldots, 0)$, where $Y_i = \{y_{ik}|y_{ik} = 1$ if $k = y_i$, 0 otherwise$\}$
4: Choose a parameter $\alpha \in [0, 1)$
5: **while** not converged to $\hat{Y}^{(\infty)}$ yet **do**
6:     Iterate $\hat{Y}^{(t+1)} := \alpha\mathcal{L}\hat{Y}^{(t)} + (1 - \alpha)\hat{Y}^{(0)}$
7: **end while**
8: Label x_i by $argmax_j\hat{Y}_i^{(\infty)}$

It has been proved that Algorithm 1 always converges [4]. Indeed, the iteration equation can be represented as follows

$$\hat{Y}^{(t+1)} = (\alpha\mathcal{L})^{t+1}\hat{Y}^{(0)} + (1 - \alpha) \sum_{i=0}^{t}(\alpha\mathcal{L})^i\hat{Y}^{(0)}. \tag{1}$$

Matrix $\mathcal{L}$ is a normalized Laplacian, its eigenvalues are known to be in [-1, 1] range. Since $\alpha < 1$, eigenvalues of $\alpha\mathcal{L}$ are in (-1,1) range. Therefore, when $t \to \infty$, $(\alpha L)^t \to 0$.

Using the matrix decomposition, we have $\sum_{i=0}^{\infty}(\alpha\mathcal{L})^i \to (I - \alpha\mathcal{L})^{-1}$, so that we obtain the following convergence:

$$\hat{Y}^{(t)} \to \hat{Y}^{(\infty)} = (1 - \alpha)(I - \alpha\mathcal{L})^{-1}\hat{Y}^{(0)}. \tag{2}$$

The convergence rate of the algorithm depends on specific properties of matrix W, in particular, the eigenvalues of its Laplacian $\mathcal{L}$. In the worst case, the convergence takes $O(kn^2)$ time, where k is the number of neighbors of a point in the graph.

On the other hand, the score matrix $\hat{Y}$ can be obtained by solving a large sparse linear system $(I - \alpha \mathcal{L})\hat{Y} = (1 - \alpha)Y^{(0)}$. This numerical problem has been intensively studied [2], and efficient algorithms, whose computational time is nearly linear in the number of non-zero entries in the matrix L. Therefore, the computation gets faster as the Laplacian matrix gets sparser.

2.1 Category Mass Regularization

Algorithm 1 generates a c-dimensional vector $\hat{Y}_i$ for each unlabeled node i, where c is the number of categories and each element $\hat{y}_{ij}$ between 0 and 1 gives a score for category j. To obtain the category for i, Algorithm 1 takes the category with the highest value, $argmax_j \, \hat{y}_{ij}$. Such a rule works well when categories are well balanced. However, in real-world data categories are often unbalanced and the categorization resulting from Algorithm 1 may not reflect the prior category distribution.

To solve this problem, we perform the category mass normalization, similarly to [6]. It rescales categories in such a way that their respective weights over unlabeled examples match the prior category distribution estimated from labeled examples.

Category mass normalization is performed in the following way. First, let p_j denote the prior probability of category j estimated from the labeled examples: $p_j = \frac{1}{l} \sum_{i=1}^{l} y_{ij}$. Second, the mass of category j as given by the average of estimated weights of j over unlabeled examples, $m_j = \frac{1}{u} \sum_{i=l+1}^{n} \hat{y}_{ij}$. Then the category mass normalization consists in scaling each category j by the factor $v_j = \frac{p_j}{m_j}$. In other words, instead of the decision function $argmax_j \, \hat{y}_{ij}$, we categorize node i in the category given by $argmax_j \, v_j \hat{y}_{ij}$. The goal is to make the scaled masses match the prior category distribution, i.e. after normalization we have that for all j

$$p_j = \frac{v_j m_j}{\sum_{i=1}^{c} v_i m_i}.$$

Generally, such a scaling gives a better categorization performance when there are enough labeled data to accurately estimate the category distribution, and when the unlabeled data come from the same distribution. Moreover, if there exists such m that each category mass is $m_j = m p_j$, i.e., the masses already reflect the prior category distribution, then the mass normalization step has no effect, since $v_j = \frac{1}{m}$ for all j.

2.2 Graph Construction

The label expansion algorithm starts with a graph G and associated weighted matrix W. To build the graph G for the Wikipedia collection, we first reuse its link structure by transforming directed links into undirected ones. We analyze the number of incoming and outcoming links for all pages in the Wikipedia collection. Figure 1 shows the In-Out frequencies for the corpus; note the log scale set for all dimensions.

In the undirected graph, we remove self-links as required by Algorithm 1. We then remove links between nodes with high weights w_{ij} having different labels in order to fits the smoothness condition. It turns out that the link graph is not totally connected.

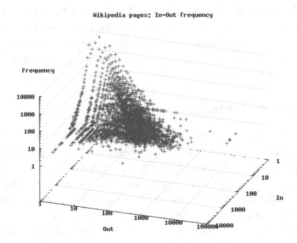

Fig. 1. Wikipedia nodes: In-Out frequencies

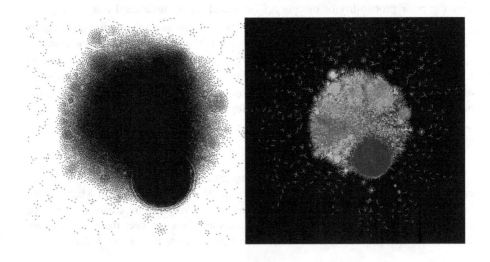

Fig. 2. Wikipedia corpus: the link graph plotted with LGL package

Figure 2 plots the link graph with the help of the Large Graph Layout package[1]. As the figure shows, the graph includes one connected component and about 160 small components covering less than 1% of collection. The right plot in Figure 2 additionally projects the category information on the link graph, where each category is shown by a particular color.

We are also interested in building graphs G which are different the original link structure. The standard approach [4] is to build the k-NN (Nearest Neighbors) graph

---

[1] http://bioinformatics.icmb.utexas.edu/lgl/

by taking the top k weights w_{ij} for each node. Unfortunately, the exhaustive k-NN procedure is infeasible even for the Wikipedia fragment used in the challenge. Thus we build a graph G' by modifying G with randomly sampling of node pairs from Wikipedia and selecting the top k=100 ones per node. Note using the content or structure similarity will produce different versions of G'. In the evaluation section, we report results of tests run on both G and G' graphs.

Content matrix. To generate a content weighted matrix W, we extract descriptor x_i for node i in the graph by using *"bag-of-words"* model and the *tf-idf* values, (term frequency-inverted document frequency) as $x_{ij} = tf_{ij} \cdot idf_i$, where

- tf_{ij} is the term frequency given by $\frac{n_{i,j}}{\sum_k n_{k,j}}$, where n_{ij} is the number of occurrences of the term in document d_j, and the denominator is the number of occurrences of all terms in document d_j.
- idf_i is the inverted document frequency $\log \frac{n}{|\{d_j : t_i \in d_j\}|}$, where n is the total number of documents and $|\{d_j : t_i \in d_j\}|$ is the number of documents where the term t_i occurs.

The tf-idf weighting scheme is often used in the vector space model together with cosine similarity to determine the similarity between two documents.

Layout matrix. In the structure graph, node descriptors x_i are generated following the *"bag-of-tags"* approach which is similar to bag-of-words used in the content graph. Instead of words, it uses elements of the page structure. In the HTML formatted pages, the presentation is guided by instructions encoded as HTML tags, attributes and their values. The HTML structure forms a nested structure. The "bag-of-tags" model might have different instantiations, below we report some of them, where the terms form one of the following sets:

1. Set of tag names, like (table) or (font),
2. Set of descendant tag pairs, like (table, span) or (tr, td),
3. Set of root-to-leaf paths in HTML page, like (html, body, table, tr, td),
4. Set of (tag,attribute) pairs, like (table, font),
5. Set of (tag,attribute,attribute_value) triples, like (table, font, times).

For each of the above sets, we extract descriptors x_i for node i according to the conventional tf-idf weights. We build the weighted matrix W using the *structure similarity* between pages evaluated with "bag-of-tags" model and one of the listed tag sets.

Similarity measures. Once we have obtained description vectors x_i for all nodes in graph G, we can get the weighted matrix W by measuring a similarity between two nodes i and j in G. Two possible measures are the following:

1. The Gaussian (RBF) kernel of width σ, $w_{ij} = exp\left(-\frac{||x_i - x_j||^2}{2\sigma^2}\right)$, where the width σ is evaluated from the variance of the descriptors x_i.
2. The standard cosine function, $w_{ij} = \frac{x_i \cdot x_j}{||x_i|| \, ||x_j||}$.

3   Evaluation

The collection used in the INEX XML Mining challenge is composed of n=114,366 pages from the Wikipedia XML Corpus; 10% of these pages have been annotated (l=11,437) with c=15 categories, 90% of pages (u=102,929) are unannotated. Some global characteristics of the corpus is given in Table 1. The word set is composed of all lexemized keywords; neither non-English words not stop words were excluded.

Table 1. Wikipedia collection: some global characteristics

Set	Size	Set	Size	Set	Size
Text words	727,667	Tag+attribute pairs	5,772	Infobox templates	602
Infobox tags	1,208	Root-to-leaf paths	110,099	Hyperlinks	636,187
Tags	1,257	Tag+attribute+value triples	943,422		

In all experiments, we measure the accuracy of a transductive categorizer using 10-fold cross validation on the training set (in the presence of unlabeled data). As the baseline method, we used the semi-supervised learning with the multi-class SVM, with x_i node descriptors being feature values. We also combine content, structure and infobox views, by concatenating the corresponding descriptors. However, direct concatenation of these alternative views brings no benefit (see 'Content+Tag+Attr+IB' line in Table 2).

For the label expansion method, we tested the link-based graph G and the sampling-enriched link graph G', with matrices W_c and W_s being generated with content or structure similarity measures, respectively. Using tag+attribute descriptors enriched with infoboxes generates a transductive categorizer whose performance is comparable to the content categorizer. Finally, the best performance is achieved by combining two graphs G' with weights w_{ij} obtained the content and structure similarity. The resulting weighted matrix is obtained as $W = \alpha W_s + (1 - \alpha)W_c$ with the optimal $\alpha = 0.34$ obtained by the cross validation. The right column in Table 2 reports the evaluation results for different (graph, similarity) comvinations and aligns them with the SVM results.

Three submissions to the INEX challenge have been done with three values of α: 0.34, 0.37 and 0.38. They yielded the accuracy values 73.71%, 73.79% and 73.47%,

Table 2. Performance evaluation for different methods

SVM Method	Accuracy(%)	LP Method	Accuracy (%)	Comment
Content	73.312	G-Content	72.104	Cosine
		G'-Content	75.03	idem
Tag+Attr	72.744	G'-Tag+Attr	72.191	Gaussian, δ=1.5
Paths	59.432	G'-Paths	64.824	idem
Tag+Attr+InfoBox	72.921	G-Tag+Attr+IB	70.287	idem
Content+Tag+Attr+IB	73.127	G'-Tag+Attr+IB	74.753	idem
		G'-Content + G'-TAIB	77.572	α=0.34

respectively. Despite the high density, these results are a clear underperformance with respect to the cross validation tests and results by the relatively simpler SVM classifiers. Nevertheless, the graph-based methods clearly represent a powerful mechanism for classifying the linked data like Wikipedia; thus we intend to conduct further studies to realize their potential.

4 Conclusion

We applied the graph-based semi-supervised methods to the categorization challenge defined on Wikipedia collection. The methods benefit from the recent advances in spectral graph analysis and offer a good scalability in the case of sparse graphs. From the series of experiments on the Wikipedia collection, we may conclude that the optimal graph construction remains the main issue. In particular, the good choice of the graph generator and node similarity distance is a key to get an accurate categorizer. The use of the Wikipedia link graph offers the baseline performance, while the sampling technique brings a clear improvement. Nevertheless, its impact remains limited as the graph smoothness requirement is satisfied only partially. To better satisfy the requirement, we would need a smarter sampling technique and an extension of the method toward the graph regularization and an advanced text analysis.

Acknowledgment

This work is partially supported by the ATASH Project co-funded by the French Association on Research and Technology (ANRT).

References

1. Riehle, D.: How and why Wikipedia works: an interview with Angela Beesley, Elisabeth Bauer, and Kizu Naoko. In: Proc. WikiSym 2006, New York, NY, USA, pp. 3–8 (2006)
2. Saad, Y.: Iterative Methods for Sparse Linear Systems, 2nd edn. SIAM, Philadelphia (2008)
3. Wu, F., Weld, D.S.: Autonomously semantifying Wikipedia. In: CIKM 2007: Proc. 16th ACM Conf. Information and Knowledge Management, pp. 41–50 (2007)
4. Zhou, D., Bousquet, O., Navin Lal, T., Weston, J., Olkopf, B.S.: Learning with local and global consistency. In: Advances in NIPS 16, pp. 321–328. MIT Press, Cambridge (2004)
5. Zhu, X.: Semi-supervised learning literature survey. In: University of Wisconsin-Madison, CD Department, Technical Report 1530 (2005)
6. Zhu, X., Ghahramani, Z., Lafferty, J.: Semisupervised learning using Gaussian fields and harmonic functions. In: Proc. 12th Intern. Conf. Machine Learning, pp. 912–919 (2003)

Document Clustering with K-tree

Christopher M. De Vries and Shlomo Geva

Faculty of Science and Technology,
Queensland University of Technology, Brisbane, Australia
chris@de-vries.id.au, s.geva@qut.edu.au

Abstract. This paper describes the approach taken to the XML Mining track at INEX 2008 by a group at the Queensland University of Technology. We introduce the K-tree clustering algorithm in an Information Retrieval context by adapting it for document clustering. Many large scale problems exist in document clustering. K-tree scales well with large inputs due to its low complexity. It offers promising results both in terms of efficiency and quality. Document classification was completed using Support Vector Machines.

Keywords: INEX, XML Mining, Clustering, K-tree, Tree, Vector Quantization, Text Classification, Support Vector Machine.

1 Introduction

The XML Mining track consists of two tasks, classification and clustering. Classification labels documents in known categories. Clustering groups similar documents without any knowledge of categories. The corpus consisted of 114,366 documents and 636,187 document-to-document links. It is a subset of the XML Wikipedia corpus [1]. Submissions were made for both tasks using several techniques.

We introduce K-tree in the Information Retrieval context. K-tree is a tree structured clustering algorithm introduced by Geva [2] in the context of signal processing. It is particularly suitable for large collections due to its low complexity. Non-negative Matrix Factorization (NMF) was also used to solve the clustering task. Applying NMF to document clustering was first described by Xu et. al. at SIGIR 2003 [3]. Negentropy has been used to measure clustering performance using the labels provided for documents. Entropy has been used by many researchers [4,5,6] to measure clustering results. Negentropy differs slightly but is fundamentally measuring the same system property.

The classification task was solved using a multi-class Support Vector Machine (SVM). Similar approaches have been taken by Joachims [7] and Tong and Koller [8]. We introduce a representation for links named Link Frequency Inverse Document Frequency (LF-IDF) and make several extensions to it.

Sections 2, 3, 4, 5, 6 and 7 discuss document representation, classification, cluster quality, K-tree, NMF and clustering respectively. The paper ends with a discussion of future research and a conclusion in Sects. 8 and 9.

S. Geva, J. Kamps, and A. Trotman (Eds.): INEX 2008, LNCS 5631, pp. 420–431, 2009.

2 Document Representation

Document content was represented with TF-IDF [9] and BM25 [10]. Stop words were removed and the remaining terms were stemmed using the Porter algorithm [11]. TF-IDF is determined by term distributions within each document and the entire collection. Term frequencies in TF-IDF were normalized for document length. BM25 works with the same concepts as TF-IDF except that is has two tuning parameters. The BM25 tuning parameters were set to the same values as used for TREC [10], $K1 = 2$ and $b = 0.75$. $K1$ influences the effect of term frequency and b influences document length.

Links were represented as a vector of LF-IDF weighted link frequencies. This resulted in a document-to-document link matrix. The row indicates the origin and the column indicates the destination of a link. Each row vector of the matrix represents a document as a vector of link frequencies to other documents. The motivation behind this representation is that documents with similar content will link to similar documents. For example, in the current Wikipedia both car manufacturers BMW and Jaguar link to the Automotive Industry document. Term frequencies were simply replaced with link frequencies resulting in LF-IDF. Link frequencies were normalized by the total number of links in a document.

All representations were culled to reduce the dimensionality of the data. This is necessary to fit the representations in memory when using a dense representation. K-tree will be extended to work with sparse representations in the future. A feature's rank is calculated by summation of its associated column vector. This is the sum of all weights for each feature in all documents. Only the top n features are kept in the matrix and the rest are discarded. TF-IDF was culled to the top 2000 and 8000 features. The selection of 2000 and 8000 features is arbitrary. BM25 and LF-IDF were only culled to the top 8000 features.

3 Classification Task

The classification task was completed using an SVM and content and link information. This approach allowed evaluation of the different document representations. It allowed the most effective representation to be chosen for the clustering task.

SVM[multiclass] [12] was trained with TF-IDF, BM25 and LF-IDF representations of the corpus. BM25 and LF-IDF feature vectors were concatenated to train on both content and link information simultaneously. Submissions were made only using BM25, LF-IDF or both because BM25 out performed TF-IDF.

3.1 Classification Results

Table 1 lists the results for the classification task. They are sorted in order of decreasing recall. Recall is simply the accuracy of predicting labels for documents not in the training set. Concatenating the link and content representations did not drastically improve performance. Further work has been subsequently performed to improve classification accuracy.

Table 1. Classification Results

Name	Recall	Name	Recall
Expe 5 tf idf T5 10000	0.7876	Expe 4 tf idf T5 100	0.7231
Expe 3 tf idf T4 10000	0.7874	Kaptein 2008NBscoresv02	0.6981
Expe 1 tf idf TA	0.7874	Kaptein 2008run	0.6979
Vries text and links	0.7849	Romero nave bayes	0.6767
Vries text only	0.7798	Expe 2.tf idf T4 100	0.6771
Boris inex tfidf1 sim 0.38.3	0.7347	Romero nave bayes links	0.6814
Boris inex tfidf sim 037 it3	0.7340	Vries links only	0.6233
Boris inex tfidf sim 034 it2	0.7310		

3.2 Improving Representations

Several approaches have been carried out
to improve classification performance.
They were completed after the end of official submissions for INEX. The same
train and test splits were used. All features were used for text and links, where
earlier representations were culled to the
top 8000 features. Links were classified without LF-IDF weighting. This was
to confirm LF-IDF was improving the
results. Document length normalization
was removed from LF-IDF. It was noticed
that many vectors in the link representation contained no features. Therefore,
inbound links were added to the repre-

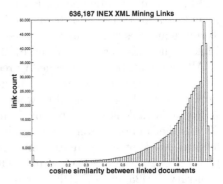

Fig. 1. Text Similarity of Links

sentation. For i, the source document and j, the destination document, a weight
of one is added to the i, j position in the document-to-document link matrix.
This represents an outbound link. To represent an inbound link, i is the destination document and j is the source document. Thus, if a pair of documents both
link to each other they receive a weight of two in the corresponding columns
in their feature vectors. Links from the entire Wikipedia were inserted into this
matrix. This allows similarity to be associated on inbound and outbound links
outside the XML Mining subset. This extends the 114,366×114,366 document-
to-document link matrix to a 114,366×486,886 matrix. Classifying links in this
way corresponds to the idea of hubs and authorities in HITS [13]. Overlap on outbound links indicates the document is a hub. Overlap on inbound links indicates
the document is an authority. The text forms a 114,366×206,868 document term
matrix when all terms are used. The link and text representation were combined
using two methods. In the first approach text and links were classified separately.
The ranking output of the SVM was used to choose the most appropriate label. We call this SVM by committee. Secondly, both text and link features were

Table 2. Classification Improvements

Dimensions	Type	Representation	Recall
114,366	Links (XML Mining subset)	unweighted	0.6874
114,366	Links (XML Mining subset)	LF-IDF	0.6906
114,366	Links (XML Mining subset)	LF-IDF no normalization	0.7095
486,886	Links (Whole Wikipedia)	unweighted	0.7480
486,886	Links (Whole Wikipedia)	LF-IDF	0.7527
486,886	Links (Whole Wikipedia)	LF-IDF no normalization	0.7920
206,868	Text	BM25	0.7917
693,754	Text, Links (Whole Wikipedia)	BM25 + LF-IDF committee	0.8287
693,754	Text, Links (Whole Wikipedia)	BM25 + LF-IDF concatenation	0.8372

converted to unit vectors and concatenated forming a 114,366×693,754 matrix. Table 2 highlights the performance of these improvements.

The new representation for links has drastically improved performance from a recall of 0.62 to 0.79. It is now performing as well as text based classification. However, the BM25 parameters have not been optimized. This could further increase performance of text classification. Interestingly, 97 percent of the correctly labeled documents for text and link classification agree. To further explain this phenomenon, a histogram of cosine similarity of text between linked documents was created. Figure 1 shows this distribution for the links in XML Mining subset. Most linked documents have a high degree of similarity based on their text content. Therefore, it is valid to assume that linked documents are highly semantically related. By combining text and link representations we can disambiguate many more cases. This leads to an increase in performance from 0.7920 to 0.8372 recall. The best results for text, links and both combined, performed the same under 10 fold cross validation using a randomized 10% train and 90% test split.

LF-IDF link weighting is motivated by similar heuristics to TF-IDF term weighting. In LF-IDF the link inverse document frequency reduces the weight of common links that associate documents poorly and increases the weight of links that associate documents well. This leads to the concept of stop-links that are not useful in classification. Stop-links bare little semantic information and occur in many unrelated documents. Consider for instance a document collection of the periodic table of the elements, where each document corresponds to an element. In such a collection a link to the "Periodic Table" master document would provide no information on how to group the elements. Noble gases, alkali metals and every other category of elements would all link to the "Periodic Table" document. However, links that exist exclusively in noble gases or alkali metals would be excellent indicators of category. Year links in the Wikipedia are a good example of a stop-link as they occur with relatively high frequency and convey no information about the semantic content of pages in which they appear.

4   Document Cluster Quality

The purity measure for the track is calculated by taking the most frequently occurring label in each cluster. Micro purity is the mean purity weighted by cluster size and macro is the unweighted arithmetic mean. Taking the most frequently occurring label in a cluster discards the rest of the information represented by the other labels. Due to this fact negentropy was defined. It is the opposite of information entropy [14]. If entropy is a measure of uncertainty associated with a random variable then negentropy is a measure of certainty. Thus, it is better when more labels of the same class occur together. When all labels are evenly distributed across all clusters the lowest possible negentropy is achieved.

Negentropy is defined in Equations (1), (2) and (3). D is the set of all documents in a cluster. X is the set of all possible labels. $l(d)$ is the function that maps a document d to its label x. $p(x)$ is the probability for label x. $H(D)$ is the negentropy for document cluster D. The negentropy for a cluster falls in the range $0 \leq H(D) \leq 1$ for any number of labels in X. Figure 2 shows the difference between entropy and negentropy. While they are exact opposites for a two class problem, this property does not hold for more than two classes. Negentropy al-

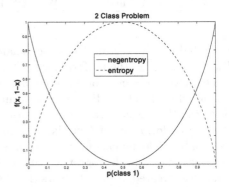

Fig. 2. Entropy Versus Negentropy

ways falls between zero and one because it is normalized. Entropy is bounded by the number of classes. The difference between the maximum value for negentropy and entropy increase when the number of classes increase.

$$l(d) = \{(d_1, x_1), (d_2, x_2), \ldots, (d_{|D|}, x_{|D|})\} \tag{1}$$

$$p(x) = \frac{|\{d \in D : x = l(d)\}|}{|D|} \tag{2}$$

$$H(D) = 1 + \frac{1}{\log_2 |X|} \sum_{\substack{x \in X \\ p(x) \neq 0}} p(x) \log_2 p(x) \tag{3}$$

The difference between purity and negentropy can easily be demonstrated with an artificial four class problem. There are six of each of the labels A, B, C and D. For each cluster in Solution 1 purity and negentropy is 0.5. For each cluster in Solution 2 the purity is 0.5 and the negentropy is 0.1038. Purity makes no differentiation between the two solutions. If the goal of document clustering is to group similar documents together then Solution 1 is clearly better because each label occurs in two clusters instead of four. The grouping of labels is better defined because they are less spread. Figures 3 and 4 show Solutions 1 and 2.

Cluster	Label Counts
1	A=3, B=3
2	A=3, C=3
3	B=3, D=3
4	C=3, D=3

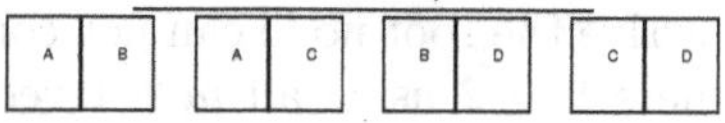

Fig. 3. Solution 1

Cluster	Label Counts
1	A=3, B=1, C=1, D=1
2	B=3, C=1, D=1, A=1
3	C=3, D=1, A=1, B=1
4	D=3, A=1, B=1, C=1

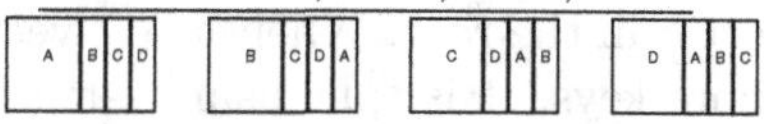

Fig. 4. Solution 2

5   K-tree

The K-tree algorithm is a height balanced cluster tree. It can be downloaded from http://ktree.sf.net. It is inspired by the B^+-tree where all data records are stored in the leaves at the lowest level in the tree and the internal nodes form a nearest neighbour search tree. The k-means algorithm is used to perform splits when nodes become full. The constraints placed on the tree are relaxed in comparison to a B^+-tree. This is due to the fact that vectors do not have a total order like real numbers.

B^+-tree of order m

1. All leaves are on the same level.
2. Internal nodes, except the root, contain between $\lceil \frac{m}{2} \rceil$ and m children.
3. Internal nodes with n children contain $n - 1$ keys, partitioning the children into a search tree.
4. The root node contains between 2 and m children. If the root is also a leaf then it can contain a minimum of 0.

K-tree of order m

1. All leaves are on the same level.
2. Internal nodes contain between one and m children. The root can be empty when the tree contains no vectors.
3. Codebook vectors (cluster representatives) act as search keys.
4. Internal nodes with n children contain n keys, partitioning the children into a nearest neighbour search tree.
5. The level immediately above the leaves form the codebook level containing the codebook vectors.
6. Leaf nodes contain data vectors.

The leaf nodes of a K-tree contain real valued vectors. The search path in the tree is determined by a nearest neighbour search. It follows the child node associated with nearest vector. This follows the same recursive definition of a B^+-tree where each tree is made up of a smaller sub tree. The current implementation of K-tree uses Euclidean distance for all measures of similarity. Future versions will have the ability to specify any distance measure.

5.1   Building a K-tree

The K-tree is constructed dynamically as data vectors arrive. Initially the tree contains a single empty root node at the leaf level. Vectors are inserted via a nearest neighbour search, terminating at the leaf level. The root of an empty tree is a leaf, so the nearest neighbour search terminates immediately, placing the vector in the root. When $m + 1$ vectors arrive the root node can not contain any more keys. It is split using k-means where $k = 2$ using all $m + 1$ vectors. The two centroids that result from k-means become the keys in a new parent. New root and child nodes are constructed and each centroid is associated with a child. The vectors associated with each centroid from k-means are placed into the associated child. This process has created a new root for the tree. It is now two levels deep. The root has two keys and two children, making a total of three nodes in the tree. Now that the tree is two levels deep, the nearest neighbour search finds the closest centroid in the root and inserts it in the associated child. When a new vector is inserted the centroids are updated along the nearest neighbour search path. They are weighted by the number of data vectors contained beneath them. This process continues splitting leaves until the root node becomes full. K-means is run on the root node containing centroids. The keys in the new root node become centroids of centroids. As the tree grows internal and leaf nodes are split in the same manner. The process can potentially propagate to a full root node and cause construction of a new root. Figure 6 shows this construction process for a K-tree of order three $(m = 3)$.

The time complexity of building a K-tree for n vectors is $O(n \log n)$. An insertion of a single vector has the time complexity of $O(\log n)$. These properties are inherent to the tree based algorithm. This allows the K-tree to scale efficiently with the number of input vectors. When a node is split, k-means is always restricted to $m + 1$ vectors and two centroids $(k = 2)$. Figure 7 compares k-means performance with K-tree where k for k-means is determined by the number of codebook vectors. This means that both algorithms produce the same

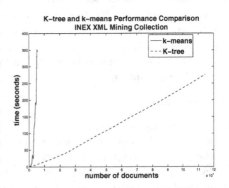

Fig. 5. K-tree Performance

number of document clusters and this is necessary for a meaningful comparison. The order, m, for K-tree was 50. Each algorithm was run on the 8000 dimension BM25 vectors from the XML mining track.

5.2   K-tree Submissions

K-tree was used to create clusters using the Wikipedia corpus. Documents were represented as 8000 dimension BM25 weighted vectors. Thus, clusters were

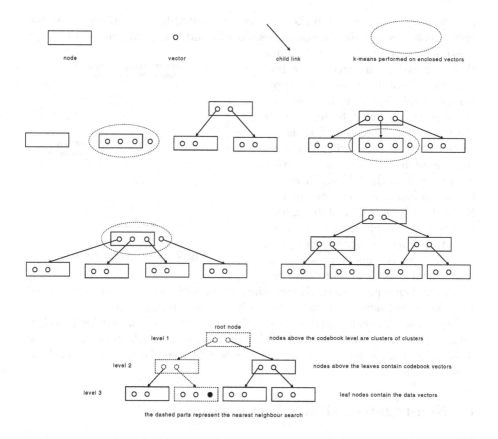

Fig. 6. K-tree Construction

formed using text only. This representation was used because it was most effective text representation in the classification task. The K-tree was constructed using the entire collection. Cluster membership was determined by comparing each document to all centroids using cosine similarity. The track required a submission with 15 clusters but K-tree does not produce a fixed number of clusters. Therefore, the codebook vectors were clustered using k-means++ where $k = 15$. The codebook vectors are the cluster centroids that exist above the leaf level. This reduces the number of vectors used for k-means++, making it quick and inexpensive. As k-means++ uses a randomised seeding process, it was run 20 times to find the solution with the lowest distortion. The k-means++ algorithm [15] improves k-means by using the D^2 weighting for seeding. Two other submission were made representing different levels of a K-tree. A tree of order 100 had 42 clusters in the first level and a tree of order 20 had 147 clusters in the second level. This made for a total of three submissions for K-tree.

Negentropy was used to determine the optimal tree order. K-tree was built using the documents in the 10% training set from the classification task. A tree was constructed with an order of 800 and it was halved each time.

Negentropy was measured in the clusters represented by the leaf nodes. As the order decreases the size of the nodes shrinks and the purity increases. If all clusters became pure at a certain size then decreasing the tree order further would not improve negentropy. However, this was not the case and negentropy continued to increase as the tree order decreased. This is expected because there will usually be some imperfection in the clustering with respect to the labels. Therefore, the sharp increase in negentropy in a K-tree below an order of 100 suggests that the natural cluster size has been observed. This can be seen in Fig. 7. The "left as is" line represents the K-tree as it is built initially. The "re-

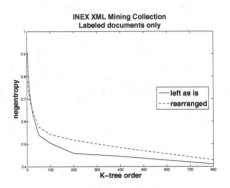

Fig. 7. K-tree Negentropy

arranged" line represents the K-tree when all the leaf nodes have been reinserted to their nearest neighbours without modifying the internal nodes.

Negentropy was calculated using the 10% training set labels provided on clusters for the whole collection. This was used to determine which order of 10, 20 or 35 fed into k-means++ with $k = 15$ was best. A tree of order 20 provided the best negentropy.

6   Non-negative Matrix Factorization

NMF factorizes a matrix into two matrices where all the elements are ≥ 0. If V is a $n \times m$ matrix and r is a positive integer where $r < min(n, m)$, NMF finds two non-negative matrices $W_{n \times r}$ and $H_{r \times m}$ such that $V \approx WH$. When applying this process to document clustering V is a term document matrix. Each column in H represents a document and the largest value represents its cluster. Each row in H is a cluster and each column is a document.

The projected gradient method was used to solve the NMF problem [16]. V was a 8000×114366 term document matrix of BM25 weighted terms. The algorithm ran for a maximum of 70 iterations. It produced the W and H matrices. Clusters membership was determined by the maximum value in the columns of H. NMF was run with r at 15, 42 and 147 to match the submissions made with K-tree.

7   Clustering Task

Every team submitted at least one solution with 15 clusters. This allows for a direct comparison between different approaches. It only makes sense to compare results where the number of clusters are the same. The K-tree performed well according to the macro and micro purity measures in comparison to the rest of

Table 3. Clustering Results Sorted by Micro Purity

Name	Size	Micro	Macro	Name	Size	Micro	Macro
K-tree	15	0.4949	0.5890	QUT LSK 1	15	0.4518	0.5594
QUT LSK 3	15	0.4928	0.5307	QUT LSK 4	15	0.4476	0.4948
QUT Entire collection 15	15	0.4880	0.5051	QUT LSK 2	15	0.4442	0.5201
NMF	15	0.4732	0.5371	Hagenbuchner	15	0.3774	0.2586

Table 4. Comparison of Different K-tree Methods

Method	Clusters	Micro	Macro	Clusters	Micro	Macro	Clusters	Micro	Macro
left as is	17	0.4018	0.5945	397	0.5683	0.6996	7384	0.6913	0.7202
rearranged	17	0.4306	0.6216	371	0.6056	0.7281	5917	0.7174	0.7792
cosine	17	0.4626	0.6059	397	0.6584	0.7240	7384	0.7437	0.7286

the field. The difference in macro and micro purity for the K-tree submissions can be explained by the uneven distribution of cluster sizes. Figure 9 shows that many of the higher purity clusters are small. Macro purity is simply the average purity for all clusters. It does not take cluster size into account. Micro purity does take size into account by weighting purity in the average by the cluster size. Three types of clusters appear when splitting the x-axis in Fig. 9 in thirds. There are very high purity clusters that are easy to find. In the middle there are some smaller clusters that have varying purity. The larger, lower purity clusters in the last third are hard to distinguish. Figure 8 shows clusters sorted by purity and size. K-tree consistently found higher purity clusters than other submissions. Even with many small high purity clusters, K-tree achieved a high micro purity score. The distribution of cluster size in K-tree was less uniform than other submissions. This can be seen in Figure 8. It found many large clusters and many small clusters, with very few in between.

The K-tree submissions were determined by cosine similarity with the centroids produced by K-tree. The tree has an internal ordering of clusters as well. A comparison between the internal arrangement and cosine similarity is listed in Table 4. This data is based on a K-tree of order 40. Levels 1, 2 and 3 produced 17, 397 and 7384 clusters respectively. Level 3 is the above leaf or codebook vector level. The "left as is" method uses the K-tree as it is initially built. The rearranged method uses the K-tree when all vectors are reinserted into the tree to their nearest neighbour. The cosine method determines cluster membership by cosine similarity with the centroids produced. Nodes can become empty when reinserting vectors. This explains why levels 2 and 3 in the rearranged K-tree contain less clusters. Using cosine similarity with the centroids improved purity in almost all cases.

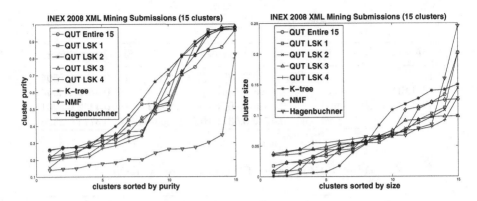

Fig. 8. All Submissions with 15 Clusters

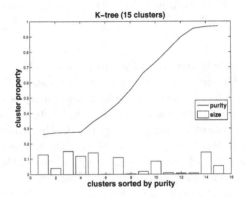

Fig. 9. K-tree Breakdown

8   Future Work

The work in this area falls into two categories, XML mining and K-tree. Further work in the XML mining area involves better representation of structure. For example, link information can be included into clustering via a modified similarity measure for documents. Future work with the K-tree algorithm will involve different strategies to improve quality of clustering results. This will require extra maintenance of the K-tree as it is constructed. For example, reinsertion of all data records can happen every time a new root is constructed.

9   Conclusion

In this paper an approach to the XML mining track was presented, discussed and analyzed. A new representation for links was introduced, extended and analyzed.

It was combined with text to further improve classification performance. The K-tree algorithm was applied to document clustering for the first time. The results show that it is well suited for the task. It produces good quality clustering solutions and provides excellent performance.

References

1. Denoyer, L., Gallinari, P.: The Wikipedia XML Corpus. SIGIR Forum (2006)
2. Geva, S.: K-tree: a height balanced tree structured vector quantizer. In: Proceedings of the 2000 IEEE Signal Processing Society Workshop Neural Networks for Signal Processing X, vol. 1, pp. 271–280 (2000)
3. Xu, W., Liu, X., Gong, Y.: Document clustering based on non-negative matrix factorization. In: SIGIR 2003: Proceedings of the 26th annual international ACM SIGIR conference on Research and development in informaion retrieval, pp. 267–273. ACM Press, New York (2003)
4. Surdeanu, M., Turmo, J., Ageno, A.: A hybrid unsupervised approach for document clustering. In: KDD 2005: Proceedings of the eleventh ACM SIGKDD international conference on Knowledge discovery in data mining, pp. 685–690. ACM Press, New York (2005)
5. Hotho, A., Staab, S., Stumme, G.: Ontologies improve text document clustering. In: Third IEEE International Conference on Data Mining. ICDM 2003, November 2003, pp. 541–544 (2003)
6. Steinbach, M., Karypis, G., Kumar, V.: A comparison of document clustering techniques 34, 35 (2000)
7. Joachims, T.: Text categorization with Support Vector Machines: Learning with many relevant features. In: Text categorization with Support Vector Machines: Learning with many relevant features, pp. 137–142 (1998)
8. Tong, S., Koller, D.: Support vector machine active learning with applications to text classification. Journal of Machine Learning Research 2, 45–66 (2002)
9. Salton, G., Fox, E.A., Wu, H.: Extended boolean information retrieval. Communications of the ACM 26(11), 1022–1036 (1983)
10. Robertson, S., Jones, K.: Simple, proven approaches to text retrieval. Update (1997)
11. Porter, M.: An algorithm for suffix stripping. Program: Electronic Library and Information Systems 40(3), 211–218 (2006)
12. Tsochantaridis, I., Joachims, T., Hofmann, T., Altun, Y.: Large Margin Methods for Structured and Interdependent Output Variables. Journal of Machine Learning Research 6, 1453–1484 (2005)
13. Kleinberg, J.M.: Authoritative sources in a hyperlinked environment. J. ACM 46(5), 604–632 (1999)
14. Shannon, C., Weaver, W.: The mathematical theory of communication. University of Illinois Press (1949)
15. Arthur, D., Vassilvitskii, S.: k-means++: the advantages of careful seeding. In: SODA 2007: Proceedings of the eighteenth annual ACM-SIAM symposium on Discrete algorithms, Philadelphia, PA, USA, pp. 1027–1035. Society for Industrial and Applied Mathematics (2007)
16. Lin, C.: Projected Gradient Methods for Nonnegative Matrix Factorization. Neural Computation 19(10), 2756–2779 (2007)

Using Links to Classify Wikipedia Pages

Rianne Kaptein[1] and Jaap Kamps[1,2]

[1] Archives and Information Studies, Faculty of Humanities, University of Amsterdam
[2] ISLA, Faculty of Science, University of Amsterdam

Abstract. This paper contains a description of experiments for the 2008 INEX XML-mining track. Our goal for the XML-mining track is to explore whether we can use link information to improve classification accuracy. Our approach is to propagate category probabilities over linked pages. We find that using link information leads to marginal improvements over a baseline that uses a Naive Bayes model. For the initially misclassified pages, link information is either not available or contains too much noise.

1 Introduction

Previous years of the XML mining track have explored the utility of using XML document structure for classification accuracy. It proved to be difficult to obtain better performance [1]. This year the data consists of a collection of wikipedia XML documents that have to be categorized into fairly high-level wikipedia categories and the link structure between these documents. Link structure has been found to be a useful additional source of information for other tasks such as ad hoc retrieval [2] and entity ranking [3]. Our aim at the XML Mining Track is to examine whether link structure can also be exploited for this classification task.

2 Classification Model

For our baseline classification model we use a classical Naive Bayes model [4]. The probability of a category given a document is:

$$P(cat|d) = \frac{P(d|cat) * P(cat)}{P(d)} \tag{1}$$

Since $P(d)$ does not change over the range of categories we can omit it. For each document the categories are ranked by their probabilities, and the category with the highest probability is assigned to the document:

$$
\begin{aligned}
ass\_cat(d) &= \arg\max_{cat \in cats} P(d|cat) * P(cat) \\
&= \arg\max_{cat \in cats} P(t_1|cat) * P(t_2|cat) * .. * P(t_n|cat) * P(cat) \tag{2}
\end{aligned}
$$

where $t_i...t_n$ are all terms in a document. The probability of a term occurring in a category is equal to its term frequency in the category divided by the total number of

S. Geva, J. Kamps, and A. Trotman (Eds.): INEX 2008, LNCS 5631, pp. 432–435, 2009.

terms in the category. Feature (term) selection is done according to document frequency. We keep 20% of the total number of features [5]. We propagate category information over links as follows:

$$P_0(cat|d) = P(cat|d)$$

$$P_1(cat|d) = \sum_{d \to d'} P(d|d')P(cat|d')$$

$$P_2(cat|d) = \sum_{d' \to d''} P(d|d'')P(cat|d'') \qquad (3)$$

where d' consist of all documents that are linked to or from d, and d'' are all documents that are linked to or from all documents d'. The probabilities are uniformly distributed among the incoming and/or outgoing links. The final probability of a category given a document is now:

$$P'(cat|d) = \mu P_0(cat|d) + (1 - \mu)(\alpha P_1(cat|d) + (1 - \alpha)P_2(cat|d)) \qquad (4)$$

The parameter μ determines the weight of the original classification versus the weight of the probabilities of the linked documents. Parameter α determines the weight of the first order links versus the weight of the second order links.

3   Experimental Results

Documents have to be categorized into one of fifteen categories. For our training experiments, we use 66% of the training data for training, and we test on the remaining 33%. Throughout this paper we use accuracy as our evaluation measure. Accuracy is defined as the percentage of documents that is correctly classified, which is equal to micro average recall. Our baseline Naive Bayes model achieves an accuracy of 67.59%. Macro average recall of the baseline run is considerably lower at 49.95%. All documents in the two smallest categories are misclassified. Balancing the training data can improve our macro average recall.

When we use the link information we try three variants: do not use category information of linked data, use category information of the training data, and always use category information of linked data. Other parameters are whether to use incoming or outgoing links, μ and α. For parameter μ we tried all values from 0 to 1 with steps of 0.1, only the best run is shown. The results are given in Table 1. The accuracy of the runs using link information is at best only marginally better than the accuracy of the baseline. This means that the difficult pages, which are misclassified in the baseline model, do not profit from the link information. The links to or from pages that do not clearly belong to a category and are misclassified in the baseline run, do not seem to contribute to classification performance. These linked pages might also be more likely to belong to a different category.

On the test data we made two runs, a baseline run that achieves an accuracy of 69.79%, and a run that uses in- and outlinks, $\alpha = 0.5$ and $\mu = 0.4$, with an accuracy of 69.81%. Again the improvement in accuracy when link information is used is only marginal.

Table 1. Training Classification results

Link info	α	μ	Inlinks Accuracy	μ	Outlinks Accuracy	μ	In- and Outlinks Accuracy
Baseline			0.6759		0.6759		0.6759
None	0.75	1.0	0.6759	1.0	0.6759	1.0	0.6759
None	1.0	1.0	0.6759	1.0	0.6759	1.0	0.6759
Training	0.5	0.5	**0.6793**	0.4	0.6777	0.4	0.6819
Training	0.75	0.5	**0.6793**	0.5	0.6777	0.5	0.6806
Training	1.0	0.6	0.6780	0.5	0.6780	0.6	0.6777
All	0.5	0.5	0.6780	0.3	0.6816	0.4	**0.6858**
All	0.75	0.6	0.6780	0.3	0.6848	0.5	0.6819
All	1.0	0.6	0.6784	0.4	**0.6858**	0.6	0.6787

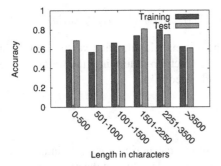

Fig. 1. Accuracy vs. document length   **Fig. 2.** Accuracy of first two categories

4   Analysis

We try to analyze on which kind of pages the most errors are made in our baseline run. Considering the length of pages, shorter pages do not tend to be more difficult than longer pages as can be seen in Fig. 1. When the output probabilities of the two highest scoring categories lie close together, the page is more likely to be misclassified. This is shown in Fig. 2 where we divided the training data over 6 bins of approximately the same size sorted by the fraction (P_{cat1}/P_{cat2}).

In our baseline run pages without links also seem to get misclassified more often than pages with in- and/or outlinks (see Fig. 3). When link information is available, and we try to use it, there are two sources of error. The first source of error, is that not all linked pages belong to the same category as the page to classify (see Table 2). However, when we classify pages that have links using only the link information, there are some cases where the accuracy on these pages is well above the accuracy of the complete set. To obtain our test data we have used both incoming and outgoing links, which means that almost half of the pages do not belong to the same category as the page to classify. Secondly, we only know the real categories of the pages in the training data, which is only 10% of all data. For all pages in the test data, we estimate the probability of

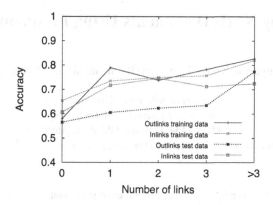

Fig. 3. Accuracy vs. number of links

Table 2. Statistics of training and test data

Data	# pages				links /page	% links with same cat.		
	total	with inlinks	with outlinks	with links		inlinks	outlinks	links
Training	11,437	2,627 (23%)	5,288 (46%)	5,925 (52%)	0.7	76.8%	41.1%	45.8%
Test	113,366	88,174 (77%)	103,781 (91%)	107,742 (94%)	5.6	77.2%	53.4%	59.0%

each category belonging to that page. With a classification accuracy of almost 70%, this means we introduce a large additional source of error.

5   Conclusion

It is difficult to use link information to improve classification accuracy. A standard Naive Bayes model achieves an accuracy of almost 70%. While link information may provide supporting evidence for the pages that are easy to classify, for the difficult pages link information is either not available or contains too much noise.

Acknowledgments. This research is funded by the Netherlands Organization for Scientific Research (NWO, grant # 612.066.513).

References

[1] Denoyer, L., Gallinari, P.: Report on the xml mining track at inex 2007 categorization and clustering of xml documents. SIGIR Forum 42(1), 22–28 (2008)

[2] Kamps, J., Koolen, M.: The importance of link evidence in Wikipedia. In: Macdonald, C., Ounis, I., Plachouras, V., Ruthven, I., White, R.W. (eds.) ECIR 2008. LNCS, vol. 4956, pp. 270–282. Springer, Heidelberg (2008)

[3] Vercoustre, A.M., Pehcevski, J., Thom, J.A.: Using wikipedia categories and links in entity ranking. In: Focused Access to XML Documents, pp. 321–335 (2007)

[4] Sebastiani, F.: Machine learning in automated text categorization. ACM Computing Surveys 34, 1–47 (2002)

[5] Williams, K.: Ai: categorizer - automatic text categorization. Perl Module (2003)

Clustering XML Documents Using Frequent Subtrees

Sangeetha Kutty, Tien Tran, Richi Nayak, and Yuefeng Li

Faculty of Science and Technology
Queensland University of Technology
GPO Box 2434, Brisbane Qld 4001, Australia
{s.kutty,t4.tran,r.nayak,y2.li}@qut.edu.au

Abstract. This paper presents an experimental study conducted over the INEX 2008 Document Mining Challenge corpus using both the structure and the content of XML documents for clustering them. The concise common substructures known as the closed frequent subtrees are generated using the structural information of the XML documents. The closed frequent subtrees are then used to extract the constrained content from the documents. A matrix containing the term distribution of the documents in the dataset is developed using the extracted constrained content. The k-way clustering algorithm is applied to the matrix to obtain the required clusters. In spite of the large number of documents in the INEX 2008 Wikipedia dataset, the proposed frequent subtree-based clustering approach was successful in clustering the documents. This approach significantly reduces the dimensionality of the terms used for clustering without much loss in accuracy.

Keywords: Clustering, XML document mining, Frequent mining, Frequent subtrees, INEX, Wikipedia, Structure and content.

1 Introduction

Due to the inherent flexibility in structure representation, eXtensible Markup Language (XML) is fast becoming a ubiquitous standard for data representation and exchange on the Internet as well as in Intranets. The self-describing nature of XML documents has resulted in their acceptance in a wide range of industries from education to entertainment and business to government sectors.

With the rapid growth of XML documents, many issues arise concerning the effective management of these documents. Clustering has been perceived as an effective solution to organize these documents. This involves grouping XML documents based on their similarity, without any prior knowledge of the taxonomy[2]. Clustering has been frequently applied to group text documents based on similarity of content. However, XML document clustering presents a new challenge as the document contains structural information as well as text data (or content). The structure of the XML documents is hierarchical in nature, representing the relationship between the elements at various levels.

Clustering of XML documents involves consideration of two document input features, namely structure and content, for determining the similarity between them.

S. Geva, J. Kamps, and A. Trotman (Eds.): INEX 2008, LNCS 5631, pp. 436–445, 2009.

Most of the existing approaches do not focus on utilizing these two features due to increased computational storage and processing. However, in order to achieve meaningful clustering results, it is essential to utilize both these XML document dimensions. This study not only combines the structure and content of XML documents effectively but also provides an approach that helps to reduce the dimensions for clustering without considerable loss in accuracy.

In this paper, we utilize the Prefix-based Closed Induced Tree Miner (PCIT-Miner)[3] algorithm to generate the closed frequent induced (CFI) subtrees. CFI is then utilized to extract the content information from the XML documents. The extracted content information which contains the document terms is represented in the Vector Space Model (VSM). A pair-wise distance based clustering method is then applied on the VSM to group the XML documents.

The assumption that we have made in this paper, based on the previous research[4], is that the structure of XML documents in the absence of their schema could be represented using frequent subtrees. Also, the content corresponding to the frequent subtrees is important, whereas the content contained within infrequent subtrees is redundant and hence can be removed.

The rest of this paper is organized as follows: Section 2 provides the overview of our approach. Section 3 covers the details about the pre-processing of structure. Section 4 details the structure mining to extract the common substructures among the XML documents. Section 5 discusses the extraction of content using the common substructures. The clustering process is covered in Section 6. In Section 7, we present the experimental results and discussion.

2 An Overview

As illustrated in Fig.1, there are four major phases in the approach that we have adopted for the INEX 2008 Document Mining Challenge corpus. The first phase is the pre-processing of XML documents to represent their structure. The structure of XML documents is modeled as a document tree. Each document tree contains nodes which represent the tag names. PCITMiner[3] is then applied to generate the CFI subtrees from the document trees in the second phase for a given support threshold.

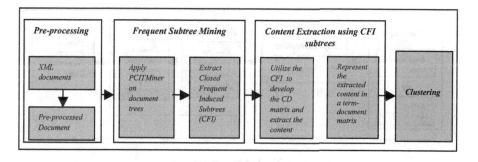

Fig. 1. Overview of the clustering method

The CFI subtrees distribution is represented as a CD Matrix. This matrix is used in the third phase to extract and preprocess the content within the CFI subtrees.. The pre-processed content is represented as a term-document matrix, TDoc|Term|x|D|, where Term represents the terms corresponding to the CFI subtrees and D represents the XML document in the given dataset. Each cell in the TDoc matrix represents the number of occurrences of the terms for the set of closed frequent subtrees in a given XML document. This matrix is used in the final phase to compute the similarity between the XML documents for the clustering of the dataset. The next section describes each phase in more detail.

3   Phase 1: Pre-processing of Structure

In the pre-processing phase, each XML document is modeled into a tree structure with nodes representing only the tag names. These tag names are then mapped to unique integers for ease of computation. The semantic and syntactic meanings of the tags are ignored since the Wikipedia documents conform to the same schema using the same tag set. Additionally, previous research has shown that the semantic variations of tags do not provide any significant contribution to the clustering process[2, 4]. We will consider the content contained within nodes to determine their semantic similarity. Other node information such as data types, attributes and constraints is also ignored as the empirical evaluation revealed that this information did not contribute to any improvement in accuracy.

The structure of the XML document has many representations depending on its usability, such as graph, tree and path. Rather than using path representation, which has been used by a number of researchers[2, 5], we have chosen to use tree representation since the tree representation preserves the sibling information of the nodes. As shown in Fig. 2, the pre-processing of XML documents involves three sub-phases:

1.   Parsing
2.   Representation
3.   Duplicate branches removal

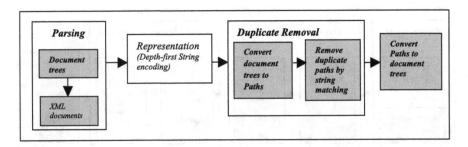

Fig. 2. The Pre-processing phase

3.1 Parsing

Each XML document in the INEX Wikipedia corpus is parsed and modeled as a rooted labeled ordered *document tree*. As the document tree has a root node and all the nodes are labeled using the tag names, it is *rooted* and *labeled*. The left-to-right *ordering* is preserved among the child nodes of a given parent in the document tree and therefore they are ordered.

3.2 Representation

The document trees need to be represented in a way that is suitable for mining in the next phase. We utilize the depth-first string format[6], a popular tree representation to represent the document trees. The *depth-first string encoding* traverses a tree in the *depth-first order* beginning from the root node. It represents the *depth-first traversal* of a given document tree in a string-like format where every node has a "–1" to represent backtracking and "#" to represent the end of the string encoding. For a document tree T with only one node r, the depth-first string of T is $S(T) = l_r\#$ where l is the integer label of the root node r. For a document tree T with multiple nodes, where r is the root node and the children nodes of r are $r_1,...,r_k$ preserving left to right ordering, the depth-first string of T is $S(T) = l_r l_{r_1}\text{-}1\ l_{r_2}\text{-}1...l_{r_k}\text{-}1\#$.

3.3 Duplicate Branches Removal

An analysis of the INEX Wikipedia dataset reveals that a large number of *document trees* contain duplicate branches. These duplicate branches are redundant information for structure mining and hence they cause additional overhead in the mining process. In order to remove the duplicate branches, the document tree is converted into a series of paths. The duplicate paths of the document trees are identified using string matching, and are removed. The remaining paths are combined together to create document trees without any duplicate branches.

4 Phase 2: Structure Mining

The structure mining phase involves mining the frequent subtrees. Instead of utilizing all the structures (nodes) for content extraction, we need to identify only the frequent or common subtrees. The assumption is that the content corresponding to infrequent subtrees should not play an important role in clustering. These nodes and their combinational structure are not common among the corpus so the content within these structures should also not be distinguishable for segmenting the corpus. Firstly document trees are mined for frequent subtrees for a given user-specified support threshold. However, there could be a very large number of frequent subtrees generated at the lower support threshold. In order to control the explosion we utilize closed frequent subtrees which are condensed representations of frequent subtrees without any information loss[7]. Frequent subtree mining on XML documents can be formally defined as follows:

Problem Definition for Frequent Subtree Mining on XML Documents

Given a collection of XML documents $D = \{D_1, D_2, D_3,...,D_n\}$ modeled as document trees $DT = \{DT_1, DT_2, DT_3,...,DT_n\}$ where n represents the number of XML documents or document trees in the corpus. There exists a subtree $DT' \subseteq DT_k$ that preserves the parent-child relationship among the nodes as that of the document tree DT_k. This type of subtree is called as an induced subtree.

Support(DT') (or frequency(DT')) is defined as the number of document trees in DT in which DT' is an induced subtree. An induced subtree DT' is frequent if its support is not less than a user-defined minimum support threshold. In other words, DT' is a frequent induced subtree in the document trees in DT such that,

$$frequency\ (DT')/|DT| \geq min\_supp \qquad (1)$$

where min_supp is the user-given support threshold and $|DT|$ is the number of document trees in the document tree dataset DT.

Due to the large number of frequent subtrees generated at lower support thresholds, recent researchers have focused on using condensed representations without any information loss [3]. The popular condensed representation is the closed frequent subtree which is defined as follows.

Problem Definition for Closed Frequent Induced (CFI) Subtree

For a given document tree dataset, $DT = \{DT_1, DT_2, DT_3,...,DT_n\}$, if there exists two frequent induced subtrees DT' and DT'', the frequent induced subtree DT' is a closed representation of DT'' iff for every $DT' \supseteq DT''$, $supp(DT') = supp(DT'')$ and there exists no superset for DT' having the same support as that of DT'. This property is called as *closure*.

	DT_1	DT_2	DT_3	DT_4
cfi_1	1	0	1	1
cfi_2	0	1	0	1
cfi_3	1	1	1	0

Fig. 3. *CD* matrix

In order to generate the closed frequent induced subtrees from the pre-processed document trees, PCITMiner[3] is utilized. This algorithm adopts the partition-based approach to determine the CFI subtrees. After CFI subtrees are generated from the corpus, their distribution in the corpus is modeled as a boolean subtree-document matrix, denoted by $CD_{|CFI|\times|DT|}$, where *CFI* subtrees represent the closed frequent induced subtrees and DT represents the document trees in the given document tree collection. Each cell in the CD matrix has a Boolean value to indicate the presence or absence of a given closed frequent induced subtree $\{cfi_1, cfi_2, ..., cfi_n\}$ in the document tree $\{DT_1, DT_2, DT_3, ..., DT_n\}$. Fig. 3 shows a CD matrix which stores the structural information of the XML documents distributed in the corpus. The CD matrix is represented as $CD_{|CFI| \times |DT|}$, with closed frequent induced subtrees $\{cfi_1, cfi_2, cfi_3\}$ in the document trees $DT = \{DT_1, DT_2, DT_3, DT_4\}$.

5   Phase 3: Content Extraction and Pre-processing

The CD matrix is used to extract the content from each of the XML documents. Using the CD matrix, the CFI subtrees present in a given XML document are listed. For every node label in the CFI subtree in the given document, its corresponding node values (or content) are extracted. Though the CD matrix does not include the occurrence of the CFI subtree in a given document, the content corresponding to every occurrence of the CFI subtree is stored.

The extracted content is a list of terms which is then pre-processed using the following steps:

1. Stop-word removal
2. Stemming
3. Integer removal
4. Shorter length words removal

5.1   Stop-Word Removal

Stop words are words that are considered poor as index terms[8], such as words that occur very frequently('the', 'of', 'and'). These words need to be removed prior to performing any natural language processing or data analysis. The most common stop list available for English text, from Christopher Fox, contains a list of 421 words[9]. Fox's stop list includes variants of the stop words, such as the word 'group' with its variants: 'groups', 'grouped' and 'grouping'.

However, there are always pitfalls in using the common stop list without any modification to suit to the domain under investigation. For example, the use of a common stop list causes the removal of the word 'back' even though 'back' (a part of the body) is a useful term in the medical domain. It is therefore essential to customize the stop list, considering the domain specific knowledge, in order to avoid removing important words. In this research, the stop word list has been customised considering the tag names of the XML documents; it contains 536 stopwords.

5.2   Stemming

Word stemming is a process to remove affixes (suffixes or prefixes) from the words and/or to replace them with the root word. For example, the word 'students' becomes 'student' and the word 'says' becomes 'say'. This research uses the most commonly used Porter stemming algorithm for affix removal[10].

5.3 Integer Removal

Due to the huge size of the Wikipedia dataset and its very large number of unique terms; it is essential to reduce the dimension of the dataset without any information loss. A careful analysis of the dataset revealed that there were a large number of integers that not contribute to the semantics of the documents and hence these were removed in the pre-processing step. The filenames of the XML documents and the links to other XML documents which were integers were retained.

5.4 Shorter Length Words Removal

Also based on the analysis of the dataset, words with fewer than 4 charactersare not meaningful, thus, they were removed. After the pre-processing of the extracted content, the content is represented as a term-document matrix where each cell in that matrix contains the term frequency of that corresponding term in the respective document.

The pre-processed content is represented as a term-document matrix, $TDoc_{|Term| \times |D|}$, where *Term* represents the terms corresponding to the CFI subtrees for the respective document in the collection D. Each cell in the *TDoc* matrix contains the term frequency for the corresponding set of closed frequent subtrees in a given XML document. This matrix contains a reduced number of terms in comparison to the entire collection, as the terms corresponding to infrequent subtrees were removed. The *TDoc* matrix is used in the next phase to compute the similarity between the XML documents for the clustering of the dataset.

6 Phase 4: Clustering

The term-document matrix generated from the previous phase becomes input to a partitional clustering algorithm. The k-way clustering algorithm[11] is used in this research as it groups the documents in the required number of clusters. The k-way clustering solution computes cluster by performing a sequence of $k-1$ repeated bisections. The input matrix is first clustered into two groups, and then one of these groups is chosen and bisected further. This process of bisection continues until the desired number of bisections is reached. During each step of bisection, the cluster is bisected so that the resulting 2-way clustering solution locally optimizes a particular criterion function[11].

7 Experiments and Discussion

We used the Microsoft Visual C++ 2005 implementation of PCITMiner[3]. The same development environment was used to implement the algorithm for content extraction using closed frequent induced subtrees. A number of experiments were conducted on the Wikipedia corpus using the INEX XML Mining Challenge 2008 testing dataset. The testing dataset contains 114,366 documents and the required number of clusters for the INEX result submission was 15 clusters. To cluster the XML documents, the k-way clustering algorithm implemented in CLUTO[11] is applied to the term-document matrix representing the constrained content of documents.

The two commonly used clustering measures, namely Micro F1 (intra-cluster purity) and Macro F1 (inter-cluster purity), are used to evaluate the accuracy of the clustering solutions which are based on the F1-measure. The F1 measure can be defined by

$$F1 = \frac{2 * Precision * Recall}{Precision + Recall} \tag{2}$$

$$Precision = \frac{TP}{TP + FP} \tag{3}$$

$$Recall = \frac{TP}{TP + FN} \tag{4}$$

where TP denotes True Positives, FP denotes False Positives, and FN denotes False Negatives. Micro-average F1 is calculated by summing up the TP, FP and FN values from all the categories individually. Macro-average F1, on the other hand, is derived from averaging the F1 values over all the categories. The best clustering solution for an input data set is the one where micro- and macro-average F1 measures are close to 1. Table 1 summarises the clustering results for INEX Wikipedia XML Mining Track 2008.

Table 1. Submitted clustering results for INEX Wikipedia XML Mining Track 2008

Approaches	No. of clusters	Micro F1	Macro F1
Hagenbuchner-01	15	0.26	0.38
QUT LSK_1	15	0.45	0.56
	30	0.53	0.57
Vries_15k_20k	15	0.49	0.59
QUT collection_15 (Our approach)	15	0.48	0.51
	30	0.53	0.58

From Table 1, it is very clear that our proposed approach performs better than the structure-only approach using Self-Organizing Map (SOM) of Hagenbuchner-01. Also, our approach showed not much loss in accuracy to other content-only methods such as QUT LSK_1 using Latent Semantic Kernel (LSK) and Vries_15k_20k using K-tree and BM25 representation. Hence, our method is suitable for combining the structure and content of XML documents without compromising on the accuracy.

In order to measure the reduction in the number of terms, a Vector Space Model (VSM) was built on all the terms of the documents and clustering was then applied on it. Table 2 summarises the dimensionality reduction in both the number of unique terms and the total number of terms.

Table 2. Dimensionality reduction

No. of Clus	Method	Micro-avg F1	Macro-avg F1	Num of Uniq. Terms	#Num of terms
	Our Approach	**0.53**	**0.58**	**442509**	**8528006**
30	On all the terms (without dimension reduction)	0.54	0.58	896050	13813559

From the experimental results summarised in Table 2 it is evident that, in spite of the reduction in the number of unique terms and the total number of terms by about 50% and 40% respectively, there is not any significant loss in accuracy of the clustering solution. This confirms that the content within the infrequent trees can be avoided as it is surplus for the clustering solution. The proposed approach is able to effectively combine the structure and content for clustering the documents, and to reduce the dimensionality.

8 Conclusion

In this paper, we have proposed and presented the results of a clustering approach using frequent subtrees for mining both the structure and content of XML documents on the INEX 2008 Wikipedia dataset. The main aim of this study is to explore and understand the importance of the content and structure of the XML documents for the clustering task. In order to cluster the XML documents, we have used content corresponding to the frequent subtrees in a given document and have generated a terms by document matrix. Using the matrix, we have computed the similarity between XML documents then clustered them based on their similarity values. We have demonstrated that by including the structure we could not only reduce the dimensionality but also provide more meaningful clusters.

References

1. Nayak, R., Witt, R., Tonev, A.: Data Mining and XML Documents. In: International Conference on Internet Computing (2002)
2. Tran, T., Nayak, R.: Evaluating the Performance of XML Document Clustering by Structure Only. In: Comparative Evaluation of XML Information Retrieval Systems, pp. 473–484 (2007)
3. Kutty, S., Nayak, R., Li, Y.: PCITMiner-Prefix-based Closed Induced Tree Miner for finding closed induced frequent subtrees. In: Sixth Australasian Data Mining Conference (AusDM 2007). ACS, Gold Coast (2007)

4. Nayak, R.: Investigating Semantic Measures in XML Clustering. In: Proceedings of the 2006 IEEE/WIC/ACM International Conference on Web Intelligence, pp. 1042–1045. IEEE Computer Society Press, Los Alamitos (2006)
5. Aggarwal, C.C., et al.: Xproj: a framework for projected structural clustering of xml documents. In: Proceedings of the 13th ACM SIGKDD international conference on Knowledge discovery and data mining, pp. 46–55. ACM, San Jose (2007)
6. Chi, Y., et al.: Frequent Subtree Mining-An Overview. In: Fundamenta Informaticae, pp. 161–198. IOS Press, Amsterdam (2005)
7. Kutty, S., Nayak, R., Li, Y.: XML Data Mining: Process and Applications. In: Song, M., Wu, Y.-F. (eds.) Handbook of Research on Text and Web Mining Technologies. Idea Group Inc., USA (2008)
8. Rijsbergen, C.J.v.: Information Retrieval. Butterworth, London (1979)
9. Fox, C.: A stop list for general text. ACM SIGIR Forum 24(1-2), 19–35 (1989)
10. Porter, M.F.: An algorithm for suffix stripping. Program 14(3), 130–137 (1980)
11. Karypis, G.: CLUTO-Software for Clustering High-Dimensional Datasets | Karypis Lab, May 25 (2007), http://glaros.dtc.umn.edu/gkhome/views/cluto

UJM at INEX 2008 XML Mining Track*

Mathias Géry, Christine Largeron, and Christophe Moulin

Université de Lyon, F-42023, Saint-Étienne, France
CNRS UMR 5516, Laboratoire Hubert Curien
Université de Saint-Étienne Jean Monnet, F-42023, France
{mathias.gery,christine.largeron,christophe.moulin}@univ-st-etienne.fr

Abstract. This paper reports our experiments carried out for the INEX XML Mining track, consisting in developing categorization (or classification) and clustering methods for XML documents. We represent XML documents as vectors of indexed terms. For our first participation, the purpose of our experiments is twofold: Firstly, our overall aim is to set up a categorization text only approach that can be used as a baseline for further work which will take into account the structure of the XML documents. Secondly, our goal is to define two criteria (CC and CCE) based on terms distribution for reducing the size of the index. Results of our baseline are good and using our two criteria, we improve these results while we slightly reduce the index term. The results are slightly worse when we sharply reduce the size of the index of terms.

1 Introduction

The INEX XML Mining Track is organized in order to identify and design machine learning algorithms suited for XML documents mining [1]. Two tasks are proposed: clustering and categorization. Clustering is an unsupervised process through which all the documents must be classified into clusters. The problem is to find meaningful clusters without any prior information. Categorization (or classification) is a supervised task for which, given a set of categories, a training set of preclassified documents is provided. Using this training set, the task consists in learning the classes descriptions in order to be able to classify a new document in one of the categories.

This second task is considered in this article. Moreover, even if the content information (the text of the documents), the structural information (the XML structure of the documents) and the links between the documents can be used for this task, we have only exploited the textual information. Indeed, this is our first participation to this track and our aim was to design a framework that could be used as a baseline for further works dealing with structured documents.

More precisely, we focus on the preprocessing step, particularly the features selection, which is an usual step of the knowledge discovery process [8,3,2]. On

---

* This work has been partly funded by the Web Intelligence project (région Rhône-Alpes, cf. http://www.web-intelligence-rhone-alpes.org).

S. Geva, J. Kamps, and A. Trotman (Eds.): INEX 2008, LNCS 5631, pp. 446–452, 2009.

textual data, this step can be essential for improving the performance of the categorization algorithm. It exists a lot of words in the natural language, including stop words, synonymous, *etc.*. These words are not equally useful for categorization. Moreover, their distribution must also be considered. For example, words that appear in a single document are not useful for the categorization task.

So, we need to extract from the text a subset of terms that can be used to efficiently represent the documents in view of their categorization. In this paper, the documents are represented according to the vector space model (VSM [5]). Our aim is to adapt some VSM principles, for example the measure of the discriminatory power of a term, to the categorization task. We propose two criteria based on terms distribution aiming at extracting the indexing terms from the training set corpora. After a brief presentation of the VSM given to introduce our notations in section 2, these criteria are defined in the following section. Our categorization approach is described in section 3 while the experiments and the obtained results are detailed in sections 4 and 5.

2   Document Model for Categorization

2.1   Vector Space Model (VSM)

Vector space model, introduced by Salton and al. [5], has been widely used for representing text documents as vectors which contain terms weights. Given a collection D of documents, an index $T = \{t_1, t_2, ..., t_{|T|}\}$, where $|T|$ denotes the cardinal of T, gives the list of terms (or features) encountered in the documents of D. A document d_i of D is represented by a vector $\boldsymbol{d_i} = (w_{i,1}, w_{i,2}, ..., w_{i,|T|})$ where $w_{i,j}$ represents the weight of the term t_j in the document d_i. In order to calculate this weight, TF.IDF formula can be used.

2.2   TF: Term Representativeness

TF (Term Frequency), the relative frequency of term t_j in a document d_i, is defined by:

$$tf_{i,j} = \frac{n_{i,j}}{\sum_l n_{i,l}}$$

where $n_{i,j}$ is the number of occurrences of t_j in document d_i normalized by the number of terms in document d_i. The more frequent the term t_j in document d_i, the higher is the $tf_{i,j}$.

2.3   IDF: Discriminatory Power of a Term

IDF (Inverse Document Frequency) measures the discriminatory power of the term t_j. It is defined by:

$$idf_j = \log \frac{|D|}{|\{d_i : t_j \in d_i\}|}$$

where $|D|$ is the total number of documents in the corpus and $|\{d_i : t_j \in d_i\}|$ is the number of documents in which the term t_j occurs at least one time. The less frequent the term t_j in the collection of documents, the higher is the idf_j.

The weight $w_{i,j}$ of a term t_j in a document d_i is then obtained by combining the two previous criteria:

$$w_{i,j} = tf_{i,j} \times idf_j$$

The more frequent the term t_j is in document d_i and the less frequent it is in the other documents, the higher is the weight $w_{i,j}$.

3   Criteria for Features Selection

This VSM is widely used for text mining and information retrieval, as well for free format document like scientific articles as for semi structured document written in markup languages like HTML or XML.

But, in the context of categorization, even for limited collections, the dimensionality of the index can be exceedingly large. For example, in INEX collection, 652 876 non trivial words have been identified. This is a real problem for categorization since the terms belonging to this bag of words are not necessarily discriminatory features of the categories. So, we introduced two criteria (CC and CCE) in order to select a subset of T providing a description of the documents belonging to the same category. We consider that these terms must be very frequent in the documents of the category and, on the contrary, that they must be infrequent in the other categories.

3.1   Category Coverage criteria (CC)

Let df_j^k be the number of documents in the category C_k where term t_j appears, and f_j^k be the frequency of documents belonging to C_k and including t_j:

$$df_j^k = |\{d_i \in C_k : t_j \in d_i\}|, k \in \{1, ...r\} \tag{1}$$

$$f_j^k = \frac{df_j^k}{|C_k|} \tag{2}$$

The higher the number of documents of C_k containing t_j, the higher is f_j^k.

On the other hand, the term t_j could be considered as a discriminant term if most of the documents, where t_j appears, belongs to the same category. Thus, a first criteria, noted CC (Category Coverage), is computed as follows:

$$CC_j^k = \frac{df_j^k}{|C_k|} * \frac{f_j^k}{\sum_k f_j^k}$$

$$CC_j^k = \frac{(f_j^k)^2}{\sum_k f_j^k}$$

If the value of CC_j^k is high, then t_j is a characteristic feature of the category C_k.

3.2   Category Coverage Entropy Criteria (CCE)

The frequency f_j^k considers the number of documents containing t_j but it does not take into account the number of occurrences of t_j in the category. It is the reason why we consider also p_j^k the frequency of t_j in the category C_k and a measure commonly used in information theory, called entropy, which evaluates the purity of the categories for the term t_j. In the context of text categorization, it measures the discriminatory power of t_j. Let n_j^k be the number of occurrences of t_j in the documents of C_k and p_j^k the corresponding frequency:

$$n_j^k = \sum_{d_i \in C_k} n_{i,j} \qquad\qquad p_j^k = \frac{n_j^k}{\sum_{k=1,r} n_j^k}$$

The Shannon entropy E_j of the term t_j is given by [6]:

$$E_j = - \sum_{k=1,r} (p_j^k) * log_2(p_j^k)$$

The entropy is minimal, equal to 0, if the term t_j appears only in one category. We consider that this term might have a good discriminatory power for the categorization task. Conversely, the entropy is maximal if t_j is not a good feature for representing the documents *i.e.* if t_j appears in all the categories with the same frequency.

We propose a second criteria, denoted CCE (Category Coverage Entropy), combining f_j^k (from CC) and entropy. CCE is defined by:

$$CCE_j^k = (alpha * f_j^k) + (1 - alpha) * (1 - \frac{E_j}{MaxE})$$

where $alpha$ is a parameter and $MaxE$ is the maximal value of E. When the term t_j is characteristic of the category C_k, the value of the criteria is high.

For each category, a subset of the terms of T corresponding to the highest values of the criterion is built. Then, the index is defined as the union of these subsets.

4   Experiments

4.1   Collection INEX XML Mining

The collection is composed of 114 336 XML documents of the Wikipedia XML Corpus. This subset of Wikipedia represents 15 categories, each corresponding to one subject or topic. Each document of the collection belongs to one category. In the XML Mining Track, the training set is composed of 10% of the collection.

4.2   Preprocessing

The first step of the categorization approach that we propose, consists in a preprocessing of the collection. It begins by the construction of the list all the terms (or features) encountered in the documents of the collection. This index of 652 876 terms is build using the LEMUR software[1]. The Porter Algorithm [4] has also been applied in order to reduce different forms of a word to a common form. This operation reduces the index to 560 209 terms. However, it still remains a large number of irrelevant terms that could degrade the categorization, e.g.: numbers (7277, -1224, 0d254c, etc.), terms with less than three characters, terms that appear less than three times, or terms that appear in almost all the documents of the training set corpus. The index obtained at this stage is denoted I. In our experiments, its size is reduced to 161 609 terms on all the documents of the collection and to 77 697 on the training set.

4.3   Features Selection

However, as explained in the previous section, the terms of I are not necessarily appropriated for the categorization task inasmuch they are not discriminatory for the categories. This is the reason why our criteria based on entropy and on frequency are used to select more suited features. The terms were sorted according to CC and CCE and only those corresponding to the highest values are retained. In our experiments, the top 100 terms by class and the top 10 000 terms by class were considered for each criteria to build four indexes, denoted respectively CC_{100} and CC_{10000} using CC and CCE_{100} and CCE_{10000} using CCE. Table 1 indicates the size of these indexes.

Table 1. Indexes sizes

Index	number of words
I	77697
CC_{100}	1 051
CC_{10000}	75 181
CCE_{100}	909
CCE_{10000}	77 580

Using one of these indexes, the content of a document is then represented by the $tf.idf$ vector model described in the first section.

The second step is the categorization step itself. Two usual methods of classification are used: Support Vector Machines (SVM) and k-nearest neighbors. Only the most promising results obtained with the SVM were submitted. SVM was introduced by Vapnik for solving two class pattern recognition problems using Structural Risk Minimization principal[7]. In our experiments, the SVM

---

[1] Lemur is available at the URL http://www.lemurproject.org

algorithm available in the Liblinear library[2] has been used. The results provided by this approach are presented in the next section.

5   Experimental Results

This work has been done with a dual purpose: firstly develop a categorization text approach usable as a baseline for further work on XML categorization taking into account the structure, and secondly evaluate performances of this method using our selection features approach.

5.1   Global Results

We have submitted 5 experiments using our 5 indexes presented in table 1. The global results of XML Mining 2008 are synthesized in table 2 (participant: LaHC).

Table 2. Summary of all XML Mining results

Rank	Participant	Run	Recall	Documents
1	LaHC	submission.expe_5.tf_idf_T5_10000.txt	0.7876	102 929
2	LaHC	submission.expe_3.tf_idf_T4_10000.txt	0.7874	102 929
3	LaHC	submission.expe_1.tf_idf_TA.txt	0.7873	102 929
4	Vries	Vries_classification_text_and_links.txt	0.7849	102 929
5	boris	boris_inex.tfidf.sim.037.it3.txt	0.7379	102 929
6	boris	boris_inex.tfidf1.sim.0.38.3.txt	0.7347	102 929
7	boris	boris_inex.tfidf.sim.034.it2.txt	0.7309	102 929
8	LaHC	submission.expe_4.tf_idf_T5_100.txt	0.7230	102 929
9	kaptein	kaptein_2008NBscoresv02.txt	0.6980	102 929
10	kaptein	kaptein_2008run.txt	0.6978	102929
11	romero	romero_naive_bayes_links.txt	0.6813	102 929
12	LaHC	submission.expe_2.tf_idf_T4_100.txt	0.6770	102 929
13	romero	romero_naive_bayes.txt	0.6767	102 929
14	Vries	Vries_classification_links_only.txt	0.6232	102 929
15	Vries	Vries_classification_text_only.txt	0.2444	92 647

5.2   Baseline Results

Our baseline corresponds to the first experiment (*expe_1*), which was ranked 3th with a quite good recall: 0.7873.

5.3   Selection Features Improves Results

When we select 10 000 terms for each class using *CCE* (*expe_3*) and *CC* (*expe_5*), we reduce the size of the index to respectively 77 580 and 75 181.

---

[2] http://www.csie.ntu.edu.tw/~cjlin/liblinear/ - L2 loss support vector machine primal

This reduction is small compared to the size of the baseline index (77 697). However, it lets us to slightly improve our baseline to 0.7874 with CCE and 0.7876 with CC. These three runs obtained the three best results of the XML Mining challenge.

5.4   Selection Features Reduces Indexes

The last two submitted runs correspond to the selection of the first 100 terms for each class using CCE (expe_2) and CC (expe_4). As presented in table 1, the size of the index is sharply reduced to 909 terms for CCE and 1 051 for CC. This reduction respectively correspond to 85% and 74% of the size of the baseline index. Even if the obtained results are lower than the results obtained with larger indexes, they are still relatively good. Indeed, the obtained recall is 0.6770 with CCE and 0.7230 with CC.

6   Conclusion

We proposed a categorization text approach for XML documents that let us obtain a good baseline for further work. For now we just used CC and CCE criteria as a threshold to select terms in order to build the index. For future work, we aim at exploiting the computed value of CC and CCE to improve the categorization. Moreover, we could use the structure information of XML documents represented by the links between document to improve even more the results.

References

1. Denoyer, L., Gallinari, P.: Report on the xml mining track at inex 2007 categorization and clustering of xml documents. SIGIR Forum 42(1), 22–28 (2008)
2. Forman, G., Guyon, I., Elisseeff, A.: An extensive empirical study of feature selection metrics for text classification. Journal of Machine Learning Research 3, 1289–1305 (2003)
3. Joachims, T.: Text categorization with support vector machines: Learning with many relevant features. In: Nédellec, C., Rouveirol, C. (eds.) ECML 1998. LNCS, vol. 1398, pp. 137–142. Springer, Heidelberg (1998)
4. Porter, M.F.: An algorithm for suffix stripping. In: Readings in information retrieval, pp. 313–316 (1997)
5. Salton, G., McGill, M.J.: Introduction to modern information retrieval. McGraw-Hill, New York (1983)
6. Shannon, C.E.: A mathematical theory of communication. Bell System Technical Journal 27, 379–423, 623–656 (1948)
7. Vapnik, V.: The Nature of Statistical Learning Theory. Springer, Heidelberg (1995)
8. Yang, Y., Pedersen, J.: A comparative study on feature selection in text categorization. In: Int. Conference on Machine Learning ICML 1997, pp. 12–420 (1997)

Probabilistic Methods for Link-Based Classification at INEX 2008

Luis M. de Campos, Juan M. Fernández-Luna,
Juan F. Huete, and Alfonso E. Romero

Departamento de Ciencias de la Computación e Inteligencia Artificial
E.T.S.I. Informática y de Telecomunicación, Universidad de Granada,
18071 – Granada, Spain
{lci,jmfluna,jhg,aeromero}@decsai.ugr.es

Abstract. In this paper we propose a new method for link-based classification using Bayesian networks. It can be used in combination with any content only probabilistic classsifier, so it can be useful in combination with several different classifiers. We also report the results obtained of its application to the XML Document Mining Track of INEX'08.

1 Introduction

The 2008 edition of the INEX Workshop was the second year that members of the research group "Uncertainty Treatment in Artificial Intelligence" at the University of Granada participate in the Document Mining track. Our aim is as in previous editions, to provide a solution to the proposed problems on the framework of Probabilistic Graphical Models (PGMs).

The corpus given for 2008 differs slightly on the one of the previous year [3]. Again, as in 2007, it is a single-label corpus (a subset of the AdHoc one [2] but using a different set of 16 categories). Moreover, this year a file with the list of links between XML documents has been added. Because the track has been substantially changed, it would be firstly interesting to check the utility of using the link information. We will show later that those links add relevant information for the categorization of documents.

On the other hand, given that the 2008 corpus is coming from the same source (Wikipedia) as the 2007 corpus, we think that it might not be worthwhile to use the structural information of the documents for categorization. In [1] we showed that even using some very intuitive XML document transformations to flat text documents, classification accuracy was not improving, getting worse in some of the cases. In this year, then, we have used a more pragmatic approach, directly ignoring the structural information by simply removing XML tags from the documents.

This work is structured as follows: firstly we perform a study of the link structure in the corpus, in order to show its importance for categorization. Secondly, we present our model for categorization, based on Bayesian networks, with some variants. Then, we make some experiments on the corpus to show how our model performs, and finally, we list some conclusions and future works.

S. Geva, J. Kamps, and A. Trotman (Eds.): INEX 2008, LNCS 5631, pp. 453–459, 2009.

2   Linked Files. Study on the Corpus

As we said before, we are given a set of links between document files as additional training information, making some explicit dependencies arise between documents. This information violates a "natural" assumption of traditional classification methods: the documents are independent of each other. This case of non independent documents can have different forms; and the graph of relationships among documents are not neccesarily regular (it can be a general directed graph, neither a tree nor a forest). •

Clearly, for the problem of document categorization, these intra-corpus dependences could be ignored, applying "traditional" text categorization algorithms but, as we will show afterwards, the information from linked files can be a very valuable data.

But, how are those links supposed to help in the final process of text categorization? Obviously, not all kinds of links are equal, because they can give different information (even none). A careful review of those different kinds of dependencies represented by hyperlinks (*regularities*) is given by Yang [6], and following her terminology we can conjecture that we are in a "encyclopedia regularity". We reproduce here her definition:

> *One of the simplest regularities is that certain documents with a class label only link with documents with the same class label. This regularity can be approximately found in encyclopedia corpus, since encyclopedia articles generally reference other articles which are topically similar.*

We have plotted, in figure 2, a matrix where the rows and columns are one of the 16 categories. Each matrix value $m_{i,j}$ represents the probability that a document of class i links a document of class j, estimated from the training document collection.

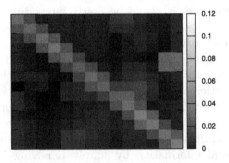

Fig. 1. Probability that a document from category i links a document from category j

As it can be seen (the matrix has a strong weight in its diagonal), documents of one category tend to link documents of the same category. Moreover, doing the same plot with the probability of a document of class i being linked by a document of class j, and another one with the probability of a document of class i links or is linked by a document of class j, we obtain a similar result (a matrix with a high weight for the diagonal values).

Thus, although we could think that only the outlinks tend to be useful, we can affirm that also inlinks are useful, and also consider the links without any direction.

3   Proposed Method

3.1   Original Method

The method proposed is an extension of a probabilistic classifier (we shall use in the experiments the Naive Bayes classifier but other probabilistic classifiers could also be employed) where the evidence is not only the document to classify, but this document together with the set of related documents. Note that, in principle, we will try to use only information which is available in a natural way for a text classifier. Considering that different documents are processed through batch processing, the information easily available to a system, given a document, is the set of documents it links (not the set of documents that link it).

Consider a document d_0 which links with documents $d_1, \ldots, d_m$. We shall consider the random variables $C_0, C_1, \ldots, C_m$, all of them taking values in the set of possible category labels. Each variable C_i represents the event "The class of document d_i is". Let e_i be the evidence available concerning the possible classification of each document d_i (the set of terms used to index the document d_i or the class label of d_i). The proposed model can be graphically represented as the Bayesian network displayed in figure 2.

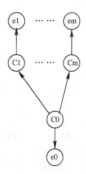

Fig. 2. Bayesian network representing the proposed model

The independencies represented by the Bayesian network are the following: given the true class of the document we want to classify, the categories of the linked documents are independent among each other. Moreover, given the true category of a linked document, the evidence about this category due to the document content is independent of the original category of the document we want to classify.

Our objective is to compute the posterior probability $p(C_0|e)$, where e is all the available evidence concerning document d_0, $e = \{e_0, e_1, \ldots, e_m\}$. It can be proven that, taking into account the independencies represented in the Bayesian network, this probability can be expressed as follows:

$$p(C_0 = c_0|e) \propto p(C_0 = c_0|e_0) \prod_{i=1}^{m} \left(\sum_{c_i} p(C_i = c_i|c_0) \frac{p(C_i = c_i|e_i)}{p(C_i = c_i)} \right) \quad (1)$$

As we can observe in equation (1), the posterior probability of C_0 has two components: a part which only depends on the evidence associated to the document d_0 to be classified ($p(C_0 = c_0|e_0)$) and another part related with the information about the class labels of each one of the documents linked with d_0, which can be obtained using its own local evidence ($p(C_i = c_i|e_i)$). This information is combined with the estimated probabilities of a linked document being of class c_i given that the document linking to it is of class c_0.

The posterior probabilities $p(C_0 = c_0|e_0)$ and $p(C_i = c_i|e_i)$ can be obtained using some standard probabilistic classifier, whereas the probabilities $p(C_i = c_i)$ and $p(C_i = c_i|c_0)$ can be estimated from the training data simply by following these formulas:

$$p(C_i = c_i) = \frac{N_i}{N}$$

and

$$p(C_i = c_i|c_0) = \frac{L_{0i} + 1}{L_{0\bullet} + |C|}$$

where N_i is the number of training documents classified by category i, N is the total number of documents, L_{0i} is the number of links from documents of category 0 to category i, $L_{0\bullet}$ is the total number of links from documents of category 0, and $|C|$ is the number of categories. Note that in the estimation of $p(C_i = c_i|c_0)$ we have used Lapace smoothing. In all our posterior experiments, using Laplace gives better results than not using it.

Therefore, we can think of the proposed model as a method to modify the results offered by a base probabilistic classifier taking into account the information available about the linked documents and the relationships between categories (the prior probabilities $p(C_i = c_i)$ and the values $p(C_i = c_i|c_j)$).

3.2   Extension to Inlinks and Undirected Links

The independencies represented by the Bayesian network are not directly related with the direction of the links. Instead of outlinks, we could think of the previous model as a model that takes into consideration the incoming links. Thus, the

e_i $(i > 0)$ variables would represent the documents that link to one (instead of the files linked by one), and the formula (1) would still be valid. In the case of the incoming links, we should reestimate the dependencies among categories as follows:

$$p(C_i = c_i|c_0) = \frac{L_{i0} + 1}{L_{\bullet 0} + |C|}$$

where L_{i0} is, as previously stated, the number of links from documents of category i to category 0, and $L_{\bullet 0}$ is the total number of links to documents of category 0.

Moreover, in the collective classification literature, the direction of the links is often not considered, so, we also could propose a model where e_i $(i > 0)$ represent the documents linked or being linked (that is to say, neighboured) by the document to classify. In that case, the probabilities would be these:

$$p(C_i = c_i|c_0) = \frac{L_{i0} + L_{0i} + 1}{L_{\bullet 0} + L_{0\bullet} + |C|}.$$

Therefore, these would be our three models: the original one (with incoming links), and the extensions using outlinks and undirected links.

4   Experimental Results

To make the values comparable with the submitted runs, we have also performed some experiments on the test set in order to show the effectiveness (recall) of our approach. First of all we study the two submitted runs, a baseline (flat text classifier) and our proposals (combined with Naive Bayes):

- A classical Naive Bayes algorithm on the flat text documents: 0.67674 of recall.
- Our model (outlinks): 0.6787 of recall.

The two aditional models which were not submitted to the track give the following results:

- Our model (inlinks): 0.67894 of recall.
- Our model (neighbours): 0.68273 of recall.

Although all our methods improve the baseline, the results achieved are not really significant. In order to justify the value of our model, we are asking now ourselves which is the predicting power of our proposal, by making some additional computations in an "ideal setting". This "ideal setting" is, for a document being classified, to be surrounded (linking, linked by or both of them) with documents whose class membership is perfectly known (and hence we can set for a related document d_k of category c_i, $P(C_k = c_i|d_k) = 1$ -the true class- and $P(C_k = c_j|d_k) = 0$ -the false categories- $\forall c_j \neq c_i$). Remember that, in previous experiments, a surrounding file whose category was not known should be first classified by Naïve Bayes, and then that estimation (the output probability values) was used in our model.

So, the procedure is the following: for each document to classify, look at the surrounding files. For each one, if it is a training file, use that information (perfect knowledge), and if it is a test file, use also its categorization information taken from the test set labels to have our file related to documents with perfect knowledge. This "acquired" knowledge is obviously removed for the next document classification.

In this "ideal setting" we have made two experiments: one combining naïve Bayes with our model (like the second one of the previous two), and one which combined a "blind classifier" (the one that gives equal probability to each category) with our model. The first should be better than the two previous ones, and the second one could give us an idea of the true contribution to the predictive power of our model, despite the underlying basic classifier used.

- Model for outlinks in an "ideal setting" using Naive Bayes as a base classifier: 0.69553 of recall.
- Model for outlinks in an "ideal setting" using a "blind classifier": 0.46500 of recall.
- Model for inlinks in an "ideal setting" using Naive Bayes as a base classifier: 0.69362 of recall.
- Model for inlinks in an "ideal setting" using a "blind classifier": 0.73278 of recall.
- Model for neighbours in an "ideal setting" using Naive Bayes as a base classifier: 0.70212 of recall.
- Model for neighbours in an "ideal setting" using a "blind classifier": 0.66271 of recall.

The first experiment provides the desired result: the recall is improved (although not so much). The small improvement could be due, in some part, to the extreme values given in this corpus by the Naive Bayes classifier (very close to 0 and 1). The introduction of these values in the final formula, as the first factor in the final posterior probability of each document, makes difficult to take into account (in the categories of the values close to 0) the information provided by the second factor (the combination of the information given by all the linked files), vanishing in some cases because of the low value of the first factor.

However, the second experiment showed us that, only using link information, and ignoring all content information of the document to classify, in this "ideal setting" of knowing the true class of each surrounding document, our method can reach 0.46500, 0.73278 or 0.66271 of recall. In the case of the inlinks, ignoring the content of the document to classify and perfectly knowing the values of the categories of the surrounding documents, gives better results than using this content. Besides, these values are clearly high, whichg gives us the idea of the predictive power of link information in this problem.

5 Conclusions and Future Works

We have proposed a new model for classification of linked documents, based on Bayesian networks. We have also justified the possibly good performance of the model in an "ideal" environment, with some promising results. Regrettably, our results in this track have been very discrete, reaching the final positions and not improving so much the naïve Bayes baseline.

To improve those poor results in the future, we could use a classifier (probabilistic) with a better performance. Such a classifier could be a logistic regression procedure, a higher dependence network or just a SVM with probabilistic output (using Platt's algorithm [5]). The probability assignments should also be "softer", in the sense that several categories should receive positive probability (naïve Bayes tended to concentrate all the probability in one category, zeroing the others and making the information provided by the links not useful, in some way).

As future work we would like to study this problem as a collaborative classification problem (see, for instance [4], and try to apply this method in one of the particular solutions (those that need a "local classifier") that are being given to it.

Acknowledgments. This work has been jointly supported by the Spanish Consejería de Innovación, Ciencia y Empresa de la Junta de Andalucía, Ministerio de Ciencia de Innovación and the research programme Consolider Ingenio 2010, under projects TIC-276, TIN2008-06566-C04-01 and CSD2007-00018, respectively.

References

1. de Campos, L.M., Fernández-Luna, J.M., Huete, J.F., Romero, A.E.: Probabilistic Methods for Structured Document Classification at INEX 2007. In: Fuhr, N., Kamps, J., Lalmas, M., Trotman, A. (eds.) INEX 2007. LNCS, vol. 4862, pp. 195–206. Springer, Heidelberg (2008)
2. Denoyer, L., Gallinari, P.: The Wikipedia XML Corpus. SIGIR Forum 40(1), 64–69 (2006)
3. Denoyer, L., Gallinari, P.: Report on the XML mining track at INEX 2007 categorization and clustering of XML documents. SIGIR Forum 42(1), 22–28 (2008)
4. Sen, P., Getoor, L.: Link-based Classification, Technical Report CS-TR-4858, University of Maryland, Number CS-TR-4858 - February 2007 (2007)
5. Platt, J.: Probabilistic Outputs for Support Vector Machines and Comparisons to Regularized Likelihood Methods. In: Smola, A., Bartlett, P., Scholkopf, B., Schuurmans, D. (eds.) Advances in Large Margin Classifiers. MIT Press, Cambridge (1999)
6. Yang, Y., Slattery, S.: A study of approaches to hypertext categorization. Journal of Intelligent Information Systems 18, 219–241 (2002)

Utilizing the Structure and Content Information for XML Document Clustering

Tien Tran, Sangeetha Kutty, and Richi Nayak

Faculty of Science and Technology
Queensland University of Technology
GPO Box 2434, Brisbane Qld 4001, Australia
{t4.tran,s.kutty,r.nayak}@qut.edu.au

Abstract. This paper reports on the experiments and results of a clustering approach used in the INEX 2008 document mining challenge. The clustering approach utilizes both the structure and content information of the Wikipedia XML document collection. A latent semantic kernel (LSK) is used to measure the semantic similarity between XML documents based on their content features. The construction of a latent semantic kernel involves the computing of singular vector decomposition (SVD). On a large feature space matrix, the computation of SVD is very expensive in terms of time and memory requirements. Thus in this clustering approach, the dimension of the document space of a term-document matrix is reduced before performing SVD. The document space reduction is based on the common structural information of the Wikipedia XML document collection. The proposed clustering approach has shown to be effective on the Wikipedia collection in the INEX 2008 document mining challenge.

Keywords: Wikipedia, clustering, LSK, INEX 2008.

1 Introduction

Most electronic data on the web, nowadays, is presented in the form of semi-structured data, such as in XML or HTML format. Semi-structured data, usually, does not follow a specific structure and its data is nested and heterogeneous. Due to the continuous growth of semi-structured data on the web, the need to efficiently manage these data becomes inevitable. Data mining techniques such as clustering has been widely used to group data based on their common features without prior knowledge [1]. Data or document clustering has become important in applications such as database indexing, data-warehouse, data integration and document engineering.

A number of clustering approaches for semi-structured documents have been proposed based on the commonality of document content information [2, 3] or of document structural information [4, 5]. The content and the structural information of semi-structured documents provide a lot of information in aiding the clustering process. The content-based approaches [6, 7] usually utilize techniques such as vector space model [8] (VSM) for representing and measuring the similarity between document for

S. Geva, J. Kamps, and A. Trotman (Eds.): INEX 2008, LNCS 5631, pp. 460–468, 2009.
© Springer-Verlag Berlin Heidelberg 2009

clustering. However, the vector space model using tf-idf weighting method [8] is associated with a number of limitations [9]. Firstly, it assigns weights to terms according to term frequency and ignores the semantic association of terms. Secondly, it does not perform well when both the structural and content features are used together to infer the similarity of the XML documents [6]. Therefore, for the INEX 2008 document mining challenge we constructed a semantic kernel based on the latent semantic analysis [10] to determine the semantic similarity between the content of the Wikipedia documents. In practice, latent semantic kernel performs better than tf-idf weighting for learning the semantic similarity between the content of documents; however, it is much more expensive to use in terms of computational time and memory requirements.

For the clustering of Wikipedia collection in the INEX 2008 document mining challenge, documents are first grouped based on the document structural information. Based on the structural groupings, the document space is then reduced and a latent semantic kernel is constructed using the document content in the Wikipedia collection. The kernel is later used to group the Wikipedia collection based on the content similarity of the Wikipedia documents. This paper is structured as follows. The next section explains the clustering approach which has been used for the INEX 2008 document mining challenge in more detail. Section 3 is the discussion of our clustering results along with other participants. The paper is then concluded in section 4.

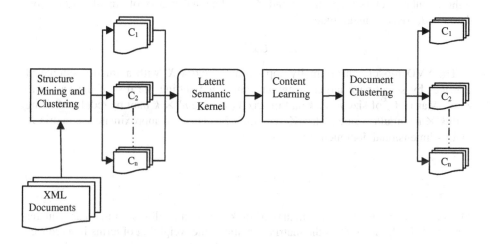

Fig. 1. Overview of the clustering approach

2 The Clustering Approach

Figure 1 illustrates the overview of the clustering approach used in the INEX 2008 document mining challenge. The Wikipedia XML document collection is first clustered based on the structural information present in the Wikipedia documents. The output of the clustering process in the groupings of the Wikipedia documents according to the commonality of their structural information. The structure-based clustering solution is not used for the final clustering solution because the structural information in collections such as the Wikipedia is not sufficient enough in classifying the documents. In this clustering approach, the structure-based clustering solution is utilized to reduce the document space of a term-document matrix which is then used to build a latent semantic kernel for learning the semantic similarity between the Wikipedia documents based on their content features.

For example, given a collection of XML documents $\{d_1, d_2 ... d_n\}$, a term-document matrix X of $m \quad n$ can be derived, where m stands for the number of unique terms and n stands for the number of documents in the collection. It is very computational expensive (and sometimes infeasible) to construct a LSK on this matrix. Thus in the INEX 2008 challenge, the matrix X is modified into a term-document matrix X' of $m \times n'$ where m stands for the number of unique terms and n' stands for the modified number of documents which is lesser than n documents. This is done based on an assumption that by reducing the document dimension and keeping the original term dimension intact allows the resulting kernel to retain the terms of the whole input collection for a better content similarity measure between the Wikipedia documents. When applying Singular Value Decomposition (SVD) on X' the matrix is decomposed into 3 matrices (equation 1), where U and V have orthonormal columns of left and right singular vectors respectively and Z is a diagonal matrix of singular values ordered in decreasing magnitude.

$$X' = UZV^T \tag{1}$$

The SVD model can optimally approximate matrix X' with a smaller sample of matrices by selecting k largest singular values and setting the rest of the values to zero. Matrix U_k of size $m \times k$ and matrix V_k of size $n' \times k$ may be redefined along with $k \times k$ singular value matrix Z_k (equation 2). This can approximate the matrix X' in a k-dimensional document space.

$$\hat{X}'_{m \times n'} = U_k Z_k V_k^T \tag{2}$$

Matrix $\hat{X}'$ is known to be the matrix of rank k, which is closest in the least squares sense to X' [11]. Since U_k is the matrix containing the weighting of terms in a reduced dimension space, it can be used as a kernel for latent learning the semantic between concepts. The U_k matrix becomes the semantic kernel that is used to group the Wikipedia documents based on their content features.

Example. Let us take a collection D that contains 4 XML documents $\{d_1, d_2, d_3, d_4\}$ as shown in figure 2; element names in the documents are shown as embraced within brackets, where $<R>$ is the root element and $<E_i>$ is the internal element or leaf element. The content of a document is denoted by T. Table 1 shows an example of a term-document matrix X. Assume that these terms are extracted after the standard text pre-processing of stop-word removal and stemming.

Table 1. Example of an X matrix

	d_1	d_2	d_3	d_4
t_1	2	1	2	2
t_2	2	2	2	0
t_3	2	2	2	0
t_4	2	2	1	4
t_5	2	0	2	0
t_6	1	0	0	0
t_7	2	2	2	1
t_8	0	1	0	1
t_9	1	1	1	0
t_{10}	0	1	0	0

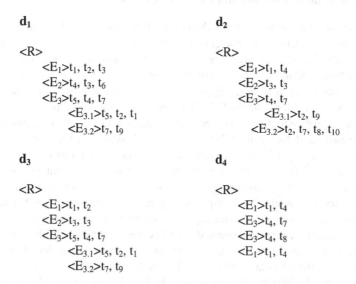

Fig. 2. Example of a collection D that contains 4 XML documents

2.1 Structure Mining and Clustering

The first step in the clustering approach is to determine the structural commonality among the Wikipedia documents. Given a collection of XML documents $\{d_1, d_2... d_n\}$, denoted by D, and a set of distinct paths $\{s_1, s_2... s_f\}$, denoted by S, extracted from the collection D. A path contains element names from the root element to the leaf element. The leaf element is the element containing the textual content. The structure of the documents in a collection is modelled as a path-document matrix $SD_{f \times n}$, where f is the number of distinct paths in collection D and n is the number of documents in collection D.

Example. Revisiting the example of the collection D in figure 2, an example of a path-document matrix SD is shown in table 2.

Table 2. Example of a *SD* matrix

	d_1	d_2	d_3	d_4
R/E$_1$	1	1	1	2
R/E$_2$	1	1	1	0
R/E$_3$/E$_{3.1}$	1	2	1	0
R/E$_3$/E$_{3.2}$	1	0	1	0
R/E$_3$	1	1	1	2

A structural clustering solution *SC* is obtained by applying a *k*-partitioning clustering method [12] on the *SD* matrix. The number of clusters for the document structure is equaled to the number of clusters in the final clustering solution which is 15 clusters.

2.2 Latent Semantic Kernel and Content Learning

The clustering solution *SC* contains documents belonging to each cluster according to their structural similarity. The contents of least important documents, determined by tf-idf, from each cluster in *SC* are merged together resulting in a smaller document space which are then used to construct a term-document matrix X' of $m \times n'$. The document space is reduced such that the computing of SVD (Singular Value Decomposition) on the matrix X' can be done successfully. The resulting latent semantic kernel P, which is the matrix U_k from equation 2, is then used to determine the semantic association between the content of each pair of documents in the Wikipedia collection.

Given a collection of XML documents D, a set of distinct terms $\{t_1, t_2... t_m\}$, denoted by T, is extracted from D after the pre-processing of the content in the Wikipedia documents. Terms are words that appear in the textual content of the leaf elements in the Wikipedia documents. The pre-processing of the content involves the removal of unimportant terms and word stemming. The removal of unimportant terms includes stop words which are terms considered not to be important such as 'the', 'of', 'and', etc. With extensive experiments, integers and terms with length lesser than 4 are found not to be important, thus they are also removed from the term set.

The content of a Wikipedia document d_i is modeled as a vector $[t_1, t_2... t_m]$ which contains the frequencies of the terms, in set T, appearing in the document, where the vector has the same dimensional space as the term space in the matrix kernel P. Given two vectors d_x and d_y, the semantic similarity between the content of the two documents is measured as shown in equation 3.

$$semanticSim(d_x, d_y) = \frac{d_x^T PP^T d_y}{|P^T d_x \| P^T d_y|} \qquad (3)$$

2.3 Document Clustering

Using equation 3 described in the previous section, a pair-wise document similarity matrix can be generated by computing the semantic similarity between each pair of the Wikipedia documents based on their content features. The Wikipedia collection in the INEX 2008 document mining track has 114,366 documents to be clustered. The dimension of a pair-wise document similarity matrix generated from this collection will be 114,366 × 114,366. However, it is not possible to store and process on such a large pair-wise document similarity matrix with the standard available memory space. Thus in the challenge, only the first 1000 closest document similarity distances associated with each document in the Wikipedia collection are used for the clustering task. By selecting r closest number of document similarity distances associated with each document in a collection, it, sometimes, can improve the clustering process by discarding the outlier similarities from the pair-wise document similarity matrix.

With a number of experiments and analysis with the 11,437 Wikipedia training documents, it has shown in figure 3 that using 1000 closest document similarity distances associated with each document in the training collection performs much better in macro F1 than using the pair-wise document similarity matrix of 11,437 × 11,437. The results in figure 3 show that when the micro F1 measure improves, the macro F1 measure degrades. Even though, the 1000 closest document similarity distances associated with each document is not the best clustering solution, however, it is the balance between micro and macro values and it is closer to the solution generated from pair-wise document similarity matrix of 11,437 × 11,437.

So, instead of having a pair-wise document similarity matrix of 114,366 × 114,366, we have a similarity matrix of 114,366 × 1000 for the clustering of the Wikipedia collection for the INEX 2008 document mining challenge. Using the similarity matrix of 114,366 × 1000, a k-partitioning clustering method [12] is used. The clustering method performs by first dividing the input collection, using the similarity matrix of 114,366 × 1000, into two groups, and then one of these two groups is chosen to be bisected further. The process is repeated until the number of bisections in the process equals the number of user-defined clusters.

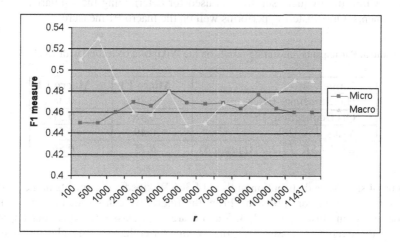

Fig. 3. Accuracy of clustering with different r values on the training collection

3 Experiments and Discussion

As mentioned before, the experiments were performed using the Wikipedia collection containing 114,366 documents from the INEX 2008 document mining challenge for the clustering task. The structure mining as described in section 2.1 is applied to find similar structured clusters within the collection. Then a selected number of documents from each cluster are used to construct the latent semantic kernel. The document dimensional space was reduced down to 257 by conducting this process. After the preprocessing of the content in the Wikipedia documents, which involved the removal of unimportant word and word stemming, 446,294 terms were extracted and used for the construction of the latent semantic kernel P. The kernel P is built by decomposing the term-document matrix of $446,294 \times 257$ using SVD and $200k$ singular values are selected for the kernel.

The required number of clusters for the Wikipedia collection was 15 clusters. Two evaluation methods were used to evaluate the performance of the clustering solutions in the INEX 2008 challenge: Micro F1 and Macro F1 measures. There were a total of 4 participants in the clustering task. Many submissions were submitted by the participants, Table 3 shows only the best result from each participant with 15 clusters. Hagenbuchner-01 approach, using a mixture of the document template, document element tag and document content features for the grouping of the Wikipedia collection, has the worse clustering solution. Vries_15k_20k approach outperforms QUT collection_15 and our approach (QUT LSK_1) but not significantly.

All these three approaches, Vries_15k_20k, QUT collection_15 and ours, use vector space model for representing and processing of the Wikipedia documents. Vries_15k_20k approach uses BM2 [13] for term weighting which has shown to perform well. QUT collection_15 approach reduces the dimension of a term-document matrix X by first mining the frequent sub-trees of the documents. Instead of using the whole content of the documents in the Wikipedia collection, only the content constrained within the frequent sub-trees is used for determining the similarity between the documents. Our clustering performs well on the macro F1 measure.

Table 3. Participants' clustering results on the Wikipedia collection with 15 clusters

Name	Micro F1	Macro F1
Hagenbuchner-01	0.26	0.38
QUT collection_15	0.48	0.51
Vries_15k_20k	0.49	0.59
QUT LSK_1 (Our Approach)	*0.45*	*0.56*

Figure 4 shows the distribution of the 114,366 Wikipedia documents in the 15 clusters discovered from our approach. From the figure, we can see that the majority of the documents in clusters 0, 1, 2, 4, 5 and 6 are from class 471. Thus, our approach can only discovered 9 classes from the Wikipedia collection. Six classes have not been able to be discovered from our approach are 323, 897, 943, 526, 1131 and 1530. Besides classes 897 and 1530, the other classes contain relatively small amount of documents which is why our approach can not discovered them.

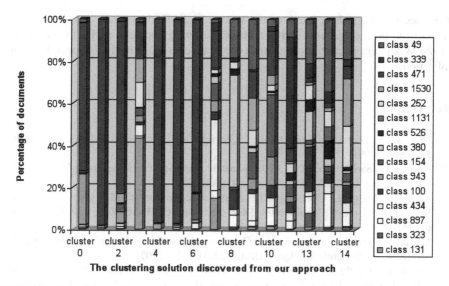

Fig. 4. The distribution of the 114,366 Wikipedia documents in the 15 clusters discovered from our approach

Based on a number of experiments and analysis, the following can be ascertained from our clustering approach: (1) Our approach performs efficiently well even though the semantic kernel is constructed with a reduced document space; (2) Instead of using a pair-wise similarity matrix of 114,366 × 114366, we used a matrix of 114,366 × 1000 for the grouping of the Wikipedia collection. Even though a lot of the document similarity distances have been discarded, our clustering result is not much difference when comparing with other participants. By selecting r closest number of document similarity distances associated with each document in a collection, it can still produce an effective clustering solution, and; (3) Even though the consideration of the structure-only similarity produced a poor clustering solution, in this case we have achieved 0.29 for the micro F1 and 0.33 for the macro F1 on the structure-based clustering solution, however it has been utilized in our approach for selecting the documents in constructing a latent semantic kernel for learning the semantic similarity between the content of the Wikipedia documents. The clustering solution with the content is much better than the clustering solution with the structure-only information or a mixture of features such as in Hagenbuchner-01 approach.

4   Conclusion and Future Work

In this paper, we have proposed using a clustering approach that utilizes both the structural and the content information of XML documents for XML document clustering. First the structure is used to group the XML documents, then, a semantic kernel is built with a selected number of documents belonging to each cluster. The kernel is then used to cluster the content of the Wikipedia XML documents. The results obtained from the experiments shows that the clustering approach performs effectively on the Wikipedia dataset. Results are comparative to other methods applied on this collection.

References

1. Han, J., Kamber, M.: Data Mining: Concepts and Techniques. Morgan Kaufmann, San Diego (2001)
2. Kurgan, L., Swiercz, W., Cios, K.J.: Semantic mapping of xml tags using inductive machine learning. In: CIKM 2002, Virginia, USA (2002)
3. Shen, Y., Wang, B.: Clustering schemaless xml document. In: Meersman, R., Tari, Z., Schmidt, D.C. (eds.) CoopIS 2003, DOA 2003, and ODBASE 2003. LNCS, vol. 2888, pp. 767–784. Springer, Heidelberg (2003)
4. Nayak, R., Tran, T.: A Progressive Clustering Algorithm to Group the XML Data by Structural and Semantic Similarity. IJPRAI 21(3), 1–21 (2007)
5. Nayak, R., Xu, S.: XCLS: A Fast and Effective Clustering Algorithm for Heterogenous XML Documents. In: Ng, W.-K., Kitsuregawa, M., Li, J., Chang, K. (eds.) PAKDD 2006. LNCS, vol. 3918, pp. 292–302. Springer, Heidelberg (2006)
6. Doucet, A., Lehtonen, M.: Unsupervised classification of text-centric xml document collections. In: INEX 2006, pp. 497–509 (2006)
7. Yao, J., Zerida, N.: Rare patterns to improve path-based clustering. In: INEX 2007, Dagstuhl Castle, Germany, December 17-19 (2007)
8. Salton, G., McGill, M.J.: Introduction to Modern Information Retrieval. McGraw-Hill Book Co., New York (1989)
9. Garcia, E.: Description, Advantages and Limitations of the Classic Vector Space Model (2006)
10. Cristianini, N., Shawe-Taylor, J., Lodhi, H.: Latent semantic kernels. JJIS 2002 18(2) (2002)
11. Landauer, T.K., Foltz, P.W., Laham, D.: An introduction to latent semantic analysis. Discourse Processes, 259–284 (1998)
12. Karypis, G.: Cluto – software for clustering high-dimensional datasets – karypis lab
13. Sparck, J.K., Walker, S., Robertson, S.E.: A probabilistic model of information retrieval: Development and comparative experiments. IP&M 36(6), 779–808, 809–840

Self Organizing Maps for the Clustering of Large Sets of Labeled Graphs

ShuJia Zhang, Markus Hagenbuchner, Ah Chung Tsoi, and Alessandro Sperduti

[1] University of Wollongong, Wollongong, Australia
{sz603,markus}@uow.edu.au
[2] Hong Kong Baptist University, Hong Kong
act@hkbu.edu.hk
[3] University of Padova, Padova, Italy
sperduti@math.unipd.it

Abstract. Data mining on Web documents is one of the most challenging tasks in machine learning due to the large number of documents on the Web, the underlying structures (as one document may refer to another document), and the data is commonly not labeled (the class in which the document belongs is not known a-priori). This paper considers latest developments in Self-Organizing Maps (SOM), a machine learning approach, as one way to classifying documents on the Web. The most recent development is called a *Probability Mapping Graph Self-Organizing Map* (PMGraphSOM), and is an extension of an earlier Graph-SOM approach; this encodes undirected and cyclic graphs in a scalable fashion. This paper illustrates empirically the advantages of the PMGraphSOM versus the original GraphSOM model in a data mining application involving graph structured information. It will be shown that the performances achieved can exceed the current state-of-the art techniques on a given benchmark problem.

1 Introduction

The Self-Organizing Map (SOM) is a popular unsupervised machine learning method which allows the projection of high dimensional data onto a low dimensional display space for the purpose of clustering or visualization [1]. Recent works proposed models of SOMs for inputs which can be expressed in terms of graphs, thus providing relational aspects (expressed as links) among objects (nodes of the graph) [2,3,4,5]. This paper describes the most recent extension to SOM which allows the processing of cyclic graphs, and demonstrates its ability on a clustering task involving 114, 366 XML formatted documents from Wikipedia. The dataset has been made available as part of a competition on XML mining and XML clustering [6]. Each document is represented by a node in the graph. The topology of the graph is defined by the hyperlink structure between documents. Textual content available for each document can be used to label the nodes.

It can be stated that most except the most exotic learning problems are suitably represented by richer structures such as trees and graphs (graphs include vectors and sequences as special cases). For example, data in molecular chemistry are more appropriately represented as a graph where nodes represent the atoms of a molecule, and links

S. Geva, J. Kamps, and A. Trotman (Eds.): INEX 2008, LNCS 5631, pp. 469–481, 2009.

between nodes represent atomic bindings. The nodes and links in such graphs may be labeled, e.g. to describe the type of atom, or the strength of an atomic binding.

Most approaches to machine learning deal with vectorial information, and hence, the traditional approach to the processing of *structured* data such as sequences and graphs involves a pre-processing step to "squash" a graph structure into a vectorial form such that the data can be processed in a conventional way. However, such pre-processing step may result in the removal of important structural relations between the atomic entities of a graph. Hence, it is preferable to design algorithms which can deal with graph structured data directly without requiring such a pre-processing step.

Early works by Kohonen [1] suggested an extension of the SOM to allow for the clustering of phonemes (sub-strings of audio signals) for speech recognition applications. More recently, SOMs have been extended to the following situation: the projection from a domain of graphs to a fixed-dimensional display space [2,3,4,5]. The pioneering work presented in [2] introduced a SOM for Structured Data (SOM-SD) which allowed the processing of labeled directed ordered acyclic graph structured data (with supersource) by individually processing the atomic components of a graph structure (the nodes, the labels, and the links). The basic idea is to present to a SOM an input vector which is a representation of a node from a given graph, a numeric data label which may be attached to the node, and information about the mappings of the offsprings of the node. A main difference of this method when compared with the traditional SOM is that the input vectors are dynamic (i.e. they can change during training). This is because the mapping of the node's offsprings can change when the codebook vectors of the SOM are updated, and hence, the input vector to a corresponding node changes. In practice, it has been shown that such dynamics in the input space do not impose a stability problem to the training algorithm [2].

When processing a node, the SOM-SD requires knowledge of the mappings of the node's offsprings. This imposes a strict causal processing order by starting from terminal nodes (which do not have any offsprings), and ending at the supersource node (which has no parent nodes or incoming links). In other words, the SOM-SD cannot, without modification, encode cyclic graphs, and cannot differentiate identical subgraphs which occur in different contextual settings.

The shortcoming of the SOM-SD was addressed through the introduction of a Contextual SOM-SD (CSOM-SD) [4] which allows for the inclusion of information about both the ancestors and the descendants of a given node, and hence, it permits the encoding of *contextual* information about nodes in a directed graph. A problem with [4] is that the improved ability to discriminate between identical substructures can create a substantially increased demand on mapping space, and hence, the computational demand can be prohibitive for large scale learning problems. A second problem is that the method cannot process cyclic graphs, as is required for the given XML clustering task.

The GraphSOM is a subsequent extension which allows for the processing of cyclic or undirected graphs, and was shown to be computationally more efficient than a CSOM-SD for large scale learning problems [5]. However, the processing time required for training a GraphSOM can remain large for the addressed learning task due to the need to use small learning rates to guarantee the stability of the approach. This stability issue is addressed in [7], allowing for much faster training time and improved performances.

This paper is structured as follows: Section 2 introduces pertinent concepts of SOM and its extensions to the graph domain. The experimental setting and experimental findings are presented in Section 3. Conclusions are drawn in Section 4.

2   Self-Organizing Maps for Graphs

In general, SOMs perform a topology preserving mapping of input data through a projection onto a low dimensional display space. For simplicity, this paper assumes the display space to be a two dimensional one. The display space is formed by a set of prototype units (neurons) which are arranged on a regular grid. There is one prototype unit associated with each element on the lattice. An input to a SOM is expected to be an k-dimensional vector, the prototype units must be of the same dimension. A codebook consisting of a k-dimensional vector with adjustable elements is associated with each prototype unit. The elements of the prototype units are adjusted during training by (1) obtaining the prototype unit which best matches a given input, and (2) adjusting the elements of the winner units and all its neighbors. These two steps are referred to as the "competitive step" and "cooperative step", respectively. An algorithmic description is given as follows:

1) Competitive step: One sample input vector $\mathbf{u}$ is randomly drawn from the input data set and its similarity to the codebook vectors is computed. When using the Euclidean distance measure, the winning neuron is obtained through:

$$r = \arg\min_i \|(\mathbf{u} - \mathbf{m}_i)^T \Lambda\|, \tag{1}$$

where $\mathbf{m}_i$ refers to the i-th prototype unit, the superscript T denotes the transpose of a vector, and Λ is a $k \times k$ dimensional diagonal matrix. For the standard SOM, all diagonal elements of Λ equal to 1.

2) Cooperative step: $\mathbf{m}_r$ itself as well as its topological neighbours are moved closer to the input vector in the input space. The magnitude of the attraction is governed by the learning rate α and by a neighborhood function $f(\Delta_{ir})$, where Δ_{ir} is the topological distance between the neurons $\mathbf{m}_r$ and $\mathbf{m}_i$. It is common to use the Euclidean distance to measure the distance topologically. The updating algorithm is given by:

$$\Delta \mathbf{m}_i = \alpha(t) f(\Delta_{ir})(\mathbf{m}_i - \mathbf{u}), \tag{2}$$

where α is the learning rate decreasing to 0 with time t, $f(\cdot)$ is a neighborhood function which controls the amount by which the codebooks are updated. Most common is the Gaussian neighborhood function: $f(\Delta_{ir}) = \exp\left((-\|\mathbf{l}_i - \mathbf{l}_r\|^2)/2\sigma(t)^2\right)$, where the spread σ is called neighborhood radius which decreases with time t, $\mathbf{l}_r$ and $\mathbf{l}_i$ are the coordinates of the winning neuron and the i-th prototype unit in the lattice respectively.

These two steps together constitute a single training step and they are repeated a given number of iterations. The number of iterations must be fixed prior to the commencement of the training process so that the rate of convergence in the neighborhood function, and the learning rate, can be calculated accordingly.

Note that this training procedure does not utilize any ground truth (target) information). This renders the algorithm to be an *unsupervised learning algorithm* which is useful to applications for which target information is not available. Note also that the computational complexity of this algorithm scales linearly with the size of the training set, and hence, this explains its suitability to data mining tasks.

When processing graphs, an input vector $\mathbf{x}$ is formed for each node in a set of graphs through concatenation of a numerical data label $\mathbf{u}$ which may be associated with the node, and the *state* information about the node's offsprings or neighbors. The literature describes two possibilities of computing the state [2,4,5]:

SOM-SD approach: The state of an offspring or neighbor can be the mapping of the offspring or the neighbor [2,4]. In this case, the input vector for the j-th node is $\mathbf{x}_j = (\mathbf{u}_j, \mathbf{y}_{ch[j]})$, where $\mathbf{u}_j$ is a numerical data vector associated with the j-th node, $\mathbf{y}_{ch[i]}$ is the concatenated list of coordinates of the winning neuron of all the children of the j-th node. Since the size of vector $\mathbf{y}_{ch[i]}$ depends on the number of offsprings, and since the SOM training algorithm requires constant sized input vectors, padding with a default value is used for nodes with less than the maximum outdegree of any graph in the training set.

GraphSOM approach: The state of an offspring or neighbor can be the activation of the SOM when mapping all the node's neighbors or offsprings [5]. In this case, the input vector is formed through $\mathbf{x}_j = (\mathbf{u}_j, \mathbf{M}_{ne[j]})$, where $\mathbf{u}_j$ is defined as before, and, $\mathbf{M}_{ne[j]}$ is a m-dimensional vector containing the activation of the map $\mathbf{M}$ when presented with the neighbors of node j. An element M_i of the map is zero if none of the neighbors are mapped at the i-th neuron location, otherwise, it is the number of neighbors that were mapped at that location. This latter approach produces fixed sized input vectors which do not require padding. Note that the latter approach requires knowledge of the mappings of all neighbors of the given node. The availability of these mappings cannot be assured when dealing with undirected or cyclic graphs. This is overcome in [5] by utilizing the mappings from a previous time step. The approximation is valid since convergence is guaranteed. The GraphSOM can process undirected or cyclic graphs.

It can be observed that the inclusion of state information in the input vector provides a local *view* of the graph structure. The iterative nature of the training algorithm ensures that local views are propagated through the nodes in a graph, and hence, structural information about the graph is passed on to all reachable nodes.

It can be observed that the concatenation of data label and state produces hybrid input vectors. The diagonal matrix Λ is used to control the influence of these two components on the mapping. The diagonal elements $\lambda_{11} \cdots \lambda_{pp}$ are set to $\mu \in (0; 1)$, all remaining diagonal elements are set to $1 - \mu$, where $p = |\mathbf{u}|$. Thus, the constant μ influences the contribution of the data label, and the state component to the Euclidean distance. Note that if $|\mathbf{u}| = |\mathbf{x}|$ and $\mu = 1$ then the algorithm reduces to Kohonen's basic SOM training algorithm.

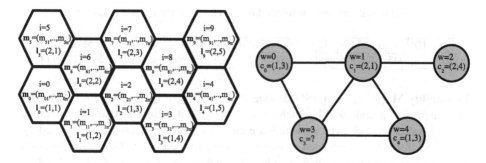

Fig. 1. A 2-dimensional map of size 5×2 (left), and an undirected graph (right). Each hexagon is a neuron. ID, codebook, and coordinate value for each neuron is shown. For each node, the node number, and coordinate of best matching codebook is shown.

After training a SOM on a set of training data it becomes possible to produce a mapping for input data from the same problem domain but which may not necessarily be contained in the training dataset. The degree of the ability of a trained SOM to properly map unseen data (data which are not part of the training set) is commonly referred to as the *generalization* performance. The generalization performance is one of the most important performance measures. However, in this paper, rather than computing the generalization performance of the SOM, we will evaluate the performance on the basis of micro purity and macro purity. This is performed to comply with guidelines set out by the INEX-XML mining competition.

The GraphSOM provides a mechanism for processing the given XML mining tasks. However, we discovered recently a stability problem with GraphSOM in [7] which we will describe by using an example. Consider the example shown in Figure 1. This figure shows a SOM of size 5×2, and an undirected graph containing 5 nodes. For simplicity, we assume that no data label is associated with any node in the graph. When processing node $w = 3$ with a GraphSOM, the network input is the k-dimensional vector $\mathbf{x}_3 = (0, 0, 2, 0, 0, 1, 0, 0, 0, 0)$. This is because two of the neighbors of node 3 are mapped to the coordinate $(1, 3)$ which refers to the 2-nd neuron, and the third neighbour of node 3 is mapped at $(2, 1)$ which refers to the 5-th neuron. The algorithm proceeds with the execution of Eq. 1 and Eq. 2. Due to the weight changes in Eq. 2 it is possible that the mapping of neighbors $w=0$, $w=1$, and $w=4$ change. Assume that there is a minor change in the mapping of node $w=0$, for example, to the nearby location $(1, 4)$. However, the Euclidean distance measure in Eq. 1 does not make a distinction as whether a mapping changed to a nearby location or to a far away location; the contribution to the Euclidean distance remains the same. This defeats the very purpose to achieve topology preserving properties, and can cause alternating states to occur. To counter this behavior it is necessary to either reduce the learning rate to a very small value (causing long training times due to an increased demand on the iterations), or to use a large value for μ (minimizing the effectiveness of the structural information). The problem can be overcome by accounting for the fact that changes in the mapping of nodes are most likely to be a location near a previous winning location. This is done with the recent introduction of the Probability Mapping GraphSOM [7] as follows:

Table 1. Document counts for different classes in the training dataset

Class-ID	471	49	339	252	1530	1542	10049	380	897	4347	9430	1310	5266	323	1131
Size	2945	1474	915	866	789	696	679	639	637	592	405	294	264	128	114

Probability Mapping GraphSOM: Due to the effects of Eq. 2 it is most likely that the mapping of a node will be unchanged at the next iteration. But since all neurons are updated, and since neurons which are close to a winner neuron (as measured by Euclidean distance) are updated more strongly (controlled by the Gaussian function), and, hence, it is more likely that any change of a mapping will be to a nearby location than to a far away location. These likelihoods are directly influenced by the neighborhood function and its spread. This is taken into account by representing a mapping in subsequent iterations as follows: $M_i = \frac{1}{\sigma(t)\sqrt{2\pi}} e^{-\frac{\|l_i - l_r\|^2}{2\sigma(t)^2}}$, where $\sigma(t)$ decreases with time t towards zero, all other quantities are as defined before. The computation is accumulative for all neighbors of the current node. Note that the term $\frac{1}{\sigma(t)\sqrt{2\pi}}$ normalizes the states such that $\sum_i M_i \approx 1.0$. It can be observed that this approach accounts for the fact that during the early stages of the training process it is likely that mappings can change significantly, whereas towards the end of the training process, as $\sigma(t) \rightarrow 0$, the state vectors become more and more similar to the hard coding method of GraphSOM. The approach is known under the acronym PMGraphSOM. It will be observed in Section 3 that the PMGraphSOM allows the use of large learning rates, and hence, reduces the required training time significantly while providing an overall improvement in the clustering performance.

3   Experiments

The experiments were conducted on a set of XML formatted documents; the task is to cluster the data. The dataset consists of $114,366$ Web documents from Wikipedia, and contains hyperlink information. The dataset produces one graph consisting of $114,366$ nodes, each of which representing one document. These nodes are connected based on the hyperlinks between documents. There are a total of $636,187$ directed links between the documents in the dataset. The maximum number of out links (outdegree) from a document is 1527, and on an average, each document has 11 neighboring documents (either linked by or linked to). 10% of the documents are labeled; each labeled document belongs to one of 15 classes. The labels are exclusively used for testing purposes, and are not used during training. The goal of the task is to associate each document with a cluster so that all documents within the same cluster are labeled alike. A closer examination of the distribution of patterns amongst the pattern classes revealed that the dataset is heavily unbalanced. This is shown in Table 1. For example, Table 1 shows that the class with ID "471" is approximately 21 times the size of class with ID "1131". Since SOM and its variants, like PMGraphSOM are unsupervised machine learning schemes, and hence, we must not use class membership information in the pre-processing phase (i.e. to balance the dataset). We expect that this imbalance will affect the quality of the clustering by the GraphSOM and PMGraphSOM. No doubt were we allowed to make use of this prior information in balancing the dataset, the clustering learning task would become simpler, and would affect the results positively.

XML structure, and textual content were available for each document. This allowed us to add a numerical data label to each node in a graph to represent some of the features of the corresponding document. We mainly considered three types of document features: XML tags, Document templates, and Document text. All three types of features were extracted in the same way. Due to the limitation on feature space dimension, we apply PCA (Principal Component Analysis) [8] and other approaches to reduce the dimension after extraction. We briefly describe the steps as follows:

- For each document in the dataset, extract only XML tags for further analysis
- Count the number of unique XML tags (denoted by n) contained in the full set, and then initialize n-dimensional tag vectors for all the documents
- For each document, update the tag vector by counting the occurrences of the tags and assigning the number of counts to the corresponding element
- Since the value of n could be large, it requires some dimension reduction. Documents are separated into 16 groups, the first 15 groups correspond to the given classes, and all unlabeled documents are covered by the 16th group.
- For each unique tag, compute the percentage of documents within each group containing this particular tag and build a $16 \times n$ matrix
- For each row in the matrix, compute the standard deviation of the percentages among different groups. The matrix can partially show whether a particular tag can contribute to differentiate document classes; based on this information, we can filter some tags according to following rules:
  1. Remove tags which are not contained in any labeled documents.
  2. Remove tags which are not contained in any unlabeled documents.
  3. Remove tags where the standard deviation is less than a given threshold.
- Use PCA to reduce the dimension by keeping the first k principal components

These k-dimensional vectors were attached to each document as the data label for training. In this dataset, there are a total of 626 unique tags and 83 of them exist in the training document set. After the application of the above described procedures and the PCA, we kept 3-dimensional tag information label. Similarly, we built a 5-dimensional text data label, and 4-dimensional template vector.

We also used the "rainbow" module from the Bag of Word model [9] as an alternative feature extraction approach to PCA. By employing the knowledge of class label information attached to the training document, we selected a total of 47 text words with top information gain associated with different document classes. Then we built a text vector for each document by counting the occurrences of each selected word.

A composite 12-dimensional vector consisting of 5-dimensional text information, 4-dimensional template information and 3-dimensional tag information is used for the experiments. For comparison purposes, we also used a 10-dimensional template-only data label (reduced by PCA), and a 47-dimensional text information by using the "rainbow" module of the Bag of Word package in separate experiments. The dimensions were the result of various trial-and-error experiments which revealed the best ratio between performance and training times.

Performances will be measured in terms of classification performance, clustering performance, macro purity, and micro purity. Macro purity, and micro purity are defined in the usual manner. Classification and clustering performance are computed as follows.

Classification: After training, we selected the neurons which are activated at least by one labeled document. For each neuron we find the largest number of documents with the same class label and associate this class label to this neuron. For all nodes in the graph we re-compute the Euclidean distance only on those activated neurons. Re-map the node to the winning neuron and assign the label of the winning neuron to this node. For each labeled document, if its original attached label matches the new label given by the network we count it as a positive result. Then we could measure the classification performance by computing the percentage of the number of positive results out of the number of all labeled documents.

Clustering: This measure is to evaluate the clustering performance. For all nodes in the graph, we compute the Euclidean distance to all neurons on the map and obtain the coordinates of the winning neurons. Then we applied K-means clustering on these coordinates. By using different values for K, we could obtain different numbers of clusters. Within each cluster, we find which class of documents are in the majority and associate the class label to this cluster. For all labeled documents, we count it as a positive result if its original attached label matches the label of the cluster which this document belongs to. The cluster performance can then be computed as the percentage of the number of positive results out of the number of all labeled documents.

3.1   Results

The training of the PMGraphSOM requires the adjustment of a number of training parameters such as $\alpha(0)$, μ, $\sigma(0)$, the network size, and the number of training iterations. The optimal value of these parameters is problem dependent, and are not known a priori. A trial and error method is used to identify a suitable set of training parameters. To extract clusters from a trained SOM, we applied K-means clustering to the mapped data of a trained PMGraphSOM. By setting K to a constant value, we can extract exactly K clusters from a SOM. The performance of these two maps when setting K to be either 15 or 512 is shown in Table 2. Table 2 presents our results as was submitted to the INEX-2008 mining challenge. In comparison, Table 3 presents the best result obtained by other participants. It can be seen that our approach is somewhat average when compared with the others. Nevertheless, a general observation was that the training of a reasonably sized PMGraphSOMs required anywhere between 13 hours and 27 hours whereas the training of a GraphSOM of similar size and with similar parameters required about 40 days (approximately one iteration per day). The time required to train the GraphSOM is long, and hence, training was interrupted, and the experiments instead focused solely on PMGraphSOM.

We later found that the main reason which held us back from producing better results was due to the unbalanced nature of the dataset. This is illustrated by Table 4 and Table 5 respectively which present the confusion matrices of the training data set and the testing data set produced by these SOMs.

Table 2. Summary of our results of the INEX'08 XML clustering task

Name	#Clusters	MacroF1	MicroF1	Classification	Clustering
hagenbuchner-01	15	0.377381	0.2586290	76.9782	40.1
hagenbuchner-02	512	0.5369338	0.464696	50.52024	47.6
hagenbuchner-03	512	0.5154928	0.470362	50.52024	48.8

Table 3. Summary of results of the INEX'08 XML clustering task by other participants

Name	#Clusters	MacroF1	MicroF1
QUT Freq_struct_30+links	30	0.551382	0.5440555
QUT collection_15	15	0.5051346	0.4879729
QUT Freq_struct_30	30	0.5854711	0.5388792
QUT LSK_7	30	0.5291029	0.5025662
QUT LSK_3	15	0.5307000	0.4927645
QUT LSK_1	15	0.5593612	0.4517649
QUT LSK_2	15	0.5201315	0.4441753
QUT LSK_8	30	0.5747859	0.5299867
QUT LSK_6	30	0.5690985	0.5261482
QUT LSK_4	15	0.4947789	0.4476466
QUT LSK_5	30	0.5158845	0.5008087
QUT Freq_struct_15	15	0.4938312	0.4833562
QUT collection_30	30	0.5766933	0.5369905
QUT Freq_struct_15+links	15	0.5158641	0.515699
Vries_20m_level2	147	0.6673795	0.5979084
Vries_100m_level1	42	0.6030854	0.5276347
Vries_nmf_15	15	0.5371134	0.4732046
Vries_nmf_147	147	0.5870940	0.5912981
Vries_15k_20m	15	0.5890083	0.4948717
Vries_nmf_42	42	0.5673114	0.5437669

Table 4 refers to the SOM of a given size which produced the best classification performance, whereas Table 5 refers to a SOM of the same size which produced the best clustering performance. In Table 4 the values on the diagonal are the number of documents correctly classified. A total of 8804 out of 11437 training patterns are classified correctly (micro purity is 68.84%). However, the results show that the generalization ability of the trained map is considerably low; the micro purity value dropped to 18.8% for the testing data set. In comparison, the SOM shown in Table 6 shows comparatively more consistent performance for both the training dataset and the testing data set.

Here a total of 48.3% training documents are classified correctly. It can be observed that for the above two experiments, the worst performing classes are the smallest classes, and hence, constitute a main contribution to the observed overall performance levels. In the former experiment, we used 12 dimension labels which combined text, template name and tags information of the documents while in the latter experiment we only attached 10 dimensional labels of template information. Even though a larger map was used for the former experiment, it cannot produce reasonable results. The combined

Table 4. Confusion Matrix for training documents generated by PMGraphSOM trained using mapsize=160x120, iteration=50, grouping=20x20, $\sigma(0)$=10, μ=0.95, $\alpha(0)$=0.9, label: text=5 + template=4 + tag=3

10049	1131	1310	1530	1542	252	323	339	380	4347	471	49	5266	897	9430	%
627	0	0	8	0	2	0	7	4	4	2	19	0	6	0	92.3417
5	**42**	1	3	5	6	1	3	6	3	3	17	1	12	6	36.8421
17	0	**169**	12	1	9	0	4	17	6	1	29	5	24	0	57.4830
29	0	4	**598**	2	6	0	6	23	3	3	66	8	41	0	75.7921
41	0	7	21	**372**	26	0	24	25	23	21	80	7	49	0	53.4483
34	0	0	31	3	**601**	0	6	31	23	1	81	8	46	1	69.3995
4	0	0	6	4	7	**61**	3	4	6	4	21	3	5	0	47.6563
36	1	9	32	4	20	0	**588**	35	23	7	103	11	46	0	64.2623
16	0	0	9	2	2	0	9	**486**	4	4	74	1	32	0	76.0563
34	0	1	16	3	6	0	6	25	**404**	3	53	10	31	0	68.2432
44	0	4	31	2	24	0	38	41	22	**2590**	90	15	39	5	87.9457
61	0	2	9	3	4	0	16	4	6	6	**1293**	1	68	1	87.7205
11	0	0	0	1	1	0	4	16	1	3	28	**185**	14	0	70.0758
39	0	1	3	2	2	0	5	3	6	6	24	0	**546**	0	85.7143
21	0	4	13	6	16	2	9	18	3	18	36	5	12	**242**	59.7531

MicroF1: 68.8489   MacroF1: 76.9782

Table 5. Cluster purity when using mapsize=160x120, iteration=50, grouping=20x20, $\sigma(0)$=10, μ=0.95, $\alpha(0)$=0.9, label: text=5 + template=4 + tag=3

	10049	1131	1310	1530	1542	252	323	339	380	4347	471	49	5266	897	9430	Macro F1	Micro F1
Train	92.3	36.8	57.5	75.8	53.4	69.4	47.7	64.3	76.1	68.2	87.9	87.7	70.1	85.7	59.8	76.978	68.848
Test	14.5	0.9	8.5	23.0	10.7	17.6	1.6	14.8	13.8	11.8	75.0	26.8	4.4	13.2	20.5	32.972	18.283

features may focus too much on the training documents and this produces a lower generalization performance than the one using the template features. It indicates that the training can be ineffective without a good set of features regardless to the size of a map.

In order to evaluate training performance by using different features, we conducted an experiment by using the 47-dimensional text vector generated via the Bag of Word model after the competition. Table 7 presents a summary of the confusion matrices for the set of training documents and set of testing documents respectively, produced by the best trained SOM so far. The corresponding map is shown in Figure 2. The results indicate that the Bag of Word based feature selection approach can provide more useful data labels which would benefit the clustering task.

Apart from selecting useful features for the training task, the size of the network can be increased to overcome issues related to imbalance in a data set. Note that the dataset

Table 6. Cluster purity when using mapsize=120x100, iteration=50, grouping=10x10, $\sigma(0)$=10, μ=0.99995, $\alpha(0)$=0.9, label: 10-dim. template

	10049	1131	1310	1530	1542	252	323	339	380	4347	471	49	5266	897	9430	Macro F1	Micro F1
Train	58.3	29.8	39.8	60.2	54.7	51.4	8.6	49.0	58.4	52.7	85.6	79.9	22.7	38.5	35.6	62.429	48.342
Test	28.8	24.7	24.2	41.5	37.8	33.8	1.4	31.2	40.2	29.5	81.7	61.5	10.5	19.5	25.0	48.131	32.752

Table 7. Cluster purity using mapsize=400x300, iteration=60, grouping=40x60, $\sigma(0)$=20, μ=0.99, $\alpha(0)$=0.9, label: 47-dim. text (BagOfWords-based approach)

	10049	1131	1310	1530	1542	252	323	339	380	4347	471	49	5266	897	9430	Macro F1	Micro F1
Train	78.1	60.5	60.5	68.8	60.3	75.4	40.6	82.8	61.5	60.5	89.7	70.1	62.5	62.2	41.5	73.096	65.012
Test	76.8	56.0	65.3	67.9	58.5	76.5	41.8	81.9	61.2	62.7	88.9	70.6	62.1	61.2	43.7	73.096	64.998

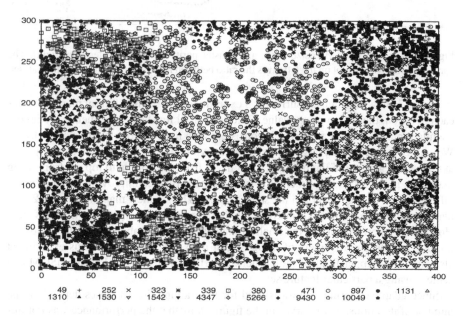

49 +	252 ×	323 ✳	339 □	380 ■	471 ○	897 ●	1131 △
1310 ▲	1530 ▽	1542 ▼	4347 ◇	5266 ◆	9430 ○	10049 ●	

Fig. 2. Mapping of all training data on a PMGraphSOM performing best in classifying the nodes

consists of 114,366 nodes which were mapped to a network of size 160 x 120 = 19200. In other words, the mapping of nodes in the dataset is compressed by at least 83%. Such high compression ratio forces the SOM to neglect less significant features and data clusters, and hence, can contribute to a poor performance in clustering performance on small classes. A study is conducted on the performance change when increasing the network size. To allow a basis of comparison, the same set of data and parameters are used while training the maps of different size. This is not generally recommended since some training parameters (most notably μ) are dependent on the network size and should be adjusted accordingly. Nevertheless, we keep these parameters constant to exclude the influence of a parameter change on the analysis of the effects of a change in network size. The results are summarized in Figure 3(left) where the x-axis indicates the proportion of the number of neurons on the map to the number of documents and the y-axis indicates the cluster purity. This shows that the clustering performance of the map can be significantly improved by increasing the training map size. However, according to the tendency of the curve, it can be predicted that it will get more and more difficult to increase the cluster purity even if the map size is increased beyond the size of the dataset.

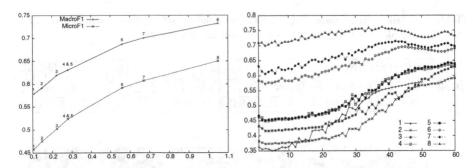

Fig. 3. Cluster purity vs. Map size (left), and Cluster purity vs. Training iteration (right). The horizontal scale of the plot on the left gives the ratio between network size and size of the dataset.

Figure 3 shows the cluster purity for 8 experiments by using different map sizes. For each iteration, the macro purity value is plotted for all finished experiments. We ignore the micro purity value here since it is consistent with macro purity in terms of the trend. We could observe that the larger maps present obvious advantages on cluster purity even at the very beginning of the training. The curves from the bottom to top are from using map sizes in the ascending order. However, the performance improving speed of using different sizes of maps is not in the same manner. The smallest map in use has 12,000 neurons while the number of neurons on the largest map is about ten times of that. However, the performance of the smallest map was increased about 0.2 at the end of the training and the improvement of using largest map was small: 0.03. This experiment also shows the upper limit of the clustering performance that the PMGraphSOM can achieve using a particular feature set.

Since adaptive learning rate was used for the training, its decreases depend on the number of iterations. Most curves in the figure show that the performance was not improved gradually during training, but instead was fast increasing over a short period. The fast increase corresponds to the steep part of the sigmoidal function, so it is possible to prolong the increase longer by stretching the sigmoidal function, This can be performed by increasing the number of iterations. In Figure 3, the results of experiments 4 and 5 are using the same map settings and parameters except the number of training iterations; these are shown in the form of hollow-square and filled-square respectively in the graph. In experiment 5, we doubled the number of iterations of the one which was set for experiment 4 (60 iterations). The curve of the results of experiment 5 is generated by plotting the macro purity value every two iterations. It can be observed that the two curves corresponding to experiments 4 and 5 are nearly identical; this confirms that the selected learning rate and number of iterations for these experiments is sufficient.

4   Conclusions

This paper presents an unsupervised machine learning approach to the clustering of a relatively large scale data mining tasks requiring the clustering of structured Web documents from Wikipedia. It was shown that this can be achieved through a suitable extension of an existing machine learning method based on the GraphSOM model. Experimental results revealed that the unbalanced nature of the training set is the main inhibiting factor in this task, and have shown that the right feature selection technique, and a sufficiently large map results in a network performance which exceeds the best of any other competitors approach. More specifically, we have shown that a map exceeding a size of 0.7 times the number of nodes in the training set outperforms any competitors approach. The task of encoding of textual information embedded within the Wikipedia documents was resolved by using a Bag of Word approach in combination with Principal Component Analysis. This extracts and compresses the information. Work on the proposed approach is ongoing with investigations on the effects of network size, and feature extraction on the clustering performance.

Acknowledgment. This work has received financial support from the Australian Research Council through Discovery Project grant DP0774168 (2007 - 2009).

References

[1] Kohonen, T.: Self-Organisation and Associative Memory, 3rd edn. Springer, Heidelberg (1990)

[2] Hagenbuchner, M., Tsoi, A., Sperduti, A.: A supervised self-organising map for structured data. In: Allison, N., Yin, H., Allison, L., Slack, J. (eds.) WSOM 2001 - Advances in Self-Organising Maps, pp. 21–28. Springer, Heidelberg (2001)

[3] Günter, S., Bunke, H.: Self-organizing map for clustering in the graph domain. Pattern Recognition Letters 23(4), 405–417 (2002)

[4] Hagenbuchner, M., Sperduti, A., Tsoi, A.: Contextual processing of graphs using self-organizing maps. In: European symposium on Artificial Neural Networks, Poster track, Bruges, Belgium, April 27-29 (2005)

[5] Hagenbuchner, M., Tsoi, A.C., Sperduti, A., Kc, M.: Efficient clustering of structured documents using graph self-organizing maps. In: Fuhr, N., Kamps, J., Lalmas, M., Trotman, A. (eds.) INEX 2007. LNCS, vol. 4862, pp. 207–221. Springer, Heidelberg (2008)

[6] Denoyer, L., Gallinari, P.: Initiative for the evaluation of xml retrieval, xml-mining track (2008), http://www.inex.otago.ac.nz/

[7] Hagenbuchner, M., Zhang, S., Tsoi, A., Sperduti, A.: Projection of undirected and non-positional graphs using self organizing maps. In: European Symposium on Artificial Neural Networks - Advances in Computational Intelligence and Learning, April 22-24 (to appear, 2009)

[8] Pearson, K.: On lines and planes of closest fit to systems of points in space. Philosophical Magazine Series 6(2), 559–572 (1901)

[9] McCallum, A.K.: Bow: A toolkit for statistical language modeling, text retrieval, classification and clustering (1996),
http://www.cs.cmu.edu/~mccallum/bow

Author Index